AF278809

# LONDON MATHEMATICAL SOCIETY LECTURE NOTE SERIES

Managing Editor: Professor M. Reid, Mathematics Institute,
University of Warwick, Coventry CV4 7AL, United Kingdom

The titles below are available from booksellers, or from Cambridge University Press
at www.cambridge.org/mathematics

London Mathematical Society Lecture
Note Series: 391

# Fusion Systems in Algebra and Topology

MICHAEL ASCHBACHER
*California Institute of Technology*

RADHA KESSAR
*University of Aberdeen*

BOB OLIVER
*Université de Paris 13*

CAMBRIDGE
UNIVERSITY PRESS

CAMBRIDGE UNIVERSITY PRESS
Cambridge, New York, Melbourne, Madrid, Cape Town,
Singapore, São Paulo, Delhi, Mexico City

Cambridge University Press
The Edinburgh Building, Cambridge CB2 8RU, UK

Published in the United States of America by Cambridge University Press, New York

www.cambridge.org
Information on this title: www.cambridge.org/9781107601000

© M. Aschbacher, R. Kessar and B. Oliver 2011

This publication is in copyright. Subject to statutory exception
and to the provisions of relevant collective licensing agreements,
no reproduction of any part may take place without the written
permission of Cambridge University Press.

First published 2011

*A catalogue record for this publication is available from the British Library*

*Library of Congress Cataloguing in Publication Data*
Aschbacher, Michael, 1944–
Fusion systems in algebra and topology / Michael Aschbacher, Radha Kessar, Bob Oliver.
p.   cm. – (London mathematical society lecture note series ; 391)
ISBN 978-1-107-60100-0 (pbk.)
1.  Combinatorial group theory.   2.  Topological groups.   3.  Algebraic topology.
I. Kessar, Radha.   II. Oliver, Robert, 1949–   III. Title.
QA182.5.A74 2011        512´.2–dc23
2011019266

ISBN 978-1-107-60100-0 Paperback

Cambridge University Press has no responsibility for the persistence or
accuracy of URLs for external or third-party internet websites referred to in
this publication, and does not guarantee that any content on such websites is,
or will remain, accurate or appropriate. Information regarding prices, travel
timetables, and other factual information given in this work is correct at
the time of first printing but Cambridge University Press does not guarantee
the accuracy of such information thereafter.

CONTENTS

# Introduction

Let $G$ be a finite group, $p$ a prime, and $S$ a Sylow $p$-subgroup of $G$. Subsets of $S$ are said to be *fused* in $G$ if they are conjugate under some element of $G$. The term "fusion" seems to have been introduced by Brauer in the fifties, but the general notion has been of interest for over a century. For example, in his text *The Theory of Groups of Finite Order* [Bu] (first published in 1897), Burnside proved that if $S$ is abelian then the normalizer in $G$ of $S$ controls fusion in $S$. (A subgroup $H$ of $G$ is said to *control fusion* in $S$ if any pair of tuples of elements of $S$ which are conjugate in $G$ are also conjugate under $H$.)

Initially, information about fusion was usually used in conjunction with transfer, as in the proof of the normal $p$-complement theorems of Burnside and Frobenius, which showed that, under suitable hypotheses on fusion, $G$ possesses a *normal $p$-complement*: a normal subgroup of index $|S|$ in $G$. But in the sixties and seventies more sophisticated results on fusion began to appear, such as Alperin's Fusion Theorem [Al1], which showed that the family of normalizers of suitable subgroups of $S$ control fusion in $S$, and Goldschmidt's Fusion Theorem [Gd3], which determined the groups $G$ possessing a nontrivial abelian subgroup $A$ of $S$ such that no element of $A$ is fused into $S \smallsetminus A$.

In the early nineties, Lluis Puig abstracted the properties of $G$-fusion in a Sylow subgroup $S$, in his notion of a *Frobenius category* on a finite $p$-group $S$, by discarding the group $G$ and focusing instead on isomorphisms between subgroups of $S$. (But even earlier in 1976 in [P1], Puig had already considered the standard example $\mathcal{F}_S(G)$ of a Frobenius category, defined below.) Puig did not publish his work until his 2006 paper [P6] and his 2009 book [P7]. Meanwhile, his approach was taken up and extended by others, and in the process, alternate terminology evolved which is now commonly used, and which we therefore have adopted here. In particular, Puig's Frobenius categories are now referred to as "saturated fusion systems" in most of the literature.

A *fusion system* $\mathcal{F}$ on a finite $p$-group $S$ is a category whose objects are the subgroups of $S$, with the set $\mathrm{Hom}_{\mathcal{F}}(P, Q)$ of morphisms from $P$ to $Q$ consisting of monomorphisms from $P$ into $Q$, and such that some weak axioms are satisfied (see Definition I.2.1 for the precise conditions). The standard example of a fusion system is the category $\mathcal{F}_S(G)$, where $G$ is a finite group, $S \in \mathrm{Syl}_p(G)$, and the morphisms are those induced by conjugation in $G$. A fusion system $\mathcal{F}$ is *saturated* if it satisfies two more axioms (Definition I.2.2), which hold in $\mathcal{F}_S(G)$ as a consequence of Sylow's Theorem.

Many classic results on fusion in a Sylow subgroup $S$ of a finite group $G$ can be interpreted as results about the fusion system $\mathcal{F} = \mathcal{F}_S(G)$. Burnside's Fusion Theorem becomes $\mathcal{F}_S(G) = \mathcal{F}_S(N_G(S))$ when $S$ is abelian. Alperin's Fusion Theorem says that $\mathcal{F}$ is generated by certain "local" subcategories of $\mathcal{F}$ (cf. Theorem I.3.5). Goldschmidt's Fusion Theorem says that an abelian subgroup $A$ of $S$ is "normal" in $\mathcal{F}$ (cf. Definition I.4.1) when no element of $A$ is fused into $S \smallsetminus A$, and goes on to use this fact to determine $G$.

Puig created his theory of Frobenius categories largely as a tool in modular representation theory, motivated in part by work of Alperin and Broue in [AB]. Later, homotopy theorists used this theory to provide a formal setting for, and prove results about, the $p$-completed classifying spaces of finite groups. As part of this process, objects called *p-local finite groups* associated to abstract fusion systems were introduced by Broto, Levi and Oliver in [BLO2]; these also possess interesting $p$-completed classifying spaces. Finally, local finite group theorists became interested in fusion systems, in part because methods from local group theory proved to be effective in the study of fusion systems, but also because certain results in finite group theory seem to be easier to prove in the category of saturated fusion systems.

These three themes — the application of fusion systems in modular representation theory, homotopy theory, and finite group theory — together with work on the foundations of the theory of saturated fusion systems, remain the focus of interest in the subject. And these are the four themes to which this volume is devoted.

This book grew out of a workshop on fusion systems at the University of Birmingham in July–August of 2007, sponsored by the London Mathematical Society and organized by Chris Parker. At that workshop there were three series of talks, one each on the role of fusion systems in modular representation theory, homotopy theory, and finite group theory, given by Kessar, Oliver, and Aschbacher, respectively. It was Chris Parker's idea to use those talks as the point of departure for this book, although he unfortunately had to pull out of the project before it was completed. We have extracted material on the foundations of the theory of fusion systems from the various series and collected them in Part I of the book, where we have also included proofs of many of the most basic results. Then the talks have been updated and incorporated in Parts II through IV of the book, which describe the state of the art of the role of fusion systems in each of the three areas.

David Craven has also written a book on fusion systems [Cr2], which also can trace its origins to the 2007 workshop in Birmingham, and which should appear at about the same time as this one. The two books are very different in style — for example, his is intended more as a textbook and ours as a survey — and also very different in the choice of topics. In this way, we expect that the two books will complement each other.

The theory of fusion systems is an emerging area of mathematics. As such, its foundations are not yet firmly established, and the frontiers of the subject are receding more rapidly than those of more established areas. With this in mind, we have two major goals for this volume: first, collect in one place the various definitions, notation, terminology, and basic results which constitute the foundation of the theory of fusion systems, but are currently spread over a number of papers in the literature. In the process we also seek to reconcile differences in notation, terminology, and even basic concepts among papers in the literature. In particular, there is a discussion of the three existing notions of a "normal subsystem" of a saturated fusion system. Second, we seek to present a snapshot of the important theorems and open problems in our four areas of emphasis at this point in time. Our hope is that the book will serve both as a basic reference on fusion systems and as an introduction to the field, particularly for students and other young mathematicians.

The book is organized as follows. Part I contains foundational material about fusion systems, including the most basic definitions, notation, concepts, and lemmas. Then Parts II, III, and IV discuss the role of fusion systems in local finite group theory, homotopy theory, and modular representation theory, respectively. Finally the book closes with an appendix which records some of the basic material on finite groups which is well known to specialists, but perhaps not to those who approach fusion systems from the point of view of representation theory or homotopy theory.

We have received help from a large number of people while working on this project. In particular, we would like to thank Kasper Andersen, David Craven, and Ellen Henke for reading large parts of the manuscript in detail and making numerous suggestions and corrections. Many of the others who have assisted us will be acknowledged in the introductions to the individual Parts.

**Notation:**    We close this introduction with a list of some of the basic notation involving finite groups used in all four Parts of the book. Almost all of this notation is fairly standard.

For $x, g \in G$, we write $^g x = gxg^{-1}$ for the *conjugate* of $x$ under $g$, and let $c_g \colon G \to G$ be *conjugation by* $g$, defined by $c_g(x) = {}^g x$. Set $x^g = g^{-1}xg$ and for $X \subseteq G$, set $^g X = c_g(X)$ and $X^g = c_{g^{-1}}(X)$. Let $N_G(X) = \{g \in G \mid {}^g X = X\}$ be the *normalizer* in $G$ of $X$ and $C_G(X) = \{g \in G \mid xg = gx$ for all $g \in G\}$ be the *centralizer* in $G$ of $X$. Write $\langle X \rangle$ for the subgroup of $G$ generated by $X$.

Similar notation will be used when conjugating by an isomorphism of (possibly distinct) groups. For example (when group homomorphisms are composed from right to left), if $\varphi\colon G \xrightarrow{\cong} H$ is an isomorphism of groups, and $\alpha \in \mathrm{Aut}(G)$ and $\beta \in \mathrm{Aut}(H)$, we write $^{\varphi}\alpha = \varphi\alpha\varphi^{-1} \in \mathrm{Aut}(H)$ and $\beta^{\varphi} = \varphi^{-1}\beta\varphi \in \mathrm{Aut}(G)$.

We write $H \leq G$, $H < G$, or $H \trianglelefteq G$ to indicate that $H$ is a subgroup, proper subgroup, or normal subgroup of $G$, respectively. Observe, for $H \leq G$, that $c\colon g \mapsto c_g$ is a homomorphism from $N_G(H)$ into $\mathrm{Aut}(H)$ with kernel $C_G(H)$; we write $\mathrm{Aut}_G(H)$ for the image $c(N_G(H))$ of $H$ under this homomorphism. Thus $\mathrm{Aut}_G(H)$ is the *automizer* in $G$ of $H$ and $\mathrm{Aut}_G(H) \cong N_G(H)/C_G(H)$. The *inner automorphism group* of $H$ is $\mathrm{Inn}(H) = \mathrm{Aut}_H(H) = c(H)$, and the *outer automorphism group* of $H$ is $\mathrm{Out}(H) = \mathrm{Aut}(H)/\mathrm{Inn}(H)$.

We write $\mathrm{Syl}_p(G)$ for the set of Sylow $p$-subgroups of $G$. When $\pi$ is a set of primes, a *$\pi$-subgroup* of $G$ is a subgroup whose order is divisible only by primes in $\pi$. We write $O_\pi(G)$ for the largest normal $\pi$-subgroup of $G$, and $O^\pi(G)$ for the smallest normal subgroup $H$ of $G$ such that $G/H$ is a $\pi$-group. We write $p'$ for the set of primes distinct from $p$; we will be particularly interested in the groups $O_p(G)$, $O_{p'}(G)$, $O^p(G)$, and $O^{p'}(G)$. Sometimes we write $O(G)$ for $O_{2'}(G)$.

As usual, when $P$ is a $p$-group for some prime $p$, we set $\Omega_1(P) = \langle g \in P \,|\, g^p = 1 \rangle$.

As for specific groups, $C_n$ denotes a (multiplicative) cyclic group of order $n$, and $D_{2^k}$, $SD_{2^k}$, $Q_{2^k}$ denote dihedral, semidihedral, and quaternion groups of order $2^k$. Also, $A_n \leq S_n$ denote alternating and symmetric groups on $n$ letters.

Throughout the book, $p$ is always understood to be a fixed prime. All $p$-groups are assumed to be finite.

## Part I. Introduction to fusion systems

This part is intended as a general introduction to the book, where we describe the properties of fusion systems which will be used throughout. We begin with the basic definitions of fusion systems of finite groups and abstract fusion systems, and give some versions of Alperin's fusion theorem in this setting. Afterwards, we discuss various topics such as normal and central subgroups of fusion systems, constrained fusion systems, normal fusion subsystems, products of fusion systems, the normalizer and centralizer fusion subsystems of a subgroup, and fusion subsystems of $p$-power index or of index prime to $p$.

### 1. THE FUSION CATEGORY OF A FINITE GROUP

For any group $G$ and any pair of subgroups $H, K \leq G$, we define

$$\mathrm{Hom}_G(H, K) = \big\{ \varphi \in \mathrm{Hom}(H, K) \,\big|$$

$$\varphi = c_g \text{ for some } g \in G \text{ such that } {}^g H \leq K \big\}.$$

In other words, $\mathrm{Hom}_G(H, K)$ is the set of all (injective) homomorphisms from $H$ to $K$ which are induced by conjugation in $G$. Similarly, we write $\mathrm{Iso}_G(H, K)$ for the set of elements of $\mathrm{Hom}_G(H, K)$ which are isomorphisms of groups.

**Definition 1.1** Fix a finite group $G$, a prime $p$, and a Sylow $p$-subgroup $S \in \mathrm{Syl}_p(G)$. The *fusion category* of $G$ over $S$ is the category $\mathcal{F}_S(G)$ whose objects are the subgroups of $S$, and which has morphism sets

$$\mathrm{Mor}_{\mathcal{F}_S(G)}(P, Q) = \mathrm{Hom}_G(P, Q).$$

Many concepts and results in finite group theory can be stated in terms of this category. We list some examples of this here. In all cases, $G$ is a finite group and $S \in \mathrm{Syl}_p(G)$.

- Alperin's fusion theorem [Al1], at least in some forms, is the statement that $\mathcal{F}_S(G)$ is generated by automorphism groups of certain subgroups of $S$, in the sense that each morphism in $\mathcal{F}_S(G)$ is a composite of restrictions of automorphisms of those subgroups.

- Glauberman's $Z^*$-theorem [Gl1] says that when $p = 2$ and $O_{2'}(G) = 1$, then $Z(G) = Z(\mathcal{F}_S(G))$, where $Z(\mathcal{F}_S(G))$ is the "center" of the fusion category in a sense which will be made precise later (Definition 4.3).

- A subgroup $H \leq G$ which contains $S$ *controls fusion in $S$* if $\mathcal{F}_S(H) = \mathcal{F}_S(G)$. Thus Burnside's fusion theorem states that when $S$ is abelian, $\mathcal{F}_S(G) = \mathcal{F}_S(N_G(S))$, and that every morphism in $\mathcal{F}_S(G)$ extends to an automorphism of $S$.

- By a theorem of Frobenius (cf. [A4, 39.4]), $G$ has a normal $p$-complement (a subgroup $H \trianglelefteq G$ of $p$-power index and order prime to $p$) if and only if $\mathcal{F}_S(G) = \mathcal{F}_S(S)$.

- The focal subgroup theorem says that

$$S \cap [G,G] = \mathfrak{foc}(\mathcal{F}_S(G)) \stackrel{\text{def}}{=} \langle x^{-1}y \,|\, x, y \in S,\ x = {}^g y \text{ for some } g \in G \rangle,$$

  and is thus described in terms of the category $\mathcal{F}_S(G)$. By the hyperfocal subgroup theorem of Puig [P5, § 1.1], $S \cap O^p(G)$ can also be described in terms of the fusion category $\mathcal{F}_S(G)$ (see Section 7).

The following lemma describes some of the properties of these fusion categories, properties which help to motivate the definition of abstract fusion systems in the next section.

**Lemma 1.2** *Fix a finite group $G$ and a Sylow $p$-subgroup $S \in \mathrm{Syl}_p(G)$.*

(a) *For each $P \leq S$, there is $Q \leq S$ such that $Q$ is $G$-conjugate to $P$ and $N_S(Q) \in \mathrm{Syl}_p(N_G(Q))$. For any such $Q$, $C_S(Q) \in \mathrm{Syl}_p(C_G(Q))$ and $\mathrm{Aut}_S(Q) \in \mathrm{Syl}_p(\mathrm{Aut}_G(Q))$.*

(b) *Fix $P, Q \leq S$ and $g \in G$ such that ${}^g P = Q$. Assume $N_S(Q) \in \mathrm{Syl}_p(N_G(Q))$. Set*

$$N = \left\{ x \in N_S(P) \,\big|\, {}^g x \in N_S(Q) \cdot C_G(Q) \right\} .$$

*Then there is $h \in C_G(Q)$ such that ${}^{hg}N \leq S$.*

*Proof.* **(a)** Fix $T \in \mathrm{Syl}_p(N_G(P))$. Since $T$ is a $p$-subgroup of $G$, there is $g \in G$ such that ${}^g T \leq S$. Set $Q = {}^g P$. Then $Q \trianglelefteq {}^g T \leq S$, and ${}^g T \in \mathrm{Syl}_p(N_G(Q))$ since $c_g \in \mathrm{Aut}(G)$. Also, ${}^g T \leq N_S(Q)$, and since ${}^g T \in \mathrm{Syl}_p(N_G(Q))$, ${}^g T = N_S(Q)$. The last statement now holds by Lemma A.3, upon identifying $\mathrm{Aut}_X(Q)$ with $N_X(Q)/C_X(Q)$ for $X = G, S$.

**(b)** Since $N_S(Q)$ normalizes $C_G(Q)$, $N_S(Q) \cdot C_G(Q)$ is a subgroup of $G$. By assumption, ${}^g N \leq N_S(Q) \cdot C_G(Q)$. Since $N_S(Q) \in \mathrm{Syl}_p(N_G(Q))$ and $N_S(Q) \cdot C_G(Q) \leq N_G(Q)$, $N_S(Q)$ is also a Sylow subgroup of $N_S(Q) \cdot C_G(Q)$. Hence there is $h \in C_G(Q)$ such that ${}^h({}^g N) \leq N_S(Q)$; i.e., ${}^{hg}N \leq N_S(Q)$. $\qquad\square$

## 2. ABSTRACT FUSION SYSTEMS

The notion of an abstract fusion system is due to Puig. The definitions we give here are modified versions of Puig's definitions (given in [P6]), but equivalent to them. The following is what he calls a "divisible $S$-category".

**Definition 2.1** ([P6], [BLO2])  A fusion system over a $p$-group $S$ is a category $\mathcal{F}$, where $\mathrm{Ob}(\mathcal{F})$ is the set of all subgroups of $S$, and which satisfies the following two properties for all $P, Q \leq S$:

- $\mathrm{Hom}_S(P, Q) \subseteq \mathrm{Mor}_{\mathcal{F}}(P, Q) \subseteq \mathrm{Inj}(P, Q)$; and

- each $\varphi \in \mathrm{Mor}_{\mathcal{F}}(P, Q)$ is the composite of an $\mathcal{F}$-isomorphism followed by an inclusion.

Composition in a fusion system $\mathcal{F}$ is always given by composition of homomorphisms. We usually write $\mathrm{Hom}_{\mathcal{F}}(P, Q) = \mathrm{Mor}_{\mathcal{F}}(P, Q)$ to emphasize that the morphisms in $\mathcal{F}$ actually are group homomorphisms, and also set $\mathrm{Aut}_{\mathcal{F}}(P) = \mathrm{Hom}_{\mathcal{F}}(P, P)$. Note that a fusion system over $\mathcal{F}$ contains all inclusions $\mathrm{incl}_P^Q$, for $P \leq Q \leq S$, by the first condition (it is conjugation by $1 \in S$). The second condition means that for each $\varphi \in \mathrm{Hom}_{\mathcal{F}}(P, Q)$, $\varphi \colon P \longrightarrow \varphi(P)$ and $\varphi^{-1} \colon \varphi(P) \longrightarrow P$ are both morphisms in $\mathcal{F}$.

Fusion systems as defined above are too general for most purposes, and additional conditions are needed for them to be very useful. This leads to the concept of what we call a "saturated fusion system": a fusion system satisfying certain axioms which are motivated by properties of fusion in finite groups. The following version of these axioms is due to Roberts and Shpectorov [RS].

**Definition 2.2**  Let $\mathcal{F}$ be a fusion system over a $p$-group $S$.

- Two subgroups $P, Q \leq S$ are *$\mathcal{F}$-conjugate* if they are isomorphic as objects of the category $\mathcal{F}$. Let $P^{\mathcal{F}}$ denote the set of all subgroups of $S$ which are $\mathcal{F}$-conjugate to $P$.

- A subgroup $P \leq S$ is *fully automized* in $\mathcal{F}$ if $\mathrm{Aut}_S(P) \in \mathrm{Syl}_p(\mathrm{Aut}_{\mathcal{F}}(P))$.

- A subgroup $P \leq S$ is *receptive* in $\mathcal{F}$ if it has the following property: for each $Q \leq S$ and each $\varphi \in \mathrm{Iso}_{\mathcal{F}}(Q, P)$, if we set

$$N_\varphi = N_\varphi^{\mathcal{F}} = \{g \in N_S(Q) \mid {}^{\varphi}c_g \in \mathrm{Aut}_S(P)\},$$

  then there is $\bar{\varphi} \in \mathrm{Hom}_{\mathcal{F}}(N_\varphi, S)$ such that $\bar{\varphi}|_Q = \varphi$.

- A fusion system $\mathcal{F}$ over a $p$-group $S$ is *saturated* if each subgroup of $S$ is $\mathcal{F}$-conjugate to a subgroup which is fully automized and receptive.

We also say that two elements $x, y \in S$ are $\mathcal{F}$-conjugate if there is an isomorphism $\varphi \in \mathrm{Iso}_{\mathcal{F}}(\langle x \rangle, \langle y \rangle)$ such that $\varphi(x) = y$, and let $x^{\mathcal{F}}$ denote the $\mathcal{F}$-conjugacy class of $x$.

The fusion category $\mathcal{F}_S(G)$ of a finite group $G$ clearly satisfies the conditions in Definition 2.1, and thus is a fusion system. It also satisfies the saturation conditions by Lemma 1.2.

**Theorem 2.3** (Puig)   *If $G$ is a finite group and $S \in \mathrm{Syl}_p(G)$, then $\mathcal{F}_S(G)$ is a saturated fusion system.*

*Proof.* By Lemma 1.2(a), each subgroup $P \leq S$ is $G$-conjugate to a subgroup $Q \leq S$ such that $N_S(Q) \in \mathrm{Syl}_p(N_G(Q))$, and each such subgroup $Q$ is fully automized in $\mathcal{F}_S(G)$. By Lemma 1.2(b), $Q$ is also receptive in $\mathcal{F}_S(G)$, and thus $\mathcal{F}_S(G)$ is saturated. $\qquad\square$

A saturated fusion system $\mathcal{F}$ over a $p$-group $S$ will be called *realizable* if $\mathcal{F} = \mathcal{F}_S(G)$ for some finite group $G$ with $S \in \mathrm{Syl}_p(G)$, and will be called *exotic* otherwise. Examples of exotic fusion systems will be described in Section III.6.

There are several, equivalent definitions of saturated fusion systems in the literature. We discuss here the definition of saturation which was given in [BLO2]. Two other definitions, the original one given by Puig and another one by Stancu, will be described and shown to be equivalent to these in Section 9 (Proposition 9.3).

In order to explain the definition in [BLO2], and compare it with the one given above, we first need to define two more concepts.

**Definition 2.4**   Let $\mathcal{F}$ be a fusion system over a $p$-group $S$.

- A subgroup $P \leq S$ is *fully centralized* in $\mathcal{F}$ if $|C_S(P)| \geq |C_S(Q)|$ for all $Q \in P^{\mathcal{F}}$.

- A subgroup $P \leq S$ is *fully normalized* in $\mathcal{F}$ if $|N_S(P)| \geq |N_S(Q)|$ for all $Q \in P^{\mathcal{F}}$.

For example, when $\mathcal{F} = \mathcal{F}_S(G)$ for some finite group $G$ and some $S \in \mathrm{Syl}_p(G)$, then by Lemma 1.2(a), a subgroup $P \leq S$ is fully normalized (centralized) in $\mathcal{F}_S(G)$ if and only if $N_S(P) \in \mathrm{Syl}_p(N_G(P))$ ($C_S(P) \in \mathrm{Syl}_p(C_G(P))$).

Definition 2.4 is different from the definition of "fully normalized" and "fully centralized" in [P6, 2.6], but it is equivalent to Puig's definition when working in saturated fusion systems. This will be discussed in much more detail in Section 9.

The following equivalent condition for a fusion system to be saturated, stated in terms of fully normalized and fully centralized subgroups, was given as the definition of saturation in [BLO2, Definition 1.2].

**Proposition 2.5** ([RS, Theorem 5.2])   *Let $\mathcal{F}$ be a fusion system over a $p$-group $S$. Then $\mathcal{F}$ is saturated if and only if the following two conditions hold.*

   (I)   *(Sylow axiom) Each subgroup $P \leq S$ which is fully normalized in $\mathcal{F}$ is also fully centralized and fully automized in $\mathcal{F}$.*

   (II)   *(Extension axiom) Each subgroup $P \leq S$ which is fully centralized in $\mathcal{F}$ is also receptive in $\mathcal{F}$.*

Proposition 2.5 is an immediate consequence of the following lemma.

**Lemma 2.6** ([RS])   *The following hold for any fusion system $\mathcal{F}$ over a $p$-group $S$.*

(a)   *Every receptive subgroup of $S$ is fully centralized.*

(b)   *Every subgroup of $S$ which is fully automized and receptive is fully normalized.*

(c)   *Assume $P \leq S$ is fully automized and receptive. Then for each $Q \in P^{\mathcal{F}}$, there is a morphism $\varphi \in \mathrm{Hom}_{\mathcal{F}}(N_S(Q), N_S(P))$ such that $\varphi(Q) = P$. Furthermore, $Q$ is fully centralized if and only if it is receptive, and is fully normalized if and only if it is fully automized and receptive.*

*Proof.* **(a)**   Assume $P \leq S$ is receptive. Fix any $Q \in P^{\mathcal{F}}$ and any $\varphi \in \mathrm{Iso}_{\mathcal{F}}(Q, P)$. Since $P$ is receptive, $\varphi$ extends to some $\bar{\varphi} \in \mathrm{Hom}_{\mathcal{F}}(N_\varphi, S)$, where $N_\varphi$ contains $C_S(Q)$ by definition. Thus $\bar{\varphi}$ sends $C_S(Q)$ injectively into $C_S(P)$, and so $|C_S(P)| \geq |C_S(Q)|$. Since this holds for all $Q \in P^{\mathcal{F}}$, $P$ is fully centralized in $\mathcal{F}$.

**(b)** Now assume $P$ is fully automized and receptive, and fix $Q \in P^{\mathcal{F}}$. Then $|C_S(Q)| \leq |C_S(P)|$ by (a), and $|\mathrm{Aut}_S(Q)| \leq |\mathrm{Aut}_S(P)|$ since $\mathrm{Aut}_{\mathcal{F}}(Q) \cong \mathrm{Aut}_{\mathcal{F}}(P)$ and $\mathrm{Aut}_S(P) \in \mathrm{Syl}_p(\mathrm{Aut}_{\mathcal{F}}(P))$. Thus

$$|N_S(Q)| = |C_S(Q)| \cdot |\mathrm{Aut}_S(Q)| \leq |C_S(P)| \cdot |\mathrm{Aut}_S(P)| = |N_S(P)|.$$

Since this holds for all $Q \in P^{\mathcal{F}}$, $P$ is fully normalized in $\mathcal{F}$.

**(c)** Assume $P$ is fully automized and receptive, and fix $Q \in P^{\mathcal{F}}$. Choose $\psi \in \mathrm{Iso}_{\mathcal{F}}(Q, P)$. Then ${}^{\psi}\mathrm{Aut}_S(Q)$ is a $p$-subgroup of $\mathrm{Aut}_{\mathcal{F}}(P)$, and hence is $\mathrm{Aut}_{\mathcal{F}}(P)$-conjugate to a subgroup of $\mathrm{Aut}_S(P)$ since $P$ is fully automized. Fix $\alpha \in \mathrm{Aut}_{\mathcal{F}}(P)$ such that ${}^{\alpha\psi}\mathrm{Aut}_S(Q)$ is contained in $\mathrm{Aut}_S(P)$. Then $N_{\alpha\psi} = N_S(Q)$ (see Definition 2.2), and so $\alpha\psi$ extends to some homomorphism $\varphi \in \mathrm{Hom}_{\mathcal{F}}(N_S(Q), S)$. Since $\varphi(Q) = \alpha\psi(Q) = P$, $\mathrm{Im}(\varphi) \leq N_S(P)$.

If $Q$ is fully centralized, then $\varphi(C_S(Q)) = C_S(P)$. Fix $R \leq S$ and $\beta \in$ $\mathrm{Iso}_{\mathcal{F}}(R, Q)$. For $g \in N_\beta$, ${}^\beta c_g \in \mathrm{Aut}_S(Q)$ implies ${}^{\alpha\psi\beta} c_g \in \mathrm{Aut}_S(P)$ since $\alpha\psi$ extends to a homomorphism defined on $N_S(Q)$, and thus $g \in N_{\alpha\psi\beta}$. Since $P$ is receptive, $\alpha\psi\beta$ extends to a homomorphism $\chi \in \mathrm{Hom}_{\mathcal{F}}(N_\beta, N_S(P))$. For each $g \in N_\beta$, ${}^\beta c_g = c_h$ for some $h \in N_S(Q)$ by definition of $N_\beta$, so $c_{\varphi(h)} = {}^\varphi c_h = c_{\chi(g)}$, and thus $\chi(g) \in \mathrm{Im}(\varphi) \cdot C_S(P) = \mathrm{Im}(\varphi)$. Thus $\mathrm{Im}(\chi) \leq \mathrm{Im}(\varphi)$, so $\chi$ factors through some $\bar\beta \in \mathrm{Hom}_{\mathcal{F}}(N_\beta, N_S(Q))$ with $\bar\beta|_R = \beta$, and this proves that $Q$ is receptive.

If $Q$ is fully normalized, then $\varphi$ is an isomorphism. Hence $\varphi$ sends $C_S(Q)$ onto $C_S(P)$, so $Q$ is fully centralized and hence receptive. Also, $\mathrm{Aut}_S(Q) \cong N_S(Q)/C_S(Q)$ is isomorphic to $\mathrm{Aut}_S(P) \cong N_S(P)/C_S(P)$, and $\mathrm{Aut}_{\mathcal{F}}(Q) \cong \mathrm{Aut}_{\mathcal{F}}(P)$ since $Q \in P^{\mathcal{F}}$. So $\mathrm{Aut}_S(Q) \in \mathrm{Syl}_p(\mathrm{Aut}_{\mathcal{F}}(Q))$ since $\mathrm{Aut}_S(P) \in \mathrm{Syl}_p(\mathrm{Aut}_{\mathcal{F}}(P))$, and $Q$ is fully automized. $\qquad\square$

We end this section with an example, which describes how to list all possible saturated fusion systems over one very small 2-group.

**Example 2.7**  Assume $S \cong D_8$: the dihedral group of order 8. Fix generators $a, b \in S$, where $|a| = 4$, $|b| = 2$, and $|ab| = 2$. Set $T_0 = \langle a^2, b \rangle$ and $T_1 = \langle a^2, ab \rangle$: these are the only subgroups of $S$ isomorphic to $C_2 \times C_2$. Then the following hold for any saturated fusion system $\mathcal{F}$ over $S$.

(a)  Since $S$ is fully automized and $\mathrm{Aut}(S)$ is a 2-group, $\mathrm{Aut}_{\mathcal{F}}(S) = \mathrm{Inn}(S)$.

(b)  If $P = \langle a \rangle$ and $Q = P$ or $S$, then $\mathrm{Hom}_{\mathcal{F}}(P, Q) = \mathrm{Hom}_S(P, Q) = \mathrm{Hom}(P, Q)$.

(c)  The subgroups $T_0$ and $T_1$ are both fully normalized in $\mathcal{F}$, and hence are fully automized and receptive. So if $T_1 \in T_0^{\mathcal{F}}$, then by Lemma 2.6(c), there is $\alpha \in \mathrm{Aut}_{\mathcal{F}}(S)$ such that $\alpha(T_0) = T_1$. Since this contradicts (a), $T_0$ and $T_1$ cannot be $\mathcal{F}$-conjugate.

(d)  Set $P = \langle a^2 \rangle$, and let $Q \leq S$ be any subgroup of order 2. Since $P \trianglelefteq S$, $P$ is fully normalized in $\mathcal{F}$ and hence fully automized and receptive. So if $Q \in P^{\mathcal{F}}$ (and $Q \neq P$), then by Lemma 2.6(c) again, there is some $\varphi \in \mathrm{Hom}_{\mathcal{F}}(T_i, S)$, where $T_i = N_S(Q)$ ($i = 0$ or $1$), such that $\varphi(Q) = P$. Also, $\varphi(T_i) = T_i$ by (c).

(e)  By (a–d), $\mathcal{F}$ is completely determined by $\mathrm{Aut}_{\mathcal{F}}(T_0)$ and $\mathrm{Aut}_{\mathcal{F}}(T_1)$. Also, for each $i$, $\mathrm{Aut}_S(T_i) \leq \mathrm{Aut}_{\mathcal{F}}(T_i) \leq \mathrm{Aut}(T_i)$, and hence $\mathrm{Aut}_{\mathcal{F}}(T_i)$ has order 2 or 6.

Thus there are at most four saturated fusion systems over $S$. Denote these fusion systems $\mathcal{F}_{ij}$, where $i = 0$ if $|\mathrm{Aut}_{\mathcal{F}}(T_0)| = 2$, $i = 1$ if $|\mathrm{Aut}_{\mathcal{F}}(T_0)| = 6$, and similarly $j = 0, 1$ depending on $|\mathrm{Aut}_{\mathcal{F}}(T_1)|$. Then $\mathcal{F}_{00}$ is the fusion system of $S \cong D_8$ itself, $\mathcal{F}_{01} \cong \mathcal{F}_{10}$ are isomorphic to the fusion system of $S_4$, and $\mathcal{F}_{11}$ is isomorphic to the fusion system of $A_6$. In particular, all four of the $\mathcal{F}_{ij}$ are saturated.

## 3. ALPERIN'S FUSION THEOREM

We next prove a version of the Alperin-Goldschmidt fusion theorem for abstract fusion systems. Before doing this, we must list some important classes of subgroups in a given fusion system. All of these are modelled on analogous definitions for $p$-subgroups of groups.

**Definition 3.1** Let $\mathcal{F}$ be a fusion system over a $p$-group $S$.

- For each $P \leq S$, set $\mathrm{Out}_{\mathcal{F}}(P) = \mathrm{Aut}_{\mathcal{F}}(P)/\mathrm{Inn}(P)$ and $\mathrm{Out}_S(P) = \mathrm{Aut}_S(P)/\mathrm{Inn}(P)$. Thus $\mathrm{Out}_S(P) \leq \mathrm{Out}_{\mathcal{F}}(P) \leq \mathrm{Out}(P)$.

- A subgroup $P$ of $S$ is $\mathcal{F}$-*centric* if $C_S(Q) = Z(Q)$ for all $Q \in P^{\mathcal{F}}$. Equivalently, $P$ is $\mathcal{F}$-centric if $P$ is fully centralized in $\mathcal{F}$ and $C_S(P) = Z(P)$.

- A subgroup $P$ of $S$ is $\mathcal{F}$-*radical* if $\mathrm{Out}_{\mathcal{F}}(P)$ is $p$-reduced; i.e., if $O_p(\mathrm{Out}_{\mathcal{F}}(P)) = 1$. We say $P$ is $\mathcal{F}$-*centric-radical* if it is $\mathcal{F}$-centric and $\mathcal{F}$-radical.

- Let $\mathcal{F}^{cr} \subseteq \mathcal{F}^c \subseteq \mathcal{F}$ denote the full subcategories whose objects are the $\mathcal{F}$-centric-radical, and $\mathcal{F}$-centric, subgroups of $S$, respectively.

- Let $\mathcal{F}^c = \mathrm{Ob}(\mathcal{F}^c)$ and $\mathcal{F}^{cr} = \mathrm{Ob}(\mathcal{F}^{cr})$ denote the sets of $\mathcal{F}$-centric and $\mathcal{F}$-centric-radical subgroups of $S$, respectively.

When $G$ is a finite group, then a $p$-subgroup $P \leq G$ is called $p$-*centric* if $Z(P) \in \mathrm{Syl}_p(C_G(P))$. If $S \in \mathrm{Syl}_p(G)$ and $\mathcal{F} = \mathcal{F}_S(G)$, then $P \leq S$ is easily seen to be $\mathcal{F}$-centric if and only if $P$ is $p$-centric in $G$. However, $P$ being $\mathcal{F}$-radical is *not* the same as being a radical $p$-subgroup of $G$: $P$ is radical in $G$ (by definition) if $O_p(N_G(P)/P) = 1$, while $P$ is $\mathcal{F}$-radical if and only if $O_p(N_G(P)/(P{\cdot}C_G(P))) = 1$. For example, if $S$ is abelian and normal in $G$, then every subgroup of $S$ is $\mathcal{F}$-radical, but the only radical $p$-subgroup of $G$ is $S$ itself. As a second example, if $G$ is dihedral of order 24, $S \in \mathrm{Syl}_2(G)$, and $P = O_2(G)$ is the normal cyclic subgroup of order 4, then $P$ is a radical 2-subgroup of $G$ but is not $\mathcal{F}$-radical (since $\mathrm{Out}_G(P)$ has order 2).

**Definition 3.2** Let $\mathcal{F}$ be a fusion system over a $p$-group $S$. A subgroup $P$ of $S$ is $\mathcal{F}$-*essential* if $P$ is $\mathcal{F}$-centric and fully normalized in $\mathcal{F}$, and $\mathrm{Out}_{\mathcal{F}}(P)$ contains a strongly $p$-embedded subgroup (Definition A.6).

Note that Puig's definition of "$\mathcal{F}$-essential" in [P6, § 5.4] is different from this one. It is, however, equivalent (except for our requirement that the subgroup be fully normalized) when the fusion system is saturated.

The next proposition describes the key property of essential subgroups.

**Proposition 3.3**  *Let $\mathcal{F}$ be a saturated fusion system over a $p$-group $S$.*

(a)  *Each $\mathcal{F}$-essential subgroup of $S$ is $\mathcal{F}$-centric-radical and fully normalized in $\mathcal{F}$.*

(b)  *Fix a proper subgroup $P$ of $S$ which is fully normalized, and let $H_P \leq \mathrm{Aut}_{\mathcal{F}}(P)$ be the subgroup generated by those $\alpha \in \mathrm{Aut}_{\mathcal{F}}(P)$ which extend to $\mathcal{F}$-isomorphisms between strictly larger subgroups of $S$. Then either $P$ is not $\mathcal{F}$-essential and $H_P = \mathrm{Aut}_{\mathcal{F}}(P)$; or $P$ is $\mathcal{F}$-essential and $H_P/\mathrm{Inn}(P)$ is strongly $p$-embedded in $\mathrm{Out}_{\mathcal{F}}(P)$.*

*Proof.* **(a)**  By definition, each $\mathcal{F}$-essential subgroup $P \leq S$ is $\mathcal{F}$-centric and fully normalized. Since $\mathrm{Out}_{\mathcal{F}}(P)$ contains a strongly $p$-embedded subgroup, $O_p(\mathrm{Out}_{\mathcal{F}}(P)) = 1$ by Proposition A.7(c), and hence $P$ is also $\mathcal{F}$-radical.

**(b)**  Fix a proper subgroup $P$ of $S$ which is fully normalized in $\mathcal{F}$. In particular, $\mathrm{Aut}_S(P) \in \mathrm{Syl}_p(\mathrm{Aut}_{\mathcal{F}}(P))$. If $P$ is not $\mathcal{F}$-centric, then each $\alpha \in \mathrm{Aut}_{\mathcal{F}}(P)$ extends to some $\mathcal{F}$-automorphism of $C_S(P){\cdot}P > P$ since $P$ is receptive, and so $H_P = \mathrm{Aut}_{\mathcal{F}}(P)$ in this case. So we can assume $P$ is $\mathcal{F}$-centric. Also, $N_S(P) > P$ since $P < S$, so $\mathrm{Out}_S(P) \neq 1$ since $P$ is $\mathcal{F}$-centric, and thus $p\,\big|\,|H_P/\mathrm{Inn}(P)|$ since $H_P \geq \mathrm{Aut}_S(P)$.

We claim that

$$H_P = \big\langle \alpha \in \mathrm{Aut}_{\mathcal{F}}(P) \,\big|\, {}^{\alpha}\mathrm{Aut}_S(P) \cap \mathrm{Aut}_S(P) > \mathrm{Inn}(P) \big\rangle. \tag{1}$$

To see this, fix any $\alpha \in \mathrm{Aut}_{\mathcal{F}}(P)$ such that ${}^{\alpha}\mathrm{Aut}_S(P) \cap \mathrm{Aut}_S(P) > \mathrm{Inn}(P)$, and recall from Definition 2.2 that

$$N_\alpha = \big\{ g \in N_S(P) \,\big|\, {}^{\alpha}c_g \in \mathrm{Aut}_S(P) \big\}.$$

Then ${}^{\alpha}\mathrm{Aut}_{N_\alpha}(P) = {}^{\alpha}\mathrm{Aut}_S(P) \cap \mathrm{Aut}_S(P) > \mathrm{Inn}(P)$, so $N_\alpha > P$. By the extension axiom, $\alpha$ extends to a morphism in $\mathrm{Hom}_{\mathcal{F}}(N_\alpha, S)$, and thus $\alpha \in H_P$. Conversely, if $\alpha \in \mathrm{Aut}_{\mathcal{F}}(P)$ extends to $\bar{\alpha} \in \mathrm{Hom}_{\mathcal{F}}(Q, S)$ for some $Q > P$, ${}^{\alpha}\mathrm{Aut}_Q(P) \leq \mathrm{Aut}_S(P)$, and

$${}^{\alpha}\mathrm{Aut}_S(P) \cap \mathrm{Aut}_S(P) \geq {}^{\alpha}\mathrm{Aut}_Q(P) > \mathrm{Inn}(P).$$

This proves (1). Hence by Proposition A.7(b), either $H_P = \mathrm{Aut}_{\mathcal{F}}(P)$, in which case $\mathrm{Out}_{\mathcal{F}}(P)$ contains no strongly $p$-embedded subgroup and $P$ is not $\mathcal{F}$-essential; or $H_P < \mathrm{Aut}_{\mathcal{F}}(P)$, in which case $H_P/\mathrm{Inn}(P)$ is strongly $p$-embedded in $\mathrm{Out}_{\mathcal{F}}(P)$ and $P$ is $\mathcal{F}$-essential.  $\square$

There is always a "universal" fusion system $\mathcal{U}$ over any $p$-group $S$ which contains all other fusion systems over $S$: for each $P, Q \leq S$, $\mathrm{Hom}_{\mathcal{U}}(P, Q)$

is the set of all monomorphisms from $P$ to $Q$. Also, the intersection of two fusion systems over $S$ is clearly again a fusion system over $S$. This allows us, in certain cases, to talk about the "smallest fusion system" over a given $S$ which satisfies certain given conditions.

**Definition 3.4** Fix a $p$-group $S$.

(a) Fix another $p$-group $T$, and an isomorphism $\varphi\colon S \xrightarrow{\;\cong\;} T$. For any fusion system $\mathcal{F}$ over $S$, let $^{\varphi}\mathcal{F}$ be the fusion system over $T$ defined by setting

$$\mathrm{Hom}_{\,^{\varphi}\mathcal{F}}(P,Q) = \{\,^{\varphi}\psi \in \mathrm{Hom}(P,Q) \mid \psi \in \mathrm{Hom}_{\mathcal{F}}(\varphi^{-1}(P), \varphi^{-1}(Q))\}$$

for all $P, Q \leq T$.

(b) For any set $\mathfrak{X}$ of monomorphisms between subgroups of $S$ and/or fusion systems over subgroups of $S$, the *fusion system generated by* $\mathfrak{X}$, denoted $\langle\mathfrak{X}\rangle_S$, is the smallest fusion system over $S$ (not necessarily saturated) which contains $\mathfrak{X}$. Thus $\langle\mathfrak{X}\rangle_S$ is the intersection of all fusion systems over $S$ which contain $\mathfrak{X}$, and the morphisms in $\langle\mathfrak{X}\rangle_S$ are the composites of restrictions of homomorphisms in the set $\mathfrak{X}\cup\mathrm{Inn}(S)$ and their inverses. We write $\langle\mathfrak{X}\rangle = \langle\mathfrak{X}\rangle_S$ when the choice of $S$ is clear.

In these terms, the "Alperin-Goldschmidt fusion theorem" for fusion systems says that a saturated fusion system is generated by certain automorphism groups. This result is originally due to Puig (see [P6, Corollary 5.10] for a linearlized version of the theorem). It is modeled on the original fusion theorems of Alperin [Al1] and Goldschmidt [Gd1].

**Theorem 3.5** *Fix a $p$-group $S$ and a saturated fusion system $\mathcal{F}$ over $S$. Then*

$$\mathcal{F} = \langle \mathrm{Aut}_{\mathcal{F}}(P) \mid P = S \text{ or } P \text{ is } \mathcal{F}\text{-essential}\rangle_S.$$

*Proof.* Set $\mathcal{E} = \langle \mathrm{Aut}_{\mathcal{F}}(P) \mid P = S \text{ or } P \text{ is } \mathcal{F}\text{-essential}\rangle$, and assume $\mathcal{E} \subsetneq \mathcal{F}$. Fix a morphism $\varphi \in \mathrm{Hom}_{\mathcal{F}}(P,Q) \smallsetminus \mathrm{Hom}_{\mathcal{E}}(P,Q)$, chosen such that $|P|$ is maximal subject to this constraint. Since $\mathrm{Aut}_{\mathcal{E}}(S) = \mathrm{Aut}_{\mathcal{F}}(S)$, $P < S$. Since $\varphi$ factors as an isomorphism in $\mathcal{F}$ followed by an inclusion (which lies in $\mathcal{E}$), we can assume $\varphi \in \mathrm{Iso}_{\mathcal{F}}(P,Q)$.

Choose $R \in P^{\mathcal{F}}$ which is fully normalized in $\mathcal{F}$. By Lemma 2.6(c), there are $\psi_1 \in \mathrm{Hom}_{\mathcal{F}}(N_S(P), N_S(R))$ and $\psi_2 \in \mathrm{Hom}_{\mathcal{F}}(N_S(Q), N_S(R))$ such that $\psi_1(P) = R$ and $\psi_2(Q) = R$. Since $P, Q < S$, $P < N_S(P)$ and $Q < N_S(Q)$ (see Lemma A.1), and so $\psi_1, \psi_2 \in \mathrm{Mor}(\mathcal{E})$ by the maximality of $|P|$. Upon replacing $\varphi$ by $(\psi_2|_Q)\varphi(\psi_1|_P)^{-1}$, we can assume $\varphi \in \mathrm{Aut}_{\mathcal{F}}(R)$ where $R$ is fully normalized. If $R$ is $\mathcal{F}$-essential, then $\mathrm{Aut}_{\mathcal{F}}(R) = \mathrm{Aut}_{\mathcal{E}}(R)$ by definition of $\mathcal{E}$. If $R$ is not $\mathcal{F}$-essential, then by Proposition 3.3(b), $\mathrm{Aut}_{\mathcal{F}}(R)$ is generated by automorphisms which extend to strictly larger subgroups, and $\mathrm{Aut}_{\mathcal{F}}(R) = \mathrm{Aut}_{\mathcal{E}}(R)$ by the maximality of $|P|$. So in either case, $\varphi \in \mathrm{Mor}(\mathcal{E})$, contradicting the original assumption on $\varphi$. Thus $\mathcal{E} = \mathcal{F}$.  $\square$

As an immediate consequence of Theorem 3.5, we get the Alperin-Goldschmidt fusion theorem in its more explicit form.

**Theorem 3.6**  *Fix a p-group $S$ and a saturated fusion system $\mathcal{F}$ over $S$. For each $P, Q \leq S$ and each $\varphi \in \mathrm{Hom}_{\mathcal{F}}(P, Q)$, there are subgroups $P = P_0, P_1, \ldots, P_k = Q$, subgroups $R_i \geq \langle P_{i-1}, P_i \rangle$ ($i = 1, \ldots, k$) which are $\mathcal{F}$-essential or equal to $S$, and automorphisms $\varphi_i \in \mathrm{Aut}_{\mathcal{F}}(R_i)$, such that $\varphi_i(P_{i-1}) \leq P_i$ for each $i$, and $\varphi = (\varphi_k|_{P_{k-1}}) \circ \cdots \circ (\varphi_1|_{P_0})$.*

The following is one very elementary consequence of Theorem 3.5 or 3.6.

**Corollary 3.7**  *Let $\mathcal{F}$ be a saturated fusion system over the p-group $S$. Assume, for each $P \leq S$, that $\mathrm{Aut}_{\mathcal{F}}(P)$ is a p-group. Then $\mathcal{F}$ is the fusion system of the group $S$.*

*Proof.*  Since all automorphism groups are $p$-groups, there are no $\mathcal{F}$-essential subgroups. Hence by Theorem 3.5 or 3.6, $\mathcal{F}$ is generated by $\mathrm{Aut}_{\mathcal{F}}(S) = \mathrm{Inn}(S)$, and thus $\mathcal{F} = \mathcal{F}_S(S)$.  $\square$

We ended the last section by laboriously describing, almost directly from the definitions, all saturated fusion systems over a dihedral group of order 8. With the help of Theorem 3.5, we can now do the same thing much more quickly for arbitrary 2-groups which are dihedral, semidihedral, or quaternion.

**Example 3.8**  Assume $S$ is a dihedral, semidihedral, or quaternion group of order $2^n \geq 16$. Fix generators $a, b \in S$, where $|a| = 2^{n-1}$, and where $|b| = 2$ if $S$ is semidihedral. For each $i \in \mathbb{Z}$, set $T_i = \langle a^{2^{n-2}}, a^i b \rangle \cong C_2^2$ if $|a^i b| = 2$, or $T_i = \langle a^{2^{n-3}}, a^i b \rangle \cong Q_8$ if $|a^i b| = 4$. Then the following hold for any saturated fusion system $\mathcal{F}$ over $S$.

(a)  If $\alpha \in \mathrm{Aut}(S)$ has odd order, then $\alpha|_{\langle a \rangle} = \mathrm{Id}$, and hence $\alpha = \mathrm{Id}_S$ by Lemma A.2. Thus $\mathrm{Aut}(S)$ is a 2-group, and hence $\mathrm{Aut}_{\mathcal{F}}(S) = \mathrm{Inn}(S)$ since $S$ is fully automized.

(b)  By similar arguments, the only subgroups of $S$ whose automorphism groups are not 2-groups are the $T_i$. Hence these are the only subgroups which could be $\mathcal{F}$-essential.

(c)  For each $i$, $\mathrm{Out}(T_i) \cong S_3$ and $\mathrm{Out}_S(T_i) \cong C_2$. Hence $\mathrm{Out}_{\mathcal{F}}(T_i)$ must be one of these two groups.

(d)  If $i \equiv j \pmod 2$, then $T_i$ and $T_j$ are $S$-conjugate, and hence $\mathrm{Out}_{\mathcal{F}}(T_i) \cong \mathrm{Out}_{\mathcal{F}}(T_j)$.

Thus there are at most four distinct saturated fusion systems over $S$. We denote these $\mathcal{F}_{ij}$, where $i = 0$ if $|\mathrm{Out}_{\mathcal{F}}(T_0)| = 2$, $i = 1$ if $|\mathrm{Out}_{\mathcal{F}}(T_0)| = 6$, and simiarly for $j = 0, 1$ depending on $|\mathrm{Out}_{\mathcal{F}}(T_1)|$. All of these are fusion systems of finite groups. For example, when $S$ is semidihedral of order $2^n$ with $n \geq 4$ (hence $T_0 \cong C_2^2$ and $T_1 \cong Q_8$), then for any odd prime power $q \equiv 2^{n-2} - 1 \pmod{2^{n-1}}$, $\mathcal{F}_{00}$ is the fusion system of $S$ itself, $\mathcal{F}_{01}$ is the fusion system of $GL_2(q)$, $\mathcal{F}_{10}$ is the fusion system of a certain extension of $PSL_2(q^2)$ by $C_2$ (the extension by the product of a field and a diagonal automorphism), and $\mathcal{F}_{11}$ is the fusion system of $PSL_3(q)$.

In many situations, when we want to prove that a fusion system over $S$ is saturated, it is very useful to know that we need prove the saturation axioms only for certain subgroups of $S$. Before stating a theorem, we first need some terminology for formulating this.

**Definition 3.9** Let $\mathcal{F}$ be a fusion system over a $p$-group $S$, and let $\mathcal{H}$ be a set of subgroups of $S$.

- Let $\mathcal{F}|_{\mathcal{H}} \subseteq \mathcal{F}$ be the full subcategory with object set $\mathcal{H}$.

- The fusion system $\mathcal{F}$ is $\mathcal{H}$-*generated* if $\mathcal{F} = \langle \mathcal{F}|_{\mathcal{H}} \rangle_S$.

- Assume $\mathcal{H}$ is closed under $\mathcal{F}$-conjugacy. Then $\mathcal{F}$ is $\mathcal{H}$-*saturated* if each $\mathcal{F}$-conjugacy class in $\mathcal{H}$ contains a subgroup which is fully automized and receptive in $\mathcal{F}$.

By Lemma 2.6, a fusion system $\mathcal{F}$ is $\mathcal{H}$-saturated (when $\mathcal{H}$ is closed under $\mathcal{F}$-conjugacy) if and only if the subgroups in $\mathcal{H}$ all satisfy the Sylow axiom and the extension axiom (see Proposition 2.5). Hence the above definition is equivalent to that in [5a1, Definition 2.1].

Theorem 3.5 (together with Proposition 3.3(a)) implies that each saturated fusion system $\mathcal{F}$ is $\mathcal{H}$-generated, when $\mathcal{H}$ is the set of subgroups which are $\mathcal{F}$-centric-radical and fully normalized in $\mathcal{F}$. The next theorem can be thought of as a partial converse to this. It was first shown by Puig (see [P6, Theorem 3.8]) in the special case where $\mathcal{H} = \mathcal{F}^c$.

**Theorem 3.10** ([5a1, Theorem 2.2])  *Let $\mathcal{F}$ be a fusion system over a $p$-group $S$. Let $\mathcal{H}$ be a set of subgroups of $S$ closed under $\mathcal{F}$-conjugacy such that $\mathcal{F}$ is $\mathcal{H}$-saturated and $\mathcal{H}$-generated. Assume also that each $\mathcal{F}$-centric subgroup of $S$ not in $\mathcal{H}$ is $\mathcal{F}$-conjugate to some subgroup $P \leq S$ such that*

$$\mathrm{Out}_S(P) \cap O_p(\mathrm{Out}_{\mathcal{F}}(P)) \neq 1. \tag{$*$}$$

*Then $\mathcal{F}$ is saturated.*

Theorem 3.10 can be very useful, for example, when one is working with a fusion system $\mathcal{F}$ over a $p$-group $S$, where morphisms are known explicitly only among subgroups in a certain family $\mathcal{H}$. For example, in certain cases, morphisms are defined explicitly between subgroups in $\mathcal{H}$, and then among other subgroups by taking composites of restrictions. In this situation (where $\mathcal{F}$ is $\mathcal{H}$-generated by construction), if $\mathcal{H}$ contains all $\mathcal{F}$-centric subgroups (or satisfies the slightly weaker hypothesis in the theorem), then to prove $\mathcal{F}$ is saturated, it suffices to check the saturation axioms for subgroups in $\mathcal{H}$.

It is natural to ask whether condition $(*)$ in Theorem 3.10 can be replaced by the condition that $\mathcal{H}$ contain all subgroups which are $\mathcal{F}$-centric-radical. An example was given in [5a1] to show that this is not the case, but Kasper Andersen recently pointed out to us that this example was in error. So we give a different example here of a fusion system which is not saturated, but which is $\mathcal{H}$-generated and $\mathcal{H}$-saturated for some set $\mathcal{H}$ of subgroups which includes all of those which are $\mathcal{F}$-centric-radical.

**Example 3.11**   Consider the following matrices in $SL_3(\mathbb{Z}/4)$:

$$Z = \begin{pmatrix} -1 & 0 & -1 \\ 0 & -1 & 0 \\ 0 & 0 & 1 \end{pmatrix}, \quad X_1 = \begin{pmatrix} -1 & 0 & 0 \\ 2 & 1 & 1 \\ 0 & 0 & -1 \end{pmatrix}, \quad X_2 = \begin{pmatrix} -1 & -1 & 0 \\ 0 & 1 & 0 \\ 0 & 0 & -1 \end{pmatrix},$$

$$A_1 = \begin{pmatrix} 0 & 1 & 0 \\ -1 & -1 & 0 \\ 2 & 0 & 1 \end{pmatrix}, \quad A_2 = \begin{pmatrix} 1 & 0 & 0 \\ 0 & 0 & 1 \\ 2 & -1 & -1 \end{pmatrix}.$$

Set

$$T = \langle Z, X_1, X_2 \rangle, \quad T_i = \langle Z, X_i \rangle, \quad H_i = \langle T, A_i \rangle \quad (i = 1, 2), \quad G = \langle H_1, H_2 \rangle.$$

Let pr: $SL_3(\mathbb{Z}/4) \longrightarrow GL_3(2)$ be the natural projection, and set

$$N = \mathrm{Ker}(\mathrm{pr}) = O_2(SL_3(\mathbb{Z}/4)) \cong C_2^8$$

(the group of matrices $I + 2M$ for $M \in M_3(\mathbb{Z}/4)$ of trace zero). Straightforward computations show that $T \cong D_8$, $T_i \cong C_2^2$, and $T_i \trianglelefteq H_i \cong S_4$. Thus pr sends each of these subgroups injectively into $GL_3(\mathbb{Z}/2)$. In contrast, one easily sees that $\mathrm{pr}(G) = GL_3(2)$, but the element $(A_2 X_2 A_1)^3$ is a nontrivial element of $N$ and hence of $O_2(G) = G \cap N$. Also, $O_2(G) \trianglelefteq \langle G, N \rangle = SL_3(\mathbb{Z}/4)$, no proper nontrivial subgroup of $N$ is normal in $SL_3(\mathbb{Z}/4)$, and we conclude that $O_2(G) = N$ and $G = SL_3(\mathbb{Z}/4)$.

Set $A = C_4^3$, with the canonical action of $G = SL_3(\mathbb{Z}/4)$, and let $\Gamma = A \rtimes G$ be the semidirect product. Set $S = AT$, and set $\mathcal{F} = \langle \mathrm{Inn}(S), \mathrm{Aut}_{AH_1}(AT_1), \mathrm{Aut}_{AH_2}(AT_2) \rangle$ as a fusion system over $S$. Set $\mathcal{H} = \{S, AT_1, AT_2\}$. Then $\mathcal{F}$ is $\mathcal{H}$-generated by construction. Also, $\mathcal{F}$ is $\mathcal{H}$-saturated: all three subgroups in $\mathcal{H}$ are fully automized in $\mathcal{F}$, and the $AT_i$ are receptive since each $\alpha \in N_{\mathrm{Aut}_{\mathcal{F}}(AT_i)}(\mathrm{Aut}_S(AT_i)) = \mathrm{Aut}_S(AT_i)$ extends to $S$ (note that $\mathrm{Out}_{\mathcal{F}}(AT_i) \cong N_{H_i}(T_i)/T_i \cong S_3$). However, $\mathcal{F}$ is not saturated, since $A^{\mathcal{F}} = \{A\}$ but $A$ is not fully automized ($\mathrm{Aut}_S(A)$ is not a

Sylow 2-subgroup of $\mathrm{Aut}_{\mathcal{F}}(A) = G$). Theorem 3.10 does not apply in this case since condition $(*)$ fails to hold for the $\mathcal{F}$-centric subgroup $A$:

$$O_2(\mathrm{Aut}_{\mathcal{F}}(A)) \cap \mathrm{Aut}_S(A) = N \cap T = 1.$$

We claim that $\mathcal{F}^{cr} \subseteq \mathcal{H}$. If $P \leq S$ is such that $PA/A$ is cyclic of order 2 or 4 or $PA = S$, then $\mathrm{Aut}_{\mathcal{F}}(P) = \mathrm{Aut}_S(P)$ since no element of order 3 in $\mathrm{Aut}_{\mathcal{F}}(AT_i)$ (for $i = 1$ or 2) restricts to an automorphism of $P$. If $PA = AT_i$ and $P < AT_i$, then $\mathrm{Aut}_{\mathcal{F}}(P) = \mathrm{Aut}_{AH_i}(P)$, and either $P \notin \mathcal{F}^c$, or $1 \neq \mathrm{Out}_{AT_i}(P) \leq O_2(\mathrm{Out}_{\mathcal{F}}(P))$ (hence $P$ is not $\mathcal{F}$-radical). If $P \leq A$, then either $P < A$ and $P \notin \mathcal{F}^c$, or $P = A$, $\mathrm{Aut}_{\mathcal{F}}(P) = \langle H_1, H_2 \rangle = G$, and $P$ is not $\mathcal{F}$-radical since $O_2(G) = N \neq 1$. Thus $\mathcal{F}^{cr} \subseteq \mathcal{H}$ (and in fact, the two sets are equal).

## 4. NORMAL AND CENTRAL SUBGROUPS OF A FUSION SYSTEM

We next look at normal and central subgroups of a fusion system, as well as strongly and weakly closed subgroups.

**Definition 4.1**   Let $\mathcal{F}$ be a fusion system over a $p$-group $S$. Fix a subgroup $Q$ of $S$.

- $Q$ is *central* in $\mathcal{F}$ if $Q \trianglelefteq S$, and for all $P, R \leq S$ and all $\varphi \in \mathrm{Hom}_{\mathcal{F}}(P, R)$, $\varphi$ extends to a morphism $\bar{\varphi} \in \mathrm{Hom}_{\mathcal{F}}(PQ, RQ)$ such that $\bar{\varphi}|_Q = \mathrm{Id}_Q$.

- $Q$ is *normal* in $\mathcal{F}$ (denoted $Q \trianglelefteq \mathcal{F}$) if $Q \trianglelefteq S$, and for all $P, R \leq S$ and all $\varphi \in \mathrm{Hom}_{\mathcal{F}}(P, R)$, $\varphi$ extends to a morphism $\bar{\varphi} \in \mathrm{Hom}_{\mathcal{F}}(PQ, RQ)$ such that $\bar{\varphi}(Q) = Q$.

- $Q$ is *strongly closed* in $\mathcal{F}$ if no element of $Q$ is $\mathcal{F}$-conjugate to an element of $S \setminus Q$. More generally, for any $P \leq S$ which contains $Q$, $Q$ is strongly closed in $P$ with respect to $\mathcal{F}$ if no element of $Q$ is $\mathcal{F}$-conjugate to an element of $P \setminus Q$.

- $Q$ is *weakly closed* in $\mathcal{F}$ if $Q^{\mathcal{F}} = \{Q\}$. More generally, for any $P \leq S$ which contains $Q$, $Q$ is weakly closed in $P$ with respect to $\mathcal{F}$ if $Q$ is the only subgroup of $P$ in $Q^{\mathcal{F}}$.

The following implications, for any $\mathcal{F}$ and any $Q \leq S$, follow immediately from Definition 4.1:

$$Q \text{ central} \implies Q \text{ normal} \implies Q \text{ strongly closed}$$

$$\implies Q \text{ weakly closed} \implies Q \trianglelefteq S.$$

Note that when $G$ is a finite group, $S \in \mathrm{Syl}_p(G)$, and $Q \trianglelefteq S$ is normal (central) in $\mathcal{F}_S(G)$, then this need *not* imply that $Q$ is normal (central) in

$G$. What it does imply is that $N_G(Q)$ $(C_G(Q))$ controls $p$-fusion in $G$ (i.e., has the same fusion system over $S$).

The following lemma gives another, simpler criterion for a subgroup to be central — at least, when the fusion system is saturated.

**Lemma 4.2**  *Let $\mathcal{F}$ be a saturated fusion system over a $p$-group $S$. For any subgroup $Q$ in $S$, $Q$ is central in $\mathcal{F}$ if and only if $x^{\mathcal{F}} = \{x\}$ for each $x \in Q$.*

*Proof.* Assume $x^{\mathcal{F}} = \{x\}$ for each $x \in Q$. In particular, $Q \leq Z(S)$. Fix $P, R \leq S$ and $\varphi \in \mathrm{Hom}_{\mathcal{F}}(P, R)$, and set $R_0 = \varphi(P)$. Choose $T \in P^{\mathcal{F}}$ which is fully centralized in $\mathcal{F}$, and fix some $\chi \in \mathrm{Iso}_{\mathcal{F}}(R_0, T)$. Since $Q \leq Z(S)$ and $T$ is receptive in $\mathcal{F}$, $\psi = \chi\varphi$ extends to $\bar{\psi} \in \mathrm{Hom}_{\mathcal{F}}(PQ, S)$, and $\chi$ extends to $\bar{\chi} \in \mathrm{Iso}_{\mathcal{F}}(R_0 Q, S)$. For each $x \in Q$, since $x^{\mathcal{F}} = \{x\}$, $\bar{\chi}(x) = x = \bar{\psi}(x)$. Hence $\bar{\chi}(R_0 Q) = TQ = \bar{\psi}(PQ)$, and $\bar{\chi}^{-1} \circ \bar{\psi} \in \mathrm{Hom}_{\mathcal{F}}(PQ, RQ)$ extends $\varphi$ and is the identity on $Q$. Thus $Q$ is central in $\mathcal{F}$.

The converse is clear: if $Q$ is central in $\mathcal{F}$, then $x^{\mathcal{F}} = \{x\}$ for each $x \in Q$. $\qquad\square$

For example, when $S$ is a cyclic 2-group, or a quaternion 2-group of order $\geq 8$, then $S$ has a unique involution $z$. Hence $z^{\mathcal{F}} = \{z\}$, and $\langle z \rangle$ is central in $\mathcal{F}$, for *any* saturated fusion system $\mathcal{F}$ over $S$.

It follows immediately from the definitions of normal and central subgroups that if $Q_1, Q_2 \leq S$ are normal (central) in $\mathcal{F}$, then so is $Q_1 Q_2$. So it makes sense to talk about the maximal normal or central subgroup in $\mathcal{F}$.

**Definition 4.3**  For any fusion system $\mathcal{F}$ over a $p$-group $S$,

- $O_p(\mathcal{F}) \trianglelefteq S$ denotes the largest subgroup of $S$ which is normal in $\mathcal{F}$; and

- $Z(\mathcal{F}) \leq Z(S)$ denotes the largest subgroup of $S$ which is central in $\mathcal{F}$.

Recall that for any finite group $G$, $Z^*(G) \leq G$ is the subgroup such that $Z^*(G)/O_{2'}(G) = Z(G/O_{2'}(G))$. So Glauberman's $Z^*$-theorem [Gl1, Corollary 1] can be interpreted to say that for $S \in \mathrm{Syl}_2(G)$, $Z(\mathcal{F}_S(G)) = Z^*(G) \cap S$.

The next proposition is very elementary.

**Proposition 4.4**  *If $Q$ is normal in a fusion system $\mathcal{F}$ over a $p$-group $S$, then each characteristic subgroup of $Q$ is also normal in $\mathcal{F}$.*

*Proof.* Assume $U$ is characteristic in $Q$, where $Q \trianglelefteq \mathcal{F}$. Then for each $P, R \leq S$ and each $\varphi \in \mathrm{Hom}_{\mathcal{F}}(P, R)$, there is $\bar{\varphi} \in \mathrm{Hom}_{\mathcal{F}}(PQ, RQ)$ such that $\bar{\varphi}|_P = \varphi$ and $\bar{\varphi}(Q) = Q$, and $\bar{\varphi}(U) = U$ since $U$ char $Q$. Thus $U \trianglelefteq \mathcal{F}$. $\qquad\square$

One easy consequence of Alperin's fusion theorem is the following useful criterion for a subgroup to be normal in a fusion system. Conditions (a) and (b) in the following proposition were shown in [5a1, Proposition 1.6] to be equivalent.

**Proposition 4.5**   *Let $\mathcal{F}$ be a saturated fusion system over a p-group $S$. Then for any $Q \leq S$, the following conditions are equivalent:*

(a)   *$Q$ is normal in $\mathcal{F}$.*

(b)   *$Q$ is strongly closed in $\mathcal{F}$, and $Q \leq P$ for each $P \in \mathcal{F}^{cr}$.*

(c)   *If $P \leq S$ is $\mathcal{F}$-essential or $P = S$, then $P \geq Q$ and $Q$ is $\mathrm{Aut}_{\mathcal{F}}(P)$-invariant.*

*Proof.* (**a $\Rightarrow$ b**)   Assume $Q \trianglelefteq \mathcal{F}$. Clearly, $Q$ is strongly closed in $\mathcal{F}$.

For each $P \in \mathcal{F}^{cr}$ and each $\alpha \in \mathrm{Aut}_{\mathcal{F}}(P)$, $\alpha$ extends to some $\bar{\alpha} \in \mathrm{Aut}_{\mathcal{F}}(PQ)$ such that $\bar{\alpha}(Q) = Q$, and hence $\alpha$ normalizes $\mathrm{Aut}_{PQ}(P)$. Thus $\mathrm{Aut}_{PQ}(P) \trianglelefteq \mathrm{Aut}_{\mathcal{F}}(P)$, and so $\mathrm{Aut}_{PQ}(P) = \mathrm{Inn}(P)$ since $P$ is $\mathcal{F}$-radical. Since $P$ is also $\mathcal{F}$-centric, it follows that $N_{PQ}(P) = P$, and hence that $PQ = P$ (Lemma A.1). So $P \geq Q$.

(**b $\Rightarrow$ c**)   If $P$ is $\mathcal{F}$-essential or $P = S$, then $P \in \mathcal{F}^{cr}$ by Proposition 3.3(a). Hence $P \geq Q$, and $Q$ is $\mathrm{Aut}_{\mathcal{F}}(P)$-invariant since it is strongly closed in $\mathcal{F}$.

(**c $\Rightarrow$ a**)   If (c) holds for $Q$, then by Theorem 3.6, every morphism in $\mathcal{F}$ is a composite of restrictions of automorphisms of subgroups which contain $Q$ and which leave it invariant. Hence each $\varphi \in \mathrm{Hom}_{\mathcal{F}}(P, R)$ in $\mathcal{F}$ extends to some $\bar{\varphi} \in \mathrm{Hom}_{\mathcal{F}}(PQ, RQ)$ which sends $Q$ to itself, and thus $Q \trianglelefteq \mathcal{F}$.   $\square$

As one special case of Proposition 4.5, if $\mathcal{F}$ is a saturated fusion system over an *abelian* p-group $S$, then the only $\mathcal{F}$-centric subgroup is $S$ itself, and hence $S \trianglelefteq \mathcal{F}$. Translated to groups, this is just Burnside's fusion theorem: if $G$ has an abelian Sylow $p$-subgroup $S$, then $N_G(S)$ controls $p$-fusion in $G$.

The next proposition gives a different characterization of normal subgroups in a fusion system; one which is also very useful.

**Proposition 4.6**   *Fix a saturated fusion system $\mathcal{F}$ over a p-group $S$. Then for any subgroup $Q$ of $S$, $Q \trianglelefteq \mathcal{F}$ if and only if there exists a series $1 = Q_0 \leq Q_1 \leq \cdots \leq Q_n = Q$ of subgroups such that*

(a)   *for each $1 \leq i \leq n$, $Q_i$ is strongly closed in $\mathcal{F}$; and*

(b)   *for each $1 \leq i \leq n$, $[Q, Q_i] \leq Q_{i-1}$.*

*Proof.* Assume first there is a series of subgroups $Q_i \leq Q$ which are strongly closed in $\mathcal{F}$ and satisfy (b). We claim that $Q \trianglelefteq \mathcal{F}$. By Proposition 4.5, it suffices to show $Q$ is contained in each member $U$ of $\mathcal{F}^{cr}$. Fix such a $U$, choose $k$ maximal subject to $Q_k \leq U$, and assume $k < n$. Set $R = U \cap Q_{k+1} < Q_{k+1}$, and consider the series

$$1 = Q_0 \leq Q_1 \leq \cdots \leq Q_k \leq R \leq U. \tag{1}$$

Since the $Q_i$ are strongly closed, each term in (1) is $\mathrm{Aut}_{\mathcal{F}}(U)$-invariant (i.e., $\mathrm{Aut}_{\mathcal{F}}(U)$ acts on each $Q_i$ and on $R$). Also, $[N_{Q_{k+1}}(U), U] \leq U \cap Q_{k+1} = R$, $[N_{Q_{k+1}}(U), R] \leq [Q_{k+1}, Q_{k+1}] \leq Q_k$, and $[N_{Q_{k+1}}(U), Q_i] \leq [Q, Q_i] \leq Q_{i-1}$ for each $0 < i \leq k$. In other words, $\mathrm{Aut}_{Q_{k+1}}(U)$ centralizes each factor group in (1), so $\mathrm{Aut}_{Q_{k+1}}(U) \leq O_p(\mathrm{Aut}_{\mathcal{F}}(U))$ by Lemma A.2. Hence $\mathrm{Aut}_{Q_{k+1}}(U) \leq \mathrm{Inn}(U)$ since $U$ is $\mathcal{F}$-radical, so $N_{Q_{k+1}}(U) \leq UC_S(U)$, and $UC_S(U) = U$ since $U$ is $\mathcal{F}$-centric. Thus $N_{Q_{k+1}U}(U) = U$, $Q_{k+1}U = U$ by Lemma A.1, and this contradicts the original assumption that $Q_{k+1} \not\leq U$. We conclude that $k = n$, $Q \leq U$ for each $U \in \mathcal{F}^{cr}$, and hence $Q \trianglelefteq \mathcal{F}$.

Conversely, assume $Q \trianglelefteq \mathcal{F}$, and let $1 = Q_0 \leq \cdots \leq Q_n = Q$ be the ascending central series for $Q$. Then each $Q_i$ is characteristic in $Q$, hence is normal in $\mathcal{F}$ by Proposition 4.4, and hence is strongly closed in $\mathcal{F}$. The series $\{Q_i\}$ thus satisfies (a) and (b), completing the proof. $\qquad\square$

We now list some easy corollaries of Proposition 4.6. Recall (Definition 4.3) that we write $O_p(\mathcal{F})$ for the largest normal subgroup in a fusion system $\mathcal{F}$.

**Corollary 4.7**  *The following hold for any saturated fusion system $\mathcal{F}$ over a $p$-group $S$.*

(a)  *An abelian subgroup $Q$ of $S$ is normal in $\mathcal{F}$ if and only if $Q$ is strongly closed in $\mathcal{F}$.*

(b)  *$O_p(\mathcal{F}) \neq 1$ if and only if there is a nontrivial abelian subgroup of $S$ which is strongly closed in $\mathcal{F}$.*

*Proof.* **(a)**  Let $Q$ be an abelian subgroup of $S$. If $Q \trianglelefteq \mathcal{F}$, then $Q$ is strongly closed by Proposition 4.5. Conversely if $Q$ is strongly closed then $Q \trianglelefteq \mathcal{F}$ by Proposition 4.6, since conditions (a) and (b) of 4.6 hold with respect to the series $1 = Q_0 \leq Q_1 = Q$.

**(b)**  If there is a nontrivial strongly closed abelian subgroup $Q$, then $Q \trianglelefteq \mathcal{F}$ by (a), so $O_p(\mathcal{F}) \neq 1$. Conversely, if $Q = O_p(\mathcal{F}) \neq 1$, then $Z(Q) \neq 1$, and hence $Z(Q)$ is a nontrivial abelian subgroup of $S$ which is normal in $\mathcal{F}$ (hence strongly closed) by Proposition 4.4. $\qquad\square$

We next look at constrained fusion systems, which will play an important role in Parts II and IV.

**Definition 4.8**  Fix a saturated fusion system $\mathcal{F}$ over a $p$-group $S$.

- $\mathcal{F}$ is *constrained* if there is a normal subgroup $Q \trianglelefteq \mathcal{F}$ which is $\mathcal{F}$-centric; equivalently, if $O_p(\mathcal{F}) \in \mathcal{F}^c$.

- If $\mathcal{F}$ is constrained, then a *model* for $\mathcal{F}$ is a finite group $G$ such that $S \in \mathrm{Syl}_p(G)$, $\mathcal{F}_S(G) = \mathcal{F}$, and $C_G(O_p(G)) \leq O_p(G)$.

- $\mathcal{G}(\mathcal{F})$ denotes the class of all finite groups which are models for $\mathcal{F}$.

The definition of a constrained fusion system is motivated by the analogous terminology for groups: a finite group $G$ is $p$-constrained if $O_p(G/O_{p'}(G))$ contains its centralizer in $G/O_{p'}(G)$. We say that $G$ is *strictly $p$-constrained* if $G$ is $p$-constrained and $O_{p'}(G) = 1$; equivalently, $C_G(O_p(G)) \leq O_p(G)$. Thus the models for $\mathcal{F}$ are the finite, strictly $p$-constrained groups which realize $\mathcal{F}$.

The next theorem says that each constrained fusion system has models, and that they are unique up to isomorphism in a strong sense. This result was originally conjectured by Puig in unpublished notes.

**Theorem 4.9** (Model theorem for constrained fusion systems) *Let $\mathcal{F}$ be a constrained saturated fusion system over a $p$-group $S$. Fix $Q \trianglelefteq S$ which is $\mathcal{F}$-centric and normal in $\mathcal{F}$. Then the following hold.*

(a) *There are models for $\mathcal{F}$; i.e., $\mathcal{G}(\mathcal{F}) \neq \varnothing$.*

(b) *If $G_1$ and $G_2$ are two models for $\mathcal{F}$, then there is an isomorphism $\varphi \colon G_1 \xrightarrow{\cong} G_2$ such that $\varphi|_S = \mathrm{Id}_S$.*

(c) *For any finite group $G$ containing $S$ as a Sylow $p$-subgroup such that $Q \trianglelefteq G$, $C_G(Q) \leq Q$, and $\mathrm{Aut}_G(Q) = \mathrm{Aut}_{\mathcal{F}}(Q)$, there is $\beta \in \mathrm{Aut}(S)$ such that $\beta|_Q = \mathrm{Id}_Q$ and $\mathcal{F}_S(G) = {}^{\beta}\mathcal{F}$. Thus there is a model for $\mathcal{F}$ which is isomorphic to $G$.*

*Proof.* This is shown in [5a1, Proposition C], except for the strong uniqueness property (b), which is shown in [A5, 2.5]. A different proof (of all three parts of the theorem) is given in Part III (Theorem III.5.10). $\square$

## 5. Normalizer fusion systems

Let $\mathcal{F}$ be a fusion system over a $p$-group $S$. For any subgroup $Q \leq S$ and any group of automorphisms $K \leq \mathrm{Aut}(Q)$, set $\mathrm{Aut}_{\mathcal{F}}^{K}(Q) = K \cap \mathrm{Aut}_{\mathcal{F}}(Q)$, $\mathrm{Aut}_S^{K}(Q) = K \cap \mathrm{Aut}_S(Q)$, and

$$N_S^{K}(Q) = \{x \in N_S(Q) \mid c_x \in K\}.$$

Thus $N_S^{\mathrm{Aut}(Q)}(Q) = N_S(Q)$ is the usual normalizer, and $N_S^{\{1\}}(Q) = C_S(Q)$ is the centralizer. Also, for any monomorphism $\varphi \in \mathrm{Hom}(Q, R)$, we write

$$^\varphi K = \{^\varphi\chi \mid \chi \in K\} \leq \mathrm{Aut}(\varphi(Q)).$$

**Definition 5.1**  Let $\mathcal{F}$ be any fusion system over $S$. For any subgroup $Q \leq S$ and any group of automorphisms $K \leq \mathrm{Aut}(Q)$, $Q$ is *fully $K$-normalized* in $\mathcal{F}$ if $|N_S^K(Q)| \geq |N_S^{\varphi K}(\varphi(Q))|$ for all $\varphi \in \mathrm{Hom}_{\mathcal{F}}(Q, S)$.

Thus $Q$ is fully normalized or fully centralized in $\mathcal{F}$ if and only if it is fully $K$-normalized, for $K = \mathrm{Aut}(Q)$ or $K = 1$, respectively. The following proposition, most of which was proven in [BLO2, Proposition A.2], describes the key properties which characterize fully $K$-normalized subgroups in a saturated fusion system. Part of the proposition has already been established in Lemma 2.6(c): the case where $K = \mathrm{Aut}(Q)$.

**Proposition 5.2**  *Let $\mathcal{F}$ be a saturated fusion system over a p-group $S$. Fix a subgroup $Q \leq S$ and a group of automorphisms $K \leq \mathrm{Aut}(Q)$. Then the following three conditions are equivalent:*

(a)  *$Q$ is fully $K$-normalized in $\mathcal{F}$.*

(b)  *$Q$ is fully centralized in $\mathcal{F}$ and $\mathrm{Aut}_S^K(Q) \in \mathrm{Syl}_p(\mathrm{Aut}_{\mathcal{F}}^K(Q))$.*

(c)  *For each $P \leq S$ and each $\varphi \in \mathrm{Iso}_{\mathcal{F}}(P, Q)$, there are homomorphisms $\chi \in \mathrm{Aut}_{\mathcal{F}}^K(Q)$ and $\bar{\varphi} \in \mathrm{Hom}_{\mathcal{F}}(P\cdot N_S^{K^\varphi}(P), S)$ such that $\bar{\varphi}|_P = \chi \circ \varphi$.*

*Proof.* **(a $\Rightarrow$ b)**  Assume $Q$ is fully $K$-normalized in $\mathcal{F}$. Choose $P \leq S$ which is $\mathcal{F}$-conjugate to $Q$, and fully automized and receptive in $\mathcal{F}$, and fix $\psi \in \mathrm{Iso}_{\mathcal{F}}(Q, P)$. By Lemma 2.6(a), $P$ is fully centralized.

Fix $T \in \mathrm{Syl}_p(\mathrm{Aut}_{\mathcal{F}}^{\psi K}(P))$ such that $T \geq {}^\psi\mathrm{Aut}_S^K(Q)$. Since $\mathrm{Aut}_S(P) \in \mathrm{Syl}_p(\mathrm{Aut}_{\mathcal{F}}(P))$, there is $\alpha \in \mathrm{Aut}_{\mathcal{F}}(P)$ such that ${}^\alpha T \leq \mathrm{Aut}_S(P)$. Set $\varphi = \alpha \circ \psi \in \mathrm{Iso}_{\mathcal{F}}(Q, P)$. Then

$$\mathrm{Aut}_S^{\varphi K}(P) = \mathrm{Aut}_S(P) \cap \mathrm{Aut}_{\mathcal{F}}^{\varphi K}(P) \geq {}^\alpha T \in \mathrm{Syl}_p(\mathrm{Aut}_{\mathcal{F}}^{\varphi K}(P)),$$

and hence $\mathrm{Aut}_S^{\varphi K}(P) = {}^\alpha T$ is also a Sylow subgroup. We thus have

$$|C_S(Q)|\cdot|\mathrm{Aut}_S^K(Q)| = |N_S^K(Q)| \geq |N_S^{\varphi K}(P)| = |C_S(P)|\cdot|\mathrm{Aut}_S^{\varphi K}(P)|;$$

since $Q$ is fully $K$-normalized, while

$$|C_S(Q)| \leq |C_S(P)| \qquad \text{and} \qquad |\mathrm{Aut}_S^K(Q)| \leq |\mathrm{Aut}_S^{\varphi K}(P)|$$

since $P$ is fully centralized and $\mathrm{Aut}_S^{\varphi K}(P)$ is a Sylow $p$-subgroup. So all of these inequalities are equalities, $Q$ is fully centralized, and $\mathrm{Aut}_S^K(Q) \in \mathrm{Syl}_p(\mathrm{Aut}_{\mathcal{F}}^K(Q))$.

**(b $\Rightarrow$ c)**  Assume $\mathrm{Aut}_S^K(Q) \in \mathrm{Syl}_p(\mathrm{Aut}_{\mathcal{F}}^K(Q))$ where $Q$ is fully centralized in $\mathcal{F}$. Fix $\varphi \in \mathrm{Iso}_{\mathcal{F}}(P, Q)$, and set $L = K^\varphi \leq \mathrm{Aut}(P)$. Then $^\varphi\mathrm{Aut}_S^L(P)$ is a

$p$-subgroup of $\operatorname{Aut}_{\mathcal{F}}^K(Q)$, so there is $\chi \in \operatorname{Aut}_{\mathcal{F}}^K(Q)$ such that ${}^{\chi\varphi}\operatorname{Aut}_S^L(P) \leq \operatorname{Aut}_S^K(Q)$. Since $Q$ is fully centralized, it is receptive, and hence $\chi\varphi$ extends to a morphism $\bar{\varphi} \in \operatorname{Hom}_{\mathcal{F}}(P{\cdot}N_S^L(P), S)$. Thus $Q$ satisfies the condition in (c).

**(c $\Rightarrow$ a)** Assume $Q$ satisfies the extension condition in (c). Thus for each $\varphi \in \operatorname{Hom}_{\mathcal{F}}(Q, S)$, there exist

$$\chi \in \operatorname{Aut}_{\mathcal{F}}^K(Q) \quad \text{and} \quad \bar{\varphi} \in \operatorname{Hom}_{\mathcal{F}}(\varphi(Q){\cdot}N_S^{\varphi K}(\varphi(Q)), S)$$

with $\bar{\varphi}|_{\varphi(Q)} = \chi \circ \varphi^{-1}$. In particular, $\bar{\varphi}(N_S^{\varphi K}(\varphi(Q))) \leq N_S^K(Q)$, and so $|N_S^{\varphi K}(\varphi(Q))| \leq |N_S^K(Q)|$. Thus $Q$ is fully $K$-normalized in $\mathcal{F}$. $\qquad\square$

We can now define the normalizer fusion subsystems in a fusion system.

**Definition 5.3** Let $\mathcal{F}$ be a fusion system over a $p$-group $S$, and fix $Q \leq S$ and $K \leq \operatorname{Aut}(Q)$. Let $N_{\mathcal{F}}^K(Q) \subseteq \mathcal{F}$ be the fusion system over $N_S^K(Q)$ where for $P, R \leq N_S^K(Q)$,

$$\operatorname{Hom}_{N_{\mathcal{F}}^K(Q)}(P, R) = \big\{\varphi \in \operatorname{Hom}_{\mathcal{F}}(P, R) \,\big|\, \exists\, \bar{\varphi} \in \operatorname{Hom}_{\mathcal{F}}(PQ, RQ)$$
$$\text{with } \bar{\varphi}|_P = \varphi, \; \bar{\varphi}(Q) = Q, \text{ and } \bar{\varphi}|_Q \in K\big\}.$$

As special cases, set $N_{\mathcal{F}}(Q) = N_{\mathcal{F}}^{\operatorname{Aut}(Q)}(Q)$ and $C_{\mathcal{F}}(Q) = N_{\mathcal{F}}^{\{1\}}(Q)$: the normalizer and centralizer fusion systems, respectively, of $Q$.

It follows immediately from Definitions 4.1 and 5.3 and Lemma 4.2 that for any saturated fusion system $\mathcal{F}$ over a $p$-group $S$, and any $Q \leq S$, $Q \trianglelefteq \mathcal{F}$ if and only if $N_{\mathcal{F}}(Q) = \mathcal{F}$, and $Q \leq Z(\mathcal{F})$ ($Q$ is central in $\mathcal{F}$) if and only if $C_{\mathcal{F}}(Q) = \mathcal{F}$.

Definition 5.3 is motivated by the following proposition, which relates normalizers in fusion systems with those in finite groups.

**Proposition 5.4** *Assume $\mathcal{F} = \mathcal{F}_S(G)$ for some finite group $G$ with $S \in \operatorname{Syl}_p(G)$, and fix $Q \leq S$ and $K \leq \operatorname{Aut}(Q)$. Then $Q$ is fully $K$-normalized in $\mathcal{F}$ if and only if $N_S^K(Q) \in \operatorname{Syl}_p(N_G^K(Q))$. If this is the case, then $N_{\mathcal{F}}^K(Q) = \mathcal{F}_{N_S^K(Q)}(N_G^K(Q))$.*

*Proof.* By Proposition 5.2, $Q$ is fully $K$-normalized in $\mathcal{F}$ if and only if $Q$ is fully centralized and $\operatorname{Aut}_S^K(Q) \in \operatorname{Syl}_p(\operatorname{Aut}_{\mathcal{F}}^K(Q))$. Also, by Lemma 1.2, $Q$ is fully centralized in $\mathcal{F}$ if and only if $C_S(Q) \in \operatorname{Syl}_p(C_G(Q))$. Since we can identify $\operatorname{Aut}_S^K(Q) = N_S^K(Q)/C_S(Q)$ and $\operatorname{Aut}_G^K(Q) = N_G^K(Q)/C_G(Q)$, Lemma A.3 now implies that $Q$ is fully $K$-normalized if and only if $N_S^K(Q) \in \operatorname{Syl}_p(N_G^K(Q))$.

Now assume $N_S^K(Q) \in \operatorname{Syl}_p(N_G^K(Q))$. For $P, R \leq N_S^K(Q)$, a morphism $\varphi \in \operatorname{Hom}_G(P, R)$ has the form $\varphi = c_g$ for some $g \in N_G^K(Q)$ if and only if $\varphi$

extends to some $\bar{\varphi} \in \mathrm{Hom}_G(PQ, RQ)$ such that $\bar{\varphi}(Q) = Q$ and $\bar{\varphi}|_Q \in K$. Thus $N_{\mathcal{F}}^K(Q)$ is the fusion system of $N_G^K(Q)$ over $N_S^K(Q)$. $\qquad\square$

Of course, for the normalizer fusion systems to be very useful, they must be saturated.

**Theorem 5.5** ([P6, Proposition 2.15], [BLO2, Proposition A.6]) *Fix a saturated fusion system $\mathcal{F}$ over a p-group $S$. Assume $Q \leq S$ and $K \leq \mathrm{Aut}(Q)$ are such that $Q$ is fully $K$-normalized in $\mathcal{F}$. Then $N_{\mathcal{F}}^K(Q)$ is a saturated fusion system over $N_S^K(Q)$.*

*Proof.* Set $S_0 = N_S^K(Q)$ and $\mathcal{F}_0 = N_{\mathcal{F}}^K(Q)$ for short. For each $P \leq S_0$, set

$$K_P = \{\alpha \in \mathrm{Aut}(PQ) \,|\, \alpha(P) = P, \ \alpha(Q) = Q, \ \alpha|_Q \in K\}.$$

We will show that

(a)   each subgroup of $S_0$ is $\mathcal{F}_0$-conjugate to a subgroup $P$ such that $PQ$ is fully $K_P$-normalized in $\mathcal{F}$;

(b)   if $P \leq S_0$ and $PQ$ is fully $K_P$-normalized in $\mathcal{F}$, then $P$ is fully automized in $\mathcal{F}_0$; and

(c)   if $P \leq S_0$ and $PQ$ is fully $K_P$-normalized in $\mathcal{F}$, then $P$ is receptive in $\mathcal{F}_0$.

The theorem then follows immediately from Definition 2.2.

**(a)**   Fix $R \leq S_0$. Let $\varphi \in \mathrm{Hom}_{\mathcal{F}}(RQ, S)$ be such that $\varphi(RQ)$ is fully ${}^{\varphi}K_R$-normalized. Set $R_1 = \varphi(R)$ and $Q_1 = \varphi(Q)$. Since $Q$ is fully $K$-normalized by assumption, by Proposition 5.2, there are $\chi \in \mathrm{Aut}_{\mathcal{F}}^K(Q)$ and $\psi \in \mathrm{Hom}_{\mathcal{F}}(N_S^{\varphi K}(Q_1)\cdot Q_1, S)$ such that $\chi \circ \psi|_{Q_1} = (\varphi|_Q)^{-1}$. Since $R \leq S_0 = N_S^K(Q)$, $\varphi(R) \leq N_S^{\varphi K}(Q_1)$. Thus $\psi \circ \varphi \in \mathrm{Hom}_{\mathcal{F}}(RQ, S)$, $\psi \circ \varphi|_Q = \chi^{-1} \in K$, and so $\psi \circ \varphi|_R \in \mathrm{Hom}_{\mathcal{F}_0}(R, S_0)$. Set $P = \psi(R_1) = \psi \circ \varphi(R)$.

Set $L = {}^{\varphi}K_R$, so that $K_P = {}^{\psi\varphi}K_R = {}^{\psi}L$. Then

$$N_S^{K_P}(PQ) = N_S(P) \cap N_S^K(Q) \quad \text{and} \quad N_S^L(\varphi(RQ)) = N_S(R_1) \cap N_S^{\varphi K}(Q_1).$$

Hence $\psi(N_S^L(\varphi(RQ))) \leq N_S^{K_P}(PQ)$; and $PQ$ is fully $K_P$-normalized in $\mathcal{F}$ since $\varphi(RQ)$ is fully $L$-normalized in $\mathcal{F}$.

**(b)**   Assume $P \leq S_0$, and $PQ$ is fully $K_P$-normalized in $\mathcal{F}$. By Proposition 5.2, $\mathrm{Aut}_S^{K_P}(PQ) \in \mathrm{Syl}_p(\mathrm{Aut}_{\mathcal{F}}^{K_P}(PQ))$. The images of these two subgroups under restriction to $P$ are $\mathrm{Aut}_{S_0}(P)$ and $\mathrm{Aut}_{\mathcal{F}_0}(P)$, respectively, so $\mathrm{Aut}_{S_0}(P) \in \mathrm{Syl}_p(\mathrm{Aut}_{\mathcal{F}_0}(P))$, and $P$ is fully automized in $\mathcal{F}_0$.

**(c)**   Assume $P \leq S_0$, and $PQ$ is fully $K_P$-normalized in $\mathcal{F}$. Fix $R \leq S_0$ and $\varphi \in \mathrm{Iso}_{\mathcal{F}_0}(R, P)$, and let $N_{\varphi} = N_{\varphi}^{\mathcal{F}_0}$ be as in Definition 2.2. Set

$$L = \{\alpha \in \mathrm{Aut}(PQ) \,|\, \alpha|_P \in \mathrm{Aut}_{S_0}(P), \ \alpha|_Q \in K\} \leq K_P.$$

For each $g \in N_S^{K_P}(PQ)$, $g \in S_0$ since $c_g|_Q \in K$, and thus $g \in N_{S_0}(P)$. So $\mathrm{Aut}_S^L(PQ) = \mathrm{Aut}_S^{K_P}(PQ)$. Since $PQ$ is fully $K_P$-normalized in $\mathcal{F}$, it is also fully $L$-normalized.

By definition of $\mathcal{F}_0 = N_{\mathcal{F}}^K(Q)$, $\varphi$ extends to some $\bar{\varphi} \in \mathrm{Iso}_{\mathcal{F}}(RQ, PQ)$ such that $\bar{\varphi}|_Q \in K$. By Proposition 5.2, there are $\chi \in \mathrm{Aut}_{\mathcal{F}}^L(PQ)$ and $\psi \in \mathrm{Hom}_{\mathcal{F}}(N_\varphi{\cdot}Q, N_S^K(Q){\cdot}Q)$ such that $\psi|_{RQ} = \chi \circ \bar{\varphi}$. Since $\chi|_P \in \mathrm{Aut}_{S_0}(P)$, there is $g \in N_{S_0}(P)$ such that $\chi|_P = c_g|_P$. Then $c_g^{-1} \circ \psi|_{N_\varphi} \in \mathrm{Hom}_{\mathcal{F}_0}(N_\varphi, S_0)$ and $c_g^{-1} \circ \psi|_R = \varphi$; and this finishes the proof that $P$ is receptive in $\mathcal{F}_0$. $\qquad\square$

The following is one easy application of Theorem 5.5.

**Lemma 5.6** *Let $\mathcal{F}$ be a saturated fusion system over a $p$-group $S$. Assume $Q \leq P \leq S$ are such that $Q \in \mathcal{F}^c$. Let $\varphi, \varphi' \in \mathrm{Hom}_{\mathcal{F}}(P, S)$ be such that $\varphi|_Q = \varphi'|_Q$. Then there is $x \in Z(Q)$ such that $\varphi' = \varphi \circ c_x$.*

*Proof.* This was shown in [BLO2, Proposition A.8], but we give here a different, shorter proof. Since $\varphi \circ c_g = c_{\varphi(g)} \circ \varphi$ for each $g \in Q$, it suffices to show that $\varphi' = c_y \circ \varphi$ for some $y \in Z(\varphi(Q))$. Upon replacing $P$ by $\varphi'(P)$, $Q$ by $\varphi(Q) = \varphi'(Q)$, and $\varphi$ by $\varphi \circ (\varphi')^{-1}$, we can assume that $\varphi' = \mathrm{incl}_P^S$ and $\varphi|_Q = \mathrm{Id}_Q$. We must show that $\varphi = c_x$ for some $x \in Z(Q)$. Since $Q$ is subnormal in $P$, it suffices to prove this when $Q \trianglelefteq P$.

Set $K = \mathrm{Aut}_P(Q)$. Since $Q$ is $\mathcal{F}$-centric, it is fully centralized. Since $\mathrm{Aut}_{\mathcal{F}}^K(Q) = \mathrm{Aut}_S^K(Q) = K$, $Q$ is fully $K$-normalized by Proposition 5.2. Hence by Theorem 5.5, the normalizer subsystem $N_{\mathcal{F}}^K(Q)$ over $N_S^K(Q) = P{\cdot}C_S(Q) = P$ is saturated. Also, since $\varphi|_Q = \mathrm{Id}$, $\mathrm{Aut}_{\varphi(P)}(Q) = \mathrm{Aut}_P(Q) = K$. Thus $\varphi(P) \leq N_S^K(Q)$, and $\varphi \in \mathrm{Mor}(N_{\mathcal{F}}^K(Q))$.

It thus suffices to prove that $N_{\mathcal{F}}^K(Q) = \mathcal{F}_P(P)$. To show this, by Corollary 3.7, it suffices to show that all automorphism groups in $N_{\mathcal{F}}^K(Q)$ are $p$-groups. Assume otherwise: then there are $R \leq P$ and $\alpha \in \mathrm{Aut}_{\mathcal{F}}(R)$ such that $Q \leq R$, $\alpha(Q) = Q$, $\alpha|_Q \in K$, and $\alpha \neq \mathrm{Id}_R$ has order prime to $p$. Since $\alpha|_Q \in K$ and $K$ is a $p$-group, $\alpha|_Q = \mathrm{Id}_Q$. Hence for $g \in R$, $g$ and $\alpha(g)$ have the same conjugation action on $Q$, and $g^{-1}\alpha(g) \in C_R(Q) \leq Q$. Thus $\alpha$ induces the identity on $R/Q$, so $\alpha = \mathrm{Id}_R$ by Lemma A.2, and this is a contradiction. $\qquad\square$

## 6. Normal Fusion Subsystems and Products

Several different definitions of normal fusion subsystems are found in the literature. That there are differences is not surprising, since the conditions one wants to impose depend to a great extent on how one needs to use them. The most restrictive definition is due to Aschbacher [A5], and is what we call here a normal fusion subsystem. The definitions of Linckelmann [Li3] and Oliver [O4] are equivalent to what we call here a weakly normal subsystem, while that of Puig [P6] is equivalent to our invariant subsystem.

**Definition 6.1**   Fix a prime $p$, and a fusion system $\mathcal{F}$ over a $p$-group $S$.

- A (saturated) fusion subsystem of $\mathcal{F}$ is a subcategory $\mathcal{E} \subseteq \mathcal{F}$ which is itself a (saturated) fusion system over a subgroup $T \leq S$.

- A fusion subsystem $\mathcal{E} \subseteq \mathcal{F}$ over $T \trianglelefteq S$ is $\mathcal{F}$-*invariant* if $T$ is strongly closed in $\mathcal{F}$, and the two following conditions hold:
  - *(invariance condition)* $^{\alpha}\mathcal{E} = \mathcal{E}$ for each $\alpha \in \mathrm{Aut}_{\mathcal{F}}(T)$, and

  - *(Frattini condition)* for each $P \leq T$ and each $\varphi \in \mathrm{Hom}_{\mathcal{F}}(P,T)$, there are $\alpha \in \mathrm{Aut}_{\mathcal{F}}(T)$ and $\varphi_0 \in \mathrm{Hom}_{\mathcal{E}}(P,T)$ such that $\varphi = \alpha \circ \varphi_0$.

- A fusion subsystem $\mathcal{E} \subseteq \mathcal{F}$ over $T \trianglelefteq S$ is *weakly normal* in $\mathcal{F}$ ($\mathcal{E} \mathrel{\dot{\trianglelefteq}} \mathcal{F}$) if $\mathcal{E}$ and $\mathcal{F}$ are both saturated and $\mathcal{E}$ is $\mathcal{F}$-invariant.

- A fusion subsystem $\mathcal{E} \subseteq \mathcal{F}$ over $T \trianglelefteq S$ is *normal* ($\mathcal{E} \trianglelefteq \mathcal{F}$) if $\mathcal{E}$ is weakly normal, and
  - *(Extension condition)* each $\alpha \in \mathrm{Aut}_{\mathcal{E}}(T)$ extends to some $\bar{\alpha} \in \mathrm{Aut}_{\mathcal{F}}(TC_S(T))$ such that $[\bar{\alpha}, C_S(T)] \leq Z(T)$.

- $\mathcal{F}$ is *simple* if it contains no proper nontrivial normal fusion subsystem.

Note that the fusion system of a simple group need not be simple. For example, the fusion systems of the group $A_5$ (at any prime $p = 2, 3, 5$) are not simple.

The Frattini condition in Definition 6.1 is motivated by the Frattini argument (see Proposition A.4(a)), which says that when $H \trianglelefteq G$ and $T \in \mathrm{Syl}_p(H)$, then $G = N_G(T) \cdot H$. This is one step in the proof of the following proposition, which says that when $H \trianglelefteq G$ are finite groups, the fusion system of $H$ is normal in the fusion system of $G$.

**Proposition 6.2**   *Fix a finite group* $G$, *a normal subgroup* $H \trianglelefteq G$, *and a Sylow subgroup* $S \in \mathrm{Syl}_p(G)$, *and set* $T = S \cap H \in \mathrm{Syl}_p(H)$. *Then* $\mathcal{F}_T(H) \trianglelefteq \mathcal{F}_S(G)$.

*Proof.* Since $H \trianglelefteq G$, no element of $T$ is $G$-conjugate to any element of $S \setminus T$. Hence $T$ is strongly closed in $\mathcal{F}_S(G)$. The invariance condition holds for $\mathcal{F}_T(H) \subseteq \mathcal{F}_S(G)$ since conjugation by $g \in N_G(T)$ (or by $c_g$) sends $\mathrm{Hom}_H(P,Q)$ to $\mathrm{Hom}_H({}^g P, {}^g Q)$. By the Frattini argument (see Proposition A.4(a)), $G = N_G(T) \cdot H$, so each morphism in $\mathcal{F}_S(G)$ between subgroups of $T$ factors as a morphism in $\mathcal{F}_T(H)$ followed by one in $\mathrm{Aut}_G(T)$. This proves the Frattini condition in Definition 6.1, and finishes the proof that $\mathcal{F}_T(H)$ is weakly normal in $\mathcal{F}_S(G)$.

By the Frattini argument again, this time applied to $C_H(T){\cdot}C_S(T) \trianglelefteq N_H(T){\cdot}C_S(T)$ and the Sylow $p$-subgroup $C_S(T)$,

$$N_H(T){\cdot}C_S(T) = C_H(T){\cdot}C_S(T){\cdot}N_{N_H(T)}(C_S(T)) \, .$$

Thus for each $g \in N_H(T)$, $g = ah$ for some $a \in C_{HS}(T)$ and some $h \in N_{N_H(T)}(C_S(T))$, and hence $c_g \in \operatorname{Aut}_H(T)$ extends to $c_h \in \operatorname{Aut}_G(T{\cdot}C_S(T))$ where $[c_h, C_S(T)] \leq H \cap C_S(T) = Z(T)$. This proves the extension condition, and finishes the proof that $\mathcal{F}_T(H) \trianglelefteq \mathcal{F}_S(G)$. $\qquad\square$

As another example, when $\mathcal{F}$ is a saturated fusion system over $S$ and $Q \trianglelefteq S$, then $Q \trianglelefteq \mathcal{F}$ if and only if $\mathcal{F}_Q(Q) \trianglelefteq \mathcal{F}$.

The following is one example of a pair of saturated fusion systems $\mathcal{E} \subseteq \mathcal{F}$, where $\mathcal{E}$ is weakly normal in $\mathcal{F}$ but not normal. It is the smallest member of the class of examples appearing after Proposition 6.7.

**Example 6.3**  Set $p = 3$, $G_1 = G_2 = S_3$, and $G = G_1 \times G_2$, where we regard $G_1$ and $G_2$ as subgroups of $G$. Fix $S_i \in \operatorname{Syl}_3(G_i)$ $(i = 1, 2)$, and set $S = S_1 \times S_2 \in \operatorname{Syl}_3(G)$. Let $G_0 < G$ be the (unique) subgroup of index 2 which contains neither $G_1$ nor $G_2$. Set $\mathcal{F} = \mathcal{F}_S(G_0)$ and $\mathcal{E} = \mathcal{F}_{S_1}(G_1)$. Then $\mathcal{E} \subseteq \mathcal{F}$: it is the full subcategory whose objects are the subgroups of $S_1$. Also, $\mathcal{E}$ is saturated since $\mathcal{E} = \mathcal{F}_{S_1}(G_1)$, and $\mathcal{E}$ is $\mathcal{F}$-invariant since $\operatorname{Aut}_{\mathcal{E}}(S_1) = \operatorname{Aut}_{\mathcal{F}}(S_1)$. Thus $\mathcal{E} \trianglelefteq \mathcal{F}$. However, $\mathcal{E}$ is not normal in $\mathcal{F}$, since for $g \in G_1$ of order 2, there is no extension of $c_g \in \operatorname{Aut}_{G_1}(S_1)$ to $\bar{\alpha} \in \operatorname{Aut}_{G_0}(S)$ such that $[\bar{\alpha}, S] \leq S_1$.

The fusion systems of Example 6.3 are a special case of the more general situation which will be described in Proposition 6.7(c). A still more general theorem of David Craven, describing the difference between normal and weakly normal subsystems, will be stated in the next section (Theorem 7.8).

We next list some of the different, but equivalent, definitions of what we here call $\mathcal{F}$-invariant fusion subsystems. In most cases (see [Li3, §3], [P6, §6.4], and [A5, §3]), some version of the strong invariance condition (c) or (d) below was used.

When $\mathcal{F}$ is a fusion system over $S$ and $T \leq S$, we let $\mathcal{F}|_{\leq T}$ denote the full subcategory of $\mathcal{F}$ with objects the subgroups of $T$. This is always a fusion system over $T$, but not, in general, saturated.

**Proposition 6.4** ([A5, Theorem 3.3])  *Let $\mathcal{F}$ be a saturated fusion system over a $p$-group $S$, and let $\mathcal{E} \subseteq \mathcal{F}$ be a fusion subsystem (not necessarily saturated) over a subgroup $T$ of $S$. Assume $T$ is strongly closed in $\mathcal{F}$. Then the following conditions are equivalent:*

(a)  $\mathcal{E}$ *is* $\mathcal{F}$-*invariant.*

(b)  $^{\alpha}\mathcal{E} = \mathcal{E}$ *for each* $\alpha \in \mathrm{Aut}_{\mathcal{F}}(T)$, *and* $\mathcal{F}|_{\leq T} = \langle \mathrm{Aut}_{\mathcal{F}}(T), \mathcal{E} \rangle$.

(c)  $^{\alpha}\mathcal{E} = \mathcal{E}$ *for each* $\alpha \in \mathrm{Aut}_{\mathcal{F}}(T)$, *and* $\mathrm{Aut}_{\mathcal{E}}(P) \trianglelefteq \mathrm{Aut}_{\mathcal{F}}(P)$ *for each* $P \leq T$.

(d)  (strong invariance condition) *For each pair of subgroups* $P \leq Q \leq T$, *each* $\varphi \in \mathrm{Hom}_{\mathcal{E}}(P, Q)$, *and each* $\psi \in \mathrm{Hom}_{\mathcal{F}}(Q, T)$, $\psi\varphi(\psi|_P)^{-1} \in \mathrm{Hom}_{\mathcal{E}}(\psi(P), \psi(Q))$.

*Proof.* **(a $\Rightarrow$ d)**  Fix $\varphi$ and $\psi$ as in (d). Since $\mathcal{E}$ is $\mathcal{F}$-invariant, we can write $\psi = \alpha \circ \psi_0$ for some $\psi_0 \in \mathrm{Hom}_{\mathcal{E}}(Q, T)$ and some $\alpha \in \mathrm{Aut}_{\mathcal{F}}(T)$. Then $\varphi' \overset{\mathrm{def}}{=} \psi_0\varphi(\psi_0|_P)^{-1} \in \mathrm{Hom}_{\mathcal{E}}(\psi_0(P), \psi_0(Q))$, and so $\psi\varphi(\psi|_P)^{-1} = {}^{\alpha}\varphi' \in \mathrm{Mor}(^{\alpha}\mathcal{E}) = \mathrm{Mor}(\mathcal{E})$.

**(d $\Rightarrow$ c)**  Clear.

**(c $\Rightarrow$ b)**  Assume $\mathcal{E}$ satisfies the hypothesis in (c), but $\langle \mathcal{E}, \mathrm{Aut}_{\mathcal{F}}(T) \rangle \subsetneq \mathcal{F}|_{\leq T}$. Fix a morphism $\varphi \in \mathrm{Hom}_{\mathcal{F}}(P, Q)$ (for $P, Q \leq T$) which is not in $\langle \mathcal{E}, \mathrm{Aut}_{\mathcal{F}}(T) \rangle$, chosen such that $|P|$ is maximal subject to these conditions. By Theorem 3.6 (and since $T$ is strongly closed), we can assume $P = Q = R \cap T$ for some $\mathcal{F}$-essential subgroup $R \leq S$, and $\varphi = \bar{\varphi}|_P$ for some $\bar{\varphi} \in \mathrm{Aut}_{\mathcal{F}}(R)$. Set $G = \mathrm{Aut}_{\mathcal{F}}(R)$, and let $H$ be the set of all $\alpha \in G$ such that $\alpha|_P \in \mathrm{Aut}_{\mathcal{E}}(P)$. Then $H \trianglelefteq G$, since $\mathrm{Aut}_{\mathcal{E}}(P) \trianglelefteq \mathrm{Aut}_{\mathcal{F}}(P)$ by assumption. Set $U = \mathrm{Aut}_S(R) \cap H \in \mathrm{Syl}_p(H)$.

By the Frattini argument (see Proposition A.4(a)), $G = H \cdot N_G(U)$. Hence $\bar{\varphi} = \bar{\varphi}_0 \circ \beta$, where $\varphi_0 \overset{\mathrm{def}}{=} \bar{\varphi}_0|_P \in \mathrm{Aut}_{\mathcal{E}}(P)$, and $\beta$ normalizes $\mathrm{Aut}_S(R) \cap H$. Set $N = \{ g \in N_S(R) \,|\, c_g \in H \}$. Then $\beta$ normalizes $\mathrm{Aut}_N(R)$, and hence $\beta$ extends to an element of $\mathrm{Aut}_{\mathcal{F}}(N)$ by the extension axiom (and since $R$ is fully normalized). So $\beta|_P$ extends to an $\mathcal{F}$-automorphism of $N \cap T \geq N_{TR}(R) \cap T > R \cap T = P$ (Lemma A.1), so $\beta|_{N \cap T}$ is in $\langle \mathcal{E}, \mathrm{Aut}_{\mathcal{F}}(T) \rangle$ by the maximality assumption on $P$, and thus $\varphi \in \langle \mathcal{E}, \mathrm{Aut}_{\mathcal{F}}(T) \rangle$.

**(b $\Rightarrow$ a)**  Assume $^{\alpha}\mathcal{E} = \mathcal{E}$ for all $\alpha \in \mathrm{Aut}_{\mathcal{F}}(T)$. Then for any $\varphi \in \mathrm{Hom}_{\mathcal{E}}(P, Q)$ and any $\alpha \in \mathrm{Aut}_{\mathcal{F}}(T)$, $\varphi \circ \alpha|_{\alpha^{-1}(P)} = \alpha|_{\alpha^{-1}(Q)} \circ (\varphi^{\alpha})$ where $\varphi^{\alpha} = {}^{\alpha^{-1}}\varphi \in \mathrm{Hom}_{\mathcal{E}}(\alpha^{-1}(P), \alpha^{-1}(Q))$.

Thus any composite of $\mathcal{E}$-morphisms and restrictions of elements of $\mathrm{Aut}_{\mathcal{F}}(T)$ can be rearranged so that the $\mathcal{E}$-morphisms come first and are followed by the restriction of one $\mathcal{F}$-automorphism of $T$. So the Frattini condition holds if $\mathcal{F}|_{\leq T} = \langle \mathcal{E}, \mathrm{Aut}_{\mathcal{F}}(T) \rangle$. $\square$

We next look at products of fusion systems. These are defined in the obvious way.

**Definition 6.5** For any pair $\mathcal{F}_1$ and $\mathcal{F}_2$ of fusion systems over $p$-groups $S_1$ and $S_2$, $\mathcal{F}_1 \times \mathcal{F}_2$ is the fusion system over $S_1 \times S_2$ generated by the set of all $(\varphi_1, \varphi_2) \in \mathrm{Hom}(P_1 \times P_2, Q_1 \times Q_2)$ for $\varphi_i \in \mathrm{Hom}_{\mathcal{F}_i}(P_i, Q_i)$.

The first thing to check is that the product of two saturated fusion systems is saturated.

**Theorem 6.6** ([BLO2, Lemma 1.5]) *Assume $\mathcal{F}_1$ and $\mathcal{F}_2$ are fusion systems over $p$-groups $S_1$ and $S_2$, respectively. Then for all $P, Q \le S_1 \times S_2$, if $P_i$ and $Q_i$ denote the images of $P$ and $Q$ under projection to $S_i$,*

$$\mathrm{Hom}_{\mathcal{F}_1 \times \mathcal{F}_2}(P, Q) = \left\{ (\varphi_1, \varphi_2)|_P \mid \varphi_i \in \mathrm{Hom}_{\mathcal{F}_i}(P_i, Q_i), \ (\varphi_1, \varphi_2)(P) \le Q \right\}. \tag{1}$$

*If $\mathcal{F}_1$ and $\mathcal{F}_2$ are both saturated, then so is $\mathcal{F}_1 \times \mathcal{F}_2$.*

*Proof.* Set $S = S_1 \times S_2$ and $\mathcal{F} = \mathcal{F}_1 \times \mathcal{F}_2$ for short. For each subgroup $P \le S$, let $P_i \le S_i$ be the image of $P$ under projection, and set $\widehat{P} = P_1 \times P_2$. Thus $P \le \widehat{P}$ in all cases.

Let $\mathcal{F}_*$ be the category whose objects are the subgroups of $S$, and whose morphisms are those given by (1). This is a fusion system: $\mathrm{Mor}(\mathcal{F}_*)$ contains $\mathrm{Inn}(S)$ and is closed under restrictions. By definition, it contains all morphisms $(\varphi_1, \varphi_2) \in \mathrm{Hom}(P_1 \times P_2, Q_1 \times Q_2)$ for $\varphi_i \in \mathrm{Hom}_{\mathcal{F}_i}(P_i, Q_i)$, and every morphism in $\mathcal{F}_*$ is the restriction of such a morphism. Thus $\mathcal{F}_* = \mathcal{F}$ by definition of $\mathcal{F} = \mathcal{F}_1 \times \mathcal{F}_2$.

Now assume $\mathcal{F}_1$ and $\mathcal{F}_2$ are both saturated. To prove that $\mathcal{F}$ is saturated, we must show that each subgroup $P$ of $S$ is $\mathcal{F}$-conjugate to one which is fully automized and receptive in $\mathcal{F}$. It suffices to do this when each $P_i$ is fully automized and receptive in $\mathcal{F}_i$ $(i = 1, 2)$.

Since $\mathrm{Aut}_{\mathcal{F}}(\widehat{P}) = \mathrm{Aut}_{\mathcal{F}_1}(P_1) \times \mathrm{Aut}_{\mathcal{F}_2}(P_2)$ and similarly for $\mathrm{Aut}_S(\widehat{P})$, and since each $P_i$ is fully automized in $\mathcal{F}_i$, $\widehat{P}$ is fully automized in $\mathcal{F}$. Each $\alpha \in \mathrm{Aut}_{\mathcal{F}}(P)$ extends to a unique automorphism $\widehat{\alpha} \in \mathrm{Aut}_{\mathcal{F}}(\widehat{P})$ by (1), and we use this to identify $\mathrm{Aut}_{\mathcal{F}}(P)$ with a subgroup of $\mathrm{Aut}_{\mathcal{F}}(\widehat{P})$. Choose $\beta \in \mathrm{Aut}_{\mathcal{F}}(\widehat{P})$ such that $\mathrm{Aut}_S(\widehat{P})$ contains a Sylow $p$-subgroup of $^{\beta}\mathrm{Aut}_{\mathcal{F}}(P) = \mathrm{Aut}_{\mathcal{F}}(\beta(P))$. Set $P^* = \beta(P)$. Thus $\mathrm{Aut}_S(P^*) = \mathrm{Aut}_{\mathcal{F}}(P^*) \cap \mathrm{Aut}_S(\widehat{P})$ is a Sylow $p$-subgroup of $\mathrm{Aut}_{\mathcal{F}}(P^*)$, and so $P^*$ is fully automized in $\mathcal{F}$.

It remains to show that $P^*$ is receptive. Fix $Q \le S$ and $\varphi \in \mathrm{Iso}_{\mathcal{F}}(Q, P^*)$, let $\varphi_i \in \mathrm{Iso}_{\mathcal{F}_i}(Q_i, P_i^*)$ be such that $\varphi = (\varphi_1, \varphi_2)|_Q$, and set $\widehat{\varphi} = (\varphi_1, \varphi_2) \in \mathrm{Iso}_{\mathcal{F}}(\widehat{Q}, \widehat{P}^*)$. Let $N_{\varphi} = N_{\varphi}^{\mathcal{F}} \le N_S(Q)$ and $N_{\widehat{\varphi}} \le N_S(\widehat{Q})$ be as in Definition 2.2. Then $N_{\widehat{\varphi}} = N_{\varphi_1}^{\mathcal{F}_1} \times N_{\varphi_2}^{\mathcal{F}_2}$, each $\varphi_i$ extends to $N_{\varphi_i}^{\mathcal{F}_i}$ since $P_i^*$ is receptive in $\mathcal{F}_i$, and thus $\widehat{\varphi}$ (and $\varphi$) extend to a morphism defined on $N_{\widehat{\varphi}}$. But $N_{\varphi} \le N_{\widehat{\varphi}}$, so $\varphi$ extends to a morphism defined on $N_{\varphi}$; and this proves $P^*$ is receptive. $\qquad\square$

The following proposition helps to illustrate the difference between normal and weakly normal subgroups. It can be thought of as a version for fusion systems (though much more elementary) of a theorem of Goldschmidt [Gd2]. Goldschmidt's theorem says that if $S = S_1 \times S_2 \in \mathrm{Syl}_2(G)$, where $S_1$ and $S_2$ are strongly closed in $S$ with respect to $G$ and $O(G) = 1$, then there are $H_1, H_2 \trianglelefteq G$ such that $S_i \in \mathrm{Syl}_2(H_i)$ and $H_1 \cap H_2 = 1$.

**Proposition 6.7** ([AOV, Proposition 3.3])  *Let $\mathcal{F}$ be a saturated fusion system over a $p$-group $S = S_1 \times S_2$, where $S_1$ and $S_2$ are strongly closed in $\mathcal{F}$. Set $\mathcal{F}_i = \mathcal{F}|_{\leq S_i}$ $(i = 1, 2)$, regarded as a fusion system over $S_i$. For each $i$, let $\mathcal{F}'_i \subseteq \mathcal{F}_i$ be the fusion subsystem over $S_i$ where for $P, Q \leq S_i$,*

$$\mathrm{Hom}_{\mathcal{F}'_i}(P, Q) = \big\{ \varphi \in \mathrm{Hom}_{\mathcal{F}_i}(P, Q) \,\big|$$
$$(\varphi, \mathrm{Id}_{S_{3-i}}) \in \mathrm{Hom}_{\mathcal{F}}(PS_{3-i}, QS_{3-i}) \big\}.$$

*Then for each $i = 1, 2$,*

(a)  *$\mathcal{F}'_i$ and $\mathcal{F}_i$ are saturated fusion systems;*

(b)  *$\mathcal{F}'_i$ is normal in $\mathcal{F}$ and in $\mathcal{F}_i$; and*

(c)  *$\mathcal{F}_i$ is weakly normal in $\mathcal{F}$, and is normal in $\mathcal{F}$ only if $\mathcal{F}_i = \mathcal{F}'_i$.*

*Furthermore,*

(d)  *$\mathcal{F}'_1 \times \mathcal{F}'_2 \subseteq \mathcal{F} \subseteq \mathcal{F}_1 \times \mathcal{F}_2$, and all three are equal if $\mathrm{Aut}_{\mathcal{F}}(S) = \mathrm{Aut}_{\mathcal{F}'_1 \times \mathcal{F}'_2}(S)$.*

*Proof.* Points (a) and (d) were shown in [AOV, Proposition 3.3]. Points (b) and (c) were pointed out to us by David Craven.

We first claim that for $i = 1, 2$,

$$P, Q \leq S_i, \ \varphi \in \mathrm{Hom}_{\mathcal{F}_i}(P, Q) \quad \Longrightarrow \quad \exists\, \psi \in \mathrm{Aut}_{\mathcal{F}}(S_{3-i})$$
$$\text{and } \chi \in \mathrm{Aut}_{\mathcal{F}}(S_i) \quad \text{s.t.} \quad (\varphi, \psi) \in \mathrm{Hom}_{\mathcal{F}}(PS_{3-i}, QS_{3-i}) \qquad (2)$$
$$\text{and } \quad \chi|_Q \circ \varphi \in \mathrm{Hom}_{\mathcal{F}'_i}(P, S_i).$$

If $\varphi(P)$ is fully centralized in $\mathcal{F}$, the existence of $\psi$ follows by the extension axiom, and since the $S_i$ are all strongly closed in $\mathcal{F}$. The general case then follows upon choosing $\alpha \in \mathrm{Iso}_{\mathcal{F}}(\varphi(P), R)$ where $R \leq S_i$ is fully centralized in $\mathcal{F}$, and applying the extension axiom to $\alpha \circ \varphi$ and to $\alpha$. By the extension axiom again, this time applied to $\psi$, there is $\chi$ such that $(\chi^{-1}, \psi) \in \mathrm{Aut}_{\mathcal{F}}(S)$, and hence $\chi|_Q \circ \varphi \in \mathrm{Hom}_{\mathcal{F}'_i}(P, S_i)$. This finishes the proof of (2).

(a)  Fix $i = 1, 2$. Fix $P \leq S_i$, and choose $Q$ which is $\mathcal{F}$-conjugate to $P$ and fully normalized in $\mathcal{F}$. By (2), there is $Q^*$ in the $\mathrm{Aut}_{\mathcal{F}}(S_i)$-orbit of $Q$ which is $\mathcal{F}'_i$-conjugate to $P$. Then $Q^*$ is also fully normalized in $\mathcal{F}$, and upon replacing $Q$ by $Q^*$, we can assume $Q$ is $\mathcal{F}'_i$-conjugate to $P$.

Since $\mathrm{Aut}_S(Q) \leq \mathrm{Aut}_{\mathcal{F}'_i}(Q) \leq \mathrm{Aut}_{\mathcal{F}_i}(Q) = \mathrm{Aut}_{\mathcal{F}}(Q)$ (and since $Q$ is fully automized in $\mathcal{F}$), $Q$ is fully automized in $\mathcal{F}_i$ and in $\mathcal{F}'_i$. In the notation of Definition 2.2, for all $R \leq S_i$ and $\varphi \in \mathrm{Iso}_{\mathcal{F}}(R, Q)$, $N_{\varphi}^{\mathcal{F}_i} = N_{\varphi}^{\mathcal{F}} \cap S_i$, and so $Q$ is receptive in $\mathcal{F}_i$ since it is receptive in $\mathcal{F}$. Finally, if $\varphi \in \mathrm{Iso}_{\mathcal{F}'_i}(R, Q)$, then since $QS_{3-i}$ is also fully normalized and hence receptive in $\mathcal{F}$, $(\varphi, \mathrm{Id}_{S_{3-i}}) \in \mathrm{Iso}_{\mathcal{F}}(RS_{3-i}, QS_{3-i})$ extends to an $\mathcal{F}$-homomorphism on $N_{\varphi}^{\mathcal{F}'_i} S_{3-i}$, this restricts to an $\mathcal{F}'_i$-homomorphism on $N_{\varphi}^{\mathcal{F}'_i}$, and thus $Q$ is also receptive in $\mathcal{F}'_i$. We now conclude that $\mathcal{F}_i$ and $\mathcal{F}'_i$ are both saturated fusion systems over $S_i$.

**(b)** By construction, for all $\varphi \in \mathrm{Aut}_{\mathcal{F}_i}(S_i) = \mathrm{Aut}_{\mathcal{F}}(S_i)$ $(i = 1, 2)$, $^{\varphi}\mathcal{F}'_i = \mathcal{F}'_i$. Since $\mathcal{F}_i = \langle \mathrm{Aut}_{\mathcal{F}}(S_i), \mathcal{F}'_i \rangle$ by (2) (and since $\mathcal{F}'_i$ is saturated and $S_i$ is strongly closed in $\mathcal{F}$), $\mathcal{F}'_i$ is weakly normal in $\mathcal{F}$ and in $\mathcal{F}_i$.

By definition of $\mathcal{F}'_i$, each $\alpha \in \mathrm{Aut}_{\mathcal{F}'_i}(S_i)$ extends to some $\bar{\alpha} \in \mathrm{Aut}_{\mathcal{F}}(S)$ such that $\bar{\alpha}|_{S_{3-i}} = \mathrm{Id}$. Hence $\mathcal{F}'_i$ is normal in $\mathcal{F}$ and in $\mathcal{F}_i$ by Proposition 6.4.

**(c)** Since $\mathcal{F}_i$ is a full subcategory of $\mathcal{F}$ and $S_i$ is strongly closed, $\mathcal{F}_i$ is $\mathcal{F}$-invariant. Thus $\mathcal{F}_i$ is weakly normal in $\mathcal{F}$ since it is saturated by (a).

If $\mathcal{F}_i$ is normal in $\mathcal{F}$, then each $\alpha \in \mathrm{Aut}_{\mathcal{F}}(S_i)$ extends to some $\bar{\alpha} \in \mathrm{Aut}_{\mathcal{F}}(S)$ such that $[\bar{\alpha}, Z(S_i)S_{3-i}] \leq Z(S_i)$. Since $S_{3-i}$ is strongly closed in $\mathcal{F}$, this means that $\bar{\alpha}$ acts trivially on $S_{3-i}$, and hence that $\alpha \in \mathrm{Aut}_{\mathcal{F}'_i}(S_i)$. So in this case, $\mathrm{Aut}_{\mathcal{F}'_i}(S_i) = \mathrm{Aut}_{\mathcal{F}}(S_i)$, and $\mathcal{F}'_i = \mathcal{F}_i$ by what was shown in (b).

**(d)** Clearly, $\mathcal{F}$ contains $\mathcal{F}'_1 \times \mathcal{F}'_2$. We must show that $\mathcal{F} \subseteq \mathcal{F}_1 \times \mathcal{F}_2$. Assume otherwise. Then by Theorem 3.5, there is a subgroup $P \leq S$ such that $P = S$ or $P$ is $\mathcal{F}$-essential, and an automorphism $\alpha \in \mathrm{Aut}_{\mathcal{F}}(P)$ such that $\alpha \notin \mathrm{Aut}_{\mathcal{F}_1 \times \mathcal{F}_2}(P)$. Assume $P$ is maximal subject to this constraint. For $i = 1, 2$, let $\mathrm{pr}_i \colon S \longrightarrow S_i$ be the projection, and set $P_i = \mathrm{pr}_i(P)$. We regard $P$ as a subgroup of $\widehat{P} \stackrel{\mathrm{def}}{=} P_1 \times P_2$. Since $\alpha(S_i \cap P) = S_i \cap P$ for $i = 1, 2$ (the $S_i$ being strongly closed), there are unique automorphisms $\alpha_i \in \mathrm{Aut}(P_i)$, defined by setting $\alpha_i(\mathrm{pr}_i(g)) = \mathrm{pr}_i(\alpha(g))$ for $g \in P$. Thus, if $g = (g_1, g_2)$ (i.e., $g_i = \mathrm{pr}_i(g)$), then $\alpha(g) = (\alpha_1(g_1), \alpha_2(g_2))$; and we conclude that $\alpha = (\alpha_1, \alpha_2)|_P$.

Since $P$ is fully normalized in $\mathcal{F}$ (hence receptive), $\alpha$ extends to an $\mathcal{F}$-automorphism of $N_{\widehat{P}}(P)$. By the maximality assumption on $P$, this implies $N_{\widehat{P}}(P) = P$, and so $\widehat{P} = P$ by Lemma A.1. Hence $\alpha = (\alpha_1, \alpha_2)$, so $\alpha_i = \alpha|_{P_i} \in \mathrm{Aut}_{\mathcal{F}_i}(P_i)$, and this contradicts the original assumption that $\alpha \notin \mathrm{Aut}_{\mathcal{F}_1 \times \mathcal{F}_2}(P)$. Thus $\mathcal{F} \subseteq \mathcal{F}_1 \times \mathcal{F}_2$.

If $\mathrm{Aut}_{\mathcal{F}}(S) = \mathrm{Aut}_{\mathcal{F}'_1 \times \mathcal{F}'_2}(S)$, then $\mathrm{Aut}_{\mathcal{F}_i}(S_i) = \mathrm{Aut}_{\mathcal{F}'_i}(S_i)$ $(i = 1, 2)$ by construction, and so $\mathcal{F}_i = \langle \mathcal{F}'_i, \mathrm{Aut}_{\mathcal{F}_i}(S_i) \rangle = \mathcal{F}'_i$ by (b). Hence $\mathcal{F} = \mathcal{F}'_1 \times \mathcal{F}'_2 = \mathcal{F}_1 \times \mathcal{F}_2$. $\qquad \square$

As an example, fix a prime $q \neq p$, and two pairs of finite groups $H_i \trianglelefteq G_i$ ($i = 1, 2$) such that $|G_i/H_i| = q$, and such that for $S_i \in \mathrm{Syl}_p(G_i)$, $C_{G_i}(S_i) \leq H_i$. Let $\Gamma \leq G_1 \times G_2$ be a subgroup of index $q$ which contains $H_1 \times H_2$, and contains neither $G_1$ nor $G_2$. Set $S = S_1 \times S_2 \in \mathrm{Syl}_p(\Gamma)$ and $\mathcal{F} = \mathcal{F}_S(\Gamma)$, and let $\mathcal{F}_i$ and $\mathcal{F}'_i$ be as in Proposition 6.7. Then $\mathcal{F}_i = \mathcal{F}_{S_i}(G_i)$, $\mathcal{F}'_i = \mathcal{F}_{S_i}(H_i)$, and $\mathcal{F}'_i \subsetneqq \mathcal{F}_i$. The subgroups $H_i$ are normal in $\Gamma$, corresponding to $\mathcal{F}'_i$ being normal in $\mathcal{F}$, while $G_i$ is not even contained in $\Gamma$ (and $\mathcal{F}_i$ is only weakly normal in $\mathcal{F}$).

## 7. Fusion subsystems of $p$-power index or of index prime to $p$

We next look at certain fusion subsystems of a saturated fusion system which play a role analogous to that of subgroups of a finite group $G$ which contain $O^p(G)$ or $O^{p'}(G)$. Before we can define exactly what we mean by this, we need to define the focal and hyperfocal subgroups of a fusion system. These definitions are also motivated by those for the focal and hyperfocal subgroup of a finite group.

**Definition 7.1**   For any saturated fusion system $\mathcal{F}$ over a $p$-group $S$, the *focal subgroup* $\mathfrak{foc}(\mathcal{F})$ and the *hyperfocal subgroup* $\mathfrak{hyp}(\mathcal{F})$ are defined by setting

$$\mathfrak{foc}(\mathcal{F}) = \langle g^{-1}h \mid g, h \in S \text{ are } \mathcal{F}\text{-conjugate} \rangle$$
$$= \langle g^{-1}\alpha(g) \mid g \in P \leq S,\ \alpha \in \mathrm{Aut}_{\mathcal{F}}(P) \rangle$$

$$\mathfrak{hyp}(\mathcal{F}) = \langle g^{-1}\alpha(g) \mid g \in P \leq S,\ \alpha \in O^p(\mathrm{Aut}_{\mathcal{F}}(P)) \rangle.$$

The two definitions of $\mathfrak{foc}(\mathcal{F})$ are equivalent by Theorem 3.5. One easily sees that $\mathfrak{foc}(\mathcal{F})$ and $\mathfrak{hyp}(\mathcal{F})$ are not only normal in $S$, but are also invariant under the action of each $\alpha \in \mathrm{Aut}_{\mathcal{F}}(S)$. The next lemma describes the relationship between these two subgroups.

**Lemma 7.2**   *For any saturated fusion system $\mathcal{F}$ over a $p$-group $S$,*

$$\mathfrak{foc}(\mathcal{F}) = \mathfrak{hyp}(\mathcal{F}) \cdot [S, S].$$

*In particular, $\mathfrak{foc}(\mathcal{F}) = S$ if and only if $\mathfrak{hyp}(\mathcal{F}) = S$.*

*Proof.* By Theorem 3.5,

$$\mathfrak{foc}(\mathcal{F}) = \langle s^{-1}\alpha(s) \mid s \in P \leq S,\ P \text{ is } \mathcal{F}\text{-essential},\ \alpha \in \mathrm{Aut}_{\mathcal{F}}(P) \rangle.$$

By the Frattini argument (Proposition A.4(a)),

$$\mathrm{Aut}_{\mathcal{F}}(P) = O^p(\mathrm{Aut}_{\mathcal{F}}(P)) \cdot \mathrm{Aut}_S(P)$$

whenever $\mathrm{Aut}_S(P) \in \mathrm{Syl}_p(\mathrm{Aut}_\mathcal{F}(P))$; in particular, whenever $P$ is $\mathcal{F}$-essential. Also, $s^{-1}\alpha(s) \in [S,S]$ when $s \in P$ and $\alpha \in \mathrm{Aut}_S(P)$, and thus $\mathfrak{foc}(\mathcal{F}) = \mathfrak{hyp}(\mathcal{F}){\cdot}[S,S]$.

If $\mathfrak{hyp}(\mathcal{F}) < S$, then it is contained in a subgroup $P < S$ of index $p$. But then $[S,S] \le P$, and hence $\mathfrak{foc}(\mathcal{F}) \le P < S$. $\qquad\square$

The hyperfocal subgroup is also called $H_\mathcal{F}$ by Puig [P6, §7.1], and is denoted $O_\mathcal{F}^p(S)$ in [5a2]. If $\mathcal{F} = \mathcal{F}_S(G)$ is the fusion system of a finite group $G$ with respect to $S \in \mathrm{Syl}_p(G)$, then the focal subgroup theorem (cf. [G1, Theorem 7.3.4]) says that $\mathfrak{foc}(\mathcal{F}) = S \cap [G,G]$, while the hyperfocal subgroup theorem of Puig [P5, §1.1] says that $\mathfrak{hyp}(\mathcal{F}) = S \cap O^p(G)$.

**Definition 7.3**   Fix a saturated fusion system $\mathcal{F}$ over a $p$-group $S$, and a fusion subsystem $\mathcal{E}$ in $\mathcal{F}$ over $T \le S$.

- The subsystem $\mathcal{E}$ has *$p$-power index* in $\mathcal{F}$ if $T \ge \mathfrak{hyp}(\mathcal{F})$, and $\mathrm{Aut}_\mathcal{E}(P) \ge O^p(\mathrm{Aut}_\mathcal{F}(P))$ for each $P \le S$.

- The subsystem $\mathcal{E}$ has *index prime to $p$* in $\mathcal{F}$ if $T = S$, and $\mathrm{Aut}_\mathcal{E}(P) \ge O^{p'}(\mathrm{Aut}_\mathcal{F}(P))$ for each $P \le S$.

The saturated fusion subsystems of $p$-power index are described by the following theorem.

**Theorem 7.4**   *Let $\mathcal{F}$ be a saturated fusion system over a $p$-group $S$. Then for each subgroup $T \le S$ which contains $\mathfrak{hyp}(\mathcal{F})$, there is a unique saturated fusion subsystem*

$$\mathcal{F}_T = \big\langle \mathrm{Inn}(T), O^p(\mathrm{Aut}_\mathcal{F}(P)) \,\big|\, P \le T \big\rangle \qquad (1)$$

*over $T$ of $p$-power index in $\mathcal{F}$. Also, $\mathcal{F}_T \trianglelefteq \mathcal{F}$ if $T \trianglelefteq S$, and $\mathcal{F}_T \subseteq \mathcal{F}_U$ if $T \le U \le S$. In particular, there is a unique minimal saturated fusion subsystem $O^p(\mathcal{F}) \trianglelefteq \mathcal{F}$ of $p$-power index, over the subgroup $\mathfrak{hyp}(\mathcal{F}) \trianglelefteq S$.*

*Proof.* Except for the precise description of $\mathcal{F}_T$, and the claim that $\mathcal{F}_T \trianglelefteq \mathcal{F}$ if $T \trianglelefteq S$, this was shown in [5a2, Theorem 4.3]. A different proof of this was given in [A6, §7].

Let $\mathcal{F}_T \subseteq \mathcal{F}$ denote the unique saturated fusion subsystem over $T$ of $p$-power index. When $P$ is fully automized in $\mathcal{F}_T$, then $\mathrm{Aut}_{\mathcal{F}_T}(P)$ must be generated by $O^p(\mathrm{Aut}_\mathcal{F}(P))$ and $\mathrm{Aut}_T(P)$, since by assumption, it contains the former and contains the latter as Sylow $p$-subgroup. Hence $\mathcal{F}_T$ is generated by the $O^p(\mathrm{Aut}_\mathcal{F}(P))$ and $\mathrm{Aut}_T(P)$ for all $P \le S$ by Theorem 3.5 (Alperin's fusion theorem). This proves (1).

If $T \trianglelefteq S$, then $T$ is strongly closed in $\mathcal{F}$ by definition of $\mathfrak{hyp}(\mathcal{F})$, and $^\alpha(\mathcal{F}_T) = \mathcal{F}_{\alpha(T)} = \mathcal{F}_T$ for all $\alpha \in \mathrm{Aut}_\mathcal{F}(T)$ by (1). Also, for $P \le S$, restriction sends $\mathrm{Aut}_{\mathcal{F}}(P)$ into $\mathrm{Aut}_{\mathcal{F}}(P \cap T)$ and hence sends $O^p(\mathrm{Aut}_{\mathcal{F}}(P))$ into $O^p(\mathrm{Aut}_{\mathcal{F}}(P \cap T))$. So by (1) again,

$$\mathcal{F}|_{\leq T} = \langle \mathrm{Aut}_S(T), O^p(\mathrm{Aut}_{\mathcal{F}}(P)) \,|\, P \leq T \rangle$$
$$= \langle \mathrm{Aut}_S(T), \mathcal{F}_T \rangle = \langle \mathrm{Aut}_{\mathcal{F}}(T), \mathcal{F}_T \rangle.$$

Thus $\mathcal{F}_T$ is weakly normal in $\mathcal{F}$ by Proposition 6.4(b $\Rightarrow$ a).

To show that $\mathcal{F}_T \trianglelefteq \mathcal{F}$, it remains to prove the extension condition. Fix $\alpha \in \mathrm{Aut}_{\mathcal{F}_T}(T)$. Then $\alpha = \beta \circ \gamma$, where $\beta$ has order prime to $p$ and $\gamma$ has $p$-power order, and $\beta, \gamma \in \langle \alpha \rangle$. Since $\mathrm{Inn}(T)$ is a normal Sylow $p$-subgroup of $\mathrm{Aut}_{\mathcal{F}_T}(T)$, $\gamma = c_g$ for some $g \in T$. Since $T$ is receptive in $\mathcal{F}$ (since it is strongly closed), there is $\bar{\beta} \in \mathrm{Aut}_{\mathcal{F}}(T \cdot C_S(T))$ such that $\bar{\beta}|_T = \beta$, and (upon replacing $\bar{\beta}$ by an appropriate power, if necessary) we can assume $\bar{\beta}$ has order prime to $p$. Thus $\bar{\beta} \circ c_g \in \mathrm{Aut}_{\mathcal{F}}(T \cdot C_S(T))$ extends $\alpha$, $[g, C_S(T)] = 1$, so

$$[\bar{\beta} \circ c_g, C_S(T)] \leq [\bar{\beta}, C_S(T)] \leq \mathfrak{hyp}(\mathcal{F}) \cap C_S(T) \leq Z(T),$$

and the extension condition holds. $\qquad\square$

The following corollary follows immediately from Theorem 7.4 and Lemma 7.2.

**Corollary 7.5** *For any saturated fusion system $\mathcal{F}$ over a $p$-group $S$,*

$$O^p(\mathcal{F}) = \mathcal{F} \iff \mathfrak{hyp}(\mathcal{F}) = S \iff \mathfrak{foc}(\mathcal{F}) = S.$$

We now turn our attention to fusion subsystems of index prime to $p$. In the following lemma, by the "Frattini condition" on a pair of fusion systems $\mathcal{E} \subseteq \mathcal{F}$ over the same $p$-group $S$, we mean the condition of Definition 6.1: for each $P \leq S$ and $\varphi \in \mathrm{Hom}_{\mathcal{F}}(P, S)$, there are $\varphi_0 \in \mathrm{Hom}_{\mathcal{E}}(P, S)$ and $\alpha \in \mathrm{Aut}_{\mathcal{F}}(S)$ such that $\varphi = \alpha \circ \varphi_0$.

**Lemma 7.6** *Fix a saturated fusion system $\mathcal{F}$ over a $p$-group $S$, and let $\mathcal{E} \subseteq \mathcal{F}$ be a fusion subsystem over $S$. Then the following hold.*

(a) *If $\mathrm{Aut}_{\mathcal{E}}(P) \geq O^{p'}(\mathrm{Aut}_{\mathcal{F}}(P))$ for each $P \in \mathcal{F}^c$, then $\mathcal{E}$ has index prime to $p$ in $\mathcal{F}$, the Frattini condition holds for $\mathcal{E} \subseteq \mathcal{F}$, and $\mathcal{F}^c = \mathcal{E}^c$.*

(b) *Assume $\mathcal{E}$ has index prime to $p$ in $\mathcal{F}$. Then $\mathcal{E}$ is saturated if and only if $\mathcal{E}$ is $\mathcal{F}^c$-generated, and for each $P, Q \in \mathcal{F}^c$ and $\varphi \in \mathrm{Hom}_{\mathcal{F}}(Q, S)$ such that $P \leq Q$ and $\varphi|_P \in \mathrm{Hom}_{\mathcal{E}}(P, S)$, $\varphi \in \mathrm{Hom}_{\mathcal{E}}(Q, S)$.*

*Proof.* **(a)** Set $\mathcal{E}_0 = \langle O^{p'}(\mathrm{Aut}_{\mathcal{F}}(P)) \,|\, P \in \mathcal{F}^c \rangle$; i.e., the smallest fusion system over $S$ which contains the automorphism groups $O^{p'}(\mathrm{Aut}_{\mathcal{F}}(P))$ for $\mathcal{F}$-centric subgroups $P \leq S$. For each $\mathcal{F}$-essential subgroup $P \leq S$, $\mathrm{Aut}_{\mathcal{F}}(P) = O^{p'}(\mathrm{Aut}_{\mathcal{F}}(P)) \cdot N_{\mathrm{Aut}_{\mathcal{F}}(P)}(\mathrm{Aut}_S(P))$ by the Frattini argument (Proposition A.4(a)). Since $P$ is receptive, each morphism

$\varphi \in N_{\mathrm{Aut}_{\mathcal{F}}(P)}(\mathrm{Aut}_S(P))$ extends to some $\bar{\varphi} \in \mathrm{Aut}_{\mathcal{F}}(N_S(P))$. So by Theorem 3.5 and a downwards induction on $|P|$, we see that

$$\mathcal{F} = \langle \mathrm{Aut}_{\mathcal{F}}(S), O^{p'}(\mathrm{Aut}_{\mathcal{F}}(P)) \mid P \ \mathcal{F}\text{-essential} \rangle = \langle \mathrm{Aut}_{\mathcal{F}}(S), \mathcal{E}_0 \rangle.$$

Also, $^{\alpha}\mathcal{E}_0 = \mathcal{E}_0$ for each $\alpha \in \mathrm{Aut}_{\mathcal{F}}(S)$ by construction of $\mathcal{E}_0$. So by Proposition 6.4($\mathrm{b} \Rightarrow \mathrm{a,c}$), $\mathcal{E}_0$ is $\mathcal{F}$-invariant (hence the Frattini condition holds for $\mathcal{E}_0 \subseteq \mathcal{F}$), and $\mathrm{Aut}_{\mathcal{E}_0}(P) \trianglelefteq \mathrm{Aut}_{\mathcal{F}}(P)$ for each $P \leq S$. The last condition in turn implies that $\mathrm{Aut}_{\mathcal{E}_0}(P) \geq O^{p'}(\mathrm{Aut}_{\mathcal{F}}(P))$ for each fully normalized subgroup $P \leq S$ (since $\mathrm{Aut}_{\mathcal{E}_0}(P)$ contains the Sylow subgroup $\mathrm{Aut}_S(P)$), and hence for all $P \leq S$. Thus $\mathcal{E}_0$ has index prime to $p$ in $\mathcal{F}$.

By assumption, $\mathrm{Aut}_{\mathcal{E}}(P) \geq O^{p'}(\mathrm{Aut}_{\mathcal{F}}(P))$ for all $P \in \mathcal{F}^c$, and hence $\mathcal{E} \supseteq \mathcal{E}_0$. So $\mathcal{E}$ has index prime to $p$ in $\mathcal{F}$, and the Frattini condition holds for $\mathcal{E} \subseteq \mathcal{F}$.

Clearly, $\mathcal{F}^c \subseteq \mathcal{E}^c$: each $\mathcal{F}$-centric subgroup $P$ is also $\mathcal{E}$-centric since $P^{\mathcal{E}} \subseteq P^{\mathcal{F}}$. Conversely, if $P \in \mathcal{E}^c$, then by the Frattini condition, each subgroup in $P^{\mathcal{F}}$ has the form $\alpha(Q)$ for some $Q \in P^{\mathcal{E}}$ and $\alpha \in \mathrm{Aut}_{\mathcal{F}}(S)$, and $C_S(\alpha(Q)) \leq \alpha(Q)$ since $C_S(Q) \leq Q$. Thus $\mathcal{F}^c = \mathcal{E}^c$.

**(b1)** Assume $\mathcal{E}$ is saturated. Then $\mathcal{E}$ is $\mathcal{E}^c$-generated by Theorem 3.5, and hence is $\mathcal{F}^c$-generated by (a).

Assume $P, Q \in \mathcal{F}^c$ and $\varphi \in \mathrm{Hom}_{\mathcal{F}}(Q, S)$ are such that $P \leq Q$ and $\varphi|_P \in \mathrm{Hom}_{\mathcal{E}}(P, S)$; we must show that $\varphi \in \mathrm{Hom}_{\mathcal{E}}(Q, S)$. Since $P$ is a subnormal subgroup of $Q$, it suffices by iteration to do this when $P \trianglelefteq Q$. Set $\varphi_0 = \varphi|_P$. Since there is some extension of $\varphi_0$ to a homomorphism defined on $Q$, $N_{\varphi_0}^{\mathcal{E}} \geq Q$ in the notation of Definition 2.2. Also, $P \in \mathcal{F}^c = \mathcal{E}^c$, and hence $P$ is receptive in $\mathcal{E}$. So there is $\varphi' \in \mathrm{Hom}_{\mathcal{E}}(Q, S)$ such that $\varphi'|_P = \varphi|_P$. By Lemma 5.6, there is $x \in Z(P)$ such that $\varphi = \varphi' \circ c_x$, and so $\varphi \in \mathrm{Hom}_{\mathcal{E}}(Q, S)$.

**(b2)** Now assume $\mathcal{E}$ is $\mathcal{F}^c$-generated, and for each $P, Q \in \mathcal{F}^c$ and $\psi \in \mathrm{Hom}_{\mathcal{F}}(Q, S)$ such that $P \leq Q$ and $\psi|_P \in \mathrm{Hom}_{\mathcal{E}}(P, S)$, $\psi \in \mathrm{Hom}_{\mathcal{E}}(Q, S)$. We must show that $\mathcal{E}$ is saturated.

If $P \leq S$ is fully automized and receptive in $\mathcal{F}$, then the same holds for $\alpha(P)$ for each $\alpha \in \mathrm{Aut}_{\mathcal{F}}(S)$. So by the Frattini condition (which holds by (a)), each subgroup of $S$ is $\mathcal{E}$-conjugate to one which is fully automized and receptive in $\mathcal{F}$.

Assume $P \in \mathcal{F}^c$, and choose $Q$ which is $\mathcal{E}$-conjugate to $P$ and fully automized and receptive in $\mathcal{F}$. Thus $\mathrm{Aut}_S(Q) \in \mathrm{Syl}_p(\mathrm{Aut}_{\mathcal{F}}(Q))$, so $\mathrm{Aut}_S(Q) \in \mathrm{Syl}_p(\mathrm{Aut}_{\mathcal{E}}(Q))$, and $Q$ is also fully automized in $\mathcal{E}$. Fix an isomorphism $\varphi \in \mathrm{Iso}_{\mathcal{E}}(R, Q)$, and let $N_{\varphi} = N_{\varphi}^{\mathcal{E}} = N_{\varphi}^{\mathcal{F}}$ be as in Definition 2.2. Since $Q$ is receptive in $\mathcal{F}$, $\varphi$ extends to some $\widehat{\varphi} \in \mathrm{Hom}_{\mathcal{F}}(N_{\varphi}, S)$. Thus $\widehat{\varphi}|_R \in \mathrm{Hom}_{\mathcal{E}}(R, Q)$, $R \in \mathcal{F}^c$, so by assumption, $\widehat{\varphi} \in \mathrm{Hom}_{\mathcal{E}}(N_{\varphi}, S)$. Thus $Q$ is receptive in $\mathcal{E}$.

We have now shown, for each $P \in \mathcal{F}^c$, that there is $Q \in P^{\mathcal{E}}$ which is fully automized and receptive in $\mathcal{E}$. Thus $\mathcal{E}$ is $\mathcal{F}^c$-saturated in the sense of Definition 3.9. Since $\mathcal{E}$ is $\mathcal{F}^c$-generated by assumption, and $\mathcal{F}^c = \mathcal{E}^c$ by (a), $\mathcal{E}$ is a saturated fusion system by Theorem 3.10. $\qquad\square$

We are now ready to describe all saturated fusion subsystems of index prime to $p$ in a saturated fusion system. For any fusion system $\mathcal{F}$ over a $p$-group $S$, let $\Gamma_{p'}(\mathcal{F})$ be the free group on the set $\mathrm{Mor}(\mathcal{F}^c)$, modulo the relations induced by composition, and by dividing out by $O^{p'}(\mathrm{Aut}_{\mathcal{F}}(P))$ for all $P \in \mathcal{F}^c$. In other words, the natural map $\mathrm{Mor}(\mathcal{F}^c) \longrightarrow \Gamma_{p'}(\mathcal{F})$ is universal among all maps from $\mathrm{Mor}(\mathcal{F}^c)$ to a group which send composites to products, and send $O^{p'}(\mathrm{Aut}_{\mathcal{F}}(P))$ to the identity for each $P \in \mathcal{F}^c$. We will see later (Theorem III.4.19) that $\Gamma_{p'}(\mathcal{F})$ is the fundamental group of the geometric realization of the category $\mathcal{F}^c$.

The following theorem is essentially the same as [5a2, Theorem 5.4], although formulated somewhat differently. See also [P6, Theorem 6.11].

**Theorem 7.7** *Fix a saturated fusion system $\mathcal{F}$ over a $p$-group $S$. Let $\theta \colon \mathrm{Mor}(\mathcal{F}^c) \longrightarrow \Gamma_{p'}(\mathcal{F})$ be the canonical map. Then the following hold.*

(a) *$\Gamma_{p'}(\mathcal{F})$ is finite of order prime to $p$, and $\theta(\mathrm{Aut}_{\mathcal{F}}(S)) = \Gamma_{p'}(\mathcal{F})$.*

(b) *For each saturated fusion subsystem $\mathcal{E} \subseteq \mathcal{F}$ of index prime to $p$, there is $H \leq \Gamma_{p'}(\mathcal{F})$ such that $\mathcal{E} = \langle \theta^{-1}(H) \rangle$. Also, $\mathrm{Hom}_{\mathcal{E}}(P, Q) = \mathrm{Hom}_{\mathcal{F}}(P, Q) \cap \theta^{-1}(H)$ for each $P, Q \in \mathcal{F}^c$.*

(c) *For each $H \leq \Gamma_{p'}(\mathcal{F})$, set $\mathcal{E}_H = \langle \theta^{-1}(H) \rangle$. Then $\mathcal{E}_H$ is a saturated fusion subsystem of index prime to $p$ in $\mathcal{F}$, and $\mathcal{E}_H \trianglelefteq \mathcal{F}$ if and only if $H \trianglelefteq \Gamma_{p'}(\mathcal{F})$.*

(d) *Set $\mathcal{E}_0 = \langle O^{p'}(\mathrm{Aut}_{\mathcal{F}}(P)) \mid P \leq S \rangle$, as a fusion system over $S$. Then $\theta$ induces an isomorphism $\Gamma_{p'}(\mathcal{F}) \cong \mathrm{Aut}_{\mathcal{F}}(S)/\mathrm{Aut}^0_{\mathcal{F}}(S)$, where*

$$\mathrm{Aut}^0_{\mathcal{F}}(S) \overset{\mathrm{def}}{=} \mathrm{Ker}(\theta|_{\mathrm{Aut}_{\mathcal{F}}(S)}) =$$
$$\langle \alpha \in \mathrm{Aut}_{\mathcal{F}}(S) \mid \alpha|_P \in \mathrm{Hom}_{\mathcal{E}_0}(P, S),\ \text{some } P \in \mathcal{F}^c \rangle.$$

*In particular, there is a unique minimal saturated fusion subsystem $O^{p'}(\mathcal{F}) = \langle \theta^{-1}(1) \rangle \trianglelefteq \mathcal{F}$ of index prime to $p$, and $\mathrm{Aut}_{O^{p'}(\mathcal{F})}(S) = \mathrm{Aut}^0_{\mathcal{F}}(S)$.*

*Proof.* Let $\mathcal{E}_0 \subseteq \mathcal{F}$ be the fusion system defined in (d). Since $\theta$ sends inclusions to the identity, $\theta(\varphi) = \theta(\psi)$ whenever $\psi$ is a restriction of $\varphi$. Since $\theta(O^{p'}(\mathrm{Aut}_{\mathcal{F}}(P))) = 1$ for each $P \in \mathcal{F}^c$ by definition of $\Gamma_{p'}(\mathcal{F})$, $\theta(\mathrm{Mor}(\mathcal{E}_0)) = 1$.

By Lemma 7.6(a), $\mathcal{E}_0$ is the minimal fusion subsystem of index prime to $p$ in $\mathcal{F}$, and the Frattini condition holds for $\mathcal{E}_0 \subseteq \mathcal{F}$.

**(a)** By the Frattini condition, for each $\varphi \in \mathrm{Mor}(\mathcal{F}^c)$, $\varphi = \alpha \circ \varphi_0$ for some $\alpha \in \mathrm{Aut}_{\mathcal{F}}(S)$ and some $\varphi_0 \in \mathrm{Mor}(\mathcal{E}_0)$. Then $\theta(\varphi_0) = 1$, so $\theta(\varphi) = \theta(\alpha) \in \theta(\mathrm{Aut}_{\mathcal{F}}(S))$. Thus $\theta(\mathrm{Aut}_{\mathcal{F}}(S)) = \theta(\mathrm{Mor}(\mathcal{F}^c)) = \Gamma_{p'}(\mathcal{F})$.

In particular, since $\theta(O^{p'}(\mathrm{Aut}_{\mathcal{F}}(S))) = 1$, $\Gamma_{p'}(\mathcal{F})$ has order prime to $p$.

**(d)** Set $\Gamma = \mathrm{Aut}_{\mathcal{F}}(S)$ for short, and set

$$\Gamma_0 = \langle \alpha \in \Gamma \mid \alpha|_P \in \mathrm{Hom}_{\mathcal{E}_0}(P, S), \text{ some } P \in \mathcal{F}^c \rangle.$$

Thus $\theta(\Gamma_0) = 1$ since $\theta(\mathrm{Mor}(\mathcal{E}_0)) = 1$. We must show that $\Gamma_0 = \mathrm{Ker}(\theta|_\Gamma)$. To see this, define

$$\widehat{\theta} \colon \mathrm{Mor}(\mathcal{F}^c) \longrightarrow \Gamma/\Gamma_0$$

as follows. For each $P \in \mathcal{F}^c$ and each $\varphi \in \mathrm{Hom}_{\mathcal{F}}(P, S)$, $\varphi = \alpha \circ \varphi_0$ for some $\varphi_0 \in \mathrm{Hom}_{\mathcal{E}_0}(P, S)$ and some $\alpha \in \Gamma$ (Lemma 7.6), and we set $\widehat{\theta}(\varphi) = \alpha\Gamma_0$. To see this is well defined, assume $\varphi = \beta \circ \varphi_1$ is another such factorization, and set $Q_i = \varphi_i(P)$ ($i = 0, 1$). Thus $\varphi_1 \circ \varphi_0^{-1} = (\beta \circ \alpha^{-1})|_{Q_0} \in \mathrm{Iso}_{\mathcal{F}}(Q_0, Q_1)$, so $\beta \circ \alpha^{-1} \in \Gamma_0$, and $\beta\Gamma_0 = \alpha\Gamma_0$.

Clearly, $\widehat{\theta}$ sends composites to products, and $\widehat{\theta}(O^{p'}(\mathrm{Aut}_{\mathcal{F}}(P))) = 1$ for each $P \in \mathcal{F}^c$. Since $\theta$ was defined to be universal among such maps, $\widehat{\theta}$ factors through $\theta$, and thus $\mathrm{Ker}(\theta|_\Gamma) \leq \Gamma_0$. We conclude that $\mathrm{Aut}_{\mathcal{F}}^0(S) \overset{\mathrm{def}}{=} \mathrm{Ker}(\theta|_\Gamma) = \Gamma_0$. Since $\theta|_\Gamma$ is onto by (a), this also proves that $\Gamma_{p'}(\mathcal{F}) \cong \mathrm{Aut}_{\mathcal{F}}(S)/\mathrm{Aut}_{\mathcal{F}}^0(S)$.

**(b)** Fix a saturated fusion subsystem $\mathcal{E}$ of index prime to $p$ in $\mathcal{F}$, and set $H = \theta(\mathrm{Aut}_{\mathcal{E}}(S))$. If $\alpha \in \mathrm{Aut}_{\mathcal{F}}(S)$ is such that $\alpha|_P \in \mathrm{Hom}_{\mathcal{E}_0}(P, S)$ for some $P \in \mathcal{F}^c$, then $\alpha|_P \in \mathrm{Mor}(\mathcal{E})$, and hence $\alpha \in \mathrm{Aut}_{\mathcal{E}}(S)$ by Lemma 7.6. Thus $\Gamma_0 \leq \mathrm{Aut}_{\mathcal{E}}(S)$. Hence for arbitrary $\alpha \in \mathrm{Aut}_{\mathcal{F}}(S)$, $\alpha \in \mathrm{Aut}_{\mathcal{E}}(S)$ if and only if $\theta(\alpha) \in H$.

Fix any $\varphi \in \mathrm{Mor}(\mathcal{F}^c)$, and let $\varphi_0 \in \mathrm{Mor}(\mathcal{E}_0)$ and $\alpha \in \mathrm{Aut}_{\mathcal{F}}(S)$ be such that $\varphi = \alpha \circ \varphi_0$. Thus $\varphi_0 \in \mathrm{Mor}(\mathcal{E})$ and $\theta(\varphi_0) = 1$. So $\theta(\varphi) = \theta(\alpha)$, and $\varphi \in \mathrm{Mor}(\mathcal{E})$ if and only if $\alpha \in \mathrm{Aut}_{\mathcal{E}}(S)$ by Lemma 7.6. We just saw that $\alpha \in \mathrm{Aut}_{\mathcal{E}}(S)$ if and only if $\theta(\alpha) \in H$, and hence $\varphi \in \mathrm{Mor}(\mathcal{E})$ if and only if $\theta(\varphi) \in H$. This proves the last statement in (b).

By Lemma 7.6 again, $\mathcal{E}^c = \mathcal{F}^c$. Hence $\theta^{-1}(H) = \mathrm{Mor}(\mathcal{E}^c)$, and so $\mathcal{E} = \langle \theta^{-1}(H) \rangle$ by Theorem 3.5.

**(c)** Now fix $H \leq \Gamma_{p'}(\mathcal{F})$, and set $\mathcal{E} = \langle \theta^{-1}(H) \rangle \supseteq \mathcal{E}_0$. Thus $\mathcal{E}$ has index prime to $p$ in $\mathcal{F}$, is $\mathcal{F}^c$-generated, satisfies the Frattini condition, and satisfies the extension condition of Lemma 7.6(b). Hence by the lemma, $\mathcal{E}$ is a saturated fusion system of index prime to $p$ in $\mathcal{F}$.

If $H \trianglelefteq \Gamma_{p'}(\mathcal{F})$, then for $\alpha \in \mathrm{Aut}_{\mathcal{F}}(S)$, $^\alpha\mathcal{E} = \mathcal{E}$, since $\theta^{-1}(H)$ is invariant under conjugation by $\alpha$. Thus $\mathcal{E}$ is weakly normal in $\mathcal{F}$, and is normal since the extension condition of Definition 6.1 is vacuous when the fusion subsystem is over the same $p$-group. Conversely, if $\mathcal{E} \trianglelefteq \mathcal{F}$, then $\mathrm{Aut}_{\mathcal{E}}(S) \trianglelefteq \mathrm{Aut}_{\mathcal{F}}(S)$, and hence $H \trianglelefteq \Gamma_{p'}(\mathcal{F})$. $\hfill \square$

We can now state the following theorem of David Craven, which gives a new way of viewing the difference between weakly normal and normal fusion subsystems.

**Theorem 7.8** ([Cr3])  *Let $\mathcal{F}$ be a saturated fusion system over a $p$-group $S$, and let $\mathcal{E} \trianglelefteq \mathcal{F}$ be a weakly normal subsystem. Then $O^{p'}(\mathcal{E})$ is normal in $\mathcal{F}$.*

As one immediate consequence of Theorem 7.8, a saturated fusion system $\mathcal{F}$ is simple (has no proper nontrivial normal subsystems) if and only if it has no proper nontrivial weakly normal subsystems.

## 8. THE TRANSFER HOMOMORPHISM FOR SATURATED FUSION SYSTEMS

We describe here an injective transfer homomorphism, defined from $S/\mathfrak{foc}(\mathcal{F})$ to $S/[S,S]$ for any saturated fusion system $\mathcal{F}$ over a $p$-group $S$, which is useful for getting information about the focal subgroup of $\mathcal{F}$. This is, in fact, a special case of transfer homomorphisms in homology and in cohomology, which can be defined (at least in the case of mod $p$ cohomology) using [BLO2, Proposition 5.5]. However, to keep the exposition elementary, we deal here only with the special case $H_1(S) = S/[S,S]$ and $H_1(\mathcal{F}) = S/\mathfrak{foc}(\mathcal{F})$. Some brief remarks on the general case are given at the end of the section.

For any group $G$, we write $G^{\mathrm{ab}} = G/[G,G]$ to denote the abelianization of $G$. We take as starting point the transfer as defined, for example, in [A4, §37] and [G1, §7.3]. In both of those references, when $H \leq G$ is a pair of finite groups, the transfer is defined as a map from $\mathrm{Hom}(H,A)$ to $\mathrm{Hom}(G,A)$ for any abelian group $A$. When $A = H^{\mathrm{ab}}$, this map sends the canonical surjection $H \twoheadrightarrow A$ to a homomorphism from $G$ to $A = H^{\mathrm{ab}}$, and that in turn factors through a homomorphism $\mathrm{trf}_H^G \colon G^{\mathrm{ab}} \longrightarrow H^{\mathrm{ab}}$. This is what we refer to here as the transfer homomorphism for the pair $H \leq G$. In fact, if we identify $\mathrm{Hom}(G,A) \cong \mathrm{Hom}(G^{\mathrm{ab}},A)$ and $\mathrm{Hom}(H,A) \cong \mathrm{Hom}(H^{\mathrm{ab}},A)$ (now for arbitrary abelian $A$), then the transfer defined in [A4] and [G1] is just composition with $\mathrm{trf}_H^G$.

When $G$ is any finite group, $S \in \mathrm{Syl}_p(G)$, and $m = [G{:}S]$, then $\mathrm{trf}_S^G$ is an injective homomorphism from $G^{\mathrm{ab}}$ to $S^{\mathrm{ab}}$, with the property that for each $g \in S$, $\mathrm{trf}_S^G([g]) = [h^m]$ for some $h$ such that $g \equiv h \pmod{[G,G]}$ (cf. formula (6) in [CE, §XII.8]). Since $G \cap [S,S] = \mathfrak{foc}(\mathcal{F})$ when $\mathcal{F} = \mathcal{F}_S(G)$ by the focal subgroup theorem, $S/\mathfrak{foc}(\mathcal{F})$ is the $p$-power torsion subgroup of

$G^{\mathrm{ab}}$. Hence it is natural, when generalizing $\mathrm{trf}_S^G$ to abstract fusion systems, to construct a homomorphism from $S/\mathfrak{foc}(\mathcal{F})$ to $S^{\mathrm{ab}}$.

To construct such a map, we must first look at bisets. For any pair of groups $G$ and $H$, an $(H, G)$-*biset* is a set $X$ on which $H$ acts on the left, $G$ acts on the right, and the two actions commute. In other words, for each $g \in G$, $h \in H$, and $x \in X$, $(hx)g = h(xg)$. This is equivalent to a set with left action of $H \times G$, where a pair $(h, g)$ acts by sending $x \in X$ to $hxg^{-1}$. If the left action of $H$ is free (i.e., $hx = x$ implies $h = 1$), and $\{t_i \mid i \in I\}$ is a set of representatives for the $H$-orbits in $X$, then each element of $X$ has the form $ht_i$ for some unique $h \in H$ and $i \in I$.

When $X$ is an $(H, G)$-biset and $Y$ is a $(K, H)$-biset, we define $Y \times_H X = (Y \times X)/H$: the set of orbits of the $H$-action on $Y \times X$ defined by $(y, x)h = (yh, h^{-1}x)$. Let $[y, x] \in Y \times_H X$ denote the orbit of $(y, x) \in Y \times X$. We make $Y \times_H X$ into a $(K, G)$-biset by setting $k[y, x]g = [ky, xg]$ for $k \in K$ and $g \in G$.

**Lemma 8.1** *For any pair of finite groups $G$ and $H$, and any finite $(H, G)$-biset $X$ whose left $H$-action is free, there is a homomorphism*

$$X_*\colon G^{\mathrm{ab}} \longrightarrow H^{\mathrm{ab}}$$

*with the following property. For any set $\{t_1, \dots, t_n\} \subseteq X$ of representatives for the orbits of the left $H$-action, and any element $g \in G$, if $\sigma \in S_n$ and $h_1, \dots, h_n \in H$ are the unique permutation and elements such that $t_i g = h_i t_{\sigma(i)}$ for each $i$, then*

$$X_*([g]) = \prod_{i=1}^{n} [h_i].$$

*Furthermore, the following hold.*

(a) *For any homomorphism $\varphi \in \mathrm{Hom}(G, H)$, let $H_{G,\varphi}$ be the $(H, G)$-biset where $H$ acts on itself by left multiplication, and $G$ acts on the right via $(h, g) \mapsto h\varphi(g)$. Then $(H_{G,\varphi})_*([g]) = [\varphi(g)]$ for each $g \in G$. Thus $(H_{G,\varphi})_* = \varphi_*$ is the homomorphism induced by $\varphi$.*

(b) *If $H \leq G$, and $X = G$ is the $(H, G)$-biset with the actions defined by left and right multiplication, then $X_*$ is the usual transfer homomorphism $\mathrm{trf}_H^G$ from $G^{\mathrm{ab}}$ to $H^{\mathrm{ab}}$.*

(c) *If $X$ and $Y$ are two $(H, G)$-bisets with free left $H$-actions, then for each $g \in G$, $(X \amalg Y)_*([g]) = X_*([g]) \cdot Y_*([g])$.*

(d) *If $X$ is an $(H, G)$-biset and $Y$ is a $(K, H)$-biset, where both left actions are free, then*

$$(Y \times_H X)_* = Y_* \circ X_* \colon G^{\mathrm{ab}} \longrightarrow K^{\mathrm{ab}}.$$

(e)  *Fix $H \leq G$ of index $m$, and let $\mathrm{incl}_H^G \colon H \longrightarrow G$ be the inclusion. Then the composite*

$$G^{\mathrm{ab}} \xrightarrow{\ \mathrm{trf}_H^G\ } H^{\mathrm{ab}} \xrightarrow{\ (\mathrm{incl}_H^G)_*\ } G^{\mathrm{ab}}$$

*sends $[g]$ to $[g^m]$ for each $g \in G$.*

*Proof.* For any set $\mathcal{T} = \{t_1, \ldots, t_n\} \subseteq X$ of representatives for the orbits of the left $H$-action, define

$$X_*^{\mathcal{T}} \colon G \longrightarrow H^{\mathrm{ab}}$$

as above: $X_*^{\mathcal{T}}(g) = \prod_{i=1}^{k}[h_i]$ if $t_i g = h_i t_{\sigma(i)}$ for each $i$. If $g' \in G$ is another element, and $t_i g' = h_i' t_{\tau(i)}$ for each $i$, then $t_i g g' = h_i h_{\sigma(i)}' t_{\tau\sigma(i)}$. Thus

$$X_*^{\mathcal{T}}(gg') = \prod_{i=1}^{n}[h_i h_{\sigma(i)}'] = \prod_{i=1}^{n}[h_i] \cdot \prod_{i=1}^{n}[h_i'] = X_*^{\mathcal{T}}([g]) X_*^{\mathcal{T}}([g']).$$

This proves that $X_*^{\mathcal{T}}$ is a homomorphism.

Let $\mathcal{T}' = \{t_1', \ldots, t_n'\}$ be another set of $H$-orbit representatives. We can assume the $t_i'$ are ordered such that $t_i' \in H t_i$ for each $i$. Let $\eta_i \in H$ be such that $t_i' = \eta_i t_i$. Fix $g \in G$, and let $\sigma \in S_n$ and $h_i \in H$ be such that $t_i g = h_i t_{\sigma(i)}$ for each $i$. Then

$$t_i' g = \eta_i t_i g = \eta_i h_i t_{\sigma(i)} = \eta_i h_i \eta_{\sigma(i)}^{-1} t_{\sigma(i)}'$$

for each $i$. So

$$X_*^{\mathcal{T}'}(g) = \prod_{i=1}^{n}[\eta_i h_i \eta_{\sigma(i)}^{-1}] = \prod_{i=1}^{n}[h_i] = X_*^{\mathcal{T}}(g) \in H^{\mathrm{ab}}.$$

Thus $X_*^{\mathcal{T}'} = X_*^{\mathcal{T}}$ is independent of the choice of set of orbit representatives, and induces a unique homomorphism $X_* \colon G^{\mathrm{ab}} \longrightarrow H^{\mathrm{ab}}$.

It remains to prove points (a)–(e).

**(a)**  Fix $\varphi \in \mathrm{Hom}(G, H)$, and assume $X = H_{G,\varphi}$: the $(H, G)$-biset with underlying set $H$, and with actions defined by $(h, x, g) \mapsto hx\varphi(g)$. Then $\{1\}$ is a set of orbit representatives for the $H$-action. For $g \in G$, $1 \cdot g = \varphi(g) = \varphi(g) \cdot 1$, so $X_*([g]) = [\varphi(g)]$, and $X_*$ is just the usual homomorphism from $G^{\mathrm{ab}}$ to $H^{\mathrm{ab}}$ induced by $\varphi$.

**(b)**  Assume $H \leq G$, and let $X = G$ be the $(H, G)$-biset with the actions defined by left and right multiplication. Let $g_1, \ldots, g_n \in G$ be any set of representatives for the right cosets $Hg$. Then for each $g \in G$, $X_*([g]) = \prod_{i=1}^{n}[h_i]$, where $h_i \in H$ and $\sigma \in S_n$ are such that $g_i g = h_i g_{\sigma(i)}$ for each $i$. Thus $X_*$ is the usual transfer $\mathrm{trf}_H^G$ as defined, for example, in [A4, §37] and [G1, Theorem 7.3.2]. More precisely, in the notation of [A4] and

[G1], the composite $G \longrightarrow G^{\mathrm{ab}} \xrightarrow{X_*} H^{\mathrm{ab}}$ is the transfer of the natural homomorphism $H \longrightarrow H^{\mathrm{ab}}$.

**(c)** If $X$ and $Y$ are two $(H, G)$-bisets with free left $H$-action, then by definition of $X_*$, $(X \amalg Y)_*([g]) = X_*([g]) \cdot Y_*([g])$ for each $g \in G$.

**(d)** Assume $X$ is an $(H, G)$-biset and $Y$ is a $(K, H)$-biset, where both left actions are free. Let $t_1, \ldots, t_n$ be representatives for the left $H$-orbits in $X$, and let $u_1, \ldots, u_m$ be representatives for the left $K$-orbits in $Y$. For each $(y, x) \in Y \times X$, let $[y, x] \in Y \times_H X$ be its equivalence class. There are unique $h \in H$ and $j$ such that $x = ht_j$, so $[y, x] = [yh, t_j]$, and $yh = ku_i$ for some unique $i$ and $k \in K$. Thus the $K$-action on $Y \times_H X$ is free, and the pairs $[u_i, t_j]$ form a set of representatives for its orbits.

For fixed $g \in G$, set $t_j g = h_j t_{\sigma(j)}$ (where $h_j \in H$ and $\sigma \in S_n$), and set $u_i h_j = k_{ij} u_{\tau_j(i)}$ ($k_{ij} \in K$ and $\tau_j \in S_m$). Then $[u_i, t_j]g = k_{ij}[u_{\tau_j(i)}, t_{\sigma(j)}]$, and so

$$(Y \times_H X)_*([g]) = \prod_{j=1}^{n}\Big(\prod_{i=1}^{m} k_{ij}\Big) = \prod_{j=1}^{n} Y_*([h_j]) = Y_*(X_*([g])).$$

**(e)** Fix $H \le G$ of index $m$, and let $\{g_1, \ldots, g_m\}$ be representatives for the right cosets $Hg$. Fix $g \in G$, and let $\sigma \in S_m$ and $h_i \in H$ be such that $g_i g = h_i g_{\sigma(i)}$. Then $\prod_{i=1}^{m}[g_i g] = \prod_{i=1}^{m}[h_i g_{\sigma(i)}]$ in $G^{\mathrm{ab}}$, and hence $\prod_{i=1}^{m}[h_i] = [g]^m$ in $G^{\mathrm{ab}}$. So

$$(\mathrm{incl}_H^G)_* \circ \mathrm{trf}_H^G([g]) = \prod_{i=1}^{m}[h_i] = [g^m] \in G^{\mathrm{ab}}. \qquad \square$$

The following special case of the transfer homomorphism for groups will be needed.

**Lemma 8.2**  *Fix a prime $p$, and a $p$-group $P$. Then for each proper subgroup $Q < P$ and each element $g \in \Omega_1(Z(P))$, $\mathrm{trf}_Q^P([g]) = 1$.*

*Proof.* By Lemma 8.1(b,d), for $Q < R < P$, $\mathrm{trf}_Q^P = \mathrm{trf}_Q^R \circ \mathrm{trf}_R^P$. It thus suffices to prove the lemma when $[P{:}Q] = p$. If $g \in Q$, let $\{x_1, x_2, \ldots, x_p\}$ be any set of coset representatives. Then $x_i g = g x_i$ for each $i$, and so $\mathrm{trf}_Q^P([g]) = \prod_{i=1}^{p}[g] = [g]^p = 1$.

If $g \notin Q$, we take $\{1, g, g^2, \ldots, g^{p-1}\}$ as our set of coset representatives. Then $g^i \cdot g = 1 \cdot g^{i+1}$ if $0 \le i < p-1$, and $g^{p-1} \cdot g = 1 \cdot 1$. So $\mathrm{trf}_Q^P([g]) = [1]$ again in this case. $\qquad \square$

More generally, by similar arguments, for any pair of finite groups $H \le G$ and any $g \in Z(G)$, $\mathrm{trf}_H^G([g]) = [g^m]$ where $m = [G{:}H]$ (cf. [A4, (37.3)] or [G1, Theorem 7.3.3]).

The construction of the transfer homomorphism for saturated fusion systems over $S$ depends on the existence of an $(S, S)$-biset satisfying certain conditions. Since this biset seems to have many applications, not just for constructing transfer homomorphisms, we describe it in the following lemma.

**Lemma 8.3** ([BLO2, Proposition 5.5]) *Let $\mathcal{F}$ be a saturated fusion system over a finite p-group $S$. Then there is an $(S, S)$-biset $\Omega$ with the following properties:*

(i)   *Each $(S, S)$-orbit of $\Omega$ has the form $(S_{P,\varphi}) \times_P S$ for some $P \leq S$ and some $\varphi \in \mathrm{Hom}_{\mathcal{F}}(P, S)$. (Thus the $S$-action on each side is free.)*

(ii)  *For each $P \leq S$ and each $\varphi \in \mathrm{Hom}_{\mathcal{F}}(P, S)$, $\Omega \times_S (S_{P,\mathrm{Id}}) \cong \Omega \times_S (S_{P,\varphi})$ as $(S, P)$-bisets.*

(iii) *$|\Omega|/|S|$ is prime to p.*

Recall that an $(S, S)$-biset $X$ can be regarded as a set with left $(S \times S)$-action by setting $(s, t) \cdot x = sxt^{-1}$. In terms of this action, point (i) says that for each $x \in \Omega$, the stabilizer subgroup at $x$ of the $(S \times S)$-action has the form $\{(\varphi(g), g) \mid g \in P\}$ for some $P \leq S$ and some $\varphi \in \mathrm{Hom}_{\mathcal{F}}(P, S)$. In particular, since each such subgroup has trivial intersection with $1 \times S$ and $S \times 1$, this implies that the left and right actions of $S$ on $\Omega$ are both free.

Note, in (ii), that as $(S, P)$-bisets, $\Omega \times_S (S_{P,\mathrm{Id}})$ is isomorphic to $\Omega$ with its right action restricted to $P$, and $\Omega \times_S (S_{P,\varphi})$ is isomorphic to $\Omega$ with the right action "twisted" by $\varphi$ ($g \in P$ acts by sending $x \in \Omega$ to $x \cdot \varphi(g)$).

A biset satisfying conditions (i)–(iii) in Lemma 8.3 for the fusion system $\mathcal{F}$ will be called a *characteristic biset* for $\mathcal{F}$. These conditions were originally formulated by Linckelmann and Webb, who conjectured the existence of such a biset. As the motivating example, when $\mathcal{F} = \mathcal{F}_S(G)$ for a finite group $G$, we can take $\Omega = G$ with the left and right $S$-actions defined by multiplication. By a theorem of Puig [P7, Proposition 21.9] (discovered independently by Ragnarsson and Stancu [RSt, Theorem A]), a fusion system has a characteristic biset only if it is saturated.

As an example of another application of characteristic bisets, we note [5a2, Proposition 1.16]: for any saturated fusion system $\mathcal{F}$ over a p-group $S$, and any subgroup $P$ of $S$, the set of $S$-conjugacy classes of fully normalized subgroups in $P^{\mathcal{F}}$ has order prime to $p$. Yet another example will be given in Theorem III.4.23.

We are now ready to construct a transfer for fusion systems. This will be done, for a fusion system $\mathcal{F}$ over a p-group $S$, by applying Lemma 8.1 to a certain $(S, S)$-biset constructed in [BLO2].

**Proposition 8.4**  *Fix a $p$-group $S$, and a saturated fusion system $\mathcal{F}$ over $S$. Then there is an injective homomorphism*

$$\mathrm{trf}_{\mathcal{F}} \colon S/\mathfrak{foc}(\mathcal{F}) \longrightarrow S^{\mathrm{ab}} \overset{\mathrm{def}}{=} S/[S,S]$$

*which has the following properties.*

(a)  *There are proper subgroups $P_1, \ldots, P_m < S$, and morphisms $\varphi_i \in \mathrm{Hom}_{\mathcal{F}}(P_i, S)$ $(i = 1, \ldots, m)$, such that for $g \in S$,*

$$\mathrm{trf}_{\mathcal{F}}([g]) = \prod_{[\alpha] \in \mathrm{Out}_{\mathcal{F}}(S)} [\alpha(g)] \cdot \prod_{i=1}^{m} \varphi_{i*}(\mathrm{trf}^{S}_{P_i}([g])).$$

*Here, $[g]$ denotes the class of $g$ in $S/\mathfrak{foc}(\mathcal{F})$ or in $S^{\mathrm{ab}} = S/[S,S]$, and the terms on the right are regarded as lying in $S^{\mathrm{ab}}$.*

(b)  *There is $m$ prime to $p$ such that for arbitrary $g \in G$, $\mathrm{trf}_{\mathcal{F}}([g]) = [h^m]$ for some $h \equiv g \pmod{\mathfrak{foc}(\mathcal{F})}$.*

(c)  *If $g \in \Omega_1(Z(S))$, then $\mathrm{trf}_{\mathcal{F}}([g]) = \prod_{[\alpha] \in \mathrm{Out}_{\mathcal{F}}(S)} [\alpha(g)]$.*

(d)  *If $g \in \Omega_1(Z(S))$ is $\mathrm{Out}_{\mathcal{F}}(S)$-invariant, then $\mathrm{trf}_{\mathcal{F}}([g]) = [g]^k$ where $k = |\mathrm{Out}_{\mathcal{F}}(S)|$ is prime to $p$.*

*Proof.* Fix a characteristic biset

$$\Omega = \coprod_{i=1}^{k} (S_{P_i, \varphi_i}) \times_{P_i} S$$

for $\mathcal{F}$, as in Lemma 8.3. Thus $P_i \leq S$ and $\varphi_i \in \mathrm{Hom}_{\mathcal{F}}(P_i, S)$ for each $i$. By Lemma 8.1(a–d), for each $g \in S$,

$$\Omega_*([g]) = \prod_{i=1}^{k} \varphi_{i*}(\mathrm{trf}^{S}_{P_i}([g])).$$

Also, for each $P \leq S$, each $\varphi \in \mathrm{Hom}_{\mathcal{F}}(P, S)$, and each $g \in P$,

$$\Omega_*([g]) = \Omega_* \circ (S_{P, \mathrm{Id}})_*([g]) = \Omega_* \circ (S_{P, \varphi})_*([g]) = \Omega_*([\varphi(g)])$$

by condition (ii) and Lemma 8.1(a,d), and hence $\Omega_*([g^{-1}\varphi(g)]) = 1$. Thus $\mathfrak{foc}(\mathcal{F})/[S,S] \leq \mathrm{Ker}(\Omega_*)$, and $\Omega_*$ factors through a homomorphism $T \colon S/\mathfrak{foc}(\mathcal{F}) \longrightarrow S^{\mathrm{ab}}$.

For all $g \in G$,

$$\Omega_*([g]) = \prod_{i=1}^{k} \varphi_{i*}(\mathrm{trf}^{S}_{P_i}([g])) \equiv \prod_{i=1}^{k} (\mathrm{incl}^{S}_{P_i})_*(\mathrm{trf}^{S}_{P_i}([g]))$$

$$\pmod{\mathfrak{foc}(\mathcal{F})/[S,S]} \quad (1)$$

$$= \prod_{i=1}^{k} [g^{[S:P_i]}] = [g^{|\Omega|/|S|}],$$

where the next-to-last equality follows from Lemma 8.1(e), and the last one since

$$|S_{P_i,\varphi_i} \times_{P_i} S|/|S| = |S|/|P_i| = [S{:}P_i] \quad \text{for each } i. \tag{2}$$

In particular, if $\Omega_*([g]) = 1 \in S^{\mathrm{ab}}$, then $g^{|\Omega|/|S|} \in \mathfrak{foc}(\mathcal{F})$, and $g \in \mathfrak{foc}(\mathcal{F})$ since $|\Omega|/|S|$ is prime to $p$ by (iii). Thus $\mathrm{Ker}(\Omega_*) = \mathfrak{foc}(\mathcal{F})/[S,S]$, and so $T$ is injective.

When $P = S$ and $\alpha, \beta \in \mathrm{Aut}_{\mathcal{F}}(S)$, then an isomorphism from $S_{S,\alpha}$ to $S_{S,\beta}$ is a bijection $f\colon S \longrightarrow S$ such that $f(sx) = sf(x)$ and $f(x\alpha(s)) = f(x)\beta(s)$ for each $s, x \in S$. The first equality implies that $f(x) = xg$ for some fixed $g \in S$ (and all $x$), and the second then implies $\alpha(s)g = g\beta(s)$ and hence $\alpha = c_g \circ \beta$. In other words, $S_{S,\alpha} \cong S_{S,\beta}$ as $(S,S)$-bisets if and only if $\alpha \equiv \beta \pmod{\mathrm{Inn}(S)}$.

By (ii), for each $\alpha \in \mathrm{Aut}_{\mathcal{F}}(S)$, there is an isomorphism of bisets $f\colon \Omega \xrightarrow{\cong} \Omega \times_S S_{S,\alpha}$. For each $P \leq S$ and $\varphi \in \mathrm{Hom}_{\mathcal{F}}(P,S)$, if we set $Q = \alpha^{-1}(P)$, then

$$(S_{P,\varphi} \times_P S) \times_S S_{S,\alpha} \cong S_{Q,\varphi\alpha} \times_Q S$$

via the isomorphism which sends $[s,t,u]$ to $[s,\alpha^{-1}(tu)]$. In particular, condition (ii) in Lemma 8.3 implies that for each $\beta \in \mathrm{Aut}_{\mathcal{F}}(S)$, terms isomorphic to $S_{S,\beta}$ $(\cong S_{S,\beta} \times_S S)$ occur in $\Omega$ with the same multiplicity as those isomorphic to $S_{S,\beta\alpha}$. Since this holds for each $\alpha, \beta \in \mathrm{Aut}_{\mathcal{F}}(S)$, there is $r$ such that for each $\xi \in \mathrm{Out}_{\mathcal{F}}(S)$, there are exactly $r$ orbits in $\Omega$ isomorphic to $S_{S,\alpha}$ for $[\alpha] = \xi$. Thus if we let $I = \{1 \leq i \leq k \mid P_i < S\}$, then for all $g \in S$,

$$T([g]) = \prod_{[\alpha] \in \mathrm{Out}_{\mathcal{F}}(S)} [\alpha(g)]^r \cdot \prod_{i \in I} \varphi_{i*}\big(\mathrm{trf}^S_{P_i}([g])\big). \tag{3}$$

For each $i \in I$, $|S_{P_i,\varphi_i} \times_{P_i} S|/|S| = [S{:}P_i] \equiv 0 \pmod{p}$ by (2). Hence $|\Omega|/|S| \equiv r \cdot |\mathrm{Out}_{\mathcal{F}}(S)| \pmod{p}$, and so $p \nmid r$ by (iii). Choose $n > 0$ such that $rn \equiv 1 \pmod{|S|}$, and define $\mathrm{trf}_{\mathcal{F}}$ by setting $\mathrm{trf}_{\mathcal{F}}([g]) = T([g])^n$. By (3), $\mathrm{trf}_{\mathcal{F}}$ has the form given in (a).

Point (b) follows from (1) (with $m = n \cdot |\Omega|/|S|$). If $g \in \Omega_1(Z(S))$, then for each $P_i < S$, $\mathrm{trf}^S_{P_i}([g]) = 1$ by Lemma 8.2. Hence $\mathrm{trf}_{\mathcal{F}}([g]) = \prod_{[\alpha] \in \mathrm{Out}_{\mathcal{F}}(S)} [\alpha(g)]$, and this proves (c). Point (d) is a special case of (c). $\square$

Note that the transfer $\mathrm{trf}_{\mathcal{F}}$ as constructed in Proposition 8.4 depends a priori on the choice of characteristic biset for $\mathcal{F}$. This choice is, in fact, not unique. This uniqueness question has been studied by Ragnarsson, who described in [Rg, Proposition 5.6] a way to choose a virtual, $p$-completed biset which is unique and "canonical" in a certain sense. Also, in [DGPS, Lemma 2.6], the transfer is shown to be independent (up to a scalar multiple) of the choice of characteristic biset. But none of this is needed here:

we just need to have one transfer homomorphism with the right properties.

The following is one easy example of how Proposition 8.4 can be applied.

**Corollary 8.5** *Let $\mathcal{F}$ be a saturated fusion system over a $p$-group $S$, and set $S_0 = \Omega_1(Z(S))$ and $S_1 = S_0 \cap [S, S]$. Then no element of $S_0 \smallsetminus S_1$ whose coset in $S_0/S_1$ is fixed by the action of $\mathrm{Out}_\mathcal{F}(S)$ is in $\mathfrak{foc}(\mathcal{F})$. In particular, if $S_0 > S_1$ and $\mathrm{Out}(S)$ is a $p$-group, or if $p = 2$ and $\mathrm{rk}(S_0/S_1) = 1$, then $O^p(\mathcal{F}) \subsetneqq \mathcal{F}$ for every saturated fusion system $\mathcal{F}$ over $S$.*

*Proof.* Assume $g \in S_0 \smallsetminus S_1$ is such that $[g] \in S_0/S_1$ is fixed by the action of $\mathrm{Out}_\mathcal{F}(S)$. Set

$$g' = \prod_{[\alpha] \in \mathrm{Out}_\mathcal{F}(S)} \alpha(g) \in S_0.$$

Set $k = |\mathrm{Out}_\mathcal{F}(S)|$. Then $g' \in g^k S_1$, and $g' \notin S_1$ since $p \nmid k$. Hence $\mathrm{trf}_\mathcal{F}([g]) \neq 1$ by Proposition 8.4(c), and so $g \notin \mathfrak{foc}(\mathcal{F})$.

The other statements now follow immediately. Recall (Corollary 7.5) that $\mathfrak{foc}(\mathcal{F}) < S$ implies $O^p(\mathcal{F}) \subsetneqq \mathcal{F}$. $\square$

We finish this section with some remarks on more general forms of the transfer. When $H \leq G$ is a pair of groups such that $[G{:}H] < \infty$, then the transfer $\mathrm{trf}_H^G$ is a homomorphism from the cohomology of $H$ to that of $G$, or from the homology of $G$ to that of $H$. In other words, it goes in the opposite direction of the usual homomorphism induced by the inclusion. A special role is played by the transfer homomorphisms $\mathrm{trf}_S^G$ when $G$ is finite and $S \in \mathrm{Syl}_p(G)$: this is useful when describing the (co)homology of $G$ (after localization at $p$) in terms of that of $S$ (cf. [CE, Theorem XII.10.1]).

When $\mathcal{F}$ is a saturated fusion system over a $p$-group $S$, we define the (co)homology of $\mathcal{F}$ by setting, when $n \geq 0$ and $A$ is a $\mathbb{Z}_{(p)}$-module,

$$H^n(\mathcal{F}; A) = \varprojlim_{\mathcal{F}} H^n(-; A) \qquad \text{and} \qquad H_n(\mathcal{F}; A) = \mathrm{colim}_{\mathcal{F}} \, H_n(-; A).$$

For example, when $A$ is finite, $H^*(\mathcal{F}; A)$ is isomorphic to the cohomology of a "classifying space" for $\mathcal{F}$ by [BLO2, Theorem B] (when $A = \mathbb{F}_p$) or [5a2, Lemma 6.12] (see also Theorem III.4.23). The transfer maps for $\mathcal{F}$ are homomorphisms $H^n(S; A) \longrightarrow H^n(\mathcal{F}; A)$ or $H_n(\mathcal{F}; A) \longrightarrow H_n(S; A)$. The transfer of Proposition 8.4 is the transfer for $H_1(-; \mathbb{Z}_{(p)})$.

## 9. OTHER DEFINITIONS OF SATURATION

We now look at two other, equivalent definitions of saturation: the original definition given by Puig (who calls saturated fusion systems "Frobenius categories" in [P6, Definition 2.9]), and an alternative definition given by Stancu (see [Li3, Definition 1.4] or [KS, Definition 2.4]).

The following terminology is useful when stating Puig's definition, and will be used only in this section.

**Definition 9.1** Let $\mathcal{F}$ be any fusion system over $S$. Fix a subgroup $P \leq S$ and a group of automorphisms $K \leq \mathrm{Aut}(P)$.

- $P$ is *totally $K$-normalized* in $\mathcal{F}$ if for each $\varphi \in \mathrm{Hom}_{\mathcal{F}}(P{\cdot}N_S^K(P), S)$,
$$\varphi(N_S^K(P)) = N_S^{\varphi K}(\varphi(P)).$$

- $P$ is *$K$-receptive* in $\mathcal{F}$ if for each $Q \leq S$ and $\varphi \in \mathrm{Iso}_{\mathcal{F}}(Q, P)$, there are homomorphisms $\bar{\varphi} \in \mathrm{Hom}_{\mathcal{F}}(Q{\cdot}N_S^{K^\varphi}(Q), S)$ and $\chi \in \mathrm{Aut}_{\mathcal{F}}^K(P)$ such that $\bar{\varphi}|_Q = \chi \circ \varphi$.

Both of these concepts are equivalent to being fully $K$-normalized when the fusion system $\mathcal{F}$ is saturated. This is why we need separate names for them only when working in fusion systems which might not be saturated; in particular, when comparing different definitions of saturation. For example, "$K$-receptive" is just the condition of Proposition 5.2(c).

The term "totally $K$-normalized" is what Puig calls "fully $K$-normalized" in [P6]; we give it a different name to distinguish between these two concepts. They are equivalent in a saturated fusion system, as shown in the next lemma, but need not be equivalent in general.

**Lemma 9.2** *Fix a fusion system $\mathcal{F}$ over a p-group $S$. Let $P \leq S$ be any subgroup, and let $K \leq \mathrm{Aut}(P)$ be a group of automorphisms. If $P$ is fully $K$-normalized, then it is totally $K$-normalized. If $\mathcal{F}$ is saturated, then $P$ is fully $K$-normalized if and only if it is totally $K$-normalized.*

*Proof.* If $P$ is fully $K$-normalized, then $|N_S^K(P)| \geq |N_S^{\varphi K}(\varphi(P))|$ for each $\varphi \in \mathrm{Hom}_{\mathcal{F}}(P{\cdot}N_S^K(P), S)$, and hence $\varphi(N_S^K(P)) = N_S^{\varphi K}(\varphi(P))$. Thus $P$ is totally $K$-normalized.

Now assume $\mathcal{F}$ is saturated and $P$ is totally $K$-normalized. Choose $Q$ which is $\mathcal{F}$-conjugate to $P$, and $\varphi \in \mathrm{Iso}_{\mathcal{F}}(P, Q)$, such that $Q$ is fully $\varphi K$-normalized. Set $L = \varphi K$ for short. By Proposition 5.2, there are $\chi \in \mathrm{Aut}_{\mathcal{F}}^L(Q)$ and $\bar{\varphi} \in \mathrm{Aut}_{\mathcal{F}}(P{\cdot}N_S^K(P), S)$ such that $\bar{\varphi}|_P = \chi \circ \varphi$. In particular, $\bar{\varphi}(P) = Q$ and $^{\chi\varphi}K = L$. Since $P$ is totally $K$-normalized, $\bar{\varphi}(N_S^K(P)) = N_S^L(Q)$, so the two have the same order, and $P$ is fully $K$-normalized since $Q$ is fully $L$-normalized. $\square$

We are now ready to show the equivalence of our definitions of saturation (Definition 2.2 and Proposition 2.5) with those of Puig and of Stancu. The equivalence with Puig's definition is sketched in the appendix of [BLO2], and the equivalence with Stancu's is proven in [Li3, Propositions 1.5–1.7].

**Proposition 9.3** *The following three conditions are equivalent for a fusion system $\mathcal{F}$ over a p-group $S$.*

(a) $\mathcal{F}$ *is saturated in the sense of Definition 2.2.*

(b) *(Puig's definition)* $\mathrm{Inn}(S) \in \mathrm{Syl}_p(\mathrm{Aut}_{\mathcal{F}}(S))$; *and for each $P \leq S$ and each $K \leq \mathrm{Aut}(P)$, if $P$ is totally $K$-normalized in $\mathcal{F}$, then $P$ is $K$-receptive in $\mathcal{F}$.*

(c) *(Stancu's definition)* $\mathrm{Inn}(S) \in \mathrm{Syl}_p(\mathrm{Aut}_{\mathcal{F}}(S))$; *and for each $P \leq S$, if $P$ is fully normalized in $\mathcal{F}$, then $P$ is receptive in $\mathcal{F}$.*

*Proof.* **(a $\Rightarrow$ b)** Assume $\mathcal{F}$ is saturated. In particular, we have $\mathrm{Inn}(S) \in \mathrm{Syl}_p(\mathrm{Aut}_{\mathcal{F}}(S))$. If $P \leq S$ and $K \leq \mathrm{Aut}(P)$ are such that $P$ is totally $K$-normalized in $\mathcal{F}$, then $P$ is fully $K$-normalized by Lemma 9.2, and hence is $K$-receptive by Proposition 5.2.

**(b $\Rightarrow$ c)** Assume $\mathcal{F}$ satisfies the conditions in (b). In particular, $\mathrm{Inn}(S) \in \mathrm{Syl}_p(\mathrm{Aut}_{\mathcal{F}}(S))$.

Assume $P \leq S$ is fully normalized in $\mathcal{F}$. Then $P$ is totally $\mathrm{Aut}(P)$-normalized by Lemma 9.2, and so $P$ is $\mathrm{Aut}(P)$-receptive. Thus for each $Q$ $\mathcal{F}$-conjugate to $P$, there is $\varphi \in \mathrm{Hom}_{\mathcal{F}}(N_S(Q), N_S(P))$ such that $\varphi(Q) = P$, and in particular, $\varphi(C_S(Q)) \leq C_S(P)$. Hence $P$ is fully centralized in $\mathcal{F}$, and is totally $\{1\}$-normalized by Lemma 9.2 again.

Fix $Q \leq S$ and $\varphi \in \mathrm{Iso}_{\mathcal{F}}(Q, P)$. Set

$$N_\varphi = \{g \in N_S(Q) \,|\, {}^\varphi c_g \leq \mathrm{Aut}_S(P)\} \qquad \text{and} \qquad K = \mathrm{Aut}_{N_\varphi}(Q) \,,$$

and set $L = {}^\varphi K \leq \mathrm{Aut}_S(P)$. For each morphism $\psi \in \mathrm{Hom}_{\mathcal{F}}(P \cdot N_S^L(P), S)$, $\psi(C_S(P)) = C_S(\psi(P))$ since $P$ is totally $\{1\}$-normalized. Hence $\psi(N_S^L(P)) = N_S^{\psi L}(\psi(P))$ since

$$|N_S^L(P)/C_S(P)| = |L| \geq |N_S^{\psi L}(\psi(P))/C_S(\psi(P))|.$$

Thus $P$ is totally $L$-normalized, and hence is $L$-receptive by (b). There are thus $\chi \in L$ and $\bar{\varphi} \in \mathrm{Hom}_{\mathcal{F}}(Q \cdot N_S^K(Q), S)$ such that $\bar{\varphi}|_Q = \chi \circ \varphi$. Also, $\chi^\varphi = c_g$ for some $g \in N_\varphi$, and so $\varphi$ extends to $\bar{\varphi} \circ c_g^{-1} \in \mathrm{Hom}_{\mathcal{F}}(N_\varphi, S)$.

**(c $\Rightarrow$ a)** Assume $\mathrm{Inn}(S) \in \mathrm{Syl}_p(\mathrm{Aut}_{\mathcal{F}}(S))$, and each fully normalized subgroup of $S$ is receptive. Since each $\mathcal{F}$-conjugacy class contains a fully normalized subgroup, we need only show that each fully normalized subgroup is fully automized. Assume otherwise: let $P$ be a fully normalized subgroup which is not fully automized, chosen such that $|P|$ is maximal among such subgroups. Since $S$ is fully automized, $P < S$.

Fix a subgroup $T \in \mathrm{Syl}_p(\mathrm{Aut}_{\mathcal{F}}(P))$ such that $\mathrm{Aut}_S(P) < T$. Choose $\alpha \in N_T(\mathrm{Aut}_S(P))$ not in $\mathrm{Aut}_S(P)$ (see Lemma A.1). Since $P$ is receptive and $\alpha$ normalizes $\mathrm{Aut}_S(P)$, $\alpha$ extends to an automorphism $\bar{\alpha} \in \mathrm{Aut}_{\mathcal{F}}(N_S(P))$.

Upon replacing $\bar{\alpha}$ by $\bar{\alpha}^r$ for some appropriate $r \equiv 1 \pmod{|\alpha|}$, we can assume $\bar{\alpha}$ has $p$-power order. Choose $R \leq S$ which is $\mathcal{F}$-conjugate to $N_S(P)$ and fully normalized in $\mathcal{F}$, and fix $\varphi \in \mathrm{Iso}_{\mathcal{F}}(N_S(P), R)$. Since $|R| = |N_S(P)| > |P|$ (Lemma A.1 again), $R$ is fully automized by assumption. Hence there is $\chi \in \mathrm{Aut}_{\mathcal{F}}(R)$ such that ${}^{\chi\varphi}\bar{\alpha} \in \mathrm{Aut}_S(R) \in \mathrm{Syl}_p(\mathrm{Aut}_{\mathcal{F}}(R))$. Let $g \in N_S(R)$ be such that $c_g = {}^{\chi\varphi}\bar{\alpha}$. In particular, $g \notin R$.

Set $Q = \chi\varphi(P) \trianglelefteq R$. Then $g \in N_S(Q)$ since $\bar{\alpha}(P) = P$, and so $\langle R, g \rangle \leq N_S(Q)$. But then $|N_S(Q)| > |R| = |N_S(P)|$, which contradicts the assumption $P$ was fully normalized. Thus each fully normalized subgroup of $S$ is fully automized. $\qquad\square$

# Part II. The local theory of fusion systems

## Michael Aschbacher

Part II of the book is intended to be an introduction to the local theory of saturated fusion systems. By the "local theory of fusion systems" we mean an extension of some part of the local theory of finite groups to the setting of saturated fusion systems on finite $p$-groups.

Recall the notions of a *p-local finite group* and its associated *linking system* defined in Definitions III.4.1 and III.4.4. One can ask: Why deal with saturated fusion systems rather than $p$-local finite groups? There are two reasons for this choice. First, it is not known whether to each saturated fusion system there is associated a unique $p$-local finite group. Thus it remains possible that the class of saturated fusion systems is larger than the class of $p$-local finite groups. But more important, to date there is no accepted notion of a morphism of $p$-local finite groups, and hence no category of $p$-local groups. (The problem arises already for fusion systems and linking systems of groups, since a group homomorphism $\alpha$ from $G$ to $H$ need not send $p$-centric subgroups of $G$ to $p$-centric subgroups of $H$, so it is not clear how to associate to $\alpha$ a map which would serve as a morphism of the $p$-local group of $G$ with the $p$-local group of $H$.) The local theory of finite groups is inextricably tied to the notion of group homomorphism and factor group, so to extend the local theory of finite groups to a different category, we must at the least be dealing with an actual category.

Part I of the book records most of the basic definitions, notation, and notions from the theory of saturated fusion systems necessary for Part II. For emphasis, we repeat some of those results from Part I (which we use most often) here in Part II. For example in Theorem 4.2 in Section 4 we again record the deeper result of [5a1] that if $\mathcal{F}$ is saturated and constrained on $S$, then the set $\mathcal{G}(\mathcal{F})$ of *models* of $\mathcal{F}$ is nonempty. Here $G \in \mathcal{G}(\mathcal{F})$ if $G$ is a finite group with $S \in \mathrm{Syl}_p(G)$, $C_G(O_p(G)) \leq O_p(G)$, and $\mathcal{F} = \mathcal{F}_S(G)$. This fact is the basis for much of the local theory of fusion systems, and allows us to translate suitable statements from the local theory of groups into the setting of fusion systems.

Exercise 2.4 shows that if $\alpha\colon \mathcal{F} \to \tilde{\mathcal{F}}$ is a morphism of fusion systems, then the kernel $\ker(\alpha)$ of the group homomorphism $\alpha\colon S \to \tilde{S}$ is strongly closed in $S$ with respect to $\mathcal{F}$. In Section 5, we see how to construct a factor system $\mathcal{F}/T$ of $\mathcal{F}$ over a strongly closed subgroup $T$ of $\mathcal{F}$. Moreover when $\mathcal{F}$ is saturated, we find that there is a surjective homomorphism $\Theta\colon \mathcal{F} \to \mathcal{F}/T$, and a bijection $T \mapsto \mathcal{F}/T$ between strongly closed subgroups $T$ of $S$ and the set of isomorphism classes of homomorphic images of $\mathcal{F}$.

In [P6], Puig defines the same factor systems and proves that $\Theta$ is a surjective morphism of fusion systems, although via a different approach.

Proceeding by analogy with the situation for groups, we would like to show there exists a "normal subsystem" $\mathcal{E}$ of $\mathcal{F}$ on $T$ which is saturated, and hence realize $\mathcal{F}$ as an "extension" of $\mathcal{E}$ by $\mathcal{F}/\mathcal{E} = \mathcal{F}/T$. Unfortunately such a subsystem need not exist, but it is still possible to develop a theory of normal subsystems of saturated fusion systems which is fairly satisfactory. Section I.6 contains a discussion of various possible definitions of the notion of a "normal subsystem"; our normal subsystems are defined in Definition I.6.1, and that definition is repeated in Definition 7.1 of this Part. A theory of normal subsystems is described in Sections 6 and 7, where a few examples are also introduced to indicate some of the places where the theory diverges from the corresponding theory for groups.

We need effective conditions to verify when a subsystem of $\mathcal{F}$ on $T$ is normal, and to produce normal subsystems. Moreover in most situations, these conditions should be *local*; that is we should be able to check them in local subsystems, and indeed even in constrained local subsystems. In Section 8 we record some such conditions from [A5]. Then in Section 9 we record some of the theorems about normal subsystems from [A6] which can be proved using the conditions. In particular we define the *generalized Fitting subsystem* $F^*(\mathcal{F})$ of $\mathcal{F}$, and recall the (obvious) notion of a *simple* system, which appears already in Definition I.6.1. Of course $\mathcal{F}$ is simple if it has no nontrivial normal subsystems.

In Section 10 we define the notion of a *composition series* for saturated fusion systems, and prove a Jordan-Hölder Theorem for such systems. For example this makes possible the definition of a *solvable* system $\mathcal{F}$: All composition factors are of order $p$, or more precisely of the form $\mathcal{F}_R(R)$ for $R$ a group of order $p$. There is a second natural definition of solvability for fusion systems, due to Puig. We compare the two definitions in Section 12.

Section 13 investigates the composition factors of systems $\mathcal{F}_S(G)$, where $G$ is a finite simple group and $S \in \mathrm{Syl}_p(G)$. Often such systems are simple, but not always. Occasionally $\mathcal{F}_S(G)$ may even have an *exotic* composition factor which is not obtainable from a finite group.

The last two sections contain some speculation about the possibility of, first, classifying simple saturated fusion systems (or some suitable subclass of such systems), particularly at the prime 2, and then, second, using results on fusion systems to simplify the proof of the theorem classifying the finite simple groups. We also include a few existing results of that sort, together with some open problems which may be of interest.

Finally we note that in Part II we often abbreviate $\mathcal{F}$-*centric* by *centric*.

## 1. Notation and terminology on groups

Our convention in Part II will be to write many functions (particularly functions which may be composed, like group homomorphisms) on the right.

We adopt the notation and terminology in [A4] when discussing groups; some of this can also be found at the end of the Introduction to this volume. For example let $G$ be a group and $x, y \in G$. Then $x^y = y^{-1}xy$ is the *conjugate* of $x$ under $y$, and

$$c_y \colon G \to G$$
$$x \mapsto x^y$$

is *conjugation* by $y$. This notation differs (because we are applying maps on the right) from that in the introduction, where $c_y(x) = {}^y x$. Of course $c_y \in \operatorname{Aut}(G)$ is an automorphism of $G$. For $X \subseteq G$, write $X^y$ for the conjugate $X c_y$ of $X$ under $y$, and set $X^G = \{X^y \colon y \in G\}$, the *conjugacy class* of $X$ in $G$.

Let $K \le H \le G$. We say $K$ is *strongly closed* in $H$ with respect to $G$ if for all $k \in K$, $k^G \cap H \subseteq K$.

See Definition A.9 for the definition of the *commutator* notation, $[x, y]$, $[X, Y]$, for $x, y \in G$, $X, Y \le G$, and see Section 8 of [A4] for more discussion of these notions.

See Definition A.11 for the definition of *quasisimple groups* and the *generalized Fitting subgroup* $F^*(G)$ of a finite group $G$, and see Section 31 in [A4] for further discussion of these notions.

Our notation for the finite simple groups is defined in section 47 of [A4], and there is a much deeper discussion of the simple groups in [GLS3].

*Notation*  Suppose $\mathcal{C}$ is a category and $\alpha \colon A \to B$ is an isomorphism in $\mathcal{C}$. Let $\operatorname{Aut}_{\mathcal{C}}(A) = \operatorname{Mor}_{\mathcal{C}}(A, A)$ be the group of automorphisms of $A$ in $\mathcal{C}$. Observe that for $C$ a subobject of $A$, we have the bijection $\operatorname{Mor}_{\mathcal{C}}(C, A) \to \operatorname{Mor}_{\mathcal{C}}(C\alpha, B)$ defined by $\beta \mapsto \alpha^{-1}\beta\alpha = \beta^\alpha$, which, in the case $C = A$, induces an isomorphism of $\operatorname{Aut}_{\mathcal{C}}(A)$ with $\operatorname{Aut}_{\mathcal{C}}(B)$.

## 2. Fusion systems

Part I contains the basic notation, terminology, and concepts for fusion systems. In this short section we record a few more facts and observations.

In this section, $p$ is a prime and $\mathcal{F}$ is a fusion system on a finite $p$-group $S$. Write $\mathcal{F}^f$ for the set of nontrivial fully normalized subgroups of $S$.

Let $P \leq S$. Recall from Definition I.5.3 that $C_{\mathcal{F}}(P)$ is the category whose objects are the subgroups of $C_S(P)$, and for $U, V \leq C_S(P)$, $\mathrm{Hom}_{C_{\mathcal{F}}(P)}(U, V)$ consists of those $\phi \in \mathrm{Hom}_{\mathcal{F}}(U, V)$ which extend to $\hat{\phi} \in \mathrm{Hom}_{\mathcal{F}}(PU, PV)$ with $\hat{\phi} = 1$ on $P$. The system $C_{\mathcal{F}}(P)$ is the *centralizer* in $\mathcal{F}$ of $P$, and is a fusion system.

Similarly $N_{\mathcal{F}}(P)$ is the category on $N_S(P)$ whose objects are the subgroups of $N_S(P)$, and for $U, V \leq N_S(P)$, $\mathrm{Hom}_{N_{\mathcal{F}}(P)}(U, V)$ consists of those $\phi \in \mathrm{Hom}_{\mathcal{F}}(U, V)$ which extend to $\hat{\phi} \in \mathrm{Hom}_{\mathcal{F}}(PU, PV)$ with $\hat{\phi}$ acting on $P$.

As a consequence of Theorem I.5.5:

**Theorem 2.1**  *If $\mathcal{F}$ is saturated and $P$ is fully normalized, then $C_{\mathcal{F}}(P)$ and $N_{\mathcal{F}}(P)$ are saturated fusion systems.*

Thus the systems $N_{\mathcal{F}}(P)$, for $P \in \mathcal{F}^f$, play the role of the *local subsystems* of $\mathcal{F}$, analogous to the local subgroups in finite group theory. For example if $\mathcal{F} = \mathcal{F}_S(G)$ for some finite group $G$ with $S \in \mathrm{Syl}_p(G)$, then $N_{\mathcal{F}}(P) = \mathcal{F}_{N_S(P)}(N_G(P))$. These local subsystems are the focus of interest in the local theory of fusion systems.

Recall from Definition I.3.1 that $P \leq S$ is $\mathcal{F}$-*centric* (or just *centric*) if $C_S(Q) \leq Q$ for each $Q \in P^{\mathcal{F}}$. We write $\mathcal{F}^c$ for the set of centric subgroups of $S$.

**Definition 2.2**  A *morphism* $\alpha \colon \mathcal{F} \to \tilde{\mathcal{F}}$ of fusion systems from $\mathcal{F}$ to a system $\tilde{\mathcal{F}}$ on $\tilde{S}$, is a family $(\alpha, \alpha_{P,Q} \colon P, Q \in \mathcal{F})$ such that $\alpha \colon S \to \tilde{S}$ is a group homomorphism, and $\alpha_{P,Q} \colon \mathrm{Hom}_{\mathcal{F}}(P, Q) \to \mathrm{Hom}_{\tilde{F}}(P\alpha, Q\alpha)$ is a function, such that for all $P, Q$, $\phi\alpha = \alpha(\phi\alpha_{P,Q})$. Notice if the family *is* a morphism, then the maps $\alpha_{P,Q}$ are uniquely determined by the group homomorphism $\alpha$ and the last property.

The *kernel* $\ker(\alpha)$ of the morphism $\alpha$ is the kernel of the group homomorphism $\alpha \colon S \to \tilde{S}$. Thus $\ker(\alpha)$ is a normal subgroup of $S$.

The morphism $\alpha$ is *surjective* if $\alpha \colon S \to \tilde{S}$ is surjective, and for all $P, Q \leq S$,

$$\alpha_{P_0, Q_0} \colon \mathrm{Hom}_{\mathcal{F}}(P_0, Q_0) \to \mathrm{Hom}_{\tilde{\mathcal{F}}}(P\alpha, Q\alpha)$$

is surjective, where for $X \leq S$, $X_0$ is the preimage in $S$ of $X\alpha$ under $\alpha$.

### Exercises for Section 2

**2.1**  *Assume $G$ is a finite group, $p$ is a prime, and $S \in \mathrm{Syl}_p(G)$. Set $\bar{G} = G/O_{p'}(G)$. Prove $\mathcal{F}_S(G) \cong \mathcal{F}_{\bar{S}}(\bar{G})$.*

**2.2**  *Assume $G$ is a finite group, $p$ is a prime, and $S \in \mathrm{Syl}_p(G)$ is abelian. Let $H = N_G(S)$ and prove $\mathcal{F}_S(G) \cong \mathcal{F}_{\bar{S}}(H)$.*

**2.3**   *Assume $G$ is a finite group, $p$ is a prime, and $S \in \mathrm{Syl}_p(G)$. Prove*

(a)   *Let $T \leq S$. Then $T$ is strongly closed in $S$ with respect to $\mathcal{F}_S(G)$ iff $T$ is strongly closed in $S$ with respect to $G$.*

(b)   *Let $H \trianglelefteq G$. Then $H \cap S$ is strongly closed in $S$ with respect to $\mathcal{F}_S(G)$.*

**2.4**   *Let $\alpha\colon \mathcal{F} \to \tilde{\mathcal{F}}$ be a morphism of fusion systems. Prove $\ker(\alpha)$ is strongly closed in $S$ with respect to $\mathcal{F}$.*

## 3. Saturated fusion systems

In this section we assume $\mathcal{F}$ is a saturated fusion system on a finite $p$-group $S$.

**Lemma 3.1**   *Let $P \leq S$ and $Q \in P^{\mathcal{F}}$ with $Q$ fully normalized. Then there exists $\varphi \in \mathrm{Hom}_{\mathcal{F}}(N_S(P), N_S(Q))$ with $P\varphi = Q$.*

*Proof.* This is Lemma I.2.6.c. $\qquad\qquad\qquad\qquad\qquad\qquad\qquad\qquad\square$

Recall from Definition I.4.1 that a subgroup $R$ of $S$ is *normal* in $\mathcal{F}$ if $\mathcal{F} = N_{\mathcal{F}}(R)$, and we write $R \trianglelefteq \mathcal{F}$ to indicate that $R$ is normal in $\mathcal{F}$. Further $\mathcal{F}$ is *constrained* if there exists a normal centric subgroup of $\mathcal{F}$.

**Example 3.2**   Let $G$ be a finite group, $S \in \mathrm{Syl}_p(G)$, and $\mathcal{F} = \mathcal{F}_S(G)$. If $R$ is a normal $p$-subgroup of $G$, then by Exercise 3.1, $R \trianglelefteq \mathcal{F}$. Further if $C_G(O_p(G)) \leq O_p(G)$, then by Exercise 3.1, $\mathcal{F}$ is constrained.

**Lemma 3.3**   *Let $R \leq S$. Then the following are equivalent:*

(a)   $R \trianglelefteq \mathcal{F}$.

(b)   *$R$ is strongly closed in $S$ with respect to $\mathcal{F}$ and contained in all radical centric subgroups of $\mathcal{F}$*

(c)   *There exists a series $1 = R_0 \leq R_1 \leq \cdots \leq R_n = R$ such that*

    (i)   *for each $1 \leq i \leq n$, $R_i$ is strongly closed in $S$ with respect to $\mathcal{F}$, and*

    (ii)   *for each $1 \leq i < n$, $[R, R_{i+1}] \leq R_i$.*

*Proof.* This is a restatement of Propositions I.4.5 and I.4.6. $\qquad\qquad\square$

By Exercise 3.2, there is a largest subgroup of $S$ normal in $\mathcal{F}$. Recall from Definition I.4.3 that we write $O_p(\mathcal{F})$ for that subgroup.

**Lemma 3.4**   (a)   *An abelian subgroup $R$ of $S$ is normal in $\mathcal{F}$ iff $R$ is strongly closed in $S$ with respect to $\mathcal{F}$.*

(b)   $O_p(\mathcal{F}) \neq 1$ *iff there is a nontrivial abelian subgroup of $S$ strongly closed in $S$ with respect to $\mathcal{F}$.*

*Proof.* See I.4.7.     $\square$

**Theorem 3.5**   *(Alperin's Fusion Theorem)* $\mathcal{F} = \langle \mathrm{Aut}_{\mathcal{F}}(R) : R \in \mathcal{F}^{frc} \rangle$, *where $\mathcal{F}^{frc}$ is the set of fully normalized radical centric subgroups of $\mathcal{F}$.*

*Proof.* This special case of I.3.5 is the version of Alperin's theorem which we will use most often.     $\square$

<h2 align="center">Exercises for Section 3</h2>

**3.1**   *Let $G$ be a finite group, $S \in \mathrm{Syl}_p(G)$, and $\mathcal{F} = \mathcal{F}_S(G)$. Prove*

(a)   *If $R$ is a normal $p$-subgroup of $G$, then $R \trianglelefteq \mathcal{F}$.*

(b)   *If $C_G(O_p(G)) \leq O_p(G)$, then $\mathcal{F}$ is constrained.*

**3.2**   *Assume $\mathcal{F}$ is a saturated fusion system on the finite $p$-group $S$. Prove:*

(a)   *If $R, Q \trianglelefteq \mathcal{F}$ then $RQ \trianglelefteq \mathcal{F}$.*

(b)   *There is a largest subgroup of $S$ normal in $\mathcal{F}$.*

**3.3**   *Prove $R \trianglelefteq \mathcal{F}$ iff for each $P \leq R$ and each $\phi \in \mathrm{hom}_{\mathcal{F}}(P, S)$, $\phi$ extends to a member of $\mathrm{Aut}_{\mathcal{F}}(R)$.*

<h2 align="center">4. MODELS FOR CONSTRAINED SATURATED FUSION SYSTEMS</h2>

In this section we assume $\mathcal{F}$ is a saturated fusion system on a finite $p$-group $S$.

**Definition 4.1**   Write $\mathcal{G}(\mathcal{F})$ for the class of finite groups $G$ such that $S \in \mathrm{Syl}_p(G)$, $C_G(O_p(G)) \leq O_p(G)$, and $\mathcal{F} = \mathcal{F}_S(G)$. Call the members of $\mathcal{G}(\mathcal{F})$ the *models* of $\mathcal{F}$.

By Exercise 3.1.b, if $\mathcal{G}(\mathcal{F}) \neq \varnothing$ then $\mathcal{F}$ is constrained. By Theorem I.4.9 (see also Theorem III.5.10), the converse is true, while Lemma 4.4 says all models for a constrained system are isomorphic in a strong sense.

**Theorem 4.2**   *If $\mathcal{F}$ is constrained then $\mathcal{G}(\mathcal{F}) \neq \varnothing$.*

*Proof.* This is part (a) of Theorem I.4.9, and is proven as Theorem III.5.10.     $\square$

**Lemma 4.3**   *Assume $\mathcal{F}$ is constrained, $\tilde{\mathcal{F}}$ is a fusion system over $\tilde{S}$, and $\alpha \colon \mathcal{F} \to \tilde{\mathcal{F}}$ is an isomorphism of fusion systems. Let $G \in \mathcal{G}(\mathcal{F})$ and $\tilde{G} \in \mathcal{G}(\tilde{\mathcal{F}})$. Then*

(a)  *The set $E(\alpha)$ of isomorphisms $\breve{\alpha}\colon G \to \tilde{G}$ extending $\alpha\colon S \to \tilde{S}$ is nonempty.*

(b)  *Let $\breve{\alpha} \in E(\alpha)$. Then $E(\alpha) = \{c_z\breve{\alpha}\colon z \in Z(S)\}$, where $c_z \in \mathrm{Aut}(G)$ is the conjugation map.*

*Proof.* Part (a) is essentially I.4.9.b. Namely by transfer of structure, there is a group $H$ with $\tilde{S} \in \mathrm{Syl}_p(H)$ and $\tilde{\mathcal{F}} = \mathcal{F}_{\tilde{S}}(H)$, and an isomorphism $\beta\colon G \to H$ extending $\alpha$. By I.4.9.b, there exists an isomorphism $\gamma\colon H \to \tilde{G}$ extending the identity map $\iota$ on $\tilde{S}$. Then $\breve{\alpha} = \beta\gamma\colon G \to \tilde{G}$ is an isomorphism extending $\alpha = \alpha\iota$.

Similarly part (b) follows from III.5.10.c. Namely if $\zeta \in E(\alpha)$ then $\rho = \breve{\alpha}\zeta^{-1} \in \mathrm{Aut}(G)$ is the identity on $S$, so by III.5.10.c, $\rho = c_z$ for some $z \in Z(S)$, and hence $\zeta = c_z\breve{\alpha}$.   $\square$

**Lemma 4.4**  *Assume $\mathcal{F}$ is constrained and let $G_1, G_2 \in \mathcal{G}(\mathcal{F})$. Then there exists an isomorphism $\varphi\colon G_1 \to G_2$ which is the identity on $S$.*

*Proof.* Apply 4.3.1 with $\tilde{\mathcal{F}} = \mathcal{F}$, $\alpha$ the identity map on $\mathcal{F}$, $G = G_1$, and $\tilde{G} = G_2$. Or appeal to I.4.9.b.   $\square$

**Example 4.5**  Let $U \in \mathcal{F}^f$ with $C_S(U) \le U$, and set $\mathcal{D} = N_{\mathcal{F}}(U)$. By 2.1, $\mathcal{D}$ is a saturated fusion system on $D = N_S(U)$. By Exercise 4.1, $U \in \mathcal{F}^{fc}$. As $\mathcal{D} = N_{\mathcal{F}}(U)$, $U \trianglelefteq \mathcal{D}$. Then as $C_{\mathcal{D}}(U) = C_S(U) \le U$, $\mathcal{D}$ is constrained, so by 4.2 there exists $G = G_{\mathcal{F}}(U) \in \mathcal{G}(\mathcal{D})$. Thus $\mathcal{D} \cong \mathcal{F}_D(G)$. By 4.4, $G$ is unique up to isomorphism.

**Example 4.6**  Assume $T$ is strongly closed in $S$ with respect to $\mathcal{F}$ and let $U \le T$ with $U \in \mathcal{F}^f$ and $C_T(U) \le U$. Set $V = UC_S(U)$ and $\mathcal{D} = N_{\mathcal{F}}(V)$. By Exercise 4.2, $V \in \mathcal{F}^f$ with $C_S(V) \le V$, $\mathcal{D} \le N_{\mathcal{F}}(U)$, and $D = N_S(V) \le N_S(U)$. Hence applying Example 4.5 to $V$ in the role of $U$, we conclude that $\mathcal{D}$ is a saturated constrained fusion system on $D$, so there exist $G = G_{\mathcal{F},T}(U) \in \mathcal{G}(\mathcal{D})$, and $G$ is unique up to isomorphism.

### Exercises for Section 4

**4.1**  *Assume $U \le S$ be fully centralized with $C_S(U) \le U$. Prove $U \in \mathcal{F}^c$.*

**4.2**  *Assume $T$ is strongly closed in $S$ with respect to $\mathcal{F}$. Assume $U \le T$ with $U \in \mathcal{F}^f$ and $C_T(U) \le U$. Set $V = UC_S(U)$, and prove:*

(a)  $U = T \cap V \trianglelefteq N_{\mathcal{F}}(V)$.

(b)  $V \in \mathcal{F}^f$.

(c)  $C_S(V) \le V$.

## 5. FACTOR SYSTEMS AND SURJECTIVE MORPHISMS

In this section $\mathcal{F}$ is a fusion system over the finite $p$-group $S$.

**Definition 5.1**   Assume $S_0 \leq S$ is strongly closed in $S$ with respect to $\mathcal{F}$ and set $\mathcal{N} = N_{\mathcal{F}}(T)$.

Set $S^+ = S/S_0$ and let $\theta: S \to S^+$ be the natural map $\theta: x \mapsto x^+ = xS_0$. We define a category $\mathcal{F}^+$ and a morphism $\theta: \mathcal{N} \to \mathcal{F}^+$. The objects of $\mathcal{F}^+$ are the subgroups of $S^+$. For $P \leq S$ and $\alpha \in \mathrm{Hom}_{\mathcal{N}}(P, S)$ define $\alpha^+: P^+ \to S^+$ by $x^+\alpha^+ = (x\alpha)^+$. This is well defined as $S_0$ is strongly closed in $S$ with respect to $\mathcal{F}$. Now define

$$\mathrm{Hom}_{\mathcal{F}^+}(P^+, S^+) = \{\beta^+ : \beta \in \mathrm{Hom}_{\mathcal{N}}(PS_0, S)\},$$

and define $\theta_P: \mathrm{Hom}_{\mathcal{N}}(P, S) \to \mathrm{Hom}_{\mathcal{F}^+}(P^+, S^+)$ by $\alpha\theta_P = \alpha^+$. For $\alpha \in \mathrm{Hom}_{\mathcal{N}}(P, S)$, $\alpha$ extends to $\hat{\alpha} \in \mathrm{Hom}_{\mathcal{N}}(PS_0, S)$ and $\hat{\alpha}^+ = \alpha^+$, so $\theta_P$ is well defined and surjective.

**Lemma 5.2**   (a)   $\mathcal{F}^+$ *is a fusion system on the finite $p$-group* $S^+$.

(b)   $\theta: \mathcal{N} \to \mathcal{F}^+$ *is a surjective morphism of fusion systems.*

*Proof.* Observe $x^+ \in \ker(\alpha^+)$ iff $1 = x^+\alpha^+ = (x\alpha)^+$ iff $x\alpha \in T$ iff $x \in T$ iff $x^+ = 1$. So the members of $\mathrm{Hom}_{\mathcal{F}^+}(P^+, Q^+)$ are monomorphisms.

Suppose $\alpha \in \mathrm{Hom}_{\mathcal{N}}(P, Q)$ and $\beta \in \mathrm{Hom}_{\mathcal{N}}(Q, S)$. Then for $x \in P$,

$$(x^+\alpha^+)\beta^+ = (x\alpha)^+\beta^+ = ((x\alpha)\beta)^+ = (x(\alpha\beta))^+ = x^+(\alpha\beta)^+,$$

so

$$\alpha^+\beta^+ = (\alpha\beta)^+ \tag{!}$$

By (!), $\mathcal{F}^+$ is a category.

For $s \in S$, $c_{s^+}: S^+ \to S^+$ is the map $c_{s^+}$, so $\mathrm{Hom}_{S^+}(P^+, Q^+) \subseteq \mathrm{Hom}_{\mathcal{F}^+}(P^+, Q^+)$.

If $\phi \in \mathrm{Hom}_{\mathcal{F}^+}(P^+, Q^+)$ then $\phi = \alpha^+$ for some $\alpha \in \mathrm{Hom}_{\mathcal{F}}(PS_0, QS_0)$. As $\mathcal{F}$ is saturated, $\alpha: PS_0 \to (PS_0)\alpha = P\alpha S_0$ is in $\mathrm{Hom}_{\mathcal{F}}(PS_0, P\alpha S_0)$, so $\phi: P^+ \to P^+\phi$ is in $\mathrm{Hom}_{\mathcal{F}^+}(P^+, P^+\phi)$, since $P^+\phi = (P\alpha S_0)^+$. Suppose $\phi$ is an isomorphism. Then again as $\mathcal{F}$ is saturated, $\alpha^{-1} \in \mathrm{Hom}_{\mathcal{F}}(QS_0, PS_0)$, so $\phi^{-1} = (\alpha^{-1})^+ \in \mathrm{Hom}_{\mathcal{F}^+}(Q^+, P^+)$, completing the proof of (1).

Let $\phi \in \mathrm{Hom}_{\mathcal{F}}(P, Q)$ and $x \in P$. Then

$$x\phi\theta = (x\phi)^+ = x^+\phi^+ = x\theta(\phi\theta_{P,Q}),$$

so from 2.2, $\theta: \mathcal{F} \to \mathcal{F}^+$ is a morphism of fusion systems. We observed in 5.1 that $\theta$ is surjective. $\qquad\square$

**Definition 5.3**   Suppose $S_0$ is strongly closed in $S$ with respect to $\mathcal{F}$. Then appealing to 5.2.a, we can form the fusion system $\mathcal{F}^+$ as in 5.1. We write $\mathcal{F}/S_0$ for this fusion system, and call it the *factor system* of $\mathcal{F}$ modulo

$S_0$. By 5.2.b, $\theta\colon \mathcal{N} \to \mathcal{F}/S_0$ is a surjective morphism of fusion systems, which we denote by $\theta_{\mathcal{F}/S_0}$.

**Lemma 5.4** *Assume $\alpha\colon \mathcal{F} \to \tilde{\mathcal{F}}$ is a surjective morphism of fusion systems and $\mathcal{F}$ is saturated. Then $\tilde{\mathcal{F}}$ is saturated.*

*Proof.* This was first proved by Puig; cf. Proposition 12.3 in [P7]. See also Lemma 8.5 in [A5], where another proof appears. Here is a sketch of a different proof, based on the Roberts-Shpectorov definition of saturation.

Let $\tilde{\mathcal{F}}$ be a system on $\tilde{S}$ and let $\ker(\alpha) \leq P \leq S$ be fully automized and receptive. It suffices to show that $\tilde{P} = P\alpha$ is fully automized and receptive. Recall from 2.2 that $\alpha$ induces a surjective homomorphism $\alpha_P\colon \mathrm{Aut}_{\mathcal{F}}(P) \to \mathrm{Aut}_{\tilde{\mathcal{F}}}(\tilde{P})$ defined by $(x\alpha)(\phi\alpha_P) = x\phi\alpha$ for $\phi \in \mathrm{Aut}_{\mathcal{F}}(P)$ and $x \in P$. Check that $\mathrm{Aut}_S(P)\alpha_P = \mathrm{Aut}_{\tilde{S}}(\tilde{P})$. But as $\mathrm{Aut}_S(P)$ is Sylow in $\mathrm{Aut}_{\mathcal{F}}(P)$, its image is Sylow in the image of $\mathrm{Aut}_{\mathcal{F}}(P)$, so $\tilde{P}$ is fully automized.

Let $\tilde{\varphi} \in \mathrm{Iso}_{\tilde{\mathcal{F}}}(\tilde{Q}, \tilde{P})$. Then there is a subgroup $Q$ of $S$ containing $\ker(\alpha)$ and $\varphi \in \mathrm{Iso}_{\mathcal{F}}(Q, P)$ with $Q\alpha = \tilde{Q}$ and $\varphi\alpha_{Q,P} = \tilde{\varphi}$. Check that for $g \in N_S(Q)$, $c_g^\varphi\alpha_P = c_{g\alpha}^{\tilde{\varphi}}$. Then as $\mathrm{Aut}_S(P)\alpha_P = \mathrm{Aut}_{\tilde{S}}(\tilde{P})$, if $g \in N_\varphi$ then $c_{g\alpha}^{\tilde{\varphi}} = C_g^\varphi\alpha_P \in \mathrm{Aut}_S(P)\alpha_P = \mathrm{Aut}_{\tilde{S}}(\tilde{P})$, so that $g\alpha \in N_{\tilde{\varphi}}$. Hence $N_\varphi\alpha \leq N_{\tilde{\varphi}}$.

Next let $N$ be the preimage in $S$ of $N_{\tilde{\varphi}}$ under $\alpha$. Then

$$\mathrm{Aut}_N(Q)^\varphi\alpha_P \leq \mathrm{Aut}_{N_{\tilde{\varphi}}}(\tilde{Q})^{\tilde{\varphi}} \leq \mathrm{Aut}_{\tilde{S}}(\tilde{P}) = \mathrm{Aut}_S(P)\alpha_P,$$

and as $P$ is fully automized, $\mathrm{Aut}_S(P)$ is Sylow in $\mathrm{Aut}_S(P)\ker(\alpha_P)$, so there is $\chi \in \ker(\alpha_P)$ with $\mathrm{Aut}_N(Q)^{\varphi\chi} \leq \mathrm{Aut}_S(P)$. Then $(\varphi\chi)\alpha_{Q,P} = \tilde{\varphi}$, so replacing $\varphi$ by $\varphi\chi$ and appealing to the previous paragraph, $N_{\tilde{\varphi}} = N_\varphi\alpha$.

As $P$ is receptive, $\varphi$ lifts to $\bar{\varphi} \in \mathrm{Hom}_{\mathcal{F}}(N_\varphi, S)$. Then $\bar{\varphi}\alpha_{N_\varphi,S}$ is a lift of $\tilde{\varphi}$ to $\mathrm{Hom}_{\tilde{\mathcal{F}}}(N_{\tilde{\varphi}}, \tilde{S})$, so $\tilde{P}$ is receptive. $\square$

**Lemma 5.5** *Assume $\mathcal{F}$ is saturated and $S_0$ is strongly closed in $S$ with respect to $\mathcal{F}$. Then $\mathcal{F}/S_0$ is saturated.*

*Proof.* As $S_0$ is strongly closed in $S$, $S_0 \trianglelefteq S$, so $S_0 \in \mathcal{F}^f$. Therefore $\mathcal{N}$ is saturated by 2.1. Therefore $\mathcal{F}/S_0$ is saturated by 5.2.b and 5.4. $\square$

**Example 5.6** Assume $\mathcal{F} = \mathcal{F}_S(G)$, $H \trianglelefteq G$, and set $S_0 = S \cap H$. Let $M = N_G(S_0)$ and $M^* = M/N_H(S_0)$. By Exercise 2.3.2, $S_0$ is strongly closed in $S$. In 8.8 in [A5] is is shown that:

(a) $\mathcal{F}/S_0 \cong \mathcal{F}_{S^*}(M^*)$.

(b) $\mathcal{F}_{S^*}(M^*) \cong \mathcal{F}_{SH/H}(G/H)$.

Thus $\mathcal{F}/S_0 \cong \mathcal{F}_{SH/H}(G/H)$.

In the remainder of the section, assume $\mathcal{F}$ is saturated and $S_0$ is strongly closed in $S$ with respect to $\mathcal{F}$.

**Definition 5.7**   For $P, Q \leq S$, write $P_0$ for $P \cap S_0$ and define

$$\Phi(P,Q) = \{\phi \in \mathrm{Hom}_{\mathcal{F}}(P,Q) \colon [P,\phi] \leq S_0\},$$

where for $x \in P$, $[x,\phi] = x^{-1} \cdot x\phi \in S$, and $[P,\phi] = \langle [x,\phi] \colon x \in P \rangle \leq S$.

For $\alpha \in \mathrm{Hom}_{\mathcal{F}}(P,S)$ define $\mathfrak{F}(\alpha)$ to be the set of pairs $(\varphi, \phi)$ such that $\varphi \in \mathrm{Hom}_{\mathcal{F}}(PS_0, S)$, $\phi \in \Phi(P\varphi, S)$, and $\alpha = \varphi\phi$.

Form $\mathcal{N} = N_{\mathcal{F}}(S_0)$ and the factor system $\mathcal{F}^+ = \mathcal{F}/S_0$ on $S^+ = S/S_0$ and $\theta\colon \mathcal{N} \to \mathcal{F}^+$ as in 5.1.

**Lemma 5.8**   *Let $\alpha \in \mathrm{Hom}_{\mathcal{F}}(P,S)$, $\beta \in \mathrm{Hom}_{\mathcal{F}}(P\alpha, S)$, $(\varphi, \phi) \in \mathfrak{F}(\alpha)$, and $(\Psi, \psi) \in \mathfrak{F}(\beta)$. Then*

(a)   *If $Q, R \leq S$, $\mu \in \Phi(P,Q)$, and $\eta \in \Phi(Q,R)$, then $\mu\eta \in \Phi(P,R)$.*

(b)   $\phi^{\Psi} \in \Phi(P\varphi\Psi, S)$.

(c)   $(\varphi\Psi, \phi^{\Psi}\psi) \in \mathfrak{F}(\alpha\beta)$.

*Proof.* Exercise 5.1.   $\square$

**Theorem 5.9**   *For each $P \leq S$ and $\alpha \in \mathrm{hom}_{\mathcal{F}}(P,S)$, $\mathfrak{F}(\alpha) \neq \varnothing$.*

*Proof.* This is [A6, 12.5]. Here is the idea of the proof. Choose a counter example with $m = |S{:}P|$ minimal. Observe $P_0 \neq S_0$ as in that case $(\alpha, 1) \in \mathfrak{F}(\alpha)$; in particular $m > 1$. By minimality of $m$, $\alpha$ does not extend to a proper overgroup of $P$ in $S$. Then, using Alperin's Fusion Theorem 3.5 and 5.8, we reduce to the case where $P \in \mathcal{F}^{frc}$ and $\alpha \in \mathrm{Aut}_{\mathcal{F}}(P)$. Hence by 4.2 there is a model $G$ for $N_{\mathcal{F}}(P)$ and $\alpha = c_{g|P}$ for some $g \in G$.

Set $Q = N_S(P)$; as $P_0 \neq S_0$, $P_0 < Q_0$. Set $K = \langle Q_0^G \rangle$. As $Q_0$ centralizes $P/P_0$, so does $K$. Therefore for each $k \in K$, $c_{k|P} \in \Phi(P,P)$, so $(1, c_{k|P}) \in \mathfrak{F}(c_{k|P})$. But by a Frattini argument, $G = KN_G(Q_0)$, so $g = kh$ for some $h \in N_G(Q_0)$ and $k \in K$. As $P_0 < Q_0$, $P < PQ_0$, so $\mathfrak{F}(c_{h|P}) \neq \varnothing$ by minimality of $m$. Since also $\mathfrak{F}(c_{k|P}) \neq \varnothing$, 5.8 completes the proof.   $\square$

**Lemma 5.10**   *Let $P \leq S$, $\alpha \in \mathrm{Hom}_{\mathcal{F}}(P,S)$, and $(\varphi, \phi) \in \mathfrak{F}(\alpha)$. Then*

(a)   $\varphi \in \mathrm{Hom}_{\mathcal{N}}(P,S)$ *and* $\varphi^+ \in \mathrm{Hom}_{\mathcal{F}^+}(P^+, S^+)$.

(b)   *For $x \in P$, $(x\alpha)^+ = x^+\varphi^+$.*

*Proof.* As $(\varphi, \phi) \in \mathfrak{F}(\alpha)$, $\varphi \in \mathcal{N}$, so by definition of the $+$-notation, $x^+\varphi^+ = (x\varphi)^+$. That is (a) holds. Further

$$(x\alpha)^+ = (x\varphi\phi)^+ = (x\varphi \cdot [x\varphi, \phi])^+ = (x\varphi)^+[x\varphi, \phi]^+ = (x\varphi)^+,$$

as $\phi \in \Phi(P\varphi, S)$, so $[x\varphi, \phi] \in S_0$, the kernel of $\theta: S \to S^+$, where $\theta: x \mapsto x^+ = xS_0$. Thus (b) holds. $\square$

**Definition 5.11** For $P \leq S$ and $\alpha \in \mathrm{Hom}_{\mathcal{F}}(P, S)$, define $\alpha\Theta \in \mathrm{Hom}_{\mathcal{F}^+}(P^+, S^+)$ by $\alpha\Theta = \varphi^+$ and $(\varphi, \phi) \in \mathfrak{F}(\alpha)$. Observe that $\Theta$ is well defined: Namely by 5.10.a, $\varphi^+ \in \mathrm{Hom}_{\mathcal{F}^+}(P^+, S^+)$. Further if $(\Psi, \psi) \in \mathfrak{F}(\alpha)$ and $x \in P$, then by 5.10.b, $x^+\varphi^+ = (x\alpha)^+ = x^+\Psi^+$, so the definition of $\alpha\Theta$ is independent of the choice of $(\varphi, \phi)$ in $\mathfrak{F}(\alpha)$.

Next define $\Theta: S \to S^+$ to be the natural map $\Theta: s \mapsto s^+$.

Write $\Theta_{\mathcal{F}, S_0}$ for this map from $\mathcal{F}$ to $\mathcal{F}/S_0$.

**Theorem 5.12** (a) $\Theta = \Theta_{\mathcal{F}, S_0}: \mathcal{F} \to \mathcal{F}/S_0$ *is a surjective morphism of fusion systems.*

(b) $\theta$ *is the restriction of* $\Theta$ *to* $\mathcal{N} = N_{\mathcal{F}}(S_0)$.

*Proof.* Let $P \leq S$. For $\gamma \in \mathrm{Hom}_{\mathcal{N}}(P, S)$, $(\gamma, 1) \in \mathfrak{F}(\gamma)$, so $\gamma\Theta = \gamma^+ = \gamma\theta$. Then as $\theta = \Theta$ as a map of groups on $S$, (b) holds.

Let $\alpha \in \mathrm{Hom}_{\mathcal{F}}(P, S)$ and $\beta \in \mathrm{Hom}_{\mathcal{F}}(P\alpha, S)$. By 5.8.c, $(\alpha\beta)\Theta = \alpha\Theta \cdot \beta\Theta$. Let $(\varphi, \phi) \in \mathfrak{F}(\alpha)$ and $x \in P$. Then by 5.10.b,

$$(x\alpha)\Theta = (x\alpha)^+ = x^+\varphi^+ = (x\Theta)(\alpha\Theta),$$

so $\Theta$ is a morphism of fusion systems. By definition, $\Theta: S \to S^+$ is surjective. By (b), $\Theta$ extends $\theta$, so as $\theta$ is surjective, so is $\Theta$. $\square$

**Lemma 5.13** *Assume* $\rho: \mathcal{F} \to \tilde{F}$ *is a surjective morphism of fusion systems, with* $S_0 = \ker(\rho)$. *Then*

(a) *For* $P, Q \leq S$ *and* $\phi \in \Phi(P, Q)$, $\phi\rho = 1$.

(b) *For* $P \leq S$, $\alpha \in \mathrm{Hom}_{\mathcal{F}}(P, S)$, *and* $(\varphi, \phi) \in \mathfrak{F}(\alpha)$, $\alpha\rho = \varphi\rho$.

(c) *Define* $\pi: \mathcal{F}^+ \to \tilde{F}$ *by* $x^+\pi = x\rho$ *for* $x \in S$, *and* $\varphi^+\pi = \varphi\rho$, *for* $\varphi$ *an* $\mathcal{N}$-map. *Then* $\pi: \mathcal{F}^+ \to \tilde{F}$ *is an isomorphism of fusion systems such that* $\Theta\pi = \rho$.

*Proof.* First assume the setup of (a), and let $x \in P$. Then

$$(x\rho)(\phi\rho) = (x\phi)\rho = (x \cdot [x, \phi])\rho = x\rho \cdot [x, \phi]\rho = x\rho,$$

as $[x, \phi] \in S_0$ and $S_0\rho = 1$. Thus (a) holds.

Next assume the setup of (b). Then by (1), $\alpha\rho = (\varphi\phi)\rho = (\varphi\rho)(\phi\rho) = \varphi\rho$, establishing (b).

As $S_0 = \ker(\rho)$, $\pi: S^+ \to \tilde{S}$ is a well defined group isomorphism, with $\Theta\pi = \rho$ as a map of groups.

Let $P \leq S$ and $\eta, \mu \in \mathrm{Hom}_{\mathcal{N}}(P, S)$. Then $\eta\rho = \mu\rho$ iff for all $x \in P$, $x\eta\rho = (x\rho)(\eta\rho) = (x\rho)(\mu\rho) = (x\mu)\rho$ iff $x\eta \in x\mu S_0$. Thus if $\eta^+ = \mu^+$

then as $S_0 = \ker(\rho)$, $\eta\rho = \mu\rho$, so $\pi\colon \mathrm{Hom}_{\mathcal{N}}(P^+, S^+) \to \mathrm{Hom}_{\tilde{\mathcal{F}}}(P\rho, \tilde{S}) = \mathrm{Hom}_{\tilde{\mathcal{F}}}(P^+\pi, \tilde{S})$ is well defined.  Further $\eta^+ = \mu^+$ iff $\eta\rho = \mu\rho$, so $\pi$ is injective.

For $x \in P$,
$$(x^+\pi)(\eta^+\pi) = (x\rho)(\eta\rho) = (x\eta)\rho = (x\eta)^+\pi = (x^+\eta^+)\pi.$$

For $\xi \in \hom_{\mathcal{N}}(P\eta, S)$,
$$(\eta^+\xi^+)\pi = (\eta\xi)^+\pi = (\eta\xi)\rho = \eta\rho{\cdot}\xi\rho = \eta^+\pi{\cdot}\xi^+\pi.$$

Thus $\pi\colon \mathcal{F}^+ \to \tilde{F}$ is a morphism of fusion systems.  Further by (b),
$$\alpha\Theta\pi = \varphi^+\pi = \varphi\rho = \alpha\rho.$$

Then as $\rho$ is a surjection, so is $\pi$.  We saw $\pi\colon S^+ \to \tilde{S}$ is an isomorphism, as is $\pi\colon \mathrm{Hom}_{\mathcal{N}}(P^+S^+) \to \mathrm{Hom}_{\tilde{\mathcal{F}}}(P^+\pi, \tilde{S})$, so (c) follows.  $\square$

**Theorem 5.14**  *The map $S_0 \mapsto \mathcal{F}/S_0$ is a bijection between the set of subgroups $S_0$ of $S$, strongly closed in $S$ with respect to $\mathcal{F}$, and the set of isomorphism classes of homomorphic images of $\mathcal{F}$.*

*Proof.*  This is a consequence of 5.12.a and 5.13.c.  $\square$

**Exercises for Section 5**

**5.1**  *Prove 5.8.*

6. INVARIANT SUBSYSTEMS OF FUSION SYSTEMS

In this section $\mathcal{F}$ is a saturated fusion system over the finite $p$-group $S$.  We saw in the previous section that there is a 1-1 correspondence between the homomorphic images of $\mathcal{F}$ and strongly closed subgroups $T$ of $S$, and we may take the factor system $\mathcal{F}/T$ to be the image corresponding to $T$.  Proceeding by analogy with the situation for groups, we would like to show there exists a "normal subsystem" $\mathcal{E}$ of $\mathcal{F}$ on $T$ which is saturated, and hence realize $\mathcal{F}$ as an "extension" of $\mathcal{E}$ by $\mathcal{F}/T$.  Unfortunately such a subsystem may not exist, but in many instances one does.  We begin to investigate the situation.

Let $\mathcal{E}$ is a subsystem of $\mathcal{F}$ on a subgroup $T$ of $S$.  Write $\mathcal{F}_T^f$ for the set of nontrivial subgroups $P$ of $T$ such that $P$ is fully normalized in $\mathcal{F}$.

**Definition 6.1**  Define $\mathcal{E}$ to be *$\mathcal{F}$-invariant* if:

(I1)  $T$ is strongly closed in $S$ with respect to $\mathcal{F}$.

(I2)  For each $P \leq Q \leq T$, $\phi \in \mathrm{Hom}_{\mathcal{E}}(P, Q)$, and $\alpha \in \mathrm{Hom}_{\mathcal{F}}(Q, S)$, $\phi^\alpha \in \mathrm{Hom}_{\mathcal{E}}(P\alpha, T)$.

Notice that Definition 6.1 does not quite agree with the definition of $\mathcal{F}$-invariance in Definition I.6.1, but the two notions are equivalent by I.6.4. The following definition also appears in I.6.1:

**Definition 6.2** The subsystem $\mathcal{E}$ is $\mathcal{F}$-*Frattini* if for each $P \leq T$ and $\gamma \in \mathrm{Hom}_{\mathcal{F}}(P, S)$, there exists $\varphi \in \mathrm{Aut}_{\mathcal{F}}(T)$ and $\phi \in \mathrm{Hom}_{\mathcal{E}}(P\varphi, S)$, such that $\gamma = \varphi\phi$ on $P$.

**Lemma 6.3** *Assume $T$ is a subgroup of $S$ which is strongly closed in $S$ with respect to $\mathcal{F}$ and $\mathcal{E}$ is a subsystem of $\mathcal{F}$ on $T$. Then the following are equivalent:*

(a)  $\mathcal{E}$ *is $\mathcal{F}$-invariant.*

(b)  $\mathrm{Aut}_{\mathcal{F}}(T) \leq \mathrm{Aut}(\mathcal{E})$ *and $\mathcal{E}$ is $\mathcal{F}$-Frattini.*

(c)  $\mathrm{Aut}_{\mathcal{F}}(T) \leq \mathrm{Aut}(\mathcal{E})$ *and condition (I2A) holds:*

    (I2A) *For each $U \in \mathcal{F}_T^f$ there exists a normal subgroup $A(U)$ of $\mathrm{Aut}_{\mathcal{F}}(U)$, such that for each $U' \leq T$, and each $\beta \in \mathrm{Iso}_{\mathcal{F}}(U, U')$, $\mathrm{Aut}_T(U') \leq A(U)^{\beta} \leq \mathrm{Aut}_{\mathcal{E}}(U')$.*

*Proof.* Conditions (a) and (b) are equivalent by Proposition I.6.4. Condition (c) is shown to be equivalent to the first two conditions in [A5, 3.3]. Condition (I2A) presages the notion of an $\mathcal{F}$-invariant map in Definition 8.1, which leads in turn to the normal maps in Definition 8.6.  $\square$

Invariant subsystems are fairly natural and have many nice properties. For example:

**Lemma 6.4** *Assume $\mathcal{E}$ is $\mathcal{F}$-invariant and $\mathcal{D}$ is a subsystem of $\mathcal{F}$ on the subgroup $D$ of $S$. Then*

(a)  $\mathcal{E} \cap \mathcal{D}$ *is a $\mathcal{D}$-invariant subsystem of $\mathcal{D}$ on $T \cap D$.*

(b)  *If $\mathcal{D}$ is $\mathcal{F}$-invariant then $\mathcal{E} \cap \mathcal{D}$ is $\mathcal{F}$-invariant on $T \cap D$.*

*Proof.* Exercise 6.1.  $\square$

On the other hand invariant subsystems have the big drawback that they need not be saturated.

**Example 6.5** Assume $T$ is strongly closed in $S$ with respect to $\mathcal{F}$. Define $\mathcal{E}$ to be the subsystem of $\mathcal{F}$ on $T$ such that for each $P, Q \leq T$, $\mathrm{Hom}_{\mathcal{E}}(P, Q) = \mathrm{Hom}_{\mathcal{F}}(P, Q)$; that is $\mathcal{E}$ is the full subcategory of $\mathcal{F}$ whose objects are the subgroups of $T$. Then trivially $\mathcal{E}$ is $\mathcal{F}$-invariant. But in most circumstances, $\mathcal{E}$ is not saturated. For example for $P \leq T$, $\mathrm{Aut}_{\mathcal{E}}(P) = \mathrm{Aut}_{\mathcal{F}}(P)$. In particular if $P \in \mathcal{F}_T^f$ then as $\mathcal{F}$ is saturated, $\mathrm{Aut}_S(P) \in \mathrm{Syl}_p(\mathrm{Aut}_{\mathcal{F}}(P)) = \mathrm{Aut}_{\mathcal{E}}(T)$. However (cf. 3.4.5 in [A5]) $P \in \mathcal{E}^f$, so if $\mathrm{Aut}_S(P) \neq \mathrm{Aut}_T(P)$ then $\mathcal{E}$ is not saturated.

While we are primarily interested in saturated fusion systems, in investigating such objects it is often useful to argue in the larger category of (possibly unsaturated) fusion systems, where the notion of $\mathcal{F}$-invariance seems to be the "right" analogue of the normal subgroup relation from group theory. For example the intersection of normal subgroups $H$ and $K$ of a group $G$ is again a normal subgroup of $G$, and indeed the greatest lower bound for the pair in the lattice of subgroups of $G$. By 6.4, the same property holds for the lattice of invariant subsystems of $\mathcal{F}$. But the intersection of normal subsystems of $\mathcal{F}$ need not be normal in $\mathcal{F}$, as we find in Example 7.3. Instead the greatest lower bound for normal subsystems (in the lattice of normal subsystems of $\mathcal{F}$ partially ordered by the normality relation) is the more subtle object appearing in Theorem 9.1, whose definition is rather complicated.

## Exercises for Section 6

**6.1**   *Prove 6.4.*

## 7. NORMAL SUBSYSTEMS OF FUSION SYSTEMS

In this section $\mathcal{F}$ is a saturated fusion system over the finite $p$-group $S$ and $\mathcal{E}$ is a subsystem of $\mathcal{F}$ on a subgroup $T$ of $S$. We continue the notation and terminology from the previous sections.

We saw in the previous section that if $\mathcal{E}$ is $\mathcal{F}$-invariant, then $\mathcal{E}$ need not be saturated. One way to repair this problem is to only consider saturated $\mathcal{F}$-invariant subsystems. Recall from Definition I.6.1 that a saturated $\mathcal{F}$-invariant subsystem of $\mathcal{F}$ is said to be *weakly normal* in $\mathcal{F}$. It turns out that to obtain a class of subsystems of a saturated fusion system $\mathcal{F}$ which have properties analogous to the class of normal subgroups of a finite group, one more condition must be added to the definition of weak normality. In any event we are lead to the following definition:

**Definition 7.1**   The subsystem $\mathcal{E}$ is *normal* in $\mathcal{F}$ if $\mathcal{E}$ is $\mathcal{F}$-invariant and saturated, and the following condition holds:

(N1)   Each $\phi \in \mathrm{Aut}_{\mathcal{E}}(T)$ extends to $\hat{\phi} \in \mathrm{Aut}_{\mathcal{F}}(TC_S(T))$ such that $[\hat{\phi}, C_S(T)] \le Z(T)$.

We write $\mathcal{E} \trianglelefteq \mathcal{F}$ to indicate that $\mathcal{E}$ is normal in $\mathcal{F}$.

**Example 7.2**   Assume $\mathcal{F} = \mathcal{F}_S(G)$ for some finite group $G$ with $S \in \mathrm{Syl}_p(G)$. Let $H \trianglelefteq G$, $T = S \cap H$, and $\mathcal{E} = \mathcal{F}_T(H)$. Then from Proposition I.6.2, $\mathcal{E} \trianglelefteq \mathcal{F}$.

**Example 7.3**   Let $H = H_1 \times H_2 \times H_3$ be the direct product of three copies $H_i$, $1 \le i \le 3$, of $A_4$. Let $X_i = \langle x_i \rangle \in \mathrm{Syl}_3(H_i)$, $S_i = O_2(H_i)$, and $S = S_1 \times S_2 \times S_3 \in \mathrm{Syl}_2(H)$. Let $X = \langle x_1 x_2, x_1 x_3 \rangle \le H$ and set $G = XS$. Then $G_1 = \langle x_1 x_2, S_1, S_2 \rangle$ and $G_2 = \langle x_1 x_3, S_1, S_3 \rangle$ are normal subgroups of $G$ with Sylow 2-subgroups $T_1 = S_1 S_2$ and $T_2 = S_1 S_3$, respectively. Let $\mathcal{F}_i = \mathcal{F}_{T_i}(G_i)$, for $i = 1, 2$. As $G_i \trianglelefteq G$, $\mathcal{F}_i \trianglelefteq \mathcal{F}$ by 7.2.

Let $\mathcal{E} = \mathcal{F}_1 \cap \mathcal{F}_2$. Then $\mathcal{E} = \mathcal{F}_{S_1}(H_1)$ as

$$\phi = c_{x_1 x_2 | S_1} = c_{x_1 x_3 | S_1} = c_{x_1 | S_1} \in \mathrm{Aut}_{\mathcal{E}}(S_1).$$

In particular $\mathcal{E}$ is a saturated fusion system, and by 6.4.a, $\mathcal{E}$ is $\mathcal{F}$-invariant; that is $\mathcal{E}$ is weakly normal in $\mathcal{F}$. On the other hand $\mathcal{E}$ is *not* normal in $\mathcal{F}$, as (N1) is not satisfied. Namely $S_1 C_S(S_1) = S$, but $\phi$ does not extend to $\hat{\phi} \in \mathrm{Aut}_{\mathcal{F}}(S)$ with $[\hat{\phi}, S] \le S_1$. That is $\mathcal{E}$ is weakly normal in $\mathcal{F}$, but not normal in $\mathcal{F}$.

This shows, first, that there exist weakly normal subsystems of saturated fusion systems which are not normal, and, second, that the intersection of normal subsystems is not in general normal. Moreover this also shows, third, that (N1) is a necessary hypothesis if the converse of 7.2 is to hold for constrained fusion systems, since $\mathcal{E}$ has no model normal in $G$. Put another way, if we are to extend arguments from the local theory of finite groups to the domain of saturated fusion systems, it is crucial to have the property that, when $\mathcal{F}$ is constrained, $G \in \mathcal{G}(\mathcal{F})$, and $\mathcal{E} \trianglelefteq \mathcal{F}$, then there exist $H \in \mathcal{G}(\mathcal{E})$ with $H \trianglelefteq G$. Thus it is necessary that the definition of "normal subsystem" contain some condition such as (N1).

See also Proposition I.5.6 for more examples illustrating differences between weakly normal and normal subsystems.

Recall from Definition 4.1 that for $\mathcal{F}$ saturated, the set $\mathcal{G}(\mathcal{F})$ of models of $\mathcal{F}$ consists of groups $G$ with $S \in \mathrm{Syl}_p(G)$ and $\mathcal{F} = \mathcal{F}_S(G)$, such that $C_G(O_p(G)) \le O_p(G)$. From the discussion of the generalized Fitting subgroup $F^*(G)$ of a group $G$ in A.9 and A.10, this last condition is equivalent to the condition that $F^*(G) = O_p(G)$, and that is the form in which we usually state the condition.

**Lemma 7.4**   *Assume $\mathcal{F}$ is constrained and $T$ is strongly closed in $S$ with respect to $\mathcal{F}$. Then*

(a)   *There exists $G \in \mathcal{G}(\mathcal{F})$. Set $L = \langle T^G \rangle$.*

(b)   *If $\mathcal{E}$ is a normal subsystem of $\mathcal{F}$ on $T$, then $\mathcal{E} = \mathcal{F}_T(H)$ for the unique normal subgroup $H$ of $G$ such that $T \in \mathrm{Syl}_p(H)$, $L = O^{p'}(H)$, and $\mathrm{Aut}_H(T) = \mathrm{Aut}_{\mathcal{E}}(T)$.*

*Proof.* Part (a) is Theorem 4.2. The proof of (b) is a bit complicated and takes a number of lemmas, culminating in 6.7 in [A5]. But here is a discussion of a few of the high points.

Let $R = O_p(G)$, $Q = T \cap R$, and $X = C_{S \cap L}(Q)$. As $\mathcal{E} \trianglelefteq \mathcal{F}$, $\mathrm{Aut}_T(Q)$ is Sylow in $B = \langle \mathrm{Aut}_T(Q)^{\mathrm{Aut}_{\mathcal{F}}(Q)} \rangle$. But as $G$ is a model for $\mathcal{F}$, $B = \mathrm{Aut}_L(Q)$, so

(*) $TX \in \mathrm{Syl}_p(L)$.

Next $[T, R] \le Q$, so $L = \langle T^G \rangle$ centralizes $R/Q$. Thus as $F^*(G) = R$, $O^p(C_L(Q)) \le O^p(C_G(R/Q) \cap C_G(Q)) = R$, so $X \le R$ and hence $X \trianglelefteq L$. Now $[T, X] \le C_{T \cap R}(Q) = Z(Q)$, so $[L, X] \le Z(Q)$. Then the strong closure of $T$ together with (*) and a transfer argument show that $T \in \mathrm{Syl}_p(L)$.

As $T \in \mathrm{Syl}_p(L)$, $G = LN_G(T)$ by a Frattini argument. As $\mathcal{E} \trianglelefteq \mathcal{F}$, $\Sigma = \mathrm{Aut}_{\mathcal{E}}(T) \trianglelefteq \mathrm{Aut}_{\mathcal{F}}(T) = \mathrm{Aut}_G(T)$ and $\mathrm{Inn}(T) \in \mathrm{Syl}_p(\Sigma)$. Further by (N1), for each $p'$-element $\sigma$ in $\Sigma$, there is a $p'$-element $g \in N_G(T)$ with $c_{g|T} = \sigma$ and $[C_S(T), g] \le Z(T)$. As $\Sigma \trianglelefteq \mathrm{Aut}_G(T)$, $[R, g] \le TC_S(T)$, so as $[C_S(T), g] \le Z(T)$ and $g$ is a $p'$-element, $[R, g] \le T \cap R = Q$. Thus $g \in K = C_G(R/Q) \cap N_G(T)$. Let $K_0$ be the preimage in $K$ of $\Sigma$ and $J = O^p(K)$. Then $J \trianglelefteq N_G(T)$, so $H = LJ \trianglelefteq G$. Further argument shows that $H$ is a model for $\mathcal{E}$. The uniqueness of $H$, subject to the constraints in (b), is essentially a consequence of I.6.5.  $\square$

**Theorem 7.5**  *Assume $G$ is a finite group, $S \in \mathrm{Syl}_p(G)$, and $\mathcal{F} = \mathcal{F}_S(G)$. Assume $C_G(O_p(G)) \le O_p(G)$ and $\mathcal{E} \trianglelefteq \mathcal{F}$ on $T \le S$. Then there exists a unique normal subgroup $H$ of $G$ such that $T \in \mathrm{Syl}_p(H)$ and $\mathcal{E} = \mathcal{F}_T(H)$.*

*Proof.* As $\mathcal{F} = \mathcal{F}_S(G)$, $\mathcal{F}$ is saturated and $G \in \mathcal{G}(\mathcal{F})$. Let $R = O_p(\mathcal{F})$. As $\mathcal{F} = \mathcal{F}_S(G)$, $R = O_p(G)$. Thus $C_S(R) \le R$, and hence $\mathcal{F}$ is a constrained saturated fusion system. Now 7.4.2 completes the proof.  $\square$

**Remark 7.6**  Observe that Theorem 7.5 shows that our definition of "normal subsystem" has the desirable property discussed in Example 7.3. On the other hand our next example shows that one cannot remove the condition in 7.5 that $C_G(O_p(G)) \le O_p(G)$.

**Example 7.7**  Here is an example which shows the hypothesis in 7.5 that $F^*(G) = O_p(G)$ cannot be removed. Assume $G$ is a finite group, $S \in \mathrm{Syl}_2(G)$, and $T$ is a strongly closed abelian subgroup of $S$ contained in $Z(S)$. Then $M = N_G(T)$ controls fusion in $T$. Set $\mathcal{E} = \mathcal{F}_T(T)$. Then $\mathcal{E} \le \mathcal{F}$ and as $T$ is abelian, $\mathcal{E}$ consists only of inclusion maps. Therefore $\mathcal{E}$ is $\mathcal{F}$-invariant. Let $P \le T$. Then $\mathrm{Aut}_{\mathcal{E}}(P) = 1$, so trivially $\mathcal{E}$ is saturated and (N1) is satisfied, and hence $\mathcal{E} \trianglelefteq \mathcal{F}$.

However there are plenty of examples of this set up in which $T$ is not normal in $G$. For example take $G$ to be simple and $S = T$ an abelian Sylow group of $G$. To get examples where $T$ is proper in $S$, take $G$ simple and $S$ cyclic with $|S| > p$, and $T = \Omega_1(S)$. Or take $p = 2$, $G = Sz(2^n)$ or $U_3(2^n)$, and $T = \Omega_1(S)$.

On the other hand these examples are a bit deceiving, since $\mathcal{F}$ is isomorphic to $\mathcal{F}_S(M)$, and $T$ *is* normal in $M$. Here is a different sort of example:

**Example 7.8**  Take $\hat{G}$ to be the extension of the natural module $V$ for $G = Sz(2^n)$ or $U_3(2^n)$, take $T$ as in Example 7.7, and set $\hat{T} = TV$. Then $\hat{T}$ is strongly closed in $\hat{S} = SV$ with respect to $\hat{G}$, but is not Sylow in any normal subgroup of $\hat{G}$. Thus by 7.5, there is no normal subsystem of $\hat{\mathcal{F}} = \mathcal{F}_{\hat{S}}(\hat{G})$ on $\hat{T}$. This example shows that even when $\hat{\mathcal{F}}$ is a saturated constrained fusion system, there can be a strongly closed subgroup $\hat{T}$ of $\hat{S}$ in $\hat{\mathcal{F}}$ such that there exists no normal subsystem of $\hat{\mathcal{F}}$ on $\hat{T}$.

Further $V = O_2(\hat{G})$, so $V \trianglelefteq \hat{\mathcal{F}}$ by 7.2. Moreover $\hat{F}/V \cong \mathcal{F}_S(G)$ by 5.6, and, as we saw above, $T \trianglelefteq \mathcal{F}$. However the preimage $\hat{T}$ of $T$ under $\Theta_{\hat{\mathcal{F}},V} \colon \hat{\mathcal{F}} \to \hat{\mathcal{F}}/V$ is not normal in $\hat{\mathcal{F}}$, and indeed by 7.5, there is no normal subsystem of $\hat{\mathcal{F}}$ on $\hat{T}$. This shows that the standard result in group theory fails for morphisms of saturated fusion systems: A normal subsystem of $\hat{\mathcal{F}}/V$ need not lift under $\Theta_{\hat{\mathcal{F}},V}$ to a normal subsystem of $\hat{\mathcal{F}}$.

## 8. INVARIANT MAPS AND NORMAL MAPS

In this section $\mathcal{F}$ is a saturated fusion system over the finite $p$-group $S$, and $T$ is a subgroup of $S$ strongly closed in $S$ with respect to $\mathcal{F}$.

In order to work with the normal subsystems defined in the previous section, we need effective conditions to verify when a subsystem of $\mathcal{F}$ on $T$ is normal, and to produce normal subsystems on $T$. Moreover in most situations, these conditions should be *local*; that is we should be able to check them in local subsystems, and indeed even in local constrained subsystems.

This section contains a brief overview of some such conditions.

Recall from section 6 that $\mathcal{F}_T^f$ denotes the set of nontrivial subgroups $P$ of $T$ with $P$ fully normalized in $\mathcal{F}$. Write $\mathcal{F}_T^{fc}$ for the set of $U \in \mathcal{F}_T^f$ such that $C_T(U) \leq U$, and set

$$\mathcal{F}_T^c = \bigcup_{U \in \mathcal{F}_T^{fc}} U^{\mathcal{F}}.$$

By Exercise 8.1,

$$\mathcal{F}_T^c = \{P \leq T \colon C_T(P\phi) \leq P\phi \text{ for all } \phi \in \hom_{\mathcal{F}}(P, S)\}.$$

**Definition 8.1**   Define a $\mathcal{F}$-*invariant map* on $T$ to be a function $A$ on the set of subgroups of $T$ such that:

(IM1)   For each $P \leq T$ and $\alpha \in \operatorname{Hom}_{\mathcal{F}}(P, S)$, $A(P)^{\alpha} = A(P\alpha) \leq \operatorname{Aut}_{\mathcal{F}}(P\alpha)$.

(IM2)   For each $P \in \mathcal{F}_T^f$, $\operatorname{Aut}_T(P) \leq A(P)$.

Given an $\mathcal{F}$-invariant map $A$ on $T$, set $\mathfrak{E}(A) = \langle A(P){:}P \leq T \rangle$, regarded as a fusion system on $T$.

**Example 8.2**   Pick a set $\mathcal{U}$ of representatives in $\mathcal{F}_T^f$ for the orbits of $\mathcal{F}$ on the subgroups of $T$. For $U \in \mathcal{U}$, pick a normal subgroup $A(U)$ of $\operatorname{Aut}_{\mathcal{F}}(U)$ containing $\operatorname{Aut}_T(U)$. For $\alpha \in \operatorname{Hom}_{\mathcal{F}}(U, T)$, define $A(U\alpha) = A(U)^{\alpha}$. As $A(U) \trianglelefteq \operatorname{Aut}_{\mathcal{F}}(U)$, the function $A$ is well defined, and by construction $A$ is a $\mathcal{F}$-invariant map on $T$. Thus each such map on $\mathcal{U}$ extends uniquely to an $\mathcal{F}$-invariant map $A$, and it is easy to check that each invariant map can be obtained via this construction.

Here is a special case: Define $B(U) = \langle \operatorname{Aut}_T(U)^{\operatorname{Aut}_{\mathcal{F}}(U)} \rangle$ as in the proof of 7.4.b. Then $B$ defines an invariant map.

**Lemma 8.3**   *Let $A$ be a $\mathcal{F}$-invariant map on $T$. Then $\mathfrak{E}(A)$ is a $\mathcal{F}$-invariant subsystem of $\mathcal{F}$ on $T$.*

*Proof.* See 5.5 in [A5]. $\qquad\qquad\qquad\qquad\qquad\qquad\qquad\qquad\qquad\square$

**Example 8.4**   Pick $\mathcal{U}$ as in Example 8.2, let $\mathcal{V} = \mathcal{U} \cap \mathcal{F}_T^{fc}$ and suppose $A$ is a map on $\mathcal{V}$ satisfying $\operatorname{Aut}_T(V) \leq A(V) \trianglelefteq \operatorname{Aut}_{\mathcal{F}}(V)$ for each $V \in \mathcal{V}$. Then as in 8.2, we can extend $A$ to a map on $\mathcal{F}_T^c$ via $A(V\alpha) = A(V)^{\alpha}$. Then for $U \in \mathcal{U}$, set $A(U) = \operatorname{Aut}_{A(UC_T(U))}(U)$. One can show that $A$ satisfies the conditions of 8.2, and hence defines an invariant map.

**Definition 8.5**   Define an $\mathcal{F}$-invariant map $A$ to be *constricted* if $A(U) = \operatorname{Aut}_{A(UC_T(U))}(U)$ for each $U \in \mathcal{F}_T^f$. Thus the maps $A$ constructed in Example 8.4 are constricted.

Our object is to use invariant maps $A$ to construct normal subsystems of $\mathcal{F}$. The corresponding normal subsystem should be $\mathfrak{E}(A)$, which should satisfy $\operatorname{Aut}_{\mathfrak{E}(A)}(U) = A(U)$ for $U \leq T$. A necessary condition for $\mathfrak{E}(A)$ to be saturated is that $A$ be constricted. But still more conditions are required. This leads to the following definition:

**Definition 8.6**   Define a constricted $\mathcal{F}$-invariant map $A$ on $T$ to be *normal* if for each $U \in \mathcal{F}_T^{fc}$:

(SA1)   $\operatorname{Aut}_T(U) \in \operatorname{Syl}_p(A(U))$.

(SA2)   For each $U \leq P \leq Q = N_T(U)$ with $P$ fully normalized in $N_{\mathcal{F}}(UC_S(U))$, $\operatorname{Aut}_{A(P)}(U) = N_{A(U)}(\operatorname{Aut}_P(U))$.

(SA3) Each $\phi \in N_{A(Q)}(U)$ extends to $\hat{\phi} \in \mathrm{Aut}_{\mathcal{F}}(QC_S(Q))$ with $[C_S(Q), \hat{\phi}] \leq Z(Q)$.

Conditions (SA1) and (SA2) are necessary to insure that $\mathfrak{E}(A)$ is saturated. Of course condition (SA3) is included to insure that the pair $\mathcal{F}$, $\mathfrak{E}(A)$ satisfies condition (N1) in the definition of normal subsystem. In any event one can show:

**Theorem 8.7** *Let $A$ be a normal map on $T$. Then $\mathfrak{E}(A)$ is a normal subsystem of $\mathcal{F}$ on $T$ such that for each $P \leq T$, $\mathrm{Aut}_{\mathcal{E}(A)}(P) = A(P)$.*

Theorem 8.7 follows from a more general result which we state in a moment.

If $U \in \mathcal{F}_T^{fc}$ then by Example 4.6, $\mathcal{D}(U) = N_{\mathcal{F}}(UC_S(U))$ is saturated and constrained, so there exists $G(U) \in \mathcal{G}(\mathcal{D}(U))$; that is $G(U)$ is a finite group with $N_S(U) \in \mathrm{Syl}_p(G(U))$ and $\mathcal{D}(U) = \mathcal{F}_{N_S(U)}(G(U))$.

**Theorem 8.8** *Let $\mathcal{E}$ be a subsystem of $\mathcal{F}$ on $T$. Then the following are equivalent:*

(a) $\mathcal{E} \trianglelefteq \mathcal{F}$.

(b) *There exists a normal map $A'$ on $T$ such that $\mathcal{E} = \mathfrak{E}(A')$.*

(c) *For each $U \in \mathcal{F}_T^{fc}$ there exists a normal subgroup $H(U)$ of $G(U)$ such that $N_T(U) \in \mathrm{Syl}_p(H(U))$ and*

    (i) *for each $P \in \mathcal{D}(U)^f$ with $U \leq P$, and for each $\alpha \in \mathrm{Hom}_{\mathcal{F}}(N_S(P), S)$ with $P\alpha \in \mathcal{F}_T^{fc}$, $\mathrm{Aut}_{H(P\alpha)}(U\alpha) = \mathrm{Aut}_{N_{H(U)}(P)}(U)^{\alpha}$, and*

    (ii) $\mathcal{E} = \langle A(U\varphi) \colon U \in \mathcal{F}_T^{fc}, \ \varphi \in \mathrm{Aut}_{\mathcal{F}}(T) \rangle$, *where $A$ is the constricted invariant map defined by $A(U) = \mathrm{Aut}_{H(U)}(U)$ as in 8.4.*

*Proof.* This is 7.18 in [A5]. $\qquad\qquad\qquad\qquad\qquad\qquad\qquad\qquad\qquad\square$

**Remark 8.9** One interpretation of 8.8 is the following: Given a strongly closed subgroup $T$ of $S$, we look for normal subsystems $\mathcal{E}$ of $\mathcal{F}$ on $T$. To do so we consider the members $U$ of $\mathcal{F}_T^{fc}$, the associated constrained saturated systems $\mathcal{D}(U) = N_{\mathcal{F}}(UC_S(U))$, and their models $G(U)$. We look for a set $\{H(U) \colon U \in \mathcal{F}_T^{fc}\}$ of subgroups $H(U) \trianglelefteq G(U)$ with $N_T(U) \in \mathrm{Syl}_p(H(U))$, satisfying the compatibility conditions in 8.8.c.i. Given such a collection, $\mathcal{E}$ is essentially the subsystem of $\mathcal{F}$ generated by the systems $\mathrm{Aut}_{H(U)}(U)$, and $\mathrm{Aut}_{\mathcal{E}}(U)$ is $\mathrm{Aut}_H(U)$.

### Exercises for Section 8

**8.1** *Prove $\mathcal{F}_T^c = \{P \leq T \colon C_T(P\phi) \leq P\phi \text{ for all } \phi \in \mathrm{hom}_{\mathcal{F}}(P, S)\}$.*

## 9. Theorems on Normal Subsystems

In this section $\mathcal{F}$ is a saturated fusion system on a finite $p$-group $S$.

We list various results about normal subsystems, and extensions of theorems in the local theory of finite groups to the setting of saturated fusion systems. These result are proved in [A6]. The proofs use results from the previous section. Most of the proofs are moderately difficult.

In Example 7.3, we saw that the intersection $\mathcal{E}_1 \cap \mathcal{E}_2$ of normal subsystems $\mathcal{E}_i$ of $\mathcal{F}$ need not be normal in $\mathcal{F}$. However this is not a serious problem, since it develops that $\mathcal{E}_1 \cap \mathcal{E}_2$ is not quite the right object to consider. Rather:

**Theorem 9.1**   *Let $\mathcal{E}_i$ be a normal subsystem of $\mathcal{F}$ on a subgroup $T_i$ of $S$, for $i = 1, 2$. Then there exists a normal subsystem $\mathcal{E}_1 \wedge \mathcal{E}_2$ of $\mathcal{F}$ on $T_1 \cap T_2$ normal in $\mathcal{E}_1$ and $\mathcal{E}_2$. Moreover $\mathcal{E}_1 \wedge \mathcal{E}_2$ is the largest normal subsystem of $\mathcal{F}$ on $T_1 \cap T_2$ normal in $\mathcal{E}_1$ and $\mathcal{E}_2$.*

The next result may already appear in the literature in the special case where $\mathcal{F} = \mathcal{F}_S(G)$ is the system of a finite group $G$ on a Sylow $p$-subgroup $S$ of $G$, but we are not aware of such a result.

**Theorem 9.2**   *Assume $T_i$, $i = 1, 2$, are strongly closed in $S$ with respect to $\mathcal{F}$. Then $T_1 T_2$ is strongly closed in $S$ with respect to $\mathcal{F}$.*

If $H_1$ and $H_2$ are normal subgroups of a group $G$, then $H_1 H_2 \trianglelefteq G$. The analogue of this result may hold for saturated fusion systems, but in [A6] there is a proof only in a very special case; this case suffices for our most immediate applications.

That special case is treated in Theorem 9.3; it bears some resemblance to earlier theorems about finite groups due to Gorenstein-Harris in [GH], and Goldschmidt in [Gd2]. Namely in each of these papers, the authors prove the existence of certain normal subgroups of a group $G$ under the hypothesis that for $S \in \mathrm{Syl}_2(G)$, there are subgroups $T_i$ of $S$ for $i = 1, 2$, such that $[T_1, T_2] = 1$ and $T_i$ is strongly closed in $S$ with respect to $G$.

**Theorem 9.3**   *Assume $\mathcal{E}_i \trianglelefteq \mathcal{F}$ on $T_i$ for $i = 1, 2$, and that $[T_1, T_2] = 1$. Then there exists a normal subsystem $\mathcal{E}_1 \mathcal{E}_2$ of $\mathcal{F}$ on $T_1 T_2$. Further if $T_1 \cap T_2 \le Z(\mathcal{E}_i)$ for $i = 1, 2$, then $\mathcal{E}_1 \mathcal{E}_2$ is a central product of $\mathcal{E}_1$ and $\mathcal{E}_2$.*

Definition I.6.5 contains the definition of the *direct product* $\mathcal{F}_1 \times \mathcal{F}_2$ of fusion systems $\mathcal{F}_1$ and $\mathcal{F}_2$. A *central product* $\mathcal{F}_1 \times_Z \mathcal{F}_2$ is a factor system $(\mathcal{F}_1 \times \mathcal{F}_1)/Z$, for some $Z \le Z(\mathcal{F}_1 \times \mathcal{F}_2)$ such that $Z \cap \mathcal{F}_i = 1$ for $i = 1, 2$.

Let $\mathcal{E} \trianglelefteq \mathcal{F}$. In [A6] the *centralizer* in $\mathcal{F}$ of $\mathcal{E}$ is defined, and is denoted by $C_{\mathcal{F}}(\mathcal{E})$. Unfortunately the definition is rather complicated, so we will not reproduce it here. However it is proved that:

**Theorem 9.4** *If $\mathcal{E} \trianglelefteq \mathcal{F}$ is a fusion system on $T$, then the set of subgroups $Y$ of $C_S(T)$ such that $\mathcal{E} \leq C_{\mathcal{F}}(Y)$ has a largest member $C_S(\mathcal{E})$, $C_{\mathcal{F}}(\mathcal{E})$ is a normal subsystem of $\mathcal{F}$ on $C_S(\mathcal{E})$, and for $X \in C_{\mathcal{F}}(\mathcal{E})^{fc}$, $\mathrm{Aut}_{C_{\mathcal{F}}(\mathcal{E})}(X) = O^p(\mathrm{Aut}_{C_{\mathcal{F}}(T)}(X))\mathrm{Aut}_{C_S(\mathcal{E})}(X)$.*

Recall the *hyperfocal subgroup* $\mathfrak{hyp}(\mathcal{F})$ is defined in Definition I.7.1, and from Theorem I.7.4, there is a unique saturated subsystem $O^p(\mathcal{F})$ of $\mathcal{F}$ on $\mathfrak{hyp}(\mathcal{F})$ such that $O^p(\mathrm{Aut}_{\mathcal{F}}(P)) \leq \mathrm{Aut}_{O^p(\mathcal{F})}(P)$ for each $P \leq \mathfrak{hyp}(\mathcal{F})$. Moreover $O^p(\mathcal{F}) \trianglelefteq \mathcal{F}$. For example if $\mathcal{F} = \mathcal{F}_S(G)$ for some finite group $G$ with $S \in \mathrm{Syl}_p(G)$, then $O^p(\mathcal{F}) = \mathcal{F}_{S \cap O^p(G)}(O^p(G))$.

In [A6] there is a different proof of the existence of $O^p(\mathcal{F})$. Very roughly the proof goes as follows: First prove $T = \mathfrak{hyp}(\mathcal{F})$ is strongly closed in $S$. Then for $U \in \mathcal{F}_T^{fc}$, let $G(U)$ be the model for $\mathcal{D}(U) = N_{\mathcal{F}}(UC_S(U))$, $H^p(U) = O^p(G(U))N_T(U)$, and $A^p(U) = \mathrm{Aut}_{H^p(U)}(U)$. Let $A^p$ be the constricted $\mathcal{F}$-invariant map determined by the groups $A^p(U)$ as in Example 8.4, and set $O^p(\mathcal{F}) = \mathfrak{E}(A^p)$. Verify the compatibility conditions in 8.8.c.i, and then appeal to Theorem 8.8 to conclude $O^p(\mathcal{F}) \trianglelefteq \mathcal{F}$.

In addition it is shown in [A6] that:

**Theorem 9.5** *Let $\mathcal{E} \trianglelefteq \mathcal{F}$ on $T$, and $T \leq R \leq S$. Then there exists a unique saturated fusion subsystem $R\mathcal{E}$ of $\mathcal{F}$ on $R$ such that $O^p(R\mathcal{E}) = O^p(\mathcal{E})$. In particular $\mathcal{F} = SO^p(\mathcal{F})$.*

Recall from Definition I.6.1 that $\mathcal{F}$ is *simple* if $\mathcal{F}$ has no proper nontrivial normal subsystems. Define $\mathcal{F}$ to be *quasisimple* if $\mathcal{F} = O^p(\mathcal{F})$ and $\mathcal{F}/Z(\mathcal{F})$ is simple. Define the *components* of $\mathcal{F}$ to be the subnormal quasisimple subsystems of $\mathcal{F}$. Recall $O_p(\mathcal{F})$ is the largest subgroup of $S$ normal in $\mathcal{F}$. We can view $R = O_p(\mathcal{F})$ as the normal subsystem $\mathcal{F}_R(R)$ of $\mathcal{F}$.

Define $E(\mathcal{F})$ to be the normal subsystem of $\mathcal{F}$ generated by the set $\mathrm{Comp}(\mathcal{F})$ of components of $\mathcal{F}$ (which exists by Theorem 9.1), and set $F^*(\mathcal{F}) = E(\mathcal{F})O_p(\mathcal{F})$. We call $F^*(\mathcal{F})$ the *generalized Fitting subsystem* of $\mathcal{F}$. Of course all of these notions are similar to the analogous notions for groups (cf. Definition A.11), and the following theorem is a list of the standard properties of the generalized Fitting subgroup of a finite group, only stated for fusion systems.

**Theorem 9.6** (a) *$E(\mathcal{F})$ is a characteristic subsystem of $\mathcal{F}$.*

(b) *$E(\mathcal{F})$ is the central product of the components of $\mathcal{F}$.*

(c) *$O_p(\mathcal{F})$ centralizes $E(\mathcal{F})$.*

(d) *$C_{\mathcal{F}}(F^*(\mathcal{F})) = Z(F^*(\mathcal{F}))$.*

Next [A6] establishes a version of the Gorenstein-Walter theorem on so called *L-balance* [GW2]:

**Theorem 9.7**  *For each fully normalized subgroup $X$ of $S$, $E(N_{\mathcal{F}}(X)) \leq E(\mathcal{F})$.*

It is worth noting that the proof of $L$-balance for a group $G$ requires that the components of $G/O_{p'}(G)$ satisfy the Schreier conjecture, or when $p = 2$, a weak version of the Schreier conjecture due to Glauberman. Our proof of Theorem 9.7 requires no deep results. The theorem does not quite imply $L$-balance for groups, since there is not a nice one to one correspondence between quasisimple groups and quasisimple fusion systems. The proof can be translated into the language of groups, but even there at some point one seems to need some result like Theorem A of Goldschmidt in [Gd2], which is only proved for $p = 2$ without the classification. Still, something is going on here, which suggests that in studying fusion systems, one may be lead to new theorems or better proofs of old theorems about finite groups.

Finally by Theorem 9.1 there is a smallest normal subsystem of $\mathcal{F}$ on $S$. Denote this subsystem by $O^{p'}(\mathcal{F})$. For example if $\mathcal{F} = \mathcal{F}_S(G)$ for some finite group $G$ with $F^*(G) = O_p(G)$ and $S \in \mathrm{Syl}_p(G)$, then $O^{p'}(\mathcal{F}) = \mathcal{F}_S(O^{p'}(G))$. The system $O^{p'}(\mathcal{F})$ is the smallest subsystem of index prime to $p$, in the sense of Definition I.7.3, and as such is described in Theorems I.7.7 and III.4.19.

There is a slightly different approach to these notions in [A6]. For $P \leq S$, define $B(P) = O^{p'}(\mathrm{Aut}_{\mathcal{F}}(P))$. Then $B$ is an $\mathcal{F}$-invariant map, as defined in Section 8, so by 8.3, $\mathcal{B} = \mathfrak{E}(B)$ is an $\mathcal{F}$-invariant subsystem of $\mathcal{F}$. Essentially as in [5a2], define

$$\mathrm{Aut}^0_{\mathcal{F}}(S) = \langle \alpha \in \mathrm{Aut}_{\mathcal{F}}(S) \colon \alpha_{|P} \in \mathrm{Hom}_{\mathcal{B}}(P, S) \text{ for some } P \in \mathcal{F}^c \rangle.$$

As $\mathcal{B}$ is $\mathcal{F}$-invariant, $\mathrm{Aut}^0_{\mathcal{F}}(S) \trianglelefteq \mathrm{Aut}_{\mathcal{F}}(S)$, so we can define

$$\Gamma_{p'}(\mathcal{F}) = \mathrm{Aut}_{\mathcal{F}}(S)/\mathrm{Aut}^0_{\mathcal{F}}(S).$$

Then Theorem 8 in [A6] establishes the following version of Theorems I.7.7 and III.4.19:

**Theorem 9.8**   (a)  $\pi_1(\mathcal{F}^c) \cong \Gamma_{p'}(\mathcal{F})$.

(b)  *The map $\mathcal{E} \mapsto \mathrm{Aut}_{\mathcal{E}}(S)/\mathrm{Aut}^0_{\mathcal{F}}(S)$ is a bijection between the set of normal subsystems of $\mathcal{F}$ on $S$ and the set of normal subgroups of $\Gamma_{p'}(\mathcal{F})$.*

(c)  $\mathcal{F} = O^{p'}(\mathcal{F})$ *iff* $\mathrm{Aut}_{\mathcal{F}}(S) = \mathrm{Aut}^0_{\mathcal{F}}(S)$ *iff* $\pi_1(\mathcal{F}^c) = 1$.

(d)  *$\mathcal{F}$ is simple iff the following hold:*

    (i)  *For each normal subsystem $\mathcal{D}$ of $\mathcal{F}$ on a subgroup $D$ of $S$, we have $D = S$.*

    (ii)  $\mathrm{Aut}_{\mathcal{F}}(S) = \mathrm{Aut}^0_{\mathcal{F}}(S)$.

Recall that parts (a)-(c) of Theorem 9.8 are proved in Theorems I.7.7 and III.4.19. Observe that (d) is an easy consequence of (b), since by (b), $\mathcal{F}$ has no proper normal subsystems on $S$ iff (di) holds.

## 10. COMPOSITION SERIES

In this section $\mathcal{F}$ is a saturated fusion system over the finite $p$-group $S$.

We wish to define and explore analogues of the notions of "composition series" and "composition factors" from finite group theory in the category of saturated fusion systems. In the category of groups, a composition series is a maximal subnormal series and the simple factor groups from such a series are independent of the series and defined to be the composition factors of the group. Moreover given a finite group $G$, its composition series $1 = G_n \vartriangleleft \cdots \vartriangleleft G_0 = G$ can all be obtained from the following algorithm: Pick a maximal normal subgroup $G_1$ of $G$; then pick a maximal normal subgroup $G_2$ of $G_1$; and so on.

Unfortunately this approach does not quite work even for constrained fusion systems, as the following example shows:

**Example 10.1** Let $L$ be a finite simple group with an abelian Sylow $p$-subgroup $P$ of order $p^e > p$. Let $V$ be a faithful irreducible $\mathbf{F}_p L$-module of dimension $d > 1$. Form the semidirect product $G = LV$ and observe $S = PV \in \mathrm{Syl}_p(G)$. Let $\mathcal{F} = \mathcal{F}_S(G)$. Then $F^*(G) = V$, so $\mathcal{F}$ is constrained and from 7.2 and 7.5 the map $H \mapsto \mathcal{F}_S(H)$ is a bijection between the normal subgroups of $G$ and the normal subsystems of $\mathcal{F}$. As $L$ is simple and irreducible on $V$, $V$ is the unique nontrivial proper normal subgroup of $G$, so $\mathcal{V} = \mathcal{F}_V(V)$ is the unique nontrivial proper normal subsystem of $\mathcal{F}$. Then choosing a chain $1 = V_0 < V_1 < \cdots < V_d = V$ of subspaces with $\dim(V_i) = i$, we get a subnormal series $1 = \mathcal{V}_0 < \cdots < \mathcal{V}_d = \mathcal{V}$, where $\mathcal{V}_i = \mathcal{F}_{V_i}(V_i)$, with simple factors of order $p$; that is each factor is of the form $\mathcal{F}_U(U)$, where $U$ is a group of order $p$. Moreover, adjoining $\mathcal{F}$ to this series, we have a maximal subnormal series for $\mathcal{F}$ of length $d + 1$.

However while $\mathcal{V}$ is a maximal normal subsystem of $\mathcal{F}$, $\mathcal{F}/\mathcal{V}$ is *not* simple. Namely as $S/V \cong P$ is abelian, $\mathcal{F}/\mathcal{V} \cong \mathcal{F}_P(N_L(P))$ has a normal Sylow group $P$. Thus we obtain a series $1 = P_0 < \cdots < P_e = P$ of subgroups with $|P_i| = p^i$, giving rise to the subnormal series $1 = \mathcal{P}_0 < \cdots < \mathcal{P}_e$ of subsystems $\mathcal{P}_i = \mathcal{F}_{P_i}(P_i)$, again with factors of order $p$. As defined below in Definition 10.6, this construction gives rise to a "supranormal series" for $\mathcal{F}$, and the factors in this series are the composition factors for $\mathcal{F}$. So in this example, $\mathcal{F}$ has $e + d$ composition factors, each of order $p$.

We now begin to develop a theory of composition series and composition factors for saturated fusion systems, adjusted to allow for systems like those in the previous example and based on a more complicated algorithm than the one for groups: If $O^{p'}(\mathcal{F})$ is not simple, pick a nontrivial proper normal subsystem $\mathcal{E}$ of $O^{p'}(\mathcal{F})$ and, proceeding recursively, continue by applying this process to $\mathcal{E}$ and $O^{p'}(\mathcal{F})/\mathcal{E}$. Notice the difficulties arise from the fact that if $T$ is strongly closed in $S$ with respect to $\mathcal{F}$, and $\mathcal{D}$ is the preimage in $\mathcal{F}$ of a normal subsystem of $\mathcal{F}/T$, it need not be the case that $\mathcal{D}$ is normal in $\mathcal{F}$.

**Definition 10.2**　By 9.1, for each subgroup $T$ of $S$ such that $T$ is Sylow in a normal subsystem of $\mathcal{F}$, there is a smallest normal subsystem $[T^{\mathcal{F}}]$ of $\mathcal{F}$ on $T$. For example from Exercise 10.1, $[S^{\mathcal{F}}]$ is the system $O^{p'}(\mathcal{F})$.

**Example 10.3**　Let $G$ be a finite group with $S \in \mathrm{Syl}_p(G)$, $C_G(O_p(G)) \leq O_p(G)$, and $\mathcal{F} = \mathcal{F}_S(G)$. Then by 7.2 and 7.5 the map $H \mapsto \mathcal{F}_{S \cap H}(H)$ is a bijection between the normal subgroups $H$ of $G$ and the normal subsystems of $\mathcal{F}$. Therefore $T \leq S$ is Sylow in a normal subsystem of $\mathcal{F}$ iff $T$ is Sylow in some normal subgroup of $G$, and in that event $[T^{\mathcal{F}}]$ is $\mathcal{F}_T(\langle T^G \rangle)$. In particular $O^{p'}(\mathcal{F}) = \mathcal{F}_S(O^{p'}(G))$, since $O^{p'}(G)$ is the smallest normal subgroup $H$ of $G$ such that $G/H$ is a $p'$-group; equivalently $O^{p'}(G) = \langle S^G \rangle$.

**Definition 10.4**　Set
$$D = [0,1] \cap \mathbb{Z}[\tfrac{1}{2}] = \left\{ \tfrac{k}{2^n} \,\middle|\, n \geq 0,\ 0 \leq k \leq 2^n \right\}.$$

(a)　Define $l, u \colon D \longrightarrow D$ by $l(0) = l(1) = 0$, $u(0) = u(1) = 1$, and when $n > 0$ and $0 < k < 2^n$ is odd:
$$l\left(\tfrac{k}{2^n}\right) = \tfrac{k-1}{2^n} \quad \text{and} \quad u\left(\tfrac{k}{2^n}\right) = \tfrac{k+1}{2^n}$$

(b)　Define $h \colon D \longrightarrow \mathbb{N}$ by
$$h(0) = h(1) = 0, \qquad h\left(\tfrac{k}{2^n}\right) = n \quad \text{for all } n \geq 1,\ k \text{ odd}.$$

(c)　A finite subset $X \subseteq D$ is *admissible* if $0, 1 \in X$, $l(X) \subseteq X$, and $u(X) \subseteq X$.

(d)　Define bijections $\iota \colon D \longrightarrow D \cap [0, \tfrac{1}{2}]$ and $\kappa \colon D \longrightarrow D \cap [\tfrac{1}{2}, 1]$ by $\iota(a) = \tfrac{a}{2}$ and $\kappa(a) = \tfrac{a+1}{2}$.

(e)　For any $X \subseteq D$, set
$$X' = \iota^{-1}(X) = \{2a \mid a \in X,\ a \leq \tfrac{1}{2}\} \quad \text{and}$$
$$X'' = \kappa^{-1}(X) = \{2a - 1 \mid a \in X,\ a \geq \tfrac{1}{2}\}.$$

For $a \in D$, $h(a)$ will be called the *height* of $a$. If $0 < a < 1$, then $l(a) < a < u(a)$, and $l(a)$ and $u(a)$ are the closest elements to $a$ in $D$ with smaller height.

**Lemma 10.5**　*Let $X \subseteq D$ be a finite subset. Then $X$ is admissible if and only if either $X = \{0, 1\}$, or $\tfrac{1}{2} \in X$ and $X'$ and $X''$ are both admissible.*

*Proof.* Assume $X \neq \{0, 1\}$. If $0 \notin X$, then $X$ is not admissible and $0 \notin X'$ so $X'$ is not admissible. Similarly if $1 \notin X$, then neither $X$ nor $X''$ is admissible. So we can assume $\{0, 1\} \subsetneq X$.

Fix $a \in X$, $0 < a < 1$, such that $n = h(a)$ is minimal. If $l(a), u(a) \in X$, then since they have height less than $n$, we must have $l(a) = a - \frac{1}{2^n} = 0$ and $u(a) = a + \frac{1}{2^n} = 1$, and hence $a = \frac{1}{2}$. Thus if $X$ is admissible then $\frac{1}{2} \in X$. So we can also assume $\frac{1}{2} \in X$.

Let $X_0 = X \cap [0, \frac{1}{2}]$ and $X_1 = X \cap [\frac{1}{2}, 1]$. By Exercise 10.2 the bijections $\iota \colon X' \to X_0$ and $\kappa \colon X'' \to X_1$ commute with $l$ and $u$. Thus $l(X') \subseteq X'$ if and only $l(X_0) \subseteq X$ and $u(X') \subseteq X'$ if and only if $u(X_0) \subseteq X$, and similarly for $X''$. Thus $X$ is admissible if and only if $X'$ and $X''$ are both admissible. $\qquad\qquad\square$

Let $\mathfrak{Sub}(S)$ denote the set of subgroups of the group $S$.

**Definition 10.6**   We recursively define the set $\mathcal{S} = \mathcal{S}(\mathcal{F})$ of *supranormal series* of $\mathcal{F}$. A member of $\mathcal{S}$ is a pair $(X, \eta)$, where

(a)   $X \subseteq D$ is a finite admissible subset;

(b)   $\eta \colon X \longrightarrow \mathfrak{Sub}(S)$ is a strictly order preserving map;

(c)   $\eta(0) = 1$ and $\eta(1) = S$; and

(d)   if $\frac{1}{2} \in X$, then there is a normal subsystem of $O^{p'}(\mathcal{F})$ over $T \overset{\text{def}}{=} \eta(1/2)$, $(X', \eta')$ is a supranormal series for $[T^{\mathcal{F}}]$, and $(X'', \eta'')$ is a supranormal series for $O^{p'}(\mathcal{F})/T$ where $\eta' = \iota\eta$ and $\eta'' \colon X'' \longrightarrow \mathfrak{Sub}(S/T)$ is defined by $\eta''(a) = \eta(\kappa(a))/T$.

We call $\eta(X)$ the set of *pivots* of $(X, \eta)$ and call $T$ the *primary pivot* of the pair. The *length* of the series is $|X| - 1$. For $(X, \eta)$ and $(Y, \theta)$ in $\mathcal{S}$, define $(X, \eta) \leq (Y, \theta)$ if $\eta \subseteq \theta$, where we regard $\eta$ and $\theta$ as sets of ordered pairs. In particular this implies that $X \subseteq Y$. Define a *composition series* for $\mathcal{F}$ to be a maximal member of the partially ordered set $(\mathcal{S}, \leq)$.

In the remainder of the section let $\lambda = (X, \eta) \in \mathcal{S}(\mathcal{F})$ with $\frac{1}{2} \in X$. Set $T = \eta(\frac{1}{2})$, $\lambda' = (X', \eta')$, and $\lambda'' = (X'', \eta'')$.

**Example 10.7**   In Example 10.1, the only nontrivial proper normal subsystem of $\mathcal{F} = O^{p'}(\mathcal{F})$ is $\mathcal{F}_V(V)$. So when $n > 1$ (that is when $\frac{1}{2} \in X$), $\eta(\frac{1}{2}) = V$ is the primary pivot and $\lambda_{1/2} = \mathcal{F}_V(V)$. One series of maximal length $n = d + e$ is $1 = V_0 < \cdots < V_d < VP_1 < \cdots < VP_e = S$. For this series there are many choices for the admissible set $X$. One choice is to take
$$X = \left\{ 0, \tfrac{1}{2^i}, \tfrac{1}{2} + \tfrac{1}{2^j} \,\middle|\, 1 \leq i \leq d, \ 1 \leq j \leq e \right\},$$
and set $\eta(\frac{1}{2^i}) = V_{d+1-i}$ and $\eta(\frac{1}{2} + \frac{1}{2^j}) = VP_{1+e-j}$.

As the notion of a supranormal series $\lambda = (X, \eta)$ is defined recursively, one can view the members of the admissible set $X$ and the pivots of $\lambda$ as becoming visible over time, as measured by the height $h(a)$ of $a \in X$ and its pivot $\eta(a)$. Most results will be established by induction on the length of the series or on the height of members of $X$ and their pivots. The next few lemmas record some elementary observations useful in such proofs.

**Lemma 10.8** *Assume $a \in D$. Then $l(u(a)) \leq l(a)$; $u(a) \leq u(l(a))$; and at least one of these inequalities is an equality.*

*Proof.* This is clear if $h(a) \leq 1$ (ie. when $a \in \{0, \frac{1}{2}, 1\}$), so we assume $n = h(a) \geq 2$. Set $a = \frac{k}{2^n}$ where $k$ is odd. By Exercise 10.3, $h(l(a)) \leq n - 1 \geq h(u(a))$, so

$$l(u(a)) = a + \tfrac{1}{2^n} - \tfrac{1}{2^{h(u(a))}} \leq a - \tfrac{1}{2^n} = l(a),$$
$$u(l(a)) = a - \tfrac{1}{2^n} + \tfrac{1}{2^{h(l(a))}} \geq a + \tfrac{1}{2^n} = u(a);$$

further by Exercise 10.3, either $h(l(a))$ or $h(u(a))$ is $n - 1$, so one of the inequalities is an equality. $\square$

**Lemma 10.9** *Fix an admissible subset $X$ of $D$, and a consecutive pair of elements $a < b$ in $X$. Then either*

- $X = \{0, 1\}$, $a = 0$, *and* $b = 1$; *or*

- $h(a) > h(b)$, $b = u(a)$, *and* $l(b) < a$; *or*

- $h(a) < h(b)$, $l(b) = a$, *and* $b < u(a)$.

*Proof.* We may assume $X \neq \{0, 1\}$. Thus $a > 0$ or $b < 1$. Set $n = \max\{h(a), h(b)\} > 0$. Write $a = \frac{j}{2^n}$ and $b = \frac{k}{2^n}$ for $j < k$ in $\mathbb{Z}$. If $n = h(a)$, then $j$ is odd, $u(a) = \frac{j+1}{2^n} \in X$, and hence $b = u(a)$. Also, in this case, $h(b) < n$ since $k = j + 1$ is even, so $l(b) = b - \frac{1}{2^{h(b)}} < a$. Similarly, if $n = h(b)$, then $j = k - 1$, $a = l(b)$, $h(a) < n$, and $u(a) > b$. $\square$

Associated to each $a \in X$ and $a \leq b \leq u(a)$ are various $p$-groups $T_a$ and $T_{a,b}$ and fusion systems $\lambda_{l(a),a}$, $\lambda_{l(a),a}$, and $\lambda_{l(a),u(a)}$. We begin with the $p$-groups $T_a = \eta(a)$ and $T_{0,b} = T_b$, and the fusion systems $\lambda_{0,0} = \lambda_{1,1} = 1$, $\lambda_{0,1} = O^{p'}(\mathcal{F})$, $\lambda_{0,\frac{1}{2}} = [T^{\mathcal{F}}]$, and $\lambda_{\frac{1}{2},1} = O^{p'}(\lambda_{0,1}/T_j)$. The next few lemmas introduce groups and systems associated to pairs of elements of higher height, defined recursively.

**Lemma 10.10** *We make the following recursive definition. For $a \in X$, set $T_{0,a} = T_a = \eta(a)$, and for $a \leq b \leq u(a)$ set $T_{a,b} = T_{l(a),b}/T_{l(a),a}$. Then*

(a)  *For $a \leq b \leq u(a)$, $T_{l(a),a} \trianglelefteq T_{l(a),b}$, so $T_{a,b}$ is well defined. Moreover, $T_a \trianglelefteq T_b$ and $T_{a,b} \cong T_b/T_a$.*

(b)  *Let $T'_a$, $T'_{a,b}$ be the groups defined analogously for the series $\lambda' = (X', \eta')$. For $c \le d \le u(c)$ in $X'$, $T_{\iota(c)} = T'_c$ and $T_{\iota(c),\iota(d)} = T'_{c,d}$.*

(c)  *Let $T''_{a,b}$ be the groups defined analogously for the series $\lambda'' = (X'', \eta'')$. For $c < 1$ and $c \le d \le u(c)$ in $X''$, $T_{\kappa(d)} = T''_d$ and $T_{\kappa(c),\kappa(d)} = T''_{c,d}$.*

*Proof.* By definition of the notation, $T'_c = T_{\iota(c)}$ and $T''_d = T_{\kappa(d)}$. If $a = \frac{1}{2}$ then $T_{l(a),a} = T_{0,a} = T_a$ and $T_a \trianglelefteq S$ since $T_{1/2}$ must be strongly closed in $O^{p'}(\mathcal{F})$, hence $T_a \trianglelefteq T_b = T_{0,b}$ and $T_{0,b}/T_{0,a} = T_b/T_a$. So (a) and (c) hold in this case.

Suppose $a = \iota(c) < \frac{1}{2}$ and $b = \iota(d)$. Then $T_{l(a),a} = T'_{l(c),c}$ and $T_{l(a),b} = T'_{l(c),d}$ by induction on $h(a)$ and Exercise 10.2. Also by induction on $|X|$ and Exercise 10.2, $T'_c \trianglelefteq T'_d$ and $T'_{l(c),c} \trianglelefteq T'_{l(c),d}$. Therefore $T_a = T'_c \trianglelefteq T'_d = T_b$ and $T_{a,b} = T_{l(a),b}/T_{l(a),a} = T'_{c,d}$ is well defined. So (b) holds, as does (a) in this case.

Finally assume $a = \kappa(c) > \frac{1}{2}$ where $c \in X''$. By induction on $|X|$, for $r \le s \le u(a)$ in $X''$, $T''_{r,s}$ is well defined and isomorphic to $T''_s/T''_r$, so to complete the proof of (a) it remains to prove (c). We proved (c) when $c = 0$ in paragraph one. If $c > 0$ then by induction on $h(c)$,

$$T''_{c,d} = T''_{l(c),d}/T''_{l(c),c} = T_{\kappa(l(c)),\kappa(d)}/T_{\kappa(l(c)),\kappa(c)}$$
$$= T_{l(\kappa(c)),\kappa(d)}/T_{l(\kappa(c)),\kappa(c)} = T_{\kappa(c),\kappa(d)},$$

completing the proof. $\square$

**Lemma 10.11**  *We make the following recursive definition: Define $\lambda_{0,0} = \lambda_{1,1} = 1$, $\lambda_{0,1} = \lambda_1 = O^{p'}(\mathcal{F})$, $\lambda_{0,1/2} = [T^{\mathcal{F}}]$ and $\lambda_{1/2,1} = O^{p'}(\lambda_1/T)$. If $0 < a < \frac{1}{2}$ where $a = \iota(c)$, set $\lambda_{l(a),a} = \lambda'_{l(c),c}$ and $\lambda_{a,u(a)} = \lambda'_{c,u(c)}$. If $\frac{1}{2} < a < 1$ where $a = \kappa(d)$, set $\lambda_{l(a),a} = \lambda''_{l(d),d}$ and $\lambda_{a,u(a)} = \lambda''_{d,u(d)}$. Then for each $a \in X$, $\lambda_{l(a),u(a)}$, $\lambda_{a,u(a)}$, and $\lambda_{l(a),a}$ are well defined saturated fusion systems over $T_{l(a),u(a)}$, $T_{a,u(a)}$, and $T_{l(a),a}$, respectively, and $\lambda_{l(a),a} = [T_{l(a),a}{}^{\lambda_{l(a),u(a)}}]$ and $\lambda_{a,u(a)} = O^{p'}(\lambda_{l(a),u(a)}/T_{l(a),a})$.*

*Proof.* This is immediate if $a \in \{0, 1\}$, and follows from Definition 10.6.d if $a = \frac{1}{2}$. So we can assume $h(a) \ge 2$. Set $l = l(a)$ and $u = u(a)$. By 10.8, either $l = l(u)$ or $u = u(l)$. Thus $(l, u) = (l(u), u)$ or $(l, u(l))$, so $\lambda_{l,u}$ is well defined and saturated over $T_{l,u}$ by induction on $h(a)$.

Let $T'_{a,b}$ and $T''_{a,b}$ be the groups defined analogously to $T_{a,b}$, for the series $\lambda'$ and $\lambda''$, respectively. Thus by 10.10, $T'_{a,b} = T_{\iota(a),\iota(b)}$ and $T''_{a,b} = T_{\kappa(a),\kappa(b)}$ (when defined). We can assume by induction on $|X|$ that the lemma holds for $\lambda'$ and $\lambda''$. By this assumption and the recursive definition, $\lambda_{l(a),a}$ and $\lambda_{a,u(a)}$ are saturated fusion systems over $T_{l(a),a}$ and $T_{a,u(a)}$ for each $a \in X$.

If $0 < a < \frac{1}{2}$ and $a = \iota(c)$, then

$$\lambda_{a,u(a)} = \lambda'_{c,u(c)} = O^{p'}(\lambda'_{l(c),u(c)}/T'_{l(c),c}) = O^{p'}(\lambda_{l(a),u(a)}/T_{l(a),a})$$

(recall $\iota(l(c)) = l(a)$ and $\iota(u(c)) = u(a)$ by Exercise 10.2); and

$$\lambda_{l(a),a} = \lambda'_{l(c),c} = [T'_{l(c),c}{}^{\lambda'_{l(c),u(c)}}] = [T_{l(a),a}{}^{\lambda_{l(a),u(a)}}].$$

The proof when $\frac{1}{2} < a < 1$ is similar. $\qquad\qquad\square$

**Lemma 10.12**  *Let $a < b$ be consecutive elements in $X$. Then $\lambda_{a,b} = O^{p'}(\lambda_{a,b})$ is a saturated fusion system on $T_{a,b}$.*

*Proof.* By Lemma 10.9, either $b = u(a)$ or $a = l(b)$. The lemma is thus a special case of 10.11. $\qquad\qquad\square$

**Definition 10.13**  For $\lambda = (X, \eta) \in \mathcal{S}$, let

$$F(\lambda) = \big(\lambda_{a,b} \,\big|\, a < b \text{ are consecutive elements of } X.\big)$$

By 10.12, the above definition makes sense and the factors in $F(\lambda)$ are saturated.

**Proposition 10.14**  *For any saturated fusion system $\mathcal{F}$, a series $\lambda \in \mathcal{S}(\mathcal{F})$ is a composition series for $\mathcal{F}$ iff all factors of $\lambda$ are simple.*

*Proof.* We prove this by induction on $|X|$. Assume first $X = \{0,1\}$. If $\mu = (Y, \theta) > (X, \eta) = \lambda$, then $\frac{1}{2} \in Y$ by Lemma 10.5, $1 < \theta(\frac{1}{2}) < \theta(1) = S$, and $\mu_{0,1/2}$ is a normal subsystem of $\lambda_{0,1} = O^{p'}(\mathcal{F})$ over $\theta(\frac{1}{2})$ by Lemma 10.11. Thus $\lambda_{0,1}$ is not simple in this case.

Conversely, assume $\lambda_{0,1} = O^{p'}(\mathcal{F})$ is not simple, and let $\mathcal{E} \trianglelefteq O^{p'}(\mathcal{F})$ be a normal subsystem over a nontrivial proper subgroup $T \trianglelefteq S$. Set $Y = \{0, \frac{1}{2}, 1\}$, and define $\theta$ by setting $\theta(0) = 1$, $\theta(\frac{1}{2}) = T$, and $\theta(1) = S$. Then $(Y, \theta) \in \mathcal{S}$, $(Y, \theta) > \lambda$, and so $\lambda$ is not maximal.

Now assume $|X| > 2$, so that $\frac{1}{2} \in X$ by Lemma 10.5 again. If $\lambda$ is not maximal in $\mathcal{S}(\mathcal{F})$, then either $\lambda'$ is not maximal in $\mathcal{S}(\lambda_{0,1/2})$, or $\lambda''$ is not maximal in $\mathcal{S}(\lambda_{1/2,1})$. Hence at least one of these has a factor which is not simple, and so the same holds for $\lambda$.

Conversely, if $\lambda$ has a factor which is not simple, then the same is true of $\lambda'$ or $\lambda''$. By the induction hypothesis, there are series $\widehat{\lambda}' = (\widehat{X}', \widehat{\eta}') \geq \lambda'$ and $\widehat{\lambda}'' = (\widehat{X}'', \widehat{\eta}'') \geq \lambda''$ such that at least one of the inequalities is strict. Set $Y = \iota(\widehat{X}') \cup \kappa(\widehat{X}'')$. Then $Y' = \widehat{X}'$ and $Y'' = \widehat{X}''$, and $Y$ is admissible by 10.5. Let $\theta$ be such that $\theta' = \widehat{\eta}'$ and $\theta'' = \widehat{\eta}''$. Then $(Y, \theta) \in \mathcal{S}(\mathcal{F})$ and by Definition 10.6, $(Y, \theta) > \lambda$, so that $\lambda$ is not maximal. $\qquad\square$

**Theorem 10.15** *(Jordan-Hölder Theorem for fusion systems) If $\lambda$ and $\mu$ are composition series for a saturated fusion system $\mathcal{F}$, then $\lambda$ and $\mu$ have the same length and $F(\lambda) = F(\mu)$.*

*Proof.* Assume the theorem is false and pick a counter example $\lambda = (X, \eta)$, $\mu = (Y, \theta)$, $\mathcal{F}$ over $S$ with $\mathcal{F}$ minimal and $|X|$ of minimal order $n$. By minimality of $\mathcal{F}$, $\mathcal{F} = O^{p'}(\mathcal{F})$. If $n = 2$ then by 10.14, $\mathcal{F} = \mathcal{F}/1$ is simple, so $\mathcal{S} = \{\lambda\}$ and the theorem holds. Thus $n > 2 < |Y|$.

For any saturated fusion system $\mathcal{E}$ smaller than $\mathcal{F}$, let $CF(\mathcal{E})$ denote the set of composition factors for any composition series for $\mathcal{E}$. This set is unique by the minimality assumption on $\mathcal{F}$.

Set $T = \eta(\frac{1}{2})$ and $U = \theta(\frac{1}{2})$. Thus $\lambda_{1/2}$ and $\mu_{1/2}$ are proper nontrivial normal subsystems of $\mathcal{F}$ over $T$ and $U$, respectively. If $T \cap U \neq 1$, then by 9.1, $\mathcal{F}_0 = \lambda_{1/2} \wedge \mu_{1/2}$ is a fusion system over $S_0 = T \cap U$, and is normal in $\mathcal{F}$, $\lambda_{1/2}$, and $\mu_{1/2}$. Thus

$$
\begin{aligned}
F(\lambda) &= F(\lambda') \amalg F(\lambda'') = CF(\lambda_{1/2}) \amalg CF(\mathcal{F}/T) \\
&= CF(\mathcal{F}_0) \amalg CF(\lambda_{1/2}/S_0) \amalg CF(\mathcal{F}/T) \\
&= CF(\mathcal{F}_0) \amalg CF(\mathcal{F}/S_0) = CF(\mathcal{F}_0) \amalg CF(\mu_{1/2}/S_0) \amalg CF(\mathcal{F}/U) \\
&= CF(\mu_{1/2}) \amalg CF(\mathcal{F}/U) = F(\mu') \amalg F(\mu'') = F(\mu).
\end{aligned}
$$

Now assume $T \cap U = 1$. By 9.3, there is a normal subsystem $\mathcal{E} \trianglelefteq \mathcal{F}$ over $TU$, and $\mathcal{E} \cong \lambda_{1/2} \times \mu_{1/2}$. If $TU < S$, then $F(\lambda) = F(\mu)$ by an argument similar to that used when $T \cap U \neq 1$. Otherwise, if $TU = S$, then $\mathcal{F} = \mathcal{E} \cong \lambda_{1/2} \times \mu_{1/2}$, and

$$
F(\lambda) = F(\lambda') \amalg F(\lambda'') = CF(\lambda_{1/2}) \amalg CF(\mu_{1/2}) = F(\mu'') \amalg F(\mu') = F(\mu)
$$

for our final contradiction. $\qquad\square$

**Definition 10.16** By 10.15, we may define the family $CF(\mathcal{F})$ of *composition factors* of $\mathcal{F}$ to be the set $F(\lambda)$ of factors of any composition series $\lambda$ of $\mathcal{F}$.

**Lemma 10.17** *For each normal subsystem $\mathcal{E}$ of $\mathcal{F}$, $CF(\mathcal{F}) = CF(\mathcal{E}) \amalg CF(\mathcal{F}/\mathcal{E})$.*

*Proof.* Let $E$ be a Sylow group for $\mathcal{E}$. Replacing $\mathcal{E}$ by $O^{p'}(\mathcal{E})$ we may assume $\mathcal{E} = O^{p'}(\mathcal{E})$. Then we may choose a composition series $\lambda$ for $\mathcal{F}$ with primary pivot $E$ and associated normal system $\mathcal{E}$. Now $CF(\mathcal{F}) = F(\lambda') \amalg F(\lambda'')$ with $F(\lambda') = CF(\mathcal{E})$ and $F(\lambda'') = CF(\mathcal{F}/\mathcal{E})$. $\qquad\square$

### Exercises for Section 10

**10.1** *If $\mathcal{F}$ is a saturated fusion system on a $p$-group $S$ and $\mathcal{E}$ is a normal subsystem of $\mathcal{F}$ on $T \leq S$ then $[T^{\mathcal{F}}] = O^{p'}(\mathcal{E})$.*

**10.2** *Prove, for $a \in D$ with $0 < a < 1$, that $\iota(l(a)) = l(\iota(a))$, $\iota(u(a)) = u(\iota(a))$, $\kappa(l(a)) = l(\kappa(a))$, and $\kappa(u(a)) = u(\kappa(a))$.*

**10.3** *Prove, for $a \in D$ with $0 < a < 1$, that $h(a) = \max\{h(l(a)), h(u(a))\} + 1$.*

## 11. CONSTRAINED SYSTEMS

In this section $\mathcal{F}$ is a saturated fusion system over the finite $p$-group $S$.

In this section we consider constrained systems. If $\mathcal{F}$ is constrained, then by 4.2, $\mathcal{F} = \mathcal{F}_S(\bar{G})$ for some finite group $\bar{G}$ with $S \in \mathrm{Syl}_p(\bar{G})$ and $C_{\bar{G}}(O_p(\bar{G})) \leq O_p(\bar{G})$. (From Theorem A.13, this last condition is equivalent to $F^*(\bar{G}) = O_p(\bar{G})$.) Moreover $\bar{G}$ is unique up to isomorphism by 4.4. On the other hand there can be finite groups $G$ such that $S \in \mathrm{Syl}_p(G)$ and $\mathcal{F} = \mathcal{F}_S(G)$, but $F^*(G) \neq O_p(G)$.

**Lemma 11.1** *The following are equivalent:*

(a) *$\mathcal{F}$ is constrained.*

(b) *$F^*(\mathcal{F}) = O_p(\mathcal{F})$.*

(c) *$E(\mathcal{F}) = 1$.*

*Proof.* Let $R = O_p(\mathcal{F})$. Suppose (a) holds, so that $C_S(R) \leq R$. By 9.6.c, $E(\mathcal{F})$ centralizes $R$, so $E = S \cap E(\mathcal{F}) \leq C_S(R) = Z(R)$, so $E$ is abelian. Then $E(\mathcal{F}) = 1$ by Exercise 11.1, so (b) holds.

By definition of $F^*(\mathcal{F})$, (b) and (c) are equivalent. Assume (b) holds. By 9.6.d, $C_S(F^*(\mathcal{F})) \leq F^*(\mathcal{F})$, so $C_S(R) \leq R$, and hence (a) holds. $\square$

**Lemma 11.2** *Assume $\mathcal{F}$ is constrained. Then*

(a) *Each subnormal subsystem of $\mathcal{F}$ is constrained.*

(b) *Assume $G$ is a finite group with $S \in \mathrm{Syl}_2(G)$ and $\mathcal{F} = \mathcal{F}_S(G)$. Then:*

    (a) *For each $H \trianglelefteq\trianglelefteq G$, $\mathcal{F}_{S \cap H}(H)$ is constrained.*

    (b) *If $L$ is a component of $G$, $T = S \cap L$, and $\bar{L} = L/Z(L)$, then $\mathcal{F}_{\bar{T}}(\bar{L})$ is constrained.*

*Proof.* Exercise 11.2. $\square$

**Definition 11.3** A *Bender group* is a finite simple group which is of Lie type of characteristic 2 and Lie rank 1. A *Goldschmidt group* is a nonabelian finite simple group with a nontrivial strongly closed abelian subgroup.

The Bender groups are the groups $L_2(q)$, $Sz(q)$, $U_3(q)$, where $q$ is a suitable power of 2. By the work of Bender and Suzuki, these are the simple groups with a strongly embedded subgroup. In each case the strongly embedded subgroup is the normalizer of a Sylow 2-subgroup $T$ of the Bender group $L$. Equivalently $T$ is a *TI-subgroup* of $L$; that is distinct conjugates of $T$ in $L$ intersect trivially.

By a theorem of Goldschmidt in [Gd3], a nonabelian finite simple group $G$ is a Goldschmidt group iff $G$ is a Bender group or a Sylow 2-subgroup $S$ of $G$ is abelian. The groups in the latter case are $L_2(q)$, $q \equiv \pm 3 \bmod 8$, $^2G_2(q)$, and $J_1$. This follows from Walter's original classification of simple groups with an abelian Sylow 2-subgroup, or the later simplified treatment of Bender.

**Lemma 11.4**  *Assume $G$ is a finite group, $S \in \mathrm{Syl}_p(G)$, and $\mathcal{F} = \mathcal{F}_S(G)$. Assume in addition any one of the following hold:*

(a)  *$S$ is abelian, or*

(b)  *$S$ is a TI-subgroup of $G$, or*

(c)  *$S$ is of class 2 and $Z(S)$ is strongly closed in $S$ with respect to $G$.*

*Then $\mathcal{F} = \mathcal{F}_S(N_G(S))$ and $S = O_p(\mathcal{F})$.*

*Proof.* By Exercise 11.3, $\mathcal{F} = \mathcal{F}_S(N_G(S))$ iff $S = O_p(\mathcal{F})$ iff $N_G(S)$ controls fusion in $S$.

By Burnside's Fusion Theorem (cf. A.8), if $S$ is abelian then $N_G(S)$ controls fusion in $S$. If (b) holds then $S \cap S^g = 1$ for $g \in G - N_G(S)$, so $N_G(S)$ controls fusion in $S$. Finally suppose that (c) holds. As $Z(S)$ is strongly closed in $S$ with respect to $G$, and as $\mathcal{F} = \mathcal{F}_S(G)$, $Z(S)$ is strongly closed in $S$ with respect to $\mathcal{F}$. Then as $S$ is of class 2, the series $1 < Z(S) < S$ satisfies condition (b) of 3.3, so $S = O_p(\mathcal{F})$ by 3.3. Thus the lemma holds. $\square$

**Lemma 11.5**  *Assume $p = 2$, $G$ is a nonabelian finite simple group, $S \in \mathrm{Syl}_2(G)$, and $\mathcal{F} = \mathcal{F}_S(G)$. Then the following are equivalent:*

(a)  *$\mathcal{F}$ is constrained.*

(b)  *$S \trianglelefteq \mathcal{F}$.*

(c)  *$G$ is a Goldschmidt group. In particular either $S$ is abelian or $G$ is a Bender group.*

*Proof.* If $G$ is Bender then $S$ is a TI-subgroup of $G$. Thus (c) implies (b) by 11.4. Trivially, (b) implies (a). Finally suppose (a) holds. Then there is a nontrivial abelian subgroup of $S$, strongly closed in $S$ with respect to

$G$ by 3.4.b. Thus the first statement in (c) holds by definition of Goldschmidt groups, while the second statement follows from the theorem of Goldschmidt in [Gd3]. $\qquad\square$

## Exercises for Section 11

**11.1**  *Prove that if $1 \neq \mathcal{F}$ is quasisimple then $S$ is nonabelian.*

**11.2**  *Prove Lemma 11.2.*

**11.3**  *Assume $G$ is a finite group, $S \in \mathrm{Syl}_p(G)$, and $\mathcal{F} = \mathcal{F}_S(G)$. Prove the following are equivalent:*

(a)  $\mathcal{F} = \mathcal{F}_S(N_G(S))$.

(b)  $S = O_p(\mathcal{F})$.

(c)  $N_G(S)$ *controls fusion in* $S$.

## 12. SOLVABLE SYSTEMS

In this section $\mathcal{F}$ is a saturated fusion system over the finite $p$-group $S$.

In this section we will define the notion of a "solvable saturated fusion system". There are at least two natural definitions of such systems, which turn out to be inequivalent. Indeed the second notion is stronger and implies the first. We will focus on the weaker of the two notions.

**Definition 12.1**  Define $\mathcal{F}$ to be *solvable* if all composition factors of $\mathcal{F}$ are of the form $\mathcal{F}_G(G)$ for $G$ of order $p$. That is all composition factors of $\mathcal{F}$ are of order $p$.

As a finite group is solvable iff all its composition factors are of prime order, Definition 12.1 provides a natural extension of the notion of solvability for groups to fusion systems.

**Definition 12.2**  Following Puig in Chapter 19 of [P7], set $S_0 = S$, $\mathcal{F}_0 = \mathcal{F}$, and proceeding recursively, for $m > 0$ define $S_{2m+1} = S_{2m}$, $\mathcal{F}_{2m+1} = O^{p'}(\mathcal{F}_{2m})$, $S_{2m+2} = \mathfrak{hyp}(\mathcal{F}_{2m+1})$, and $\mathcal{F}_{2m+2} = O^p(\mathcal{F}_{2m+1})$. Then define $\mathcal{F}$ to be *Puig solvable* if $\mathcal{F}_n = 1$ for some positive integer $n$.

Notice the series $(\mathcal{F}_n : n)$ is analogous to the derived series of a group (cf. Definition A.9), and hence Puig solvability is also a natural extension of the notion of solvability for groups to fusion systems.

**Example 12.3**  Recall Example 10.1, where $L$ is a nonabelian simple group with an abelian Sylow $p$-subgroup $P$ of order $p^e > p$, $V$ is an irreducible $\mathbf{F}_p L$-module of dimension $d > 1$, $G = LV$ is the semidirect product of $L$ and $V$, $S = VP \in \mathrm{Syl}_p(G)$, and $\mathcal{F} = \mathcal{F}_S(G)$. As we saw in Example

10.7, $\mathcal{F}$ has $d + e$ composition factors, each of order $p$, so $\mathcal{F}$ is solvable. On the other hand $G = O^p(G) = O^{p'}(G)$ and $F^*(G) = V = O_p(G)$, so $\mathcal{F} = O^{p'}(\mathcal{F}) = O^p(\mathcal{F})$, and hence $\mathcal{F}$ is *not* Puig solvable.

Recall that a finite group $G$ is *p-solvable* iff each of its composition factors is of order $p$ or of order prime to $p$.

**Theorem 12.4** *(Puig) The following are equivalent:*

(a)  $\mathcal{F}$ *is Puig solvable.*

(b)  $\mathcal{F}$ *is constrained and for $G \in \mathcal{G}(\mathcal{F})$, $G$ is p-solvable.*

*Proof.* First suppose $\mathcal{F}$ is constrained and let $G \in \mathcal{G}(\mathcal{F})$. Set $G_0 = G$ and proceeding recursively, for $m > 0$ set $G_{2m+1} = O^{p'}(G)$ and $G_{2m+2} = O^p(G_{2m+1})$. Then $\mathcal{F}_n = \mathcal{F}_{S \cap G_n}(G_n)$, so $\mathcal{F}$ is Puig solvable iff $G_n = 1$ for some $n$ iff $G$ is $p$-solvable.

Thus we may assume that $\mathcal{F}$ is Puig solvable, and it remains to show that $\mathcal{F}$ is constrained. Now $\mathcal{F}_i \neq \mathcal{F}$ for $i = 1$ or 2; let $k$ be the least such $i$ and $\mathcal{E} = \mathcal{F}_k$. Then $\mathcal{E}$ is constrained, so $Q = O_p(\mathcal{E})$ is centric, so $C_{S_k}(Q) \leq Q$. As $\mathcal{E} \trianglelefteq \mathcal{F}$, also $Q \trianglelefteq \mathcal{F}$. If $k = 1$ then $S = S_k$, so $C_S(Q) \leq Q$ and hence $\mathcal{F}$ is constrained. Thus we may assume $k = 2$, so that $\mathcal{E} = O^p(\mathcal{F})$ and $S_k = \mathfrak{hyp}(\mathcal{F})$.

As $Q \trianglelefteq \mathcal{F}$, $\mathcal{C} = C_{\mathcal{F}}(Q) \trianglelefteq \mathcal{F}$ by Theorem 9.4. If $P \leq T = C_S(Q)$ and $\alpha \in \mathrm{Aut}_{\mathcal{C}}(P)$ is a $p'$-element, then by definition of $O^p(\mathcal{F})$, $[P, \alpha] \leq S_k$, so $[P, \alpha] \leq P \cap S_k = P \cap Z(Q) \leq Z(P)$, contradicting Lemma A.2. Therefore $\mathcal{C} = \mathcal{F}_T(T)$, so $T \trianglelefteq \mathcal{F}$. Thus $QT \leq O_p(\mathcal{F})$, so as $QT$ is centric, $\mathcal{F}$ is constrained. $\square$

**Lemma 12.5**  (a)  *For each normal subsystem $\mathcal{E}$ of $\mathcal{F}$, $\mathcal{F}$ is solvable iff $\mathcal{E}$ and $\mathcal{F}/\mathcal{E}$ are solvable.*

(b)  *If $\mathcal{F}$ is solvable then $\mathcal{F}$ is constrained.*

*Proof.* Part (a) follows from 10.17, which says the composition factors for $\mathcal{F}$ are the disjoint union of the factors for $\mathcal{E}$ and $\mathcal{F}/\mathcal{E}$.

Suppose $\mathcal{F}$ is solvable but not constrained. Then by 11.1, $E(\mathcal{F}) \neq 1$, so we may take $\mathcal{F} = E(\mathcal{F})$ by (a). Next each component $\mathcal{C}$ of $\mathcal{F}$ is solvable by (a), so we may take $\mathcal{F} = \mathcal{C}$. Replacing $\mathcal{F}$ by $\mathcal{F}/Z(\mathcal{F})$ and appealing to (a), we may take $\mathcal{F}$ simple. Thus $\mathcal{F}$ is of order $p$, and hence constrained. $\square$

**Definition 12.6**  Define the series $O_p^n(\mathcal{F})$ of subgroups of $S$ recursively by $O_p^0(\mathcal{F}) = 1$, and for $n > 0$, $O_p^n(\mathcal{F})$ is the preimage in $S$ of $O_p(\mathcal{F}/O_p^{n-1}(\mathcal{F}))$.

Observe that the fusion system $\mathcal{F}$ in Example 12.3 is a system with $O_p(\mathcal{F}) = V$ and $O_p^2(\mathcal{F}) = S$. In particular $O_p(\mathcal{F}) < O_p^2(\mathcal{F})$.

**Theorem 12.7**   *The following are equivalent:*

(a)   $\mathcal{F}$ *is solvable.*

(b)   $O_p^n(\mathcal{F}) = S$ *for some positive integer* $n$.

(c)   *There exists a series* $1 = S_0 \leq \cdots \leq S_m = S$ *of subgroups of* $S$ *such that for each* $0 \leq i < m$, $S_i$ *is strongly closed in* $S$ *with respect to* $\mathcal{F}$ *and* $S_{i+1}/S_i$ *is abelian.*

(d)   $\mathcal{F}$ *is constrained and for* $G \in \mathcal{G}(\mathcal{F})$, $\mathcal{F}_T(H)$ *is solvable for each composition factor* $H$ *of* $G$ *and* $T \in \mathrm{Syl}_p(H)$.

*Proof.* See 15.3 in [A6]. For example suppose (a) holds. Then $\mathcal{F}$ is constrained by 12.5.2, so $Q = O_p(\mathcal{F}) \neq 1$. By 12.5.a, $\mathcal{F}/Q$ is solvable, so by induction on $|S|$, $O_p^n(\mathcal{F}/Q) = S/Q$ for some $n$. Thus $O_p^{n+1}(\mathcal{F}) = S$, so (b) holds.                                                              $\square$

**Lemma 12.8**   *Assume* $\mathcal{F}$ *is solvable. Then each saturated subsystem of* $\mathcal{F}$ *is solvable.*

*Proof.* By 12.7 there is a series $1 = S_0 \leq \cdots \leq S_m = S$ satisfying the conditions in 12.7.c. Let $\mathcal{E}$ be a saturated subsystem of $\mathcal{F}$ on $T \leq S$ and set $T_i = T \cap S_i$. Then the series $(T_i : 0 \leq i \leq m)$ satisfies the hypotheses of 12.7.c with respect to $\mathcal{E}$, so the lemma follows from 12.7.                     $\square$

In the remainder of the section we work toward a determination of the simple groups whose fusion systems are solvable. This is achieved in the last theorem in the section. The proof uses the classification of the finite simple groups.

We begin by extending the definition of "Goldschmidt groups" to odd primes, by defining the notion of a $p$-Goldschmidt group. In this case the relevant property of Goldschmit groups $G$ with Sylow 2-group $S$ is that $S \trianglelefteq \mathcal{F}_S(G)$; recall that this is one of the three equivalent conditions defining Goldschmidt groups by 11.5, so the Goldschmidt groups are indeed the 2-Goldschmidt groups. From 12.12, when $p = 3$ the groups $G_2(q)$ with $q \equiv \pm 1 \bmod 9$ have a strongly closed abelian 3-subgroup, but are not 3-Goldschmidt. So when $p = 3$, the three conditions in 11.5 are not equivalent.

**Definition 12.9**   Define a nonabelian finite simple group $G$ with $p \in \pi(G)$ to be a *$p$-Goldschmidt group* if for $S \in \mathrm{Syl}_p(G)$, $S \trianglelefteq \mathcal{F}_S(G)$.

The following two results follow from work of Foote and Flores and Foote in [F] and [FF]; their proofs use the classification of the finite simple groups.

**Theorem 12.10**  *Let $G$ be a nonabelian finite simple group with $p \in \pi(G)$ and $S \in \mathrm{Syl}_p(G)$. Then $G$ is a $p$-Goldschmidt group iff one of the following hold:*

(a)  *$S$ is abelian.*

(b)  *$L$ is of Lie type in characteristic $p$ of Lie rank 1.*

(c)  *$p = 5$ and $L \cong Mc$.*

(d)  *$p = 11$ and $L \cong J_4$.*

(e)  *$p = 3$ and $L \cong J_2$.*

(f)  *$p = 5$ and $G \cong HS$, $Co_2$, or $Co_3$.*

(g)  *$p = 3$ and $G \cong G_2(q)$ for some prime power $q$ prime to 3 such that $q$ is not congruent to $\pm 1$ modulo 9.*

(h)  *$p = 3$ and $G \cong J_3$.*

**Remark 12.11**  In cases (b)-(d), we say $G$ is $p$-*Bender*. In those cases, $S$ is a TI-subgroup of $G$, or equivalently, strongly $p$-embedded in $G$. In cases (c)-(g), $S \cong p^{1+2}$.

**Theorem 12.12**  *(Flores-Foote) Assume $G$ is a nonabelian finite simple group and $S \in \mathrm{Syl}_p(G)$. Suppose $T$ is a nontrivial proper subgroup of $S$ strongly closed in $S$ with respect to $G$. Then either*

(a)  *$G$ is $p$-Goldschmidt, or*

(b)  *$p = 3$, $G \cong G_2(q)$ with $q \equiv \pm 1 \bmod 9$, and $T = Z(S)$ is of order 3.*

**Theorem 12.13**  *Let $G$ be a nonabelian finite simple group with $p \in \pi(G)$ and $S \in \mathrm{Syl}_p(G)$, and set $\mathcal{F} = \mathcal{F}_S(G)$. Then the following are equivalent:*

(a)  *$\mathcal{F}$ is constrained.*

(b)  *$\mathcal{F}$ is solvable.*

(c)  *$S = O_p(\mathcal{F})$.*

(d)  *$G$ is $p$-Goldschmidt.*

(e)  *$G$ satisfies one of conclusions (a)–(h) of Theorem 12.10.*

*Proof.* Trivially, (c) and (d) are equivalent and (c) implies (b). Part (b) implies (a) by 12.5.2, and (d) and (e) are equivalent by 12.10. Thus it suffices to show that (a) implies (e), so assume $\mathcal{F}$ is constrained but (e) fails. Then by 3.3, there is a nontrivial abelian subgroup $T$ of $S$ strongly closed in $S$ with respect to $\mathcal{F}$. Therefore by Theorem 12.12, conclusion (b) of that theorem holds. Then from 16.11.5 in [A6], $\mathcal{F}$ is quasisimple, contradicting $\mathcal{F}$ constrained.  $\square$

## 13. Fusion Systems in Simple Groups

In Section 14 we speculate on the possibility of classifying simple saturated $p$-fusion systems. Before undertaking such a task, one would like to have a good feeling for known examples of simple systems. An obvious source of examples are the composition factors of the systems of finite groups. If $G$ is a finite group, $S \in \mathrm{Syl}_p(G)$, $H$ is a proper normal subgroup of $G$, and $G^* = G/H$, then by 10.17 the family of composition factors of $\mathcal{F}_S(G)$ is the union of the factors of $\mathcal{F}_{S \cap H}(H)$ and $\mathcal{F}_{S^*}(G^*)$. Thus it suffices to consider the case where $G$ is a nonabelian finite simple group and $p \in \pi(G)$ is a prime divisor of the order of $G$.

So in this section $p$ is a prime, $G$ is a finite group, $1 \neq S \in \mathrm{Syl}_p(G)$, and $\mathcal{F} = \mathcal{F}_S(G)$. Thus $\mathcal{F}$ is a saturated fusion system over the finite $p$-group $S$. Our first result gives a sufficient criterion for $\mathcal{F}$ to be simple.

**Lemma 13.1** *Assume*

(a) *there exists no nontrivial proper subgroup of $S$ strongly closed in $S$ with respect to $G$, and*

(b) $\mathrm{Aut}_G(S) = \langle \mathrm{Aut}_{O^{p'}(N_G(R))}(S) : R \in \mathcal{F}^{frc} \rangle$.

*Then $\mathcal{F}_S(G)$ is simple.*

*Proof.* We apply part (d) of Theorem 9.8. Assume conditions (a) and (b) of the lemma hold. Then by 9.8.d, it suffices to verify condition (ii) of that result. However for $X \leq S$, $\mathrm{Aut}_{\mathcal{F}}(X) = \mathrm{Aut}_G(X)$, and for $R \in \mathcal{F}^{frc}$, $\mathrm{Aut}_{O^{p'}(N_G(R))}(S)$ consists of the maps $\alpha = c_g$ for $g \in O^{p'}(N_G(R) \cap N_G(S))$. Further for each such $\alpha$, $\alpha_{|R} \in \mathrm{Aut}_{O^{p'}(N_G(R))}(R) = B(R) \leq \mathrm{Aut}_{\mathcal{B}}(R)$, so $\alpha \in \mathrm{Aut}_{\mathcal{F}}^0(S)$. Therefore condition (ii) of 9.8.d does indeed hold, completing the proof. $\square$

**Remark 13.2** If $G$ is a nonabelian finite simple group which does not satisfy condition (a) of 13.1, then by 12.12, either $G$ is $p$-Goldschmidt, or $p = 3$ and $G \cong G_2(q)$ is described in 12.12.b. In the latter case $\mathcal{F}_S(G) \cong \mathcal{F}_S(H)$ is quasisimple, where $S \in \mathrm{Syl}_3(H)$ and $H \cong SL_3^\epsilon(q)$ (cf. 16.11 in [A6]). Thus in the remainder of the discussion, we may assume condition (a) of 13.1 *is* satisfied, so in particular $G$ is not $p$-Goldschmidt and $G$ is not $G_2(q)$ if $p = 3$.

We next record some results on various families of nonabelian finite simple groups from [A6].

**Lemma 13.3** *Assume $G$ is simple of Lie type and characteristic $p$, or $p = 2$ and $G = {}^2F_4(2)'$ is the Tits group. Assume the Lie rank of $G$ is at least 2. Then $\mathcal{F}$ is simple.*

*Proof.* A proof appears in 16.3 in [A6], but to illustrate how to apply 13.1, we give some indication of how the proof goes.

As the Lie rank of $G$ is at least 2, $G$ is not $p$-Goldschmidt, so from Remark 13.2 it remains to verify condition (b) of 13.1. From the Borel-Tits Theorem (cf. 3.1.3 and 3.1.5 in [GLS3]), $\mathcal{F}^{frc}$ is the set of unipotent radicals $O_p(P)$ of the proper parabolics $P$ of $G$ containing $S$. Let $B = SH$ be the Borel subgroup over $S$, with $H$ a Cartan subgroup corresponding to a root system $\Sigma$ and simple system $\pi$. For $\alpha \in \Sigma$, let $U_\alpha$ be the root subgroup of $\alpha$. The minimal parabolics over $B$ are of the form $P_\alpha = \langle B, r_\alpha \rangle$ for $\alpha \in \pi$, where $r_\alpha$ is the fundamental reflection in the Weyl group of $G$ corresponding to $\alpha$. Thus $R_\alpha = O_p(P_\alpha) \in \mathcal{F}^{frc}$ and $B = N_{P_\alpha}(S)$.

It is more convenient to work in the universal group of type $G$, so replacing $G$ by that group, we may assume $G$ is universal rather than simple. Then from the proof of Lemma 64 in [Stn],

$$H_\alpha = H \cap \langle U_\alpha, U_{-\alpha} \rangle = H \cap O^{p'}(P_\alpha).$$

Now as $G$ is universal, $H$ is the direct product of the groups $H_\alpha$, $\alpha \in \pi$ (cf. 2.4.7 in [GLS3]). Hence as $H$ is a Hall $p'$-subgroup of $B = N_G(S) = N_{P_\alpha}(S)$ for each $\alpha \in \pi$, condition (b) of 13.2 does indeed hold.  $\square$

In the next lemma we use the following notation. If $G$ is a group of permutations on a set $\Omega$ and $\Delta \subseteq \Omega$, then $G_\Delta$ denotes the pointwise stabilizer in $G$ of $\Delta$, $N_G(\Delta)$ denotes the global stabilizer in $G$ of $\Delta$, and for $X \leq N_G(\Delta)$, $X^\Delta$ denotes the image of $X$ in $Sym(\Delta)$ under the restriction map.

**Lemma 13.4**   *Assume $G \cong A_n$ is an alternating group on $\Omega = \{1, \ldots, n\}$ with $n \geq 6$ and $S \neq 1$. Write $n = ap+b$ with $0 \leq b < p$, let $X = N_G(M(S))$, and $Y = G_{M(S)}$, where $M(S)$ is the set of points of $\Omega$ moved by $S$. Then*

(a)   $\mathcal{F}_S(G) = \mathcal{F}_S(X) \cong \mathcal{F}_S(X^{M(S)})$.

(b)   $p \geq n$.

(c)   $S$ *is abelian iff* $n < p^2$.

(d)   *If* $n \geq p^2$ *and* $b \leq 1$ *then* $\mathcal{F}_S(G)$ *is simple and* $X^{M(S)} \cong A_{pa}$.

(e)   *If* $n \geq p^2$ *and* $b \geq 2$ *then* $X^{M(S)} \cong S_{pa}$, $\mathcal{F}_S(Y^{M(S)}) \triangleleft \mathcal{F}_S(X^{M(S)})$, *and* $Y^{M(S)} \cong A_{pa}$.

*Proof.* See 16.5 in [A6].  $\square$

**Lemma 13.5**   *Assume $G$ is a sporadic simple group, and let $\Pi = \Pi(G)$ be the set of odd primes $p \in \pi(G)$ such that $|G|_p > p^2$. Then*

(a)   $S$ *is nonabelian iff either:*

(i)   $p \in \Pi$ *and* $(G, p) \neq (O'N, 3)$, *or*

(ii)  $p = 2$ *and* $G$ *is not* $J_1$.

(b)   *If* $G$ *is* $M_{11}$, $M_{22}$, $M_{23}$, *or* $J_1$, *then* $\Pi = \varnothing$.

(c)   *If* $G$ *is* $M_{12}$, $M_{24}$, $J_2$, $J_3$, $Suz$, $F_{22}$, *or* $F_{23}$, *then* $\Pi = \{3\}$.

(d)   *If* $G$ *is a Conway group,* $Mc$, $Ru$, $Ly$, $F_5$, $F_3$, *or* $F_2$, *then* $\Pi = \{3, 5\}$.

(e)   *If* $G$ *is* $HS$ *then* $\Pi = \{5\}$.

(f)   *If* $G$ *is* $He$, $O'N$, *or* $F_{24}$ *then* $\Pi = \{3, 7\}$.

(g)   *If* $G$ *is* $J_4$ *then* $\Pi = \{3, 11\}$.

(h)   *If* $G$ *is* $F_1$ *then* $\Pi = \{3, 5, 7, 13\}$.

*Proof.* See 16.6 in [A6].                                            $\square$

**Lemma 13.6**   *Assume* $G$ *is a sporadic simple group, but not* $J_1$, *and* $p = 2$. *Then* $\mathcal{F}_S(G)$ *is simple.*

*Proof.* See 16.8 in [A6].                                            $\square$

**Lemma 13.7**   *Assume* $G$ *is a sporadic simple group, but not* $p$-*Goldschmidt. Then either*

(a)   $\mathcal{F}_S(G)$ *is simple, or*

(b)   $(G, p) = (Ru, 3)$, $(M_{24}, 3)$, $(Ru, 5)$, $(J_4, 3)$, *or* $(Co_1, 5)$, *and* $S \in \mathrm{Syl}_p(L)$ *where* $\mathcal{F}_S(G) \cong \mathcal{F}_S(L)$ *and* $L \cong {}^2F_4(2)$, $\mathrm{Aut}(M_{12})$, $\mathrm{Aut}(L_3(5))$, ${}^2F_4(2)$, *or* $PO_5(5)$, *respectively.*

*Proof.* See 16.10 in [A6].                                            $\square$

**Remark 13.8**   We are addressing the following question: Let $G$ be a nonabelian finite simple group, $p$ a prime divisor of $|G|$, $S \in \mathrm{Syl}_p(G)$, and $\mathcal{F} = \mathcal{F}_S(G)$. What are the composition factors of $\mathcal{F}$? In particular, when is $\mathcal{F}$ simple? The results listed above, together with the classification of the finite simple groups, have reduced us to the case where $G$ is of Lie type in characteristic $r \neq p$.

When $p = 2$ we are aware of only isolated results addressing the question. Presumably the theory of fundamental subgroups of $G$ contained in [A1] and [A3] would be useful here. In any event, during the remainder of the discussion assume that $p$ is odd.

The structure of $S$ is discussed in section 4.10 of [GLS3]. In particular one can decide when $S$ is abelian. For example if $p$ does not divide the order of the Weyl group of the associated algebraic group, then $S$ is abelian.

Let $G$ be a group over $\mathbf{F}_q$. Theorem III.1.18 tells us that the isomorphism type of $\mathcal{F}$ does not depend on $G$, but on the Lie type of $G$ and the $p$-adic evaluation of certain polynomials in $q$. Indeed there are other isomorphisms between fusion systems of groups of Lie type, particularly among the systems of classical groups. In particular from [BMO], if $G$ is classical then $\mathcal{F}$ is isomorpic to the system of $L_n(q')$ for some integer $m$ and prime power $q'$. In any event, before entering into a detailed analysis of $\mathcal{F}$ it would seem to be best to sort out such isomorphisms, which of course are of interest in their own right.

In preliminary work, Aschbacher has tentatively analyzed $\mathcal{F}$ when $G$ is an exceptional group, and in particular determined when $\mathcal{F}$ is simple. In [Rz], Ruiz analyzes the case where $G$ is $L_n(q)$. This work is discussed briefly in section III.6.5. In particular it can be determined when $\mathcal{F}$ is simple. In some cases the unique nonsolvable composition factor is exotic; see for example section 17 in [A6], where the case $G \cong L_{20}(2)$ and $p = 5$ is considered as an example, and we find that $O^{5'}(\mathcal{F})$ is simple and exotic of index 4 in $\mathcal{F}$; that is $\pi_1(\mathcal{F}^c) \cong \mathbf{C}_4$.

## 14. CLASSIFYING SIMPLE GROUPS AND FUSION SYSTEMS

Since the local theory of finite groups has proven to be a sufficiently powerful tool to classify the finite simple groups, one might hope that the local theory of fusion systems could be used to classify the simple saturated $p$-fusion systems. Evidence suggests that the behavior of simple systems for odd primes may be more wild than the behavior of simple 2-fusion systems, so perhaps the case $p = 2$ should be considered first.

There are other reasons for focusing on 2-fusion systems. For example certain arguments in local group theory seem to be easier to implement in saturated fusion systems than in groups. Thus there is some hope that the proof of the classification of the finite simple groups might be simplified by, first, proving suitable results about saturated 2-fusion systems, and then, second, using the theorems on fusion systems to prove results about finite simple groups. In this section and its sequel, we speculate a bit about the possibility of implementing these two programs.

Let $\mathcal{F}$ be a saturated fusion system on a finite $p$-group $S$.

**Definition 14.1** Define $\mathcal{F}$ be be of *characteristic p-type* if for each $U \in \mathcal{F}^f$, $N_{\mathcal{F}}(U)$ is constrained. Define $\mathcal{F}$ to be of *component type* if for some fully normalized subgroup $X$ of $S$ of order $p$, $C_{\mathcal{F}}(X)$ has a component.

**Remark 14.2**   Let $G$ be a finite group. Then $G$ is said to be of *characteristic p-type* if for each $p$-local subgroup $H$ of $G$, $C_H(O_p(H)) \leq O_p(H)$. Further $G$ is of *component type* if for some involution $t$ in $G$, $C_G(t)/O(C_G(t))$ has a component.

It is easy to see that if $G$ is of characteristic $p$-type than its $p$-fusion system is of characteristic $p$-type. If $G$ is of component type the situation is a bit more subtle. If $L$ is a component of $C_G(t)/O(C_G(t))$ with Sylow 2-group $T$, it may be the case that $\mathcal{F}_T(L)$ is solvable. Indeed from 12.13 this will be the case precisely when $L$ is a Goldschmit group. If for some involution $t$ and component $L$ of $C_G(t)/O(C_G(t))$, $L$ is not Goldschmidt, then $\mathcal{F}_S(G)$ is of component type.

Furthermore the classification of the finite simple groups proceeds by partitioning the simple groups into those of small 2-rank, those of component type, and those of characteristic 2-type. One of the first major steps in the classification of the finite simple groups is the proof that each finite simple group falls into one of these three classes (cf. [G2]). The proof of this result requires the Bender-Suzuki work on groups with a strongly embedded subgroup, signalizer functor theory, and the Gorenstein-Walter theorem on L-balance. On the other hand we see in a moment that the analogous result for fusion systems is much easier, requiring only the L-balance theorem for fusion systems. This begins to show why one might hope to be able to classify simple 2-fusion systems, and use some parts of that proof to simplify the proof of the classification of the finite simple groups.

**Theorem 14.3**   *Let $\mathcal{F}$ be a saturated p-fusion system. Then either*

(a)   *$\mathcal{F}$ is of characteristic p-type, or*

(b)   *$\mathcal{F}$ is of component type.*

*Proof.* Assume $\mathcal{F}$ is not of characteristic $p$-type. Then there exists some $U \in \mathcal{F}^f$ such that $\mathcal{N} = N_{\mathcal{F}}(U)$ is not constrained. Therefore by 11.1, $E(\mathcal{N}) \neq 1$. Let $T = N_S(U)$ and $X$ of order $p$ in $Z(T) \cap U$. Then $X$ is fully normalized in $\mathcal{N}$ and centralizes $E(\mathcal{N})$. Therefore (cf. 10.3 in [A6]) $E(\mathcal{N}) = E(C_{\mathcal{N}}(X))$. Set $\mathcal{C} = C_{\mathcal{F}}(X)$. Then $C_{\mathcal{N}}(X) = N_{\mathcal{C}}(U)$, so $E(N_{\mathcal{C}}(X)) \neq 1$. Conjugating in $\mathcal{F}$ and using 3.1, we may assume $X \in \mathcal{F}^f$ and $U \in \mathcal{C}^f$. Therefore by L-balance, Theorem 9.7, $1 \neq E(N_{\mathcal{C}}(U)) \leq E(\mathcal{C})$, completing the proof.   $\square$

**Remark 14.4**   Theorem 14.3, together with the proof of the classification of the finite simple groups, suggest that we should partition the problem of classifying simple saturated $p$-fusion systems into two subproblems:

Problem 14.4.1  Classify all simple saturated $p$-fusion systems of component type.

**Problem 14.4.2** Classify all simple saturated fusion systems of characteristic $p$-type.

However to avoid difficulties in the first problem, associated with certain wreath products and extensions of groups of Lie type of characteristic $p$ by field automorphisms of order $p$, there is reason to believe it might be better to modify the partition corresponding to the two problems. For example this is one of the guiding principals of the GLS approach to revising the classification of the finite simple groups. This leads to the following definitions, where we specialize to the case $p = 2$.

**Definition 14.5** Let $p = 2$, $P$ a 2-group, and recall the definition of the *Thompson subgroup* $J(P)$ and the *Baumann subgroup* $\mathrm{Baum}(P)$ of $P$ from Definition A.15.

Define $\mathcal{F}$ to be of *even characteristic* if $N_{\mathcal{F}}(U)$ is constrained for all $1 \neq U \trianglelefteq S$. Define $\mathcal{F}$ to be of *even component type* if $C_{\mathcal{F}}(z)$ has a component for some involution $z$ in the center of $S$,

Define $\mathcal{F}$ to be of *Baumann characteristic 2* if $N_{\mathcal{F}}(U)$ is constrained for each $U \in \mathcal{F}^f$ with $\mathrm{Baum}(S) \leq N_S(U)$. Define $\mathcal{F}$ to be of *Baumann component type* if there exists $X \in \mathcal{F}^f$ of order 2 such that $X \leq Z(\mathrm{Baum}(S))$ and $C_{\mathcal{F}}(X)$ contains a component.

**Example 14.6** Let $L$ be a group of Lie type in characteristic 2, $t$ an involutory outer automorphism of $L$, and $G = \langle t \rangle L$ the semidirect product of $L$ by $\langle t \rangle$. Let $t \in S \in \mathrm{Syl}_2(G)$ and $S_L = S \cap L$. By the Borel-Tits Theorem (cf. 3.1.3 in [GLS3]) $L$ is of characteristic 2-type. On the other hand, with a few small exceptions, $C_L(i)$ has a component for some involution $i \in tG$, so $G$ is of component type. However, as is well known, $Z(S) \leq S_L$, so $\mathcal{F} = \mathcal{F}_S(G)$ is of even characteristic. Moreover, again with some small exceptions, $\mathrm{Baum}(S) = \mathrm{Baum}(S_L)$, so $G$ is also of Baumann characteristic 2.

**Example 14.7** Let $L$ be a nonabelian finite simple group and $G$ the wreath product of $L$ by $\mathbf{C}_2$. Thus $G$ is a semidirect product $G = \langle t \rangle H$, where $H = L_1 \times L_2$ with $L_i \cong L$, and $t$ is an involution with $L_1^t = L_2$. As $G$ is a wreath product, $C_L(t) = \{ll^t : l \in L_1\} \cong L$ is a component of $C_G(t)$, so $G$ is of component type, and if $L$ is not a Goldschmidt group then $\mathcal{F} = \mathcal{F}_S(G)$ is also of component type, where $t \in S \in \mathrm{Syl}_2(G)$.

Let $S_i = S \cap L_i$, so that $S = \langle t \rangle S_1 S_2$ is also a wreath product. In particular $Z(S) = \{zz^t : z \in Z(S_i)\}$ is diagonally embedded in $L_1 \times L_2$, so $\mathcal{F}$ is of even characteristic iff $\mathcal{F}_{S_1}(L_1)$ is of even characteristic.

Finally set $J_i = J(S_i)$ be the Thompson subgroup of $S_i$ for $i = 1, 2$. As

$$m_2(C_S(t)) = 1 + m_2(L) < 2m_2(L) = m_2(H)$$

since $m_2(L) > 1$, it follows that $J(S) = J_1 \times J_2$, and then $\mathrm{Baum}(S) = B_1 B_2$, where $B_i = \mathrm{Baum}(S_i)$. In particular $G$ is of Baumann component type as $L_1$ is a component of $C_G(z_2)$ for $z_2$ an involution in $Z(S_2)$. Thus $\mathcal{F}$ is of Baumann component type unless $L$ is a Goldschmidt group.

As we will see in a while, these examples suggest that it may be best to partition the classification of the simple saturated 2-fusion systems either into those of even characteristic and those of even component type, or into those of Baumann characteristic 2 and those of Baumann component type.

In the remainder of the section assume $p = 2$. We engage in some speculation as to how to classify the simple 2-fusion systems of component type. The idea is of course to imitate the proof of the classification of the finite simple groups of component type. Here is an outline of the major steps in such a program:

**Problem 14.8**  *Translate the notion of a "classical involution" [A1] from finite group theory into the realm of 2-fusion systems, and prove an analogue of the classical involution theorem for 2-fusion systems. This result would provide a characterization of almost all the 2-fusion systems of groups of Lie type in odd characteristic.*

More specifically, consider the following class of fusion systems.

**Definition 14.9**  A *quaternion fusion packet* is a pair $(\mathcal{F}, \Omega)$ where $\mathcal{F}$ is a saturated fusion system on a finite 2-group $S$, and $\Omega$ is a collection of subgroups of $S$ such that $\Omega^{\mathcal{F}} = \Omega$ and

(a)   There exists $e \geq 3$ such that for all $K \in \Omega$, $K$ has a unique involution $z(K)$ and $K$ is nonabelian of order $2^e$.

(b)   For each pair of distinct $K, J \in \Omega$, $|K \cap J| \leq 2$.

(c)   If $K, J \in \Omega$ and $v \in J - Z(J)$, then $v^{\mathcal{F}} \cap C_S(z(K)) \subseteq N_S(K)$.

(d)   If $K, J \in \Omega$ with $z = z(K) = z(J)$, $v \in K$, and $\phi \in \mathrm{Hom}_{C_{\mathcal{F}}(z)}(\langle v \rangle, S)$, then either $v\phi \in J$ or $v\phi$ centralizes $J$.

Then try to extend the various Theorems in [A1] to results about quaternion fusion packets. In particular if $\mathcal{F}$ is simple, show that (essentially) $\mathcal{F}$ is the fusion system of some group of Lie type and odd characteristic, coming from one of the following examples:

**Example 14.10**  Let $p$ be an odd prime and $G$ a group of Lie type over a finite field of characteristic $p$, other than $L_2(q)$ or $^2G_2(q)$. Let $\Sigma$ be a root system for $G$, and for $\alpha$ a long root in $\Sigma$, write $U_\alpha$ for the center of the root group of $\alpha$ and set $L_\alpha = \langle U_\alpha, U_{-\alpha} \rangle$. The conjugates of $L_\alpha$ are the *fundamental subgroups* of $G$. Each fundamental subgroup $L$ is isomorphic to $SL_2(q)$ for some power $q$ of $p$; in particular $L$ has quaternion Sylow 2-subgroups and a unique involution $z = z(L)$. Moreover $L$ is subnormal in $C_G(z)$, so $L$ is a component of $C_G(z)$, unless $q = 3$ where $L$ is a "solvable component" in the sense that $SL_2(3)$ is almost quasisimple.

Pick a Sylow 2-subgroup $S$ of $G$ and let $\Omega$ consist of the groups $S \cap L$ such that $L$ is a fundamental subgroup and $S \cap L \in \mathrm{Syl}_2(L)$. Set $\mathcal{F} = \mathcal{F}_S(G)$. Then $(\mathcal{F}, \Omega)$ is a quaternion fusion packet and these *Lie packets* are the generic examples of quaternion fusion packets. There are however other infinite families of examples, many of which are constrained. There are also a few sporadic examples. Most of the examples are subpackets of Lie packets.

Aschbacher has done some preliminary work on this problem.

**Problem 14.11**  *Prove an analogue of the theorem of Walter* [W] *classifying finite simple groups in which some involution centralizer has a 2-component of Lie type in odd characteristic. The proof would proceed by reducing to Problem 14.8 in most cases.*

More specifically, consider the following hypothesis (or something like it):

**Hypothesis W.** $\mathcal{F}$ is a saturated fusion system on a finite 2-group $S$ with $F^*(\mathcal{F})$ quasisimple. Assume there exists an involution $i \in S$ such that $\langle i \rangle \in \mathcal{F}^f$ and $C_{\mathcal{F}}(i)$ has a component in $Chev^*(r)$ for some odd prime $r$.

Here $Chev^*(r)$ is essentially the class of fusion systems of quasisimple groups of Lie type and odd characteristic, distinct from $L_2(r^e)$ and $^2G_2(3^e)$ when $r = 3$. One would like to show that if Hypothesis W holds, then, with known exceptions, there is some collection $\Omega$ of subgroups of $S$ such that $(\mathcal{F}, \Omega)$ is a quaternion fusion packet. Note that one class of exceptions are the exotic 2-fusion systems constructed by Levi and Oliver in [LO]. These arise in [W] during the proof of Proposition 4.3 of that paper.

**Problem 14.12**  *Prove an analogue of the Component Theorem in* [A2] *from finite group theory for 2-fusion systems. The result would say that, modulo known exceptions, if $\mathcal{F}$ is a saturated fusion system on a finite 2-group $S$, and there exists an involution $i \in S$ such that $\langle i \rangle \in \mathcal{F}^f$ and $C_{\mathcal{F}}(i)$ is not constrained, then there exists a "standard component" in the centralizer of some involution.*

Let $G$ be a finite group. A subgroup $K$ of $G$ is *tightly embedded* in $G$ if $K$ is of even order but $K \cap K^g$ is of odd order for each $g \in G - N_G(K)$. A quasisimple subgroup $L$ of $G$ is a *standard component* of $G$ if $K = C_G(L)$ is tightly embedded in $G$, $N_G(K) = N_G(L)$, and $L$ commutes with none of its conjugates.

One would need to settle on the "right" definition of a "standard component" of a 2-fusion system. This would involve finding an analogue of the notion of a tightly embedded subgroup from [A2], and constructing a theory of tightly embedded subsystems of 2-fusion systems.

**Problem 14.13** *Prove the various Standard Form Theorems. That is prove that if $\mathcal{F}$ is a saturated 2-fusion system with a standard component of a given isomorphism type, then the system $\mathcal{F}$ is isomorphic to a known system.*

This is the step where it would be easier to deal with a standard component $\mathcal{L}$ either in the centralizer of an involution in the center of $S$, or normalized by the Baumann subgroup $\mathrm{Baum}(S)$ of $S$. Working with such a component would avoid treating cases leading to the following two situations:

**Situation A.** $\mathcal{F}$ is the 2-fusion system of the wreath product of a group $L$ of Lie type and characteristic 2 by $\mathbf{C}_2$.

**Situation B.** $\mathcal{F}$ is the 2-fusion system of a group of Lie type in characteristic 2 by an involutory field automorphism.

For we saw in Examples 14.6 and 14.7 that neither situation can occur in a fusion system of even component type, and only Situation A can occur in a system of Baumann component type. Even in Situation A with $\mathcal{F} = \mathcal{F}_S(G)$ for $G$ as in Example 14.7, $L_1$ is not a standard component in the group $G$ as $L_1$ commutes with its conjugate $L_2$. In addition we also have the strong condition $L_1 \trianglelefteq C_G(i)$ for each involution $i \in L_2 = C_G(L_1)$. Thus one can hope that Situation A also presents relatively few difficulties in the context of Problem 14.13.

If one takes this approach, then Problem 14.12 would also have to be modified to prove that if $\mathcal{F}$ is of even component type or of Baumann component type, then there exists a standard component in the centralizer of an involution in the center of $S$ or normalized by $\mathrm{Baum}(S)$, respectively.

**Remark 14.14** Observe that this program does *not* include any steps analogous to the treatment of the B-Conjecture or the Unbalanced Group Conjecture, two extremely difficult steps in the classification of the finite simple groups (cf. [G2]). We recall that the B-Conjecture asserts that, for a finite simple group $G$ and an involutory automorphism $t$ of $G$, $L(C_G(t)) = E(C_G(t))$, where $L(C_G(t)) = O^{p'}(\hat{E})$ and $\hat{E}$ is the preimage in $C_G(t)$ of $E(C_G(t)/O(C_G(t)))$. Similarly the Unbalanced Group Conjecture supplies a list of the simple groups $G$ and involutory automorphisms $t$ such that $O(C_G(t)) \neq 1$.

Such steps are unnecessary in the category of fusion systems since they involve the *cores* $O(H)$ of 2-local subgroups $H$ of a minimal counter example to the classification. But from Exercise 2.1, the 2-fusion systems of $H$

and $H/O(H)$ are the same, so cores of 2-locals are *not* obstructions when we work in the category of 2-fusion systems.

## 15. SYSTEMS OF CHARACTERISTIC 2-TYPE

In this section $\mathcal{F}$ is a saturated fusion system over the finite 2-group $S$. We will be primarily interested in the case where $\mathcal{F}$ is of characteristic 2-type.

Unlike the case of systems of component type, we have no grand program to offer aimed at classifying the simple saturated 2-fusion systems of characteristic 2-type. This is because the corresponding problem in finite group theory — the classification of the finite simple groups of characteristic 2-type — was achieved, in the generic case, by switching attention from 2-locals to $p$-locals, for suitable odd primes $p$. Of course such a change of point of view would appear to be impossible in the category of 2-fusion systems.

However the small groups of characteristic 2-type were treated via 2-local analysis, and there is an ongoing program of Meierfrankenfeld, Stellmacher, and Stroth [MSS] to treat all groups of characteristic 2-type 2-locally. Thus there are *some* techniques available to steal from finite group theory.

Therefore in this section we discuss some results which put in place machinery to begin the analysis of 2-fusion systems of characteristic 2-type, and then consider two theorems which classify certain special classes of systems. These theorems are analogues of results from finite simple group theory which began the study of groups of characteristic 2-type, and served as test cases for the classification of such groups. Their analogues for 2-fusion systems can thus be regarded as test cases for classifying systems of characteristic 2-type.

A finite group $H$ is a $\mathcal{K}$-*group* if each simple section of $H$ is on the list $\mathcal{K}$ of "known" simple groups appearing in the statement of the theorem classifying the finite simple groups. In a minimal counter example to that theorem, each proper subgroup is a $\mathcal{K}$-group. Presumably we also wish to classify the simple 2-fusion systems using an inductive proof, so we are lead to:

**Definition 15.1**   Define a saturated fusion system $\mathcal{D}$ on a finite $p$-group $D$ to be a *local CK-system* if for each nontrivial subgroup $P$ of $S$, $\mathrm{Aut}_{\mathcal{F}}(P)$ is a $\mathcal{K}$-group.

For example if $G$ is a minimal counter example to the classification and $S \in \mathrm{Syl}_p(G)$, then $\mathcal{F}_S(G)$ is a local CK-system. Since we hope our results can be used to simplify portions of the proof of the classification theorem, it is important not to use that theorem in their proofs, but instead to prove theorems about local CK-systems.

The following collection of 2-fusion systems share certain unusual behavior.

**Definition 15.2**  Define $\mathcal{F}$ to be an *obstruction to pushing up at the prime 2* if $\mathcal{F} = \mathcal{F}_S(G)$ for some finite group $G$ with $S \in \mathrm{Syl}_2(G)$ such that one of the following holds:

(a)  $S$ is dihedral of order at least 16 and $G \cong L_2(q)$ or $PGL_2(q)$ for some odd prime power $q$.

(b)  $S$ is semidihedral and $G \cong L_2(q)^{(1)}$ for some odd prime power $q$ which is a square.

(c)  $S$ is semidihedral of order 16 and $G \cong L_3(3)$.

(d)  $|S| = 32$ and $G \cong \mathrm{Aut}(A_6)$ or $\mathrm{Aut}(L_3(3))$.

(e)  $|S| = 2^7$ and $G \cong J_3$.

(f)  $F^*(G) \cong Sp_4(q)$, where $q = 2^e$ with $e > 1$ odd, $|G{:}O^2(G)| = 2$, and $O^2(G)$ is an extension of $F^*(G)$ by a group of field automorphisms of odd order.

(g)  $F^*(G) \cong L_3(q)$, $q = 2^e$, $e > 1$, $|O^2(G){:}F^*(G)|$ is odd, and $G$ is the extension of $O^2(G)$ by a graph or graph-field automorphism.

Here $L_2(q)^{(1)}$ is the extension of $L_2(q)$ of degree 2 with semidihedral Sylow 2-subgroups.

What is *pushing up*? The term comes from the local theory of groups of characteristic 2-type, but we will reinterpret it for 2-fusion systems. Let $\mathcal{U}$ be the set of radical centric subgroups of $S$ which are normal in $S$. By Alperin's Fusion Theorem 3.5, $\mathcal{F} = \langle \mathrm{Aut}_{\mathcal{F}}(R) : R \in \mathcal{F}^{frc} \rangle$; that is the automizers of fully normalized radical centric subgroups of $S$ control fusion. Observe that if $U \trianglelefteq S$ then $U$ is fully normalized. Moreover we would like to work, as much as possible, in local subsystems containing $S$. Thus we would like to show that $\mathcal{F} = \langle N_{\mathcal{F}}(U) : U \in \mathcal{U} \rangle$. Unfortunately this is not always the case.

In attempting to prove such a result, we are led to attempt to "push up" the local systems $N_{\mathcal{F}}(R)$, for $R \in \mathcal{F}^{frc}$ with $R$ not normal in $S$. That is we try to show $N_{\mathcal{F}}(R) \leq \langle N_{\mathcal{F}}(Q) : Q \in \mathcal{Q} \rangle$ for some suitable subset $\mathcal{Q}$ of $N_{\mathcal{F}}(R)^f$ such that $|N_S(Q)| > |N_S(R)|$ for $Q \in \mathcal{Q}$.

When $\mathcal{F}$ is of characteristic 2-type, "pushing up" techniques borrowed from local group theory allow us to show that either $\mathcal{F} = \langle N_{\mathcal{F}}(U) : U \in \mathcal{U} \rangle$ or $\mathcal{F}$ is one of the obstructions to pushing up appearing in 15.2. Then as a corollary we obtain the smaller list of finite groups $G$ of characteristic 2-type in which 2-fusion is not controlled in the normalizers of normal subgroups of a Sylow 2-subgroup of $G$. This is a new theorem for groups, giving yet another example of how it is sometime easier to prove results about fusion systems as a stepping stone to theorems about finite groups.

The theorems we just discussed appear in [A7]. Indeed it is possible to prove an even stronger result. We now state the main theorem from [A7]. Recall the *Baumann subgroup* $\mathrm{Baum}(S)$ of $S$ is defined in Definition A.15.

**Theorem 15.3** *Assume $\mathcal{F}$ is a saturated fusion system on a finite 2-group $S$, such that $\mathcal{F}$ is a local CK-system of characteristic 2-type. Then one of the following holds:*

(a) $\mathcal{F} = \langle C_{\mathcal{F}}(\Omega_1(Z(S))), N_{\mathcal{F}}(\mathrm{Baum}(S)) \rangle$.

(b) *Campbell pairs for $\mathcal{F}$ exist, and for each such pair $(C_1, C_2)$, $\mathcal{F} = \langle C_{\mathcal{F}}(C_1), N_{\mathcal{F}}(C_2) \rangle$.*

(c) *There exists a finite group $G$ with $S \in \mathrm{Syl}_2(G)$ such that $\mathcal{F} = \mathcal{F}_S(G)$ and either $G$ is $M_{23}$ or $G$ is an extension of $L_3(2^e)$ or $Sp_4(2^e)$ by a group of odd order for some positive integer $e$.*

(d) *$\mathcal{F}$ is an obstruction to pushing up at the prime 2.*

**Remark 15.4** An *$\mathcal{F}$-characteristic subgroup* of a subgroup $P$ of $S$ is a subgroup $C$ which is $\mathrm{Aut}_{\mathcal{F}}(P)$-invariant. A *Baumann pair* for $\mathcal{F}$ is a pair $(C_1, C_2)$ of nontrivial $\mathcal{F}$-characteristic subgroups of $S$, such that $C_1 \leq Z$ and $C_2$ is an $\mathcal{F}$-characteristic subgroup of $B$, where $Z = \Omega_1(Z(S))$ and $B = \mathrm{Baum}(S)$. Thus for example, $(Z, B)$ is the "largest" Baumann pair. *Campbell pairs* are certain Baumann pairs; the precise definition is technical, so we will only give a rough idea here.

Assume $\mathcal{F}$ is of even characteristic, as defined in Definition 14.5, and, as earlier, let $\mathcal{U}$ be the set of radical centric subgroups of $S$ which are normal in $S$. Pick $U \in \mathcal{U}$ and let $G$ be a model for $N_{\mathcal{F}}(U)$. A *minimal parabolic* of $G$ is an overgroup $M$ of $S$ in $G$ such that $S$ is not normal in $M$ and $S$ is contained in a unique maximal subgroup of $M$. By a result of McBride (cf. B.6.3 in [ASm]), $G$ is generated by $N_G(S)$ and its set $\mathfrak{M} = \mathfrak{M}(U)$ of minimal parabolics. Then by 2.5 and 7.12 in [A7],

$$\mathcal{N} = N_{\mathcal{F}}(U) = \langle C_{\mathcal{N}}(Z), N_{\mathcal{N}}(B), \mathcal{F}_S(M) \colon M \in \mathfrak{M} \rangle.$$

So if

(*) for each $M \in \mathfrak{M}(U)$, either $Z \leq Z(M)$ or $B \trianglelefteq M$,

then $\mathcal{N} = \langle C_{\mathcal{N}}(Z), N_{\mathcal{N}}(B) \rangle$. Therefore if $\mathcal{F} = \langle N_{\mathcal{F}}(U) \colon U \in \mathcal{U} \rangle$ and (*) holds for each $U \in \mathcal{U}$, then conclusion (1) of Theorem 15.3 holds. So suppose that (*) fails for some $U \in \mathcal{U}$ and $M \in \mathfrak{M}$. Then (cf. 8.4 in [A7]) there exists a subgroup $H$ of $M$ with $B \in \mathrm{Syl}_2(H)$ such that $H$ is a minimal parabolic in the set $\mathcal{G}(B)$ of Definition 1.3 in [A7]. Write $\mathfrak{G}(\mathcal{F})$ for the set of such groups $H$ as $U, M$ varies over pairs for which (*) fails.

A *Glauberman-Niles pair* is a Baumann pair $(C_1, C_2)$ such that for each $H \in \mathfrak{G}(\mathcal{F})$, either $C_1 \leq Z(H)$ or $C_2 \trianglelefteq H$, or $H$ has just one noncentral chief factor in $O_2(H)$. By a theorem of Glauberman and Niles in [GN], Glauberman-Niles pairs exist. Moreover in his thesis [Cm], Campbell gave a constructive proof of the existence of such a pair. Namely given $H \in \mathfrak{G}(\mathcal{F})$, set $F = \Omega_1(Z(J(O_2(H))))$ and $V = [O^2(H), V]$. Then set $C_1 = Z \cap \langle C_V(B)^{\mathrm{Aut}_{\mathcal{F}}(B)} \rangle$ and $C_2 = \langle F^{\mathrm{Aut}_{\mathcal{F}}(B)} \rangle$. The pair $(C_1, C_2)$ is a *Campbell pair*.

What is the significance of Theorem 15.3? We wish to analyze some system $\mathcal{F}$ of characteristic 2-type. Applying Theorem 15.3, we may assume we are in case (1) or (2) of that theorem, so we have a pair $(C_1, C_2)$ with $C_1$ an $\mathcal{F}$-characteristic subgroup of $S$ contained in $\Omega_1(Z(S))$, $C_2$ $\mathcal{F}$-characteristic in $\mathrm{Baum}(S)$, and $\mathcal{F} = \langle C_{\mathcal{F}}(C_1), N_{\mathcal{F}}(C_2) \rangle$. As $\mathcal{F}$ is of characteristic 2-type there are models $G_1$ for $\mathcal{F}_1 = C_{\mathcal{F}}(C_1)$ and $G_2$ for $\mathcal{F}_2 = N_{\mathcal{F}}(C_2)$. Let $\mathcal{F}_{1,2} = C_{\mathcal{F}_2}(C_1)$ and let $G_{1,2}$ be a model for $\mathcal{F}_{1,2}$. We can choose $G_{1,2} = C_{G_2}(C_1)$. Further $\mathcal{F}_{1,2} = N_{\mathcal{F}_1}(C_2)$, so $N_{G_1}(C_2)$ is also a model for $\mathcal{F}_{1,2}$, so by 4.4 there is an isomorphism $\alpha_1 : G_{1,2} \to N_{G_1}(C_2)$ extending the identity map on $S$. Form the amalgam

$$\mathcal{A} = (G_1 \xleftarrow{\alpha_1} G_{1,2} \xrightarrow{\alpha_2} G_2),$$

where $\alpha_2$ is the inclusion map. Let $G$ be the universal completion of $\mathcal{A}$; that is $G$ is the free group on $G_1 \cup G_2 \cup G_{1,2}$ modulo relations defining the groups and identifying elements of $G_{1,2}$ with those in $G_i$ via the maps $\alpha_i$, for $i = 1, 2$. Then

**Lemma 15.5**  $\mathcal{F} = \mathcal{F}_S(G)$, *so the amalgam $\mathcal{A}$ determines the fusion system $\mathcal{F}$.*

It follows from 15.5 that, given a suitable set $\Phi$ of local conditions on $\mathcal{F}$, if we can show that $\mathcal{A}$ is determined up to isomorphism by $\Phi$, then also $\mathcal{F}$ is determined up to isomorphism by $\Phi$. That is we can hope to characterize 2-fusion systems of characteristic 2-type via sets $\Phi$ of local conditions using Theorem 15.3.

Moreover Theorem 15.3 says that fusion in $\mathcal{F}$ is controlled by the normalizers of nontrivial normal subgroups $U$ of $S$. Notice each such $U$ is fully normalized, and hence $N_{\mathcal{F}}(U)$ has a model $G(U)$ which has $S$ as a Sylow 2-subgroup. There is a natural equivalence relation on the set of pairs $(U, H)$ with $1 \neq U \trianglelefteq S$ and $S \leq H \leq G(U)$; let $\mathcal{H} = \mathcal{H}(\mathcal{F})$ be the set of equivalence classes of this equivalence relation and write $[U, H]$ for the equivalence class of $(U, H)$, or sometimes just $H$. There is also a natural partial order $\leq$ on $\mathcal{H}$, making $\mathcal{H}$ into a poset. Indeed if $\mathcal{F} = \mathcal{F}_S(G)$ for some finite group $G$, then $\mathcal{H}$ can be regarded as the set $\mathcal{H}(G)$ of overgroups $H$ of $S$ in $G$ with $O_2(H) \neq 1$, and the relation $\leq$ is the subgroup relation. This theory is developed in [A8].

One would like to prove a version of Theorem 15.3 for systems of Baumann characteristic 2, or even better, for systems of even characteristic. For if one partitions systems into those of Baumann characteristic 2 and Baumann component type, in order to simplify the analysis in the "component type" case, as discussed near the end of Section 14, then one is left with the problem of dealing with systems of Baumann characteristic 2, where the first step in the analysis would be to prove some extension of Theorem 15.3.

The 2-local theory of finite groups of characteristic 2-type focuses on the poset $\mathcal{H}(G)$, and many results in the theory are theorems about this poset. We can hope to prove analogues of those theorems for the poset $\mathcal{H}(\mathcal{F})$. For example Theorem 5.4 in [A8] proves such an analogue of the Stellmacher qrc-Lemma, which appears for example as Theorem D.1.5 in [ASm].

The statement of the qrc-Lemma is technical, so we will not give it here. Interpreted in the language of fusion systems, it considers the situation where $[U_i, G_i]$, $i = 1, 2$, are equivalence classes in $\mathcal{H} = \mathcal{H}(\mathcal{F})$ such that

(qrc1)  no nontrivial subgroup of $S$ is normal in both $G_1$ and $G_2$; and

(qrc2)  $V$ is a nontrivial normal elementary abelian 2-subgroup of $G_1$ such that $O_2(\mathrm{Aut}_{G_1}(V)) = 1$ and $R = C_S(V) \trianglelefteq G_1$; and

(qrc3)  $G_2$ is a minimal parabolic; that is $S$ is not normal in $G_2$ and $S$ is contained in a unique maximal subgroup of $G_2$.

In this situation the qrc-Lemma says that the pair $(G_1, G_2)$ lives in one of five cases, with the cases distinguished by properties of the representation of $G_1$ on $V$ and the embedding of $V$ and $R$ in $G_2$.

How does one find pairs $[U_i, G_i]$, $i = 1, 2$, satisfying the hypotheses (qrc1)–(qrc3) of the qrc-Lemma? First, if $\mathcal{H}$ has a unique maximal member $[M, U]$ then (qrc1) is never satisfied, as $U \trianglelefteq H$ for each $H \in \mathcal{H}$. On the other hand in this situation, Theorem 15.3 says that either $\mathcal{F}$ is an obstruction to pushing up, or $U \trianglelefteq \mathcal{F}$. Thus one consequence of Theorem 15.3 is that we may assume that $\mathcal{H}$ has at least two maximal members.

Let $H \in \mathcal{H}$, $Z = \Omega_1(Z(S))$, and $V(H) = \langle Z^H \rangle$. Then as $\mathcal{F}$ is of characteristic 2-type, $O_2(\mathrm{Aut}_H(V(H))) = 1$ (cf. B.2.13 in [ASm]). Set $R = C_S(V(H))$. By a Frattini argument, $H = C_H(V(H))N_H(R)$ and hence $[R, N_H(R)] \in \mathcal{H}$ with $V(H) = V(N_H(R))$ and $\mathrm{Aut}_H(V(H)) = \mathrm{Aut}_{N_H(R)}(V(H))$. Moreover $R = C_S(V(H)) \trianglelefteq N_H(R)$. That is $G_1 = N_H(R)$ satisfies condition (qrc2).

Now specialize to the case where $H$ is maximal in $\mathcal{H}$ and recall we may assume there exists a second maximal member $M$ of $\mathcal{H}$. It follows that condition (qrc1) is satisfied by the pair $(H, M)$. However we wish to work with $G_1$ rather than $H$. To do so we work with a second partial order $\preceq$ on $\mathcal{H}$ and choose $H$ to also be maximal with respect to $\preceq$. As a consequence of the maximality of $H$ with respect to $\preceq$, results in [A8] show that $H$ is the unique member of $\mathcal{H}$ maximal under $<$ containing $G_1$, so the pair $(G_1, M)$ satisfies (qrc1), and indeed for each $G_2 \leq M$ such that $S \leq G_2 \not\leq H$, $(G_1, G_2)$ satisfies (qrc1). Choose $G_2$ minimal subject to these constraints. By a result of McBride (cf. B.6.3 in [ASm]), $M$ is generated by $N_M(S)$ and the minimal parabolics of $M$. Thus if $N_M(S) \leq H$ then we may choose $G_2$ to be a minimal parabolic of $M$, and we have achieved the hypotheses of the qrc-Lemma. The case where $N_{\mathcal{F}}(S)$ is not contained in $H$ is even nicer; for example in that case no nontrivial characteristic subgroup of $S$ is normal in $H$, so the structure of $H$ and its action on $V(H)$ is determined by the local $C(G, T)$-Theorem (cf. C.1.29 in [ASm]).

Using the poset $\mathcal{H}(\mathcal{F})$ and the qrc-Lemma, it is possible to prove an analogue of Glauberman's theorem classifying $S_4$-free finite simple groups, for 2-fusion systems. Indeed the theorem for fusion systems is more natural and attractive than the theorem for groups, and can be used to give a simple proof of Glauberman's theorem. We begin to describe this constellation of ideas.

Let $G$ be a finite group. A *section* of $G$ is a group of the form $H/K$, where $K \trianglelefteq H \leq G$. Given a positive integer $n$, write $S_n$ for the symmetric group on a set of order $n$. The group $G$ is $S_n$-*free* if $G$ has no section isomorphic to $S_n$.

One of the striking results in finite group theory from the decade before the classification of the finite simple groups, was the classification of the $S_4$-free finite simple groups. The first result in this direction was Thompson's proof (cf. [Th2]) that the Suzuki groups $Sz(2^{2m+1})$ are the only nonabelian finite simple groups of order prime to 3. Next in [Gl2], Glauberman proved a triple factorization theorem for constrained $S_4$-free groups, and used this result to show each $S_4$-free group has a nontrivial strongly closed abelian 2-subgroup. As a corollary to this result and Goldschmidt's theorem in [Gd3] classifying finite groups with such a subgroup, Glauberman classified the $S_4$-free and $S_3$-free nonabelian finite simple groups. Later in [St1], Stellmacher showed that in a constrained $S_4$-free group $G$, there is a nontrivial characteristic subgroup of a Sylow 2-subgroup of $G$ normal in $G$; this theorem can be used to give an alternate treatment of $S_4$-free groups.

What are the analogues of these results for 2-fusion systems?

**Definition 15.6**  $\mathcal{F}$ is $S_3$-*free* if for each subgroup $U$ of $S$, the group $\mathrm{Aut}_{\mathcal{F}}(U)$ is $S_3$-free.

**Theorem 15.7** *Let $\mathcal{F}$ be a saturated $S_3$-free fusion system on a finite 2-group $S$. Then $\mathcal{F}$ is constrained.*

*Proof.* See [A8].                                                            □

Onofrei and Stancu [OS] have proved a result on $S_3$-free 2-fusion systems stronger than Theorem 15.7, using Stellmacher's theorem on constrained $S_4$-free groups.

Once the general machinery for fusion systems of characteristic 2-type is in place, particularly Stellmacher's qrc-Lemma for fusion systems, the fusion system theoretic proof of Theorem 15.7 is only a few paragraphs, and hence to our minds, more attractive than the group theoretic proofs of Glauberman's theorem. This is another example of how certain results about finite groups are best proved fusion system theoretically.

Moreover Theorem 15.7, together with Goldschmidt's Theorem, leads almost immediately to the following corollaries for groups:

**Corollary 15.8** *Assume $G$ is a finite $S_4$-free nonabelian finite simple group. Then $G$ is a Goldschmidt group.*

**Corollary 15.9** *Assume $G$ is a finite $S_3$-free nonabelian finite simple group. Then $G$ is $Sz(2^{2m+1})$ or $L_2(3^{2m+1})$, with $m \geq 1$.*

**Corollary 15.10** *Assume $G$ is a finite nonabelian simple group of order prime to 3. Then $G \cong Sz(2^{2m+1})$ for some $m \geq 1$.*

Finally we come to the second of our seminal papers on groups of characteristic 2-type. Recall that an *N-group* is a nonabelian finite simple group $G$ such that all local subgroups of $G$ are solvable. In a monumental series of papers [Th1], Thompson determined all $N$-groups. Thompson's work on $N$-groups was a test case for the classification of the finite simple groups, and served as a blue print for the classification.

In [J], [Sm], and [GL], Janko, F. Smith, and Gorenstein and Lyons extended Thompson's work to determine all finite groups in which all 2-local subgroups are solvable. Later in [St2], Stellmacher gave an alternate approach to treating finite simple groups of characteristic 2-type in which all 2-locals are solvable, based on the amalgam method.

It should be noted that, while $N$-groups need not be of characteristic 2-type, from the modern point of view, $N$-groups which are not of characteristic 2-type are relatively easy to treat; this is basically the reduction described in Remark 14.2. Of course the machinery required to make that reduction was not available to Thompson when he did his work on $N$-groups. Rather Gorenstein and Walter were led to the reduction by an analysis of Thompson's treatment of that case in the $N$-group paper.

Our goal is to carry out an analysis of groups in which all 2-locals are solvable in the category of 2-fusion systems. This leads to the following definition:

**Definition 15.11**   Define $\mathcal{F}$ to be an *N-system* if for each $P \in \mathcal{F}^f$, $N_{\mathcal{F}}(P)$ is constrained and $\mathrm{Aut}_{\mathcal{F}}(P)$ is solvable.

For example if $G$ is a finite group in which all 2-local subgroups are solvable, and $S \in \mathrm{Syl}_2(G)$, then $\mathcal{F}_S(G)$ is an $N$-system. The converse is not true however, as we see in a moment. Such examples give yet another hint that some theorems about finite groups have more attractive statements and proofs in the language of fusion systems.

**Remark 15.12**   Observe that $\mathcal{F}$ is an $N$-system iff for each $P \in \mathcal{F}^f$, $N_{\mathcal{F}}(P)$ is Puig solvable. This follows from Theorem 12.4 and the fact that groups of odd order are solvable. Since Puig solvability is one of the two natural notions of solvability in the category of 2-fusion systems, our notion of an $N$-system is a natural translation of the notion of an $N$-group into the category of 2-fusion systems. Thus the classification of $N$-systems serves as a test case for classifying all simple 2-fusion systems of characteristic 2-type, just as the $N$-group paper was a test case for classifying simple groups of characteristic 2-type.

Of course instead of working with $N$-systems, one could work with systems $\mathcal{F}$ in which, for each $P \in \mathcal{F}^f$, $N_{\mathcal{F}}(P)$ is solvable. However this class of systems is much larger than the class of $N$-systems, and not as closely related to the 2-fusions systems of $N$-groups. For example for each group $G$ of Lie type and Lie rank 2 over a field of even order, the 2-fusion system of $G$ is in the larger class.

**Theorem 15.13**   *Assume $\mathcal{F}$ is an N-system. Then there exists a finite group $G$ with $S \in \mathrm{Syl}_2(G)$ such that $\mathcal{F} = \mathcal{F}_S(G)$, and one of the following holds:*

(a)   *$G$ is a solvable group with $F^*(G) = O_2(G)$.*

(b)   *$S$ is dihedral of order at least 16, and $G \cong L_2(q)$ or $PGL_2(q)$ for some odd prime power $q$.*

(c)   *$S$ is semidihedral and $G \cong L_2(q)^{(1)}$ for some odd prime power $q$ which is a square.*

(d)   *$S$ is semidihedral of order 16 and $G \cong L_3(3)$.*

(e)   *$|S| = 32$ and $G \cong \mathrm{Aut}(A_6)$ or $\mathrm{Aut}(L_3(3))$.*

(f)   *$G \cong L_3(2)$ or $Sp_4(2)$.*

(g)   *$G$ is isomorphic to $U_3(3)$, $G_2(2)$, $M_{12}$, $\mathrm{Aut}(M_{12})$, ${}^2F_4(2)'$, or ${}^2F_4(2)$.*

*Proof.* This appears in the preprint [A9].                    $\square$

Notice that $M_{12}$ is not a $N$-group; indeed the centralizer of a non-2-central involution of $M_{12}$ is isomorphic to $\mathbf{C}_2 \times S_5$. However the 2-fusion system of $M_{12}$ is an $N$-system. As a corollary to Theorem 15.13, we obtain an alternate approach to the treatment of finite groups in which all 2-local subgroups are solvable:

**Theorem 15.14**   *(Thompson, Janko, Smith, Gorenstein, Lyons) Assume $G$ is a finite group in which all 2-local subgroups are solvable. Let $S \in \mathrm{Syl}_2(G)$ and set $G^* = G/O(G)$ and $H^* = O^{2'}(G^*)$. Then one of the following holds:*

(a)   *$G$ is solvable.*

(b)   *$S$ is dihedral and $H^* \cong A_7$, $L_2(q)$, or $PGL_2(q)$ for some odd prime power $q$.*

(c)   *$S$ is semidihedral and $H^* \cong L_2(q)^{(1)}$ for some odd prime power $q$ which is a square.*

(d)   *$S$ is semidihedral of order 16 and $H^* \cong L_3(3)$ or $M_{11}$.*

(e)   *$|S| = 32$ and $H^* \cong \mathrm{Aut}(A_6)$ or $\mathrm{Aut}(L_3(3))$.*

(f)   *$H^* \cong Sp_4(2)$ or $S_7$.*

(g)   *$H^*$ is isomorphic to $U_3(3)$, $G_2(2)$, ${}^2F_4(2)'$, or ${}^2F_4(2)$.*

(h)   *$H^*$ is a Bender group.*

The proof of Theorem 15.14, based on Theorem 15.13, also appears in the preprint [A9]. That proof uses the classification of simple groups with dihedral Sylow 2-subgroups by Gorenstein and Walter in [GW1], or the more modern treatment of such groups by Bender and Glauberman in [Ben] and [BG]. It also requires a classification of simple groups whose Sylow 2-subgroup is semidihedral of order 16, or wreathed of order 32, in which the centralizers of involutions are solvable (cf. [ABG] or [GLS6]). The proof also appeals to Goldshmidt's theorem classifying simple groups with a strongly closed abelian 2-subgroup [Gd3], the solvable 2-signalizer functor theorem. (cf. Chapter 15 in [A4]), the Feit-Thompson Theorem [FT], and suitable characterizations of $M_{12}$ and the Tits group.

This shows the strengths and limitations of any approach to proving theorems about finite simple groups in the category of fusion systems. If the 2-local structure of a group $G$ is reasonably rich, then it seems often to be relatively easy to retrieve the group from its fusion system. On the other hand if the 2-local structure is sparse, the fusion system seems to give relatively little information. For example if $S$ is dihedral, standard elementary transfer and fusion easily pin down the fusion system $\mathcal{F}_S(G)$ (cf. Examples I.2.7 and I.3.8), but the hard work in [GW1] or [Ben] and [BG] is still required to go on and determine $G$.

## Exercises for Section 15

**15.1**   *Let $\mathcal{F}$ be a constrained, saturated 2-fusion system and $G$ a model for $\mathcal{F}$. Prove $\mathcal{F}$ is $S_3$-free iff $G/O_2(G)$ is an $S_3$-free group.*

**15.2**   *Let $G$ be a finite group with $O_{3'}(G) = 1$ such that $G$ is $S_3$-free. Prove $O^{2'}(G) = E(G)$ and $E(G) = 1$ or $E(G) = L_1 \times \cdots \times L_n$ where $L_i \cong L_2(3^{e_i})$ for some odd integer $e_i$.*

**15.3**   *Let $\mathcal{F}$ be a saturated 2-fusion system. Prove the following are equivalent:*

(a)   *$\mathcal{F}$ is $S_3$-free.*

(b)   *$\mathcal{F}$ is constrained, and if $G$ is a model for $\mathcal{F}$ and $H = O^{2'}(G)$, either $H$ is a $3'$-group or $H/O_{3'}(H) = L_1 \times \cdots \times L_n$ with $L_i \cong L_2(3^{e_i})$ for some odd integer $e_i > 1$.*

# Part III. Fusion and homotopy theory

## Bob Oliver

To each group $G$ (discrete or topological), one associates a topological space $BG$, called the classifying space of $G$. Originally, these were defined by Milnor to classify certain fiber bundles with a structure determined by the given group. Much later, the proof in the early 1980's by Lannes, Miller, and Carlsson of the Sullivan conjecture made it possible to attack certain problems involving classifying spaces which earlier had been considered impossible. One consequence was the discovery that the classifying space of a group $G$ has homotopy theoretic properties much more rigid, and much more closely connected to the structure of $G$ itself, than was previously thought possible.

For each topological space $X$ and each prime $p$, Bousfield and Kan, in the 1970's, defined the $p$-*completion* $X_p^\wedge$ of $X$: a space which allows us to focus on the properties of $X$ "at the prime $p$". There is a map $X \longrightarrow X_p^\wedge$, functorial in $X$, which in "nice" cases is universal among all maps $X \longrightarrow Y$ which induce an isomorphism in mod $p$ homology (see Section 1.4 for more details). It turns out that there is a very close connection between the homotopy theoretic properties of $BG_p^\wedge$, when $G$ is a finite group or a compact Lie group, and the $p$-*local structure* of $G$. Here, by the $p$-local structure of $G$ is meant roughly the structure of its fusion category over a Sylow $p$-subgroup $S$ (see Section I.1); i.e., the structure of $S$ together with the conjugacy relations between its subgroups.

More concretely, the *Martino-Priddy conjecture* stated that for any prime $p$ and any pair $G_1, G_2$ of finite groups, $G_1$ and $G_2$ have the same $p$-local structure (in a sense which will be made precise in Section 1.5) if and only if their $p$-completed classifying spaces $BG_{1}{}_p^\wedge$ and $BG_{2}{}_p^\wedge$ are homotopy equivalent. While working on this conjecure, and also trying to understand the group of self homotopy equivalences of $BG_p^\wedge$, Broto, Levi, and Oliver [BLO1] were led to investigate the *centric linking category* $\mathcal{L}_S^c(G)$ associated to $G$ (where $S \in \mathrm{Syl}_p(G)$). They discovered that the $p$-completed geometric realization $|\mathcal{L}_S^c(G)|_p^\wedge$ of this category has the homotopy type of $BG_p^\wedge$, and also that many of the homotopy properties of $BG_p^\wedge$ can be described in terms of properties of $\mathcal{L}_S^c(G)$. For example, the group $\mathrm{Out}(BG_p^\wedge)$ of homotopy classes of self homotopy equivalences of $BG_p^\wedge$ is isomorphic to a certain group of "outer" automorphisms of the category $\mathcal{L}_S^c(G)$.

One implication in the Martino-Priddy conjecture — $G_1$ and $G_2$ have the same $p$-local structure if their $p$-completed classifying spaces are homotopy equivalent — was shown by Martino and Priddy, via a proof based on the Sullivan conjecture and the work of Lannes [La]. As one application of this result, Broto, Møller, and Oliver [BMO] listed many examples of families of finite groups of Lie type in characteristic different from $p$ all of whose members have the same $p$-local structure (described in Section 1.7). The opposite implication in the conjecture was proven in [O2, O3], but via a proof which depends on the classification of finite simple groups. This is not very satisfactory, and there are still hopes that a "classification-free" proof could lead to a better understanding of why this relationship should be true.

The fusion category $\mathcal{F}_S(G)$ of a finite group $G$ (when $S \in \mathrm{Syl}_p(G)$ for some prime $p$), and abstract saturated fusion systems over a $p$-group, have already been described at length in Part I. Motivated by this and by work of Ron Solomon, Dave Benson (in [Be3] and in unpublished work) predicted that there should be a way of associating classifying spaces to saturated fusion systems, which would generalize the association between the $p$-fusion category of a finite group and its $p$-completed classifying space. This was carried out by Broto, Levi, and Oliver [BLO2], who defined a certain class of related categories, abstract linking systems, which can be associated to fusion systems. A classifying space for a fusion system $\mathcal{F}$ can then be defined to be the space $|\mathcal{L}|_p^\wedge$ — the $p$-completion of the geometric realization of $\mathcal{L}$ — for any linking system $\mathcal{L}$ associated to $\mathcal{F}$.

It is still not known whether or not a classifying space can be associated to each fusion system (nor whether it is unique), although this is the case in all examples which we have looked at. This problem of determining whether or not each saturated fusion system has a unique associated linking system and classifying space is closely related to the Martino-Priddy conjecture, and provides an additional reason for wanting to find a proof of that conjecture which does not use the classification of finite simple groups.

The $p$-completed classifying spaces of finite groups have some very remarkable homotopy theoretic properties, and classifying spaces of abstract fusion systems share many of these same properties. For example, the group of homotopy classes of self equivalences of a classifying space $|\mathcal{L}|_p^\wedge$ can be described explicitly in terms of certain automorphisms of the (finite) category $\mathcal{L}$; it is very unusual to find spaces whose self equivalences can be described so explicitly. The mod $p$ cohomology $H^*(|\mathcal{L}|_p^\wedge; \mathbb{F}_p)$ can be described explicitly as a "ring of stable elements" depending on the fusion system, in a way analogous to the description by Cartan and Eilenberg of $H^*(G; \mathbb{F}_p) \cong H^*(BG_p^\wedge; \mathbb{F}_p)$. These and other, similar properties are described in Section 4.6.

The fundamental group of any classifying space $|\mathcal{L}|_p^\wedge$ of a saturated fusion system $\mathcal{F}$ is always a finite $p$-group. The connected covering spaces of $|\mathcal{L}|_p^\wedge$

are all classifying spaces of fusion subsystems of $\mathcal{F}$: the subsystems of $p$-power index in $\mathcal{F}$ (Definition I.7.3). For example, the universal covering space of $|\mathcal{L}|_p^\wedge$ is the classifying space of the fusion subsystem $O^p(\mathcal{F}) \trianglelefteq \mathcal{F}$ described in Theorem I.7.4. There is a similar geometric description of $O^{p'}(\mathcal{F})$ (see Theorem I.7.7) via the universal cover of a different space associated to $\mathcal{F}$. All of these connections between covering spaces and fusion subsystems are described in Section 4.5.

A saturated fusion system is called "realizable" if it is the fusion system of a finite group, and is called "exotic" otherwise. Classifying spaces of exotic fusion systems are thus spaces which have many of the very nice properties of $p$-completed classifying spaces of finite groups, but which are not equivalent to $BG_p^\wedge$ for any finite $G$. This provides part of the motivation for describing in Section 6 some of the known examples of exotic fusion systems.

We would like to thank Kasper Andersen, Carles Broto, Paul Goerss, Jesper Grodal, Ran Levi, Kári Ragnarsson, and Albert Ruiz for their many suggestions and corrections made while preparing this survey.

**Notation:** When $\mathcal{C}$ is a (small) category, $\mathcal{C}^{\mathrm{op}}$ denotes the opposite category (morphisms going in the opposite direction), $\mathcal{N}(\mathcal{C})$ is the nerve of $\mathcal{C}$ (as a simplicial set), and $|\mathcal{C}|$ is the geometric realization of $\mathcal{C}$ (i.e., of $\mathcal{N}(\mathcal{C})$). For any functor $F\colon \mathcal{C} \longrightarrow \mathcal{D}$ between categories, we let

$$F_{c,c'}\colon \mathrm{Mor}_{\mathcal{C}}(c, c') \longrightarrow \mathrm{Mor}_{\mathcal{D}}(F(c), F(c')) \qquad \text{and} \qquad F_c = F_{c,c}$$

denote the induced maps between morphism sets.

For each $n \geq 0$, $D^n$ denotes the closed ball of radius 1 in $\mathbb{R}^n$; i.e., the space of points $x$ such that $\|x\| \leq 1$. Also, $S^{n-1} \subseteq D^n$ denotes the unit sphere: the space of those $x$ such that $\|x\| = 1$. When $X$ is any topological space and $A$ is a subspace (usually closed), $X/A$ denotes the quotient space where all points of $A$ have been identified to a point. Thus, for example, $D^n/S^{n-1}$ is homeomorphic to $S^n$ for $n \geq 1$.

Some of the basic concepts in algebraic topology, such as the fundamental group $\pi_1(X)$, covering spaces, and CW complexes and their (co)homology, are surveyed briefly in Section 1.1–1.2. However, a reader not already somewhat familiar with these topics will probably have to supplement what is written there by referring to the book of Hatcher [Ht], or that of Benson [Be2]. Note that for a pair of spaces $X$ and $Y$, $X \cong Y$ means that $X$ and $Y$ are homeomorphic, while $X \simeq Y$ means that they are homotopy equivalent (see Section 1.1).

Throughout this part, $p$ is assumed to be a fixed prime, and all $p$-groups are finite.

## 1. Classifying spaces, $p$-completion, and the Martino-Priddy conjecture

This section is intended to give an introduction and general motivation for the connection between fusion systems and topology. The main goal is to present the material needed to state precisely the Martino-Priddy conjecture (Theorem 1.17). This conjecture is what provided the original motivation for this author and some of his collaborators to begin studying the connections between fusion in finite groups and homotopy theory.

All maps between topological spaces are assumed to be continuous.

### 1.1. Homotopy and fundamental groups.

We begin with a very brief description of some of the basic concepts in algebraic topology. For more detail, but still in a concise formulation, we refer to Benson's book [Be2, Chapter 1]. Hatcher's textbook [Ht] provides a much more lengthy introduction to these topics.

Let $I = [0,1]$ denote the unit interval. When $X$ and $Y$ are two spaces, and $f, f' \colon X \longrightarrow Y$ are (continuous) maps from $X$ to $Y$, then $f$ and $f'$ are *homotopic*, written $f \simeq f'$, if there is a map $F \colon X \times I \longrightarrow Y$ such that $F(x,0) = f(x)$ and $F(x,1) = f'(x)$ for all $x \in X$. In other words, $f$ and $f'$ are homotopic if there is a "continuous deformation" of $f$ into $f'$. One easily sees that homotopy defines an equivalence relation among the set of all maps from $X$ to $Y$.

A map $f \colon X \longrightarrow Y$ is a *homotopy equivalence* if it has a homotopy inverse: a map $g \colon Y \longrightarrow X$ such that $f \circ g \simeq \mathrm{Id}_Y$ and $g \circ f \simeq \mathrm{Id}_X$. When there is a homotopy equivalence between $X$ and $Y$, one says that $X$ has the *homotopy type* of $Y$. A space $X$ is *contractible*, written $X \simeq *$, if $X$ has the homotopy type of a point; equivalently, if $\mathrm{Id}_X$ is homotopic to a constant map.

If $X$ is a space, and $x, y \in X$, then a *path* in $X$ from $x$ to $y$ is a (continuous) map $\gamma \colon I \longrightarrow X$ such that $\varphi(0) = x$ and $\gamma(1) = y$. A *loop* in $X$ based at $x$ is a path from $x$ to itself. The space $X$ is *path connected* if there is a path in $X$ between any two points. Two paths $\gamma$ and $\gamma'$ from $x$ to $y$ are homotopic (relative to their endpoints) if there is a homotopy $\Gamma \colon I \times I \longrightarrow X$ such that for all $s, t \in I$, $\Gamma(t,0) = \gamma(t)$, $\Gamma(t,1) = \gamma'(t)$, $\Gamma(0,s) = x$, and $\Gamma(1,s) = y$. Homotopy relative to endpoints defines an equivalence relation among paths between any two given points of $X$.

If $\gamma$ and $\delta$ are two paths in $X$, and $\gamma(1) = \delta(0)$, then the composite path $\delta{\cdot}\gamma$ from $\gamma(0)$ to $\delta(1)$ is defined by setting

$$(\delta{\cdot}\gamma)(t) = \begin{cases} \gamma(2t) & \text{if } 0 \le t \le \tfrac{1}{2} \\ \delta(2t - 1) & \text{if } \tfrac{1}{2} \le t \le 1. \end{cases}$$

Paths are thus composed from right to left throughout this survey (but note that composition from left to right is quite frequently found in the literature).

Composition of paths is associative up to homotopy. Also, for each path $\gamma$, we define the "inverse" path $\bar{\gamma}$ by setting $\bar{\gamma}(t) = \gamma(1 - t)$. This is an inverse up to homotopy, in the sense that $\gamma \cdot \bar{\gamma}$ and $\bar{\gamma} \cdot \gamma$ are both homotopic to constant paths.

For any choice of basepoint $x_0 \in X$, the *fundamental group of $X$ at $x_0$* is the group $\pi_1(X, x_0)$ of homotopy classes of loops based at $x_0$ (homotopy relative to the endpoints), with multiplication defined by composition of paths. When $X$ is path connected, then for any pair of points $x_0, x_1 \in X$, $\pi_1(X, x_0)$ and $\pi_1(X, x_1)$ are isomorphic. More precisely, if $\gamma$ is a path from $x_0$ to $x_1$, then there is an isomorphism

$$\gamma_\# : \pi_1(X, x_1) \xrightarrow{\;\cong\;} \pi_1(X, x_0)$$

which sends the class of a loop $\phi$ based at $x_1$ to the class of the composite loop $\bar{\gamma} \cdot \phi \cdot \gamma$ based at $x_0$ (cf. [Ht, Proposition 1.5] and [Be2, pp.6–7]). This isomorphism depends on the choice of path $\gamma$, but if $\delta$ is another path from $x_0$ to $x_1$, then $\gamma_\#$ and $\delta_\#$ differ by an inner automorphism of $\pi_1(X, x_0)$ (conjugation by $[\bar{\delta} \cdot \gamma]$).

A path connected space is *simply connected* if its fundamental group is the trivial group. We just saw that this is independent of the choice of bascpoint.

When $f \colon X \longrightarrow Y$ is a map between spaces, $x_0 \in X$, and $y_0 = f(x_0) \in Y$, then we let

$$f_\# : \pi_1(X, x_0) \longrightarrow \pi_1(Y, y_0)$$

denote the homomorphism of groups defined by setting $f_\#([\phi]) = [f \circ \phi]$. If $f$ is a homotopy equivalence, then $f_\#$ is an isomorphism of groups (cf. [Ht, Proposition 1.18]).

The fundamental group of a path connected space $X$ is closely related to its covering spaces.

**Definition 1.1** A *covering space* of a space $X$ is a space $\widetilde{X}$, together with a surjective map $f \colon \widetilde{X} \longrightarrow X$, such that each $x \in X$ has an open neighborhood $U \subseteq X$ for which $f^{-1}(U)$ is a disjoint union of open sets, each of which is sent homeomorphically to $U$. If $X$ is path connected, then $\widetilde{X}$ is a *universal covering space* if it is path connected and simply connected.

The simplest nontrivial example of a universal covering space is the map $f \colon \mathbb{R} \longrightarrow S^1$, defined by $f(t) = e^{2\pi i t}$. Here, for simplicity, we identify $S^1$ as the unit circle in the complex plane. Also, for each $n \in \mathbb{Z}$ with $n \neq 0$, the map $f_n \colon S^1 \longrightarrow S^1$ defined by $f_n(z) = z^n$ is a covering space over $S^1$.

One can show that each connected covering space $\widetilde{X} \longrightarrow S^1$, where $\widetilde{X}$ is path connected, is homeomorphic to one of these covering spaces $(\mathbb{R}, f)$ or $(S^1, f_n)$.

We refer to [Ht, §1.3] for more details about covering spaces.

Assume $f \colon \widetilde{X} \longrightarrow X$ is a universal covering space for $X$. Assume also that $X$ is locally path connected: each point has a family of arbitrarily small neighborhoods which are path connected. Let $G$ be the group of all homeomorphisms $h \colon \widetilde{X} \xrightarrow{\ \cong\ } \widetilde{X}$ such that $f \circ h = f$: the group of "deck transformations" of $\widetilde{X}$. By one of the fundamental theorems about covering spaces (cf. [Ht, Proposition 1.39]), $G \cong \pi_1(X)$, and $G$ acts freely and transitively on $f^{-1}(x)$ for each $x \in X$.

The following theorem describes a way to construct covering spaces with a given group of deck transformations.

**Proposition 1.2** *Fix a topological space $X$ which is path connected and locally path connected, and a discrete group $G$ which acts continuously on $X$. Assume $G$ acts freely and properly, in the sense that for each $x \in X$, there is an open neighborhood $U$ of $x$ such that $g(U) \cap U = \varnothing$ for each $1 \neq g \in G$. Let $X/G$ be the orbit space of the action (with the quotient topology), and let $f \colon X \longrightarrow X/G$ be the natural projection. Then $f$ is a covering space. For any $x_0 \in X$, with $G$-orbit $y_0 = f(x_0)$, the induced homomorphism $f_\#$ from $\pi_1(X, x_0)$ to $\pi_1(X/G, y_0)$ is injective, its image is normal in $\pi_1(X/G, y_0)$, and $\pi_1(X/G, y_0)/\mathrm{Im}(f_\#) \cong G$. In particular, if $X$ is simply connected, then $\pi_1(X/G, y_0) \cong G$.*

*Proof.* That $f \colon X \longrightarrow X/G$ is a covering space follows immediately from the assumption that $G$ acts freely and properly.

There is a surjective homomorphism $\chi \colon \pi_1(X/G, y_0) \longrightarrow G$ defined as follows. For each loop $\gamma$ in $X/G$ based at $y_0$, there is a unique path $\widetilde{\gamma}$ in $X$ such that $\widetilde{\gamma}(0) = x_0$ and $f \circ \widetilde{\gamma} = \gamma$ (cf. [Ht, Propositions 1.33 & 1.34]). Define $\chi([\gamma])$ to be the (unique) element $g \in G$ such that $\widetilde{\gamma}(1) = g(x_0)$. For any $g \in G$, there is a path $\phi$ from $x_0$ to $g(x_0)$, and $g = \chi([f \circ \phi])$. Thus $\chi$ is surjective. Also, $\mathrm{Ker}(\chi) = f_\#(\pi_1(X, x_0))$ by construction, and hence $\chi$ induces an isomorphism $\pi_1(X/G, y_0)/\mathrm{Im}(f_\#) \cong G$. The injectivity of $f_\#$ follows by lifting homotopies between paths, using [Ht, Propositions 1.33] again. $\qquad\square$

Returning to the example of a universal covering space $f \colon \mathbb{R} \longrightarrow S^1$ defined above, $f$ is the orbit map of the free $\mathbb{Z}$-action on $\mathbb{R}$ by translation. Hence this is the group of deck transformations of $f$, and defines an isomorphism $\mathbb{Z} \cong \pi_1(S^1)$. Under this isomorphism, each $n \in \mathbb{Z}$ is sent to the homotopy class of $f \circ \phi$, where $\phi \colon I \longrightarrow \mathbb{R}$ is any path from $0$ to $n$ (or from $k$ to $k + n$ for $k \in \mathbb{Z}$).

We will occasionally need to mention higher homotopy groups of spaces. For $n \geq 1$, let $I^n$ be the product of $n$ copies of the interval $I = [0,1]$. Let $\partial I^n$ be its boundary: the subspace of all $(t_1, \ldots, t_n) \in I^n$ such that for some $i$, $t_i \in \partial I = \{0,1\}$. When $X$ is a space and $x_0 \in X$, $\pi_n(X, x_0)$ is defined as the set of homotopy classes of maps $\phi \colon I^n \longrightarrow X$ such that $\phi(\partial I^n) = \{x_0\}$; more precisely, the homotopy classes via homotopies which are constant on $\partial I^n$. This set has the structure of a group via juxtaposition with respect to the first coordinate:

$$\phi \cdot \psi(t_1, \ldots, t_n) = \begin{cases} \phi(2t_1, t_2, \ldots, t_n) & \text{if } t_1 \leq \tfrac{1}{2} \\ \psi(2t_1 - 1, t_2, \ldots, t_n) & \text{if } t_1 \geq \tfrac{1}{2}. \end{cases}$$

Since $I^n/\partial I^n \cong S^n$, elements of $\pi_n(X, x_0)$ can also be regarded as homotopy classes of maps $S^n \longrightarrow X$ which send a basepoint to $x_0$.

When $n \geq 2$, $\pi_n(X, x_0)$ is an abelian group. Also, if $f \colon X \longrightarrow Y$ is a covering space, $x_0 \in X$, and $y_0 = f(x_0)$, then the homomorphism $f_{\#} \colon \pi_n(X, x_0) \longrightarrow \pi_n(Y, y_0)$ is an isomorphism for all $n \geq 2$. See, e.g., [Ht, pp. 338–340] or [Be2, §1.2,1.6] for more details, and for proofs of these facts.


## 1.2. CW complexes and cellular homology.

Intuitively, a *CW complex* is a space built up by starting with a discrete set (the 0-cells), and attaching higher dimensional cells (copies of the closed unit ball $D^n \subseteq \mathbb{R}^n$ for $n > 0$) via their boundaries. The following definition is one way of making this more precise.

**Definition 1.3**  A *CW complex* consists of a space $X$, together with a sequence of closed subspaces $\varnothing = X^{(-1)} \subseteq X^{(0)} \subseteq X^{(1)} \subseteq X^{(2)} \subseteq \cdots$ whose union is $X$ (the "skeleta" of $X$), indexing sets $J_n$ for all $n \geq 0$, and "characteristic maps"

$$\rho_j^n \colon D^n \longrightarrow X^{(n)} \qquad \text{for all } n \geq 0 \text{ and } j \in J_n$$

which satisfy the following conditions:

(i)  For each $n$, each $j \in J_n$, and each $t \in D^n$, $\rho_j^n(t) \in X^{(n-1)}$ if and only if $t \in S^{n-1}$.

(ii)  For each $n \geq 0$ and each $x \in X^{(n)} \setminus X^{(n-1)}$, there are unique $j \in J_n$ and $t \in D^n \setminus S^{n-1}$ such that $\rho_j^n(t) = x$.

(iii)  For each $n \geq 0$, a subset $U \subseteq X^{(n)}$ is open in $X^{(n)}$ if and only if $(\rho_j^m)^{-1}(U)$ is open in $D^m$ for each $m \leq n$ and each $j \in J_m$.

(iv)  A subset $U \subseteq X$ is open in $X$ if and only if $U \cap X^{(n)}$ is open in $X^{(n)}$ for each $n \geq 0$.

For more details (and slightly different but equivalent formulations), we refer to [Ht, pp.5–8 & Appendix] and to [Be2, §1.5].

When $X$ is a CW complex with skeleta and structural maps as above, let $C_n(X)$, for each $n \geq 0$, be the free group with basis the set $J_n$ of cells of dimension $n$. Let $x_j \in C_n(X)$ be the generator corresponding to $j \in J_n$.

A *boundary homomorphism* $\partial \colon C_n(X) \longrightarrow C_{n-1}(X)$ is defined as follows. Fix a pair of cells $j \in J_n$ and $k \in J_{n-1}$. For any $m$, let $D^m/S^{m-1}$ be the quotient space of $D^m$ where all points of the boundary $S^{m-1}$ are identified to a point. Then $D^m/S^{m-1}$ is homeomorphic to the $m$-sphere $S^m$, and we fix some identification. Consider the following map

$$\partial_{jk} \colon S^{n-1} \xrightarrow{\ \rho_j^n|_{S^{n-1}}\ } X^{(n-1)} \xrightarrow{\ \psi_k\ } D^{n-1}/S^{n-2} \cong S^{n-1},$$

where $\psi_k(x) = (\rho_k^{n-1})^{-1}(x)$ if $x \in \rho_k^{n-1}(D^{n-1}/S^{n-2})$, and $\psi_k(x)$ is the basepoint (the identification point of all elements of $S^{n-2}$) otherwise. Then $\psi_k$ is a continuous map from $X^{(n-1)}$ to $S^{n-1}$. Let $n_{jk} \in \mathbb{Z}$ be the degree of $\partial_{jk}$. By the compactness of $S^{n-1}$, for each $j \in J_n$, $\rho_j^n(S^{n-1})$ has nonempty intersection with $\rho_k^{n-1}(D^{n-1}/S^{n-2})$ for at most finitely many $k \in J_{n-1}$, so $n_{jk}$ is nonzero for at most finitely many $k$. We now define, for each $j \in J_n$, $\partial(x_j) = \sum_{k \in J_{n-1}} n_{jk} x_k$.

One can show that the composite of any two successive boundary maps is zero, and hence that $(C_*(X), \partial)$ is a chain complex. Then for any abelian group $A$ and any $q \geq 0$,

$$H_q(X; A) \cong H_q(C_*(X) \otimes_{\mathbb{Z}} A, \partial)$$
$$H^q(X; A) \cong H^q(\mathrm{Hom}_{\mathbb{Z}}(C_*(X), A), \partial^*).$$

We thus get a direct connection between the cell structure of $X$ and its homology and cohomology.

For a more detailed construction of the celluar chain complex, we refer to [Be2, § 1.5] and [Ht, § 2.2].


## 1.3. Classifying spaces of discrete groups.

We begin with the definition of an abstract classifying space of a discrete group $G$. A more explicit definition of a space which will be called *the* classifying space of $G$ will be given in Section 2.3.

**Definition 1.4**   A *classifying space* for a discrete group $G$ is a CW complex $BG$ which satisfies the following conditions:

(a)   $\pi_1(BG) \cong G$ (for any choice of basepoint); and

(b)   $BG$ has a universal covering space $EG$ which is contractible.

The assumption that a classifying space for $G$ is a CW complex has been put here for convenience, but is not really necessary. In most situations, this can be replaced by the condition that it is paracompact and has certain local properties.

The simplest example of a classifying space is the case where $G \cong \mathbb{Z}$: the circle $S^1$ is a classifying space, since its universal covering space is the real line, and its fundamental group is isomorphic to $\mathbb{Z}$.

To construct a slightly more complicated example, let $S^\infty$ be the union of the finite dimensional spheres: the space of all sequences $(x_0, x_1, x_2, \dots)$ such that $x_i = 0$ for $i$ sufficiently large, and $\sum_{i=0}^\infty x_i^2 = 1$. This space is, in fact, contractible; we describe here one way to see that. Let $\Delta^n \supseteq \partial\Delta^n$ denote the "standard $n$-simplex" and its boundary:

$$\Delta^n = \big\{(x_0, \dots, x_n) \in \mathbb{R}^{n+1} \,\big|\, x_i \geq 0 \text{ for each } i,\ \textstyle\sum_{i=0}^n x_i = 1\big\}$$
$$\partial\Delta^n = \big\{(x_0, \dots, x_n) \in \Delta^n \,\big|\, x_i = 0 \text{ for some } i\big\}.$$

Homeomorphisms $S^{n-1} \xrightarrow{\cong} \partial\Delta^n$ can be chosen for all $n$ in such a way as to be consistent with the inclusions, and hence $S^\infty$ is homeomorphic to the union of the $\partial\Delta^n$. This last space is convex, hence contractible, and so $S^\infty$ is contractible.

Since $S^\infty$ is the universal covering space of the infinite projective space $\mathbb{R}P^\infty$ (the union of the $\mathbb{R}P^n$), $\mathbb{R}P^\infty = S^\infty/(x \sim -x)$ is a classifying space for the cyclic group of order 2. More generally, if we identify $S^\infty$ with the unit sphere in $\mathbb{C}^\infty$, then for any $n > 1$, the group $\mu_n$ of $n$-th roots of unity acts freely on $S^\infty$ via scalar multiplication. So by Proposition 1.2, the quotient space $S^\infty/\mu_n$ of this action has fundamental group $\mu_n$ and universal covering space $S^\infty$, and is a classifying space for the cyclic group $\mu_n$.

We next describe a "universal property" of classifying spaces, which helps to explain the name and the origin of their interest to topologists. Very roughly, it says that for any path connected space $X$, maps from $X$ to $BG$ are "classified" by homomorphisms from $\pi_1(X)$ to $G$. We state the property here for CW complexes. A slightly different version will be given in Section 2.5 (Proposition 2.10), with a sketch of the proof. There are also versions which hold for paracompact spaces with certain local properties.

When $X$ and $Y$ have chosen basepoints $x_0 \in X$ and $y_0 \in Y$, a *pointed map* from $X$ to $Y$ is a map $f\colon X \longrightarrow Y$ such that $f(x_0) = y_0$. If $f$ and $f'$ are two pointed maps from $X$ to $Y$, then they are *pointed homotopic* if there is a homotopy $F\colon X \times I \longrightarrow Y$ such that $F(x,0) = f(x)$, $F(x,1) = f'(x)$, and $F(x_0, t) = y_0$ for all $x \in X$ and $t \in I$. The set of all pointed homotopy classes of pointed maps from $X$ to $Y$ will be denoted here $[X, Y]_*$. By comparison, the set of unpointed homotopy classes of (all) maps from $X$ to $Y$ will be denoted $[X, Y]$.

For any pair of groups $G, H$, define

$$\mathrm{Rep}(H, G) = \mathrm{Hom}(H, G)/\mathrm{Inn}(G).$$

**Proposition 1.5** *Fix a discrete group $G$, and a classifying space $BG$ for $G$ with basepoint $b_0 \in BG$. Identify $\pi_1(BG, b_0)$ with $G$. Then the following hold for any path connected CW complex $X$ with basepoint $x_0 \in X$.*

(a)  *The natural map of sets*

$$[X, BG]_* \xrightarrow{\ \cong\ } \mathrm{Hom}(\pi_1(X, x_0), G),$$

*which sends the class of a pointed map $f\colon X \longrightarrow BG$ to its induced homomorphism $f_\#$, is a bijection.*

(b)  *The natural map of sets*

$$[X, BG] \xrightarrow{\ \cong\ } \mathrm{Rep}(\pi_1(X, x_0), G)$$

*which sends the homotopy class of $f\colon X \longrightarrow BG$ to the class of $f_\#$ (modulo $\mathrm{Inn}(G)$) is a bijection. In particular, two pointed maps $f, f'\colon X \longrightarrow BG$ are (freely) homotopic if and only if the induced homomorphisms $f_\#$ and $f'_\#$ are $G$-conjugate.*

*Proof.* See, e.g., [Wh, Theorems V.4.3–4]. Point (a) is also shown in [Ht, Proposition 1B.9], and (b) follows easily from that.

To see the "only if" part of the last statement, let $F\colon X \times I \longrightarrow BG$ be a homotopy from $f$ to $f'$. Thus $F(x, 0) = f(x)$ and $F(x, 1) = f'(x)$ for all $x \in X$. Let $\phi\colon I \longrightarrow BG$ be the loop $\phi(t) = F(x_0, t)$, and set $g = [\phi] \in \pi_1(BG, b_0) = G$. Then for each loop $\gamma$ in $X$ based at $x_0$, the composite $F \circ (\gamma \times \mathrm{Id}_I)\colon I \times I \longrightarrow BG$ defines a homotopy $\phi \cdot (f \circ \gamma) \simeq (f' \circ \gamma) \cdot \phi$, so that $c_g(f_\#([\gamma])) = f'_\#([\gamma])$. $\qquad\qquad\square$

A proof of Proposition 1.5 will be sketched in Section 2.5 (Proposition 2.10), in the case where $X$ is the geometric realization of a simplicial set. In addition, some information will be given there about the individual connected components of the space of maps.

One important special case of Proposition 1.5 is that of maps between classifying spaces of discrete groups.

**Corollary 1.6** *If $G$ and $H$ are two discrete groups, with classifying spaces $BG$ and $BH$, then $[BG, BH]_* \cong \mathrm{Hom}(G, H)$ and $[BG, BH] \cong \mathrm{Rep}(G, H)$.*

In particular, Corollary 1.6 implies that two classifying spaces of the same group (or of isomorphic groups) have the same homotopy type.

There is also a close connection between classifying spaces and group (co)homology.

**Proposition 1.7** *Let $BG$ be a classifying space of the discrete group $G$. Then $H_*(BG; A) \cong H_*(G; A)$ (and $H^*(BG; A) \cong H^*(G; A)$) for all abelian groups $A$.*

*Proof.* See, e.g., [Be2, Theorem 2.2.3]. We give here a very rough sketch of a proof.

Let $EG$ be the universal covering space of $BG$, and let $C_*(EG)$ be its cellular chain complex (see Section 1.2). The sequence

$$\cdots \xrightarrow{\partial} C_1(EG) \xrightarrow{\partial} C_0(EG) \longrightarrow \mathbb{Z} \longrightarrow 0$$

is exact (the homology of $EG$ is isomorphic to that of a point since $EG$ is contractible), and each group $C_q(EG)$ has a natural basis which is permuted freely by $G$. Thus $C_*(EG)$ is a $\mathbb{Z}[G]$-free resolution of $\mathbb{Z}$. In addition, $C_*(BG) \cong C_*(EG) \otimes_{\mathbb{Z}[G]} \mathbb{Z}$, and hence for each $q \geq 0$,

$$H_q(BG; A) \cong H_q\big(C_*(EG) \otimes_{\mathbb{Z}[G]} A, \partial\big)$$
$$H^q(BG; A) \cong H^q\big(\mathrm{Hom}_{\mathbb{Z}[G]}(C_*(EG), A), \partial^*\big).$$

The result now follows from the definition of group (co)homology. $\square$

In fact, Proposition 1.7 also holds when $A$ is an arbitrary $\mathbb{Z}[G]$-module, once one defines appropriately $H_*(X, A)$ and $H^*(X; A)$ for a path connected space $X$ and a $\mathbb{Z}[\pi_1(X)]$-module $A$ ("homology with twisted coefficients").

## 1.4. **The $p$-completion functor of Bousfield and Kan.**

The Bousfield-Kan $p$-completion functor is a functor from spaces to spaces, denoted here $X \mapsto X_p^\wedge$, together with a natural transformation $\phi \colon \mathrm{Id} \longrightarrow (-)_p^\wedge$. More precisely, it is a functor from the category of simplicial sets to itself (see Section 2.1), but we suppress such details in this presentation. We refer to [BK] for the precise definition of this functor, and for most of the properties referred to here. Another source for the definition of $p$-completion and some of its properties is [GJ, § VIII.3].

A map of spaces $f \colon X \longrightarrow Y$ will be called a *$p$-equivalence* if $f$ induces an isomorphism from $H_*(X; \mathbb{F}_p)$ to $H_*(Y; \mathbb{F}_p)$. Note that since $\mathbb{F}_p$ is a field, mod $p$ cohomology is the dual of mod $p$ homology, and so $f$ is a $p$-equivalence if and only if it induces an isomorphism in mod $p$ cohomology.

Many of the important properties of $p$-completions hold only on certain classes of spaces. The next proposition describes one of the few properties which holds for all spaces.

**Proposition 1.8** ([BK, Lemma I.5.5])  *A map $f\colon X \longrightarrow Y$ induces a homotopy equivalence $f_p^\wedge\colon X_p^\wedge \longrightarrow Y_p^\wedge$ if and only if it is a p-equivalence.*

When working with $p$-completion, the classes of $p$-complete spaces and $p$-good spaces play a central role.

**Definition 1.9**  A space $X$ is

- *p-complete* if $\phi_X\colon X \longrightarrow X_p^\wedge$ is a homotopy equivalence; and is

- *p-good* if $\phi_X$ is a $p$-equivalence.

By [BK, Proposition I.5.2], a space $X$ is $p$-good if and only if $X_p^\wedge$ is $p$-complete (i.e., $X_p^\wedge \simeq (X_p^\wedge)_p^\wedge$).

One important family of examples of $p$-complete spaces are the classifying spaces of (finite) $p$-groups.

**Proposition 1.10**  *The classifying space of any p-group is p-complete.*

*Proof.* This is very well known, but does not seem to be stated explicitly anywhere in [BK]. If $P$ is an abelian $p$-group, then by [BK, Example VI.5.2], $\pi_n(BP_p^\wedge) \cong \mathbb{Z}_p \otimes_\mathbb{Z} \pi_n(BP)$ for each $n \geq 1$. Thus $BP_p^\wedge$ is again a classifying space for $P$, and so $BP$ is $p$-complete.

Now let $P$ be an arbitrary $p$-group, and set $Z = Z(P)$. We can assume by induction on $|P|$ that $B(P/Z)$ is $p$-complete. There is a fibration $BP \longrightarrow B(P/Z)$ with fiber $BZ$. By the mod-$\mathbb{F}_p$ fiber lemma of Bousfield and Kan (see Lemma II.5.1 and Example II.5.2(iv) in [BK]), and since $\pi_1(B(P/Z)) \cong P/Z$ is a finite $p$-group, the $p$-completed map $BP_p^\wedge \longrightarrow B(P/Z)_p^\wedge$ is a fibration with fiber homotopy equivalent to $BZ_p^\wedge$. (More generally, by [BK, Lemma II.5.1], if $E \longrightarrow B$ is a fibration with fiber $F$, where all three spaces are connected and the action of $\pi_1(B)$ on $H_i(F;\mathbb{F}_p)$ is nilpotent for all $i$, then $E_p^\wedge \longrightarrow B_p^\wedge$ is also a fibration and has fiber homotopy equivalent to $F_p^\wedge$.) Since $BZ$ and $B(P/Z)$ are both $p$-complete, the map $\phi_{BP}\colon BP \longrightarrow BP_p^\wedge$ must be a homotopy equivalence, and thus $BP$ is $p$-complete. $\qquad\square$

If a space is $p$-bad (i.e., not $p$-good), then in fact, all of its iterated $p$-completions are also $p$-bad [BK, Proposition I.5.2]. The simplest example of a $p$-bad space is $S^1 \vee S^1$: the one point union of two circles [Bf, Theorem 11.1]. In fact, $S^1 \vee S^n$ (the one-point union of a circle and an $n$-sphere) is $p$-bad for all $p$ and all $n \geq 2$ [Bf, Theorem 10.1]. So even a space with very nice fundamental group (the integers) can be $p$-bad.

The most important criterion for checking that a space is $p$-good, at least for the spaces of the type we work with here, is given in the following proposition. As usual, a group $G$ is called *p-perfect* if $H_1(G;\mathbb{F}_p) = 0$; equivalently, if $G$ is generated by its commutators and $p$-th powers.

**Proposition 1.11** *A connected space $X$ is $p$-good if $\pi_1(X)$ is finite; in particular, if $X$ is simply connected. More generally, $X$ is $p$-good if $\pi_1(X)$ contains a $p$-perfect subgroup of finite index. In this situation, if $K \trianglelefteq \pi_1(X)$ is the maximal $p$-perfect subgroup of $\pi_1(X)$, then $\pi_1(X_p^\wedge) \cong \pi_1(X)/K$.*

*Proof.* The first statement is shown in [BK, Proposition VII.5.1]. The other two statements are implicit in [BK], but do not seem to be proven there explicitly. So we sketch a proof here.

Since any group generated by $p$-perfect subgroups is $p$-perfect, there is a unique maximal $p$-perfect subgroup $K \trianglelefteq \pi_1(X)$ which is normal. Set $\pi = \pi_1(X)/K$ for short. Since $\pi$ is finite and contains no nontrivial $p$-perfect subgroups, it must be a $p$-group.

Let $\widetilde{X}$ be the connected covering space of $X$ with fundamental group $K$. Since $\pi_1(\widetilde{X})$ is $p$-perfect, $\widetilde{X}$ is $p$-good and $\widetilde{X}_p^\wedge$ is simply connected by [BK, VII.3.2].

Since $\pi$ is a $p$-group, the augmentation ideal in the group ring $\mathbb{F}_p[\pi]$ is nilpotent, and hence every $\mathbb{F}_p[\pi]$-module is nilpotent. In particular, $H_i(\widetilde{X}; \mathbb{F}_p)$ is nilpotent as an $\mathbb{F}_p[\pi]$-module for all $i$. Hence the mod-$\mathbb{F}_p$ fiber lemma of Bousfield and Kan ([BK, II.5.1]) implies that the fibration sequence $\widetilde{X} \longrightarrow X \longrightarrow B\pi$ is still a fibration sequence after $p$-completion. Since $B\pi$ is $p$-complete by Proposition 1.10, this means that $X_p^\wedge \longrightarrow B\pi$ is a fibration with fiber $\widetilde{X}_p^\wedge$ and total space $X_p^\wedge$, and that $\pi_1(X_p^\wedge) \cong \pi$. Upon applying the mod-$\mathbb{F}_p$ fiber lemma to this new sequence, we see that $X_p^\wedge \simeq (X_p^\wedge)_p^\wedge$ is $p$-complete, and hence that $X$ is $p$-good. Also, the universal covering space of $X_p^\wedge$ is $\widetilde{X}_p^\wedge$. $\qquad\square$

Thus, for example, when $\pi_1(X)$ is finite, $\pi_1(X_p^\wedge) \cong \pi_1(X)/O^p(\pi_1(X))$. Most importantly for our purposes here, for any finite group $G$ and any classifying space $BG$ of $G$, $BG$ is $p$-good and $\pi_1(BG_p^\wedge) \cong G/O^p(G)$.

The next proposition states that when $X$ is $p$-good, $\phi_X : X \longrightarrow X_p^\wedge$ is universal among all $p$-equivalences $X \longrightarrow Y$.

**Proposition 1.12** *For any $p$-good space $X$, and any $p$-equivalence $f : X \longrightarrow Y$, there is a map $g : Y \longrightarrow X_p^\wedge$, unique up to homotopy, such that $g \circ f \simeq \phi_X$. Thus $\phi_X : X \longrightarrow X_p^\wedge$ is a final object among homotopy classes of mod $p$ equivalences defined on $X$.*

*Proof.* By [BK, Lemma I.5.5], any $p$-equivalence $f : X \longrightarrow Y$ induces a homotopy equivalence $f_p^\wedge : X_p^\wedge \longrightarrow Y_p^\wedge$. So $g$ is defined to be $\phi_Y$ followed by a homotopy inverse to $f_p^\wedge$. $\qquad\square$

As one consequence of Proposition 1.12, we have the following condition for the $p$-completions of two spaces to be homotopy equivalent. Note that this condition does not in itself involve $p$-completion.

**Corollary 1.13**  *If $X$ and $Y$ are two spaces, and one of them is $p$-good, then their $p$-completions are homotopy equivalent if and only if there exists some space $Z$, and maps $X \xrightarrow{\ f\ } Z \xleftarrow{\ g\ } Y$, such that $f$ and $g$ are both $p$-equivalences.*

## 1.5. Equivalences between fusion systems of finite groups.

Recall that for any finite group $G$ and any $S \in \mathrm{Syl}_p(G)$, the fusion category of $G$ over $S$ is the category $\mathcal{F}_S(G)$, whose objects are the subgroups of $S$, and where for each $P, Q \leq S$,

$$\mathrm{Mor}_{\mathcal{F}_S(G)}(P,Q) = \mathrm{Hom}_G(P,Q)$$

$$\overset{\mathrm{def}}{=} \{\varphi \in \mathrm{Hom}(P,Q) \mid \varphi = c_g,\ \text{some } g \in G \text{ such that } {}^g P \leq Q\}.$$

Our goal here is to study the relationship between this category $\mathcal{F}_S(G)$ and the homotopy type of the $p$-completed classifying space $BG_p^\wedge$. In order to state precise results, we first need some terminology for describing equivalences between fusion categories.

**Definition 1.14**  Fix a prime $p$, a pair of finite groups $G_1$ and $G_2$, and Sylow $p$-subgroups $S_i \in \mathrm{Syl}_p(G_i)$. An isomorphism $\varphi \colon S_1 \xrightarrow{\ \cong\ } S_2$ is *fusion preserving* if for all $P, Q \leq S_1$ and all $\alpha \in \mathrm{Hom}(P,Q)$,

$$\alpha \in \mathrm{Hom}_{G_1}(P,Q) \quad \Longleftrightarrow \quad {}^\varphi\alpha \overset{\mathrm{def}}{=} (\varphi|_Q)\alpha(\varphi|_P)^{-1} \in \mathrm{Hom}_{G_2}(\varphi(P),\varphi(Q)).$$

In other words, in the above situation, an isomorphism $\varphi \colon S_1 \longrightarrow S_2$ is fusion preserving if and only if it induces an isomorphism of categories from $\mathcal{F}_{S_1}(G_1)$ to $\mathcal{F}_{S_2}(G_2)$ by sending $P$ to $\varphi(P)$ and $\alpha$ to ${}^\varphi\alpha$. The next proposition says, roughly, that there is a fusion preserving isomorphism between Sylow $p$-subgroups of $G_1$ and $G_2$ if and only if there is a bijection between conjugacy classes of $p$-subgroups of $G_1$ and $G_2$. Recall that for any pair of groups $H$ and $G$, $\mathrm{Rep}(H,G) = \mathrm{Hom}(H,G)/\mathrm{Inn}(G)$.

**Proposition 1.15**  *Fix a pair of finite groups $G_1$ and $G_2$, a prime $p$ and Sylow $p$-subgroups $S_i \leq G_i$. Then there is a fusion preserving isomorphism $\varphi \colon S_1 \xrightarrow{\ \cong\ } S_2$ if and only if there are bijections*

$$\beta_P \colon \mathrm{Rep}(P,G_1) \xrightarrow{\ \cong\ } \mathrm{Rep}(P,G_2),$$

*defined for all $p$-groups $P$ and natural with respect to homomorphisms.*

*Proof.* This is due mostly to Martino and Priddy [MP2]. We sketch a proof here. See also [BMO, Proposition 1.3] for more details.

Assume first $\varphi\colon S_1 \xrightarrow{\cong} S_2$ is a fusion preserving isomorphism. For each $p$-group $P$ and each $\alpha \in \mathrm{Hom}(P, G_1)$, choose $g \in G_1$ such that $c_g(\alpha(P)) \le S_1$, and set $\beta_P([\alpha]) = [\varphi \circ c_g \circ \alpha] \in \mathrm{Rep}(P, G_2)$. If $h \in G_1$ is another choice of element such that $c_h(\alpha(P)) \le S_1$, then $c_{hg^{-1}} \in \mathrm{Iso}_{G_1}({}^g(\alpha(P)), {}^h(\alpha(P)))$, so there is $x \in G_2$ such that ${}^\varphi(c_{hg^{-1}}) = c_x \in \mathrm{Iso}_{G_2}(\varphi(c_g(\alpha(P))), \varphi(c_h(\alpha(P))))$, and hence $[\varphi \circ c_g \circ \alpha] = [\varphi \circ c_h \circ \alpha]$ in $\mathrm{Rep}(P, G_2)$. Thus $\beta_P$ is well defined, and it is a bijection since an inverse can be defined via composition with $\varphi^{-1}$. Clearly, the $\beta_P$ are natural in $P$.

Now assume that $\beta_P$ are natural bijections, defined for all $p$-groups $P$. Choose $\varphi \in \mathrm{Hom}(S_1, S_2)$ and $\psi \in \mathrm{Hom}(S_2, S_1)$ such that $[\varphi] = \beta_{S_1}([\mathrm{Id}_{S_1}])$ and $[\psi] = \beta_{S_2}([\mathrm{Id}_{S_2}])$. If $\varphi$ is not injective, then it factors through $\bar{\varphi} \in \mathrm{Hom}(S_1/K, S_2)$ for some $1 \ne K \trianglelefteq S_1$, and by the naturality of the $\beta$ with respect to $S_1 \longrightarrow S_1/K$, $\mathrm{Id}_{S_1}$ must also factor through $S_1/K$. Since this is not the case, we conclude that $\varphi$ is injective; and also (by similar reasoning) that $\psi$ is injective. In particular, $|S_1| = |S_2|$, and so $\varphi$ and $\psi$ are both isomorphisms.

Now fix $P, Q \le S_1$ and $\alpha \in \mathrm{Hom}(P, Q)$. By the naturality of the $\beta$, $\beta_P$ sends $[\alpha] \in \mathrm{Rep}(P, G_1)$ to $[\varphi \circ \alpha] \in \mathrm{Rep}(P, G_2)$. Similarly, by naturality with respect to the inclusion $\mathrm{incl}_P^{S_1}$, $\beta_P([\mathrm{incl}_P^{G_1}]) = [\varphi|_P]$. Since $\alpha \in \mathrm{Hom}_{G_1}(P, Q)$ if and only if $[\alpha] = [\mathrm{incl}_P^{G_1}]$ in $\mathrm{Rep}(P, G_1)$, this is the case exactly when $[\varphi \circ \alpha] = [\varphi|_P]$ in $\mathrm{Rep}(P, G_2)$, or equivalently, ${}^\varphi\alpha \in \mathrm{Hom}_{G_2}(\varphi(P), \varphi(Q))$. Thus $\varphi$ is fusion preserving. $\qquad\square$

### 1.6. The Martino-Priddy conjecture.

The Martino-Priddy conjecture states roughly that two finite groups have the same $p$-local structure if and only if their classifying spaces have the same $p$-local structure. Here, the "$p$-local structure" of a group $G$ translates to mean its fusion at $p$, and the $p$-local structure of a space means the homotopy type of its $p$-completion.

The first hint that some such result might be true came from the following theorem of Mislin.

**Theorem 1.16** ([Ms, pp. 457–458]) *For any prime $p$, any $p$-group $Q$, and any finite group $G$, the natural map*

$$\mathrm{Rep}(Q, G) \xrightarrow{\cong} [BQ, BG_p^\wedge]$$

*is a bijection.*

Upon comparing Theorem 1.16 with Corollary 1.6, we see that $[BQ, BG_p^\wedge] \cong [BQ, BG]$ in this situation. Mislin's theorem is, however, a much deeper result than Corollary 1.6, and depends for its proof on the

Sullivan conjecture in the form shown by Carlsson [Ca], Lannes [La], and Miller [Mi].

If $G_1$ and $G_2$ are two finite groups such that $BG_{1p}^{\wedge} \simeq BG_{2p}^{\wedge}$, then by Mislin's theorem, there are bijections

$$\mathrm{Rep}(Q, G_1) \xrightarrow{\;\cong\;} \mathrm{Rep}(Q, G_2),$$

defined for all $p$-groups $Q$, and natural with respect to group homomorphisms. Hence by Proposition 1.15, there is a fusion preserving isomorphism between Sylow $p$-subgroups of $G_1$ and of $G_2$. In fact, the converse to this is also true.

**Theorem 1.17** (Martino-Priddy conjecture)  *For any prime $p$, and any pair $G_1, G_2$ of finite groups, $BG_{1p}^{\wedge} \simeq BG_{2p}^{\wedge}$ if and only if there is a fusion preserving isomorphism $S_1 \xrightarrow[\cong]{\;\varphi\;} S_2$ between Sylow $p$-subgroups $S_i \leq G_i$.*

The "if" part of Theorem 1.17 was proven for odd primes $p$ in [O2, Theorem B], and for $p = 2$ in [O3, Theorem B]. It follows from a combination of several results stated later in this survey. We describe in Theorem 3.7 how it is reduced to a question about uniqueness of linking categories (as defined in Section 3.1). The obstruction groups for answering that question are described in Proposition 5.11; and some of the techniques used to prove they vanish are discussed in Section 5.3 (see, e.g., Theorem 5.13). It is the proof that the obstruction groups vanish which fills most of [O2] and [O3], and which depends on the classification theorem for finite simple groups.

Theorem 1.17 can be thought of as a refinement of a classical theorem of Cartan and Eilenberg in group cohomology (see [CE, Theorem XII.10.1]). Their theorem describes how for any finite group $G$, the cohomology ring $H^*(BG; \mathbb{F}_p)$ is determined by $S \in \mathrm{Syl}_p(G)$ and fusion in $S$. By comparison, Theorem 1.17 says that if $G_1$ and $G_2$ have isomorphic Sylow $p$-subgroups and the same fusion at $p$, then not only is $H^*(G_1; \mathbb{F}_p) \cong H^*(BG_1; \mathbb{F}_p)$ abstractly isomorphic to $H^*(G_2; \mathbb{F}_p) \cong H^*(BG_2; \mathbb{F}_p)$, but this isomorphism is realized by maps of spaces $BG_1 \longrightarrow Z \longleftarrow BG_2$ (see Corollary 1.13).

### 1.7. An application: fusion in finite groups of Lie type.

We describe here one application of the "only if" part of the Theorem 1.17: the part proven by Martino and Priddy. So it does depend on Mislin's theorem and through that on the proofs of the Sullivan conjecture, but it does not depend on the classification of finite simple groups.

The following is the main result in a paper by Broto, Møller, and Oliver [BMO].

**Theorem 1.18** ([BMO, Theorem A])  *Fix a prime $p$, a connected reductive integral group scheme $\mathbb{G}$, and a pair of prime powers $q$ and $q'$ both prime to $p$.*

(a)  *If $\overline{\langle q \rangle} = \overline{\langle q' \rangle}$ as closed subgroups of $\mathbb{Z}_p^\times$, then there is a fusion preserving isomorphism between Sylow $p$-subgroups of $\mathbb{G}(q)$ and $\mathbb{G}(q')$.*

(b)  *If $\mathbb{G}$ is of type $A_n$, $D_n$, or $E_6$, $\tau$ is a graph automorphism of $\mathbb{G}$, and $\overline{\langle q \rangle} = \overline{\langle q' \rangle}$ as closed subgroups of $\mathbb{Z}_p^\times$, then there is a fusion preserving isomorphism between Sylow $p$-subgroups of ${}^\tau\mathbb{G}(q)$ and ${}^\tau\mathbb{G}(q')$.*

(c)  *If the Weyl group of $\mathbb{G}$ contains an element which acts on the maximal torus by inverting all elements, and $\overline{\langle -1, q \rangle} = \overline{\langle -1, q' \rangle}$ as closed subgroups of $\mathbb{Z}_p^\times$, then there is a fusion preserving isomorphism between Sylow $p$-subgroups of $\mathbb{G}(q)$ and $\mathbb{G}(q')$ (and between Sylow $p$-subgroups of ${}^\tau\mathbb{G}(q)$ and ${}^\tau\mathbb{G}(q')$ if $\tau$ is as in (b)).*

(d)  *If $\mathbb{G}$ is of type $A_n$, $D_n$ for $n$ odd, or $E_6$, $\tau$ is a graph automorphism of $\mathbb{G}$ of order 2, and $\overline{\langle -q \rangle} = \overline{\langle q' \rangle}$ as closed subgroups of $\mathbb{Z}_p^\times$, then there is a fusion preserving isomorphism between Sylow $p$-subgroups of ${}^\tau\mathbb{G}(q)$ and $\mathbb{G}(q')$.*

When $p = 2$, $q$ and $q'$ generate the same closed subgroup of $\mathbb{Z}_2^\times$ if and only if $q \equiv q' \pmod 8$, and $v_2(q^2 - 1) = v_2(q'^2 - 1)$. Thus, for example, if $q, q' \equiv 7 \pmod{16}$ and $q'' \equiv 9 \pmod{16}$, then for $n \geq 2$, $SL_n(q)$, $SL_n(q')$, and $SU_n(q'')$ all have the same 2-fusion. We refer to [BMO] for more examples (also at odd primes), and also for more precise details on how and when this theorem can be applied.

We know of no purely algebraic proof of Theorem 1.18. The proof in [BMO] is carried out by showing in each case that the $p$-completed classifying spaces of the two groups are homotopy equivalent. The starting point for doing this is a theorem of Friedlander [Fr, Theorem 12.2], which describes $B\mathbb{G}(q)_p^\wedge$ or $B{}^\tau\mathbb{G}(q)_p^\wedge$ as the "homotopy fixed space" of a certain self-equivalence of $B\mathbb{G}(\mathbb{C})_p^\wedge$. This was then combined with, among other things, a general result [BMO, Theorem 2.4] stating that under certain conditions on a space $X$, the homotopy fixed spaces of two self-equivalences are homotopy equivalent if they generate the same closed subgroup in the group $\mathrm{Out}(X)$ of homotopy classes of all self-equivalences of $X$.

## 2. THE GEOMETRIC REALIZATION OF A CATEGORY

Roughly speaking, the geometric realization of a category $\mathcal{C}$ is the space obtained starting with a disjoint set of vertices, one for each object in $\mathcal{C}$, then attaching one edge for each morphism (where the endpoints of the edge are attached to the source and target of the morphism), then one 2-simplex (triangle) for each commutative triangle of morphisms in $\mathcal{C}$, etc. Also, the edge corresponding to an identity morphism $\mathrm{Id}_c$ for $c \in$

$\mathrm{Ob}(\mathcal{C})$ is identified to the vertex corresponding to $c$, and similarly for higher dimensional simplices which involve identity morphisms.

In order to make this construction more precise, we begin by defining simplicial sets and their geometric realizations.

## 2.1. Simplicial sets and their realizations.

Let $\Delta$ be the *simplicial category*: the category whose objects are the sets $[n] = \{0, 1, \ldots, n\}$ for $n \geq 0$, and whose morphisms are the order preserving maps between objects. For each $n$, there are $n + 1$ *face morphisms* $d_n^i \in \mathrm{Mor}_\Delta([n-1], [n])$ $(0 \leq i \leq n)$, where $d_n^i$ is the (unique) injective morphism whose image does not contain $i$. Also, there are $n$ *degeneracy morphisms* $s_n^i \in \mathrm{Mor}_\Delta([n], [n-1])$ $(0 \leq i \leq n - 1)$, where $s_n^i$ is the unique surjective morphism such that $s_n^i(i) = s_n^i(i+1) = i$. One easily sees that each morphism in $\Delta$ is a composite of such face and degeneracy morphisms.

A *simplicial set* is a functor $K \colon \Delta^{\mathrm{op}} \longrightarrow \mathrm{Sets}$. If $K$ is a simplicial set, we often write $K_n = K([n])$, which is regarded as the set of "$n$-simplices" of $K$. For each $n$-simplex $\sigma \in K_n$, the codimension one faces of $\sigma$ are the $(n-1)$-simplices $d_n^{i\,*}(\sigma) \in K_{n-1}$. If $\sigma \in \mathrm{Im}(s_n^{i\,*})$ for some $i$, then $\sigma$ is called a *degenerate* simplex. Equivalently, $\sigma$ is degenerate if $\sigma \in \mathrm{Im}(\chi^*)$ for any $\chi \in \mathrm{Mor}(\Delta)$ which is not injective.

A *map of simplicial sets* or a *simplicial map* from $K$ to $L$ is a natural transformation of functors from $K$ to $L$; i.e., a sequence $\varphi = \{\varphi_n\}_{n=0}^\infty$ of maps $\varphi_n \colon K_n \longrightarrow L_n$ which commute with all face and degeneracy morphisms. We let $\mathbf{S}$ denote the category of simplicial sets, and write $\mathrm{Mor}_{\mathbf{S}}(K, L)$ for the set of all simplicial maps from $K$ to $L$.

Let $\{e_0, \ldots, e_n\}$ be the canonical basis for $\mathbb{R}^{n+1}$. Let $\Delta^n$ be the $n$-simplex spanned by these elements; i.e.,

$$\Delta^n = \left\{ (t_0, \ldots, t_n) \in \mathbb{R}^{n+1} \,\middle|\, 0 \leq t_i \leq 1, \sum_{i=0}^{n} t_i = 1 \right\}.$$

There is an obvious functor $\Delta \longrightarrow \mathrm{Top}$ which sends each object $[n]$ to $\Delta^n$, and which sends a morphism $\varphi \in \mathrm{Mor}_\Delta([n], [m])$ to the affine map $\varphi_* \colon \Delta^n \longrightarrow \Delta^m$ which sends a vertex $e_i \in \Delta^n$ to $e_{\varphi(i)} \in \Delta^m$. Equivalently, $\varphi_*(t_0, \ldots, t_n) = (s_0, \ldots, s_m)$, where $s_i = \sum_{j \in \varphi^{-1}(i)} t_j$ (and $s_i = 0$ if $\varphi^{-1}(i) = \varnothing$).

The *geometric realization* $|K|$ of a simplicial set $K$ is now defined by setting

$$|K| = \left( \coprod_{n=0}^{\infty} K_n \times \Delta^n \right) \middle/ \sim$$

(with the quotient topology), where

$$(\sigma, \varphi_*(\tau)) \sim (\varphi^*(\sigma), \tau) \qquad \forall\, \sigma \in K_m,\, \tau \in \Delta^n, \text{ and } \varphi \in \mathrm{Mor}_\Delta([n], [m]).$$

Thus $|K|$ has the structure of a CW complex with one vertex for each $\sigma \in K_0$, an edge $\Delta^1$ for each nondegenerate element of $K_1$, a 2-cell for each nondegenerate element of $K_2$, etc.

We can regard the geometric realization $|K|$ of a simplicial set $K$ as a type of "generalized" simplicial complex, but not as a simplicial complex in the strict sense. For example, the two endpoints of a (nondegenerate) edge can be equal, and there can be several different edges between the same pair of vertices.

For example, let $\Delta_m$, for any $m \geq 0$, be the simplicial set defined by setting $\Delta_m([n]) = \mathrm{Mor}_\Delta([n], [m])$. Let $\partial\Delta_m \subseteq \Delta_m$ denote its boundary: $\partial\Delta_m([n])$ is the set of all maps in $\mathrm{Mor}_\Delta([n], [m])$ which are not surjective. Let $\Delta_m/\partial\Delta_m$ be the quotient simplicial set, which sends $[n]$ to the quotient set $\Delta_m([n])/\partial\Delta_m([n])$ (i.e., all elements of the subset are identified to a point). Then $|\Delta_m|$ is an $m$-simplex, as a topological space in the usual sense, $|\partial\Delta_m|$ is its boundary (homeomorphic to $S^{m-1}$), and $|\Delta_m/\partial\Delta_m|$ is homeomorphic to $D^m/S^{m-1} \cong S^m$. Thus $\partial\Delta_{m+1}$ and $\Delta_m/\partial\Delta_m$ are two very different simplicial sets whose geometric realizations are homeomorphic to the sphere $S^m$.

For any simplicial set $K$, let $C_n(K)$, for each $n \geq 0$, be the free abelian group with basis $K_n = K([n])$. There are "boundary maps"

$$\partial\colon C_n(K) \longrightarrow C_{n-1}(K),$$

defined by setting $\partial(\sigma) = \sum_{i=0}^n (-1)^i d_n^{i\,*}(\sigma)$ for each $\sigma \in K_n$. The homology $H_*(K; A)$ of $K$ with coefficients in an abelian group $A$ is defined to be the homology of the chain complex

$$\longrightarrow C_n(K) \otimes_\mathbb{Z} A \xrightarrow{\partial \otimes \mathrm{Id}_A} C_{n-1}(K) \otimes_\mathbb{Z} A \xrightarrow{\partial \otimes \mathrm{Id}_A}$$

$$\cdots \xrightarrow{\partial \otimes \mathrm{Id}_A} C_0(K) \otimes_\mathbb{Z} A \longrightarrow 0.$$

The cohomology $H^*(K; A)$ of $K$ with coefficients in $A$ is the homology of the dual chain complex with terms $\mathrm{Hom}_\mathbb{Z}(C_n(K), A)$.

The (co)homology groups of the simplicial set $K$ are naturally isomorphic to the cellular (co)homology groups of its geometric realization $|K|$, when regarded as a CW complex. To see this, let $D_n \subseteq C_n(K)$ be the subgroup generated by all degenerate $n$-simplices. The natural map of chain complexes $(C_*(K), \partial) \longrightarrow (C_*(K)/D_*, \partial)$ induces an isomorphism in homology (cf. [McL, Theorem VIII.6.1]), and so $H_*(K; A) \cong H_*(|K|; A)$ (and likewise in cohomology) since the cells in $|K|$ as a CW complex correspond to the nondegenerate simplices in $K$.

The *singular simplicial set* $S.(X)$ of a topological space $X$ is defined by sending $[n]$ to the set $S_n(X)$ of all continuous maps $\Delta^n \longrightarrow X$, and by sending a morphism $\varphi$ to composition by $\varphi_*$. The geometric realization $|S.(X)|$ has the same homotopy and homology groups as $X$, and has the homotopy type of $X$ whenever $X$ is sufficiently "nice" (for example, when it has the structure of a CW complex). By definition, the singular (co)homology of $X$ is just the (co)homology of this simplicial set $S.(X)$.

It is an easy exercise to construct a bijection

$$\operatorname{Mor}_{\mathbf{S}}(K, S.(X)) \xrightarrow{\ \cong\ } \operatorname{map}(|K|, X),$$

for any simplicial set $K$ and any space $X$. Thus the functor $S.(-)$ from spaces to simplicial sets is right adjoint to the realization functor.

Simplicial sets behave very nicely with respect to products (and this is one way in which the "degenerate simplices" play a key role). The product $K \times L$ of two simplicial sets $K$ and $L$ is defined simply by setting $(K \times L)_n = K_n \times L_n$, and similarly for induced morphisms. One of the basic theorems in the subject says that the map of spaces

$$|K \times L| \longrightarrow |K| \times |L|$$

induced by the obvious projections of simplicial sets is always a continuous bijection, and is a homeomorphism under certain conditions (for example, if $K$ or $L$ has finitely many nondegenerate simplices). See [GZ, Section III.3] for more details about this and a proof. For our purposes here, we use this only in the case where $L = \Delta_m$ for some $m \geq 1$, and hence where $|L|$ is a simplex. Showing that $|K \times \Delta_m|$ is homeomorphic to $|K| \times \Delta^m$ is an easy exercise, and gives insight into why this also works in the general case — and also into the importance of degenerate simplices in a simplicial set.

## 2.2. The nerve of a category as a simplicial set.

Now let $\mathcal{C}$ be a small category. The *nerve* of $\mathcal{C}$ is a simplicial set $\mathcal{N}(\mathcal{C})$, defined by

$$\mathcal{N}(\mathcal{C})_n = \mathcal{N}(\mathcal{C})([n]) \overset{\text{def}}{=}$$

$$\left\{ c_0 \xrightarrow{\ \alpha_1\ } c_1 \xrightarrow{\ \alpha_2\ } \cdots \xrightarrow{\ \alpha_n\ } c_n \,\middle|\, c_i \in \operatorname{Ob}(\mathcal{C}),\ \alpha_i \in \operatorname{Mor}(\mathcal{C}) \right\},$$

and where $\varphi \in \operatorname{Mor}_\Delta([n], [m])$ sends $\mathcal{N}(\mathcal{C})_m$ to $\mathcal{N}(\mathcal{C})_n$ by composing morphisms or inserting identity morphisms as appropriate. The geometric realization $|\mathcal{C}|$ of the category $\mathcal{C}$ is defined to be the geometric realization $|\mathcal{N}(\mathcal{C})|$ of its nerve.

For example, let $(X, \leq)$ be a poset: a set with a partial ordering $\leq$. We regard $X$ as a category, with object set $X$ and with a unique morphism $x \to y$ whenever $x \leq y$. Then $\mathcal{N}(X)$ is the simplicial set where for $n \geq 0$,

$\mathcal{N}(X)_n$ is the set of all order preserving maps $[n] \longrightarrow X$. Hence there is a nondegenerate $n$-simplex for each strictly increasing chain $x_0 < x_1 < \cdots < x_n$ of elements of $X$. The faces of such a simplex are obtained by removing elements from the chain in the obvious way. Note that in this case, $|X|$ is a simplicial complex in the strict sense.

As one special case, if we regard the poset $[m] = \{0, 1, \ldots, m\}$ as a category, then its nerve $\mathcal{N}([m])$ is the simplicial set $\Delta_m$ defined above, and hence $\|[m]\|$ is an $m$-simplex.

If $f \colon \mathcal{C} \longrightarrow \mathcal{D}$ is any functor between small categories, then it follows immediately from the definition that $f$ induces a simplicial map $\mathcal{N}(f) \in \mathrm{Mor}_{\mathbf{S}}(\mathcal{N}(\mathcal{C}), \mathcal{N}(\mathcal{D}))$, and hence a map of spaces $|f| \colon |\mathcal{C}| \longrightarrow |\mathcal{D}|$ between the geometric realizations. In fact, as first shown by Segal, one can say much more.

**Proposition 2.1** ([Sg, Proposition 2.1])   *Fix two categories $\mathcal{C}$ and $\mathcal{D}$, and let $f$ and $g$ be a pair of functors from $\mathcal{C}$ to $\mathcal{D}$. Assume there is a natural transformation of functors $u$ from $f$ to $g$. Then the two maps $|f|$ and $|g|$ between the geometric realizations are homotopic.*

*Proof.* Recall that $[1]$ denotes the category with object set $\{0, 1\}$, and with a unique nonidentity morphism $0 \to 1$. By definition, the natural transformation $u$ consists of morphisms $u(c) \in \mathrm{Mor}_{\mathcal{D}}(f(c), g(c))$, for each $c \in \mathrm{Ob}(\mathcal{C})$, such that $g(\alpha) \circ u(c) = u(c') \circ f(\alpha)$ for each $\alpha \in \mathrm{Mor}_{\mathcal{C}}(c, c')$. This determines a functor
$$\widehat{u} \colon \mathcal{C} \times [1] \longrightarrow \mathcal{D},$$
on objects by setting $\widehat{u}(c, 0) = f(c)$ and $\widehat{u}(c, 1) = g(c)$, and on morphisms by setting (for each $c, c' \in \mathrm{Ob}(\mathcal{C})$ and $\alpha \in \mathrm{Mor}_{\mathcal{C}}(c, c')$):
$$\widehat{u}\big((c, 0) \xrightarrow{(\alpha, \mathrm{Id}_0)} (c', 0)\big) = f(\alpha), \qquad \widehat{u}\big((c, 1) \xrightarrow{(\alpha, \mathrm{Id}_1)} (c', 1)\big) = g(\alpha)$$
and
$$\widehat{u}\big((c, 0) \xrightarrow{(\alpha, 0 \to 1)} (c', 1)\big) = g(\alpha) \circ u(c) = u(c') \circ f(\alpha).$$

Since $\|[1]\|$ is the unit interval, $|\mathcal{C} \times [1]|$ is homeomorphic in a natural way to $|\mathcal{C}| \times I$. So the geometric realization of $\widehat{u}$ is a homotopy between $|f|$ and $|g|$. $\qquad\square$

We now list some easy consequences of Proposition 2.1. By an *equivalence* of categories we mean a functor which induces a bijection on isomorphism classes of objects, and bijections on all morphism sets. Alternatively, a functor $f \colon \mathcal{C} \longrightarrow \mathcal{D}$ is an equivalence if and only if there is a functor $g \colon \mathcal{D} \longrightarrow \mathcal{C}$ such that both composites $g \circ f$ and $f \circ g$ are naturally isomorphic to the identity functors.

**Corollary 2.2**   (a)   *If a category $\mathcal{C}$ has an initial or a final object, then $|\mathcal{C}|$ is contractible.*

(b) *If $f: \mathcal{C} \longrightarrow \mathcal{D}$ is an equivalence of categories, then $|f|$ is a homotopy equivalence from $|\mathcal{C}|$ to $|\mathcal{D}|$.*

*Proof.* If $c_0 \in \mathrm{Ob}(\mathcal{C})$ is an initial object, and $F_{c_0}$ is the constant functor which sends all objects to $c_0$ and all morphisms to the identity, then there is a natural transformation of functors from $F_{c_0}$ to $\mathrm{Id}_{\mathcal{C}}$ which sends an object $c$ to the unique morphism from $c_0$ to $c$. Hence $|F_{c_0}| \simeq \mathrm{Id}_{|\mathcal{C}|}$, and so $|\mathcal{C}|$ is contractible. The argument when $c_0$ is a final object is similar.

Point (b) follows immediately from the above remarks about equivalences of functors. $\qquad\square$

We also note the following lemma, which is among the most elementary of the many results which are useful for comparing higher limits over different categories.

**Lemma 2.3** ([BLO1, Lemma 1.3]) *Fix a prime $p$. Let $f: \mathcal{C} \longrightarrow \mathcal{D}$ be a functor between small categories which satisfies the following conditions:*

(i) *$f$ is bijective on isomorphism classes of objects and is surjective on morphism sets.*

(ii) *For each $c \in \mathrm{Ob}(\mathcal{C})$, the subgroup*
$$K(c) = \mathrm{Ker}\big[\mathrm{Aut}_{\mathcal{C}}(c) \longrightarrow \mathrm{Aut}_{\mathcal{D}}(f(c))\big]$$
*is finite of order prime to $p$.*

(iii) *For each pair of objects $c, d \in \mathrm{Ob}(\mathcal{C})$ and each $\varphi, \psi \in \mathrm{Mor}_{\mathcal{C}}(c, d)$, $f(\varphi) = f(\psi)$ if and only if there is some $\alpha \in K(c)$ such that $\psi = \varphi \circ \alpha$ (i.e., $\mathrm{Mor}_{\mathcal{D}}(f(c), f(d)) \cong \mathrm{Mor}_{\mathcal{C}}(c, d)/K(c)$).*

*Then $|f|: |\mathcal{C}| \longrightarrow |\mathcal{D}|$ is a $p$-equivalence of spaces, and hence $|f|_p^\wedge$ is a homotopy equivalence from $|\mathcal{C}|_p^\wedge$ to $|\mathcal{D}|_p^\wedge$.*

### 2.3. Classifying spaces as geometric realizations of categories.

Classifying spaces of groups and their universal covering spaces provide two very useful examples of geometric realizations of categories.

**Definition 2.4** For any discrete group $G$, define categories $\mathcal{B}(G)$ and $\mathcal{E}(G)$ as follows.

- $\mathrm{Ob}(\mathcal{B}(G)) = \{o_G\}$ and $\mathrm{Mor}_{\mathcal{B}(G)}(o_G, o_G) = G$. Thus $\mathcal{B}(G)$ has a unique object, morphisms from this object to itself are identified with the elements of $G$, and composition in $\mathcal{B}(G)$ is defined by group multiplication.

- $\mathrm{Ob}(\mathcal{E}(G)) = G$, and there is a unique morphism $(g \to h)$ for each pair $(g, h) \in G \times G$.

- $p_G \colon \mathcal{E}(G) \longrightarrow \mathcal{B}(G)$ denotes the functor which sends each object of $\mathcal{E}(G)$ to the unique object $o_G$ in $\mathcal{B}(G)$, and which sends a morphism $(g \to h)$ to the morphism $hg^{-1}$ in $\mathcal{B}(G)$.

Since each object of $\mathcal{E}(G)$ is both an initial and a final object, $|\mathcal{E}(G)|$ is contractible by Corollary 2.2. Also, $G$ acts on $\mathcal{E}(G)$ via right multiplication on objects and on morphisms, this induces a free $G$-action on the geometric realization $|\mathcal{E}(G)|$, and $|p_G|$ is the orbit map of this action. Hence by Proposition 1.2, $|\mathcal{E}(G)|$ is the universal cover of $|\mathcal{B}(G)|$, $G$ is the group of deck transformations of this covering space, and $\pi_1(|\mathcal{B}(G)|) \cong G$. In other words:

**Proposition 2.5** ([Sg, § 3])  *For any discrete group $G$, $\pi_1(|\mathcal{B}(G)|) \cong G$, $|\mathcal{B}(G)|$ is a classifying space for $G$, and $|\mathcal{E}(G)|$ is its universal covering space.*

For example, when $G = \{1, g\}$ is cyclic of order 2, then for each $n \geq 0$, $\mathcal{N}(\mathcal{E}(G))_n$ contains exactly two nondegenerate simplices, represented by the sequences of objects $(1, g, 1, g, \ldots)$ and $(g, 1, g, 1, \ldots)$ (each containing $n+1$ entries). Each of them is attached to the two nondegenerate $(n-1)$-simplices via one face each, while all of its other faces are degenerate. Using this description, it is not hard to see that $|\mathcal{E}(G)|$ is homeomorphic to $S^\infty \overset{\mathrm{def}}{=} \bigcup_{n=1}^{\infty} S^n$, with its cell structure containing $S^n$ as the $n$-skeleton for each $n$. Thus $|\mathcal{B}(G)|$ is homeomorphic to the infinite dimensional real projective space $\mathbb{R}P^\infty$.

In contrast, $|\mathcal{B}(\mathbb{Z})|$ is *not* homeomorphic to the circle $S^1$, although they have the same homotopy type. Since $\mathcal{N}(\mathcal{B}(\mathbb{Z}))$ contains nondegenerate simplices of arbitrarily large dimension, its geometric realization is an infinite dimensional space.

From now on, whenever we write *the* classifying space of $G$, or $BG$, we mean the geometric realization of the category $\mathcal{B}(G)$. Similarly, we write $EG = |\mathcal{E}(G)|$.

### 2.4. Fundamental groups and coverings of geometric realizations.

The fundamental group of the geometric realization of a category or a simplicial set is defined topologically, but it can also be described purely algebraically in terms of generators and relations. We show here one way to do this.

*Notation*   We will use the following notation for the vertices and edges of a simplex in a simplicial set $K$. Fix $n \geq 1$. For each $0 \leq i \leq n$, let $v_i \in \mathrm{Mor}_\Delta([0], [n])$ be the morphism with $v_i(0) = i \in [n]$. For each $0 \leq i < j \leq n$, let $e_{ij} \in \mathrm{Mor}_\Delta([1], [n])$ be the morphism with image $\{i, j\}$.

Thus for $\sigma \in K_n$, $v_i^*(\sigma)$ is the $i$-th vertex of $\sigma$, and for $i < j$, $e_{ij}^*(\sigma)$ is the edge from $v_i^*(\sigma)$ to $v_j^*(\sigma)$. Note that when $\sigma \in K_1$, $e_{01}^*(\sigma) = \sigma$.

**Lemma 2.6** *For any simplicial set $K$ and any discrete group $G$, restriction to 1-simplices defines a bijection*

$$R_{K,G} \colon \mathrm{Mor}_{\mathbf{S}}(K, \mathcal{N}(\mathcal{B}(G))) \xrightarrow{\;\cong\;} \{\alpha \colon K_1 \to G \mid \forall\, \sigma \in K_2,$$

$$\alpha(e_{02}^*(\sigma)) = \alpha(e_{12}^*(\sigma))\alpha(e_{01}^*(\sigma))\}.$$

*Proof.* Clearly, the condition $\varphi_1(e_{02}^*(\sigma)) = \varphi_1(e_{12}^*(\sigma))\varphi_1(e_{01}^*(\sigma))$ holds for each simplicial map $\varphi$ and each $\sigma \in K_2$. Hence $R_{K,G}$ is well defined. It is injective since a simplicial map to $\mathcal{N}(\mathcal{B}(G))$ is determined by its restriction to 1-simplices.

Now fix $\alpha \colon K_1 \longrightarrow G$ such that $\alpha(e_{02}^*(\sigma)) = \alpha(e_{12}^*(\sigma))\alpha(e_{01}^*(\sigma))$ for all $\sigma \in K_2$. In particular, if $v \in K_0$ is a vertex, and $\sigma_v^n \in K_n$ is the degenerate $n$-simplex over $v$, then $e_{ij}^*(\sigma_v^2) = \sigma_v^1$ for each $0 \le i < j \le 2$, and hence $\alpha(\sigma_v^1) = 1$. In other words, $\alpha$ sends all degenerate 1-simplices to the identity.

Let $\varphi_0 \colon K_0 \longrightarrow \mathcal{N}(\mathcal{B}(G))_0$ be the map which sends each element to the unique vertex of $\mathcal{N}(\mathcal{B}(G))$, and set $\varphi_1 = \alpha$. For $n \ge 2$, define $\varphi_n \colon K_n \longrightarrow \mathcal{N}(\mathcal{B}(G))_n$ by setting

$$\varphi_n(\sigma) = \big(\alpha(e_{01}^*(\sigma)), \alpha(e_{12}^*(\sigma)), \ldots, \alpha(e_{n-1,n}^*(\sigma))\big).$$

The $\varphi_n$ clearly commute with all face and degeneracy maps, and hence define a simplicial map $\varphi \in \mathrm{Mor}_{\mathbf{S}}(K, \mathcal{N}(\mathcal{B}(G)))$. Then $R_{K,G}(\varphi) = \alpha$, and thus $R_{K,G}$ is onto. $\qquad\square$

We are now ready to describe the fundamental group of the realization of a simplicial set. When doing this, it will be useful to consider trees as special types of simplicial sets. A (directed) graph $\Gamma$ is a simplicial set all of whose nondegenerate simplices are in $\Gamma_0$ or $\Gamma_1$. Thus the realization $|\Gamma|$ is 1-dimensional. A *circuit* in a graph $\Gamma$ is a simplicial subset whose geometric realization is homeomorphic to the circle. A *tree* is a connected graph which contains no circuits.

**Proposition 2.7** *Fix a simplicial set $K$, a vertex $x_0 \in K_0$, and a tree $T \subseteq K$ which contains all vertices of $K$ (i.e., $T_0 = K_0$). We also regard $x_0$ as a vertex in $|K|$. For each $v \in K_0$, let $\iota_v$ be a path in $|T|$ (unique up to homotopy) from $x_0$ to $v$. Define*

$$\theta \colon K_1 \longrightarrow \pi_1(|K|, x_0)$$

*by sending $e \in K_1$ to the class of the loop $\overline{\iota_{v_1^*(e)}} \cdot e \cdot \iota_{v_0^*(e)}$, where as usual we compose paths from right to left. Then $\theta$ induces an isomorphism of $\pi_1(|K|, x_0)$ with the free group on generators $[e]$ for each $e \in K_1$, modulo* *the relations $[e] = 1$ for $e \in T_1$, and $[e^*_{02}(\sigma)] = [e^*_{12}(\sigma)][e^*_{01}(\sigma)]$ for each $\sigma \in K_2$.*

*Equivalently, for any group $G$, and any map $\alpha\colon K_1 \longrightarrow G$ such that $\alpha(T_1) = 1$ and $\alpha(e^*_{02}(\sigma)) = \alpha(e^*_{12}(\sigma))\alpha(e^*_{01}(\sigma))$ for each $\sigma \in K_2$, there is a unique homomorphism*

$$\bar{\alpha}\colon \pi_1(|K|, x_0) \longrightarrow G$$

*such that $\alpha = \bar{\alpha} \circ \theta$.*

*Proof.* This is an easy consequence of standard facts about fundamental groups of cell complexes (cf. [Ht, Proposition 1.26]), and is shown explicitly in [OV1, Proposition A.3(a)]. We outline a proof here.

We first prove the last statement. For each map $\alpha\colon K_1 \longrightarrow G$ such that $\alpha(e^*_{02}(\sigma)) = \alpha(e^*_{12}(\sigma))\alpha(e^*_{01}(\sigma))$ for all $\sigma \in K_2$, there is a unique simplicial map $\varphi\colon K \longrightarrow \mathcal{N}(\mathcal{B}(G))$ such that $\varphi_1 = \alpha$ (Lemma 2.6). Then $|\varphi|$ is a map of spaces from $|K|$ to $|\mathcal{B}(G)| = BG$, and hence induces a homomorphism of groups

$$\bar{\alpha} \overset{\text{def}}{=} |\varphi|_\#\colon \pi_1(|K|, x_0) \longrightarrow \pi_1(|\mathcal{B}(G)|, *) = G.$$

If in addition, $\alpha(T_1) = 1$, then $|\varphi|$ sends $|T|$ to the basepoint by Lemma 2.6 again, and hence $\bar{\alpha} \circ \theta = \alpha$ by definition of $\theta$.

The uniqueness of $\bar{\alpha}$ is clear once we know that $\pi_1(|K|, x_0)$ is generated by $\mathrm{Im}(\theta)$. Essentially, this is a consequence of [Ht, Proposition 1.26], which implies (as a special case) that every element in the fundamental group of a cell complex $X$ can be represented by a loop following only the edges of $X$.

Thus $\theta$ is universal among all maps $K_1 \longrightarrow G$ which send $T_1$ to the identity and satisfy the relation for 2-simplices. It follows that $\pi_1(|K|, *)$ is isomorphic to the free group on $K_1$, modulo relations given by $T_1$ and 2-simplices. $\qquad\square$

When $K$ is the nerve of a category $\mathcal{C}$, Proposition 2.7 takes the following form.

**Proposition 2.8** *Fix a small category $\mathcal{C}$ and an object $c_0$ in $\mathcal{C}$. Assume we can choose morphisms $\iota_c \in \mathrm{Mor}_{\mathcal{C}}(c, c_0)$, for each $c \in \mathrm{Ob}(\mathcal{C})$, where $\iota_{c_0} = \mathrm{Id}_{c_0}$. Let $* \in |\mathcal{C}|$ be the vertex which represents the object $c_0$. Define*

$$\theta\colon \mathrm{Mor}(\mathcal{C}) \longrightarrow \pi_1(|\mathcal{C}|, *)$$

*by sending $\alpha \in \mathrm{Mor}_{\mathcal{C}}(c, d)$ to the class of the loop $\iota_d \cdot \alpha \cdot \bar{\iota}_c$. Then $\theta$ induces an isomorphism of $\pi_1(|\mathcal{C}|, *)$ with the free group on generators $[\alpha]$ for each $\alpha \in \mathrm{Mor}(\mathcal{C})$, modulo the relations $[\iota_c] = 1$ for $c \in \mathrm{Ob}(\mathcal{C})$ and $[\beta \circ \alpha] = [\beta][\alpha]$ for any composable pair $\beta, \alpha \in \mathrm{Mor}(\mathcal{C})$. Hence for any group $G$, and any* *functor $F\colon \mathcal{C} \longrightarrow \mathcal{B}(G)$ such that $F(\iota_c) = 1$ for each $c \in \mathrm{Ob}(\mathcal{C})$, there is a unique homomorphism*

$$\overline{F}\colon \pi_1(|\mathcal{C}|, *) \longrightarrow G$$

*such that $\mathrm{Mor}(F) = \overline{F} \circ \theta$.*

One can, of course, give a presentation of $\pi_1(|\mathcal{C}|)$ in terms of morphisms in $\mathcal{C}$ without the assumption that some object is the target of morphisms from all other objects, but the above formulation suffices for our purposes.

The next lemma gives a procedure, in certain cases, for constructing covering spaces (up to homotopy type) of $|\mathcal{C}|$ in terms of subcategories of $\mathcal{C}$.

**Proposition 2.9**   *Fix a small category $\mathcal{C}$, an object $c_0$ in $\mathcal{C}$, a group $G$, and a functor $F\colon \mathcal{C} \longrightarrow \mathcal{B}(G)$. Assume, for each $c \in \mathrm{Ob}(\mathcal{C})$ and each $g \in G$, there are $d \in \mathrm{Ob}(\mathcal{C})$ and $\psi \in \mathrm{Iso}_{\mathcal{C}}(c, d)$ such that $F(\psi) = g$. Let $* \in BG = |\mathcal{B}(G)|$ be the unique vertex, and let*

$$\overline{F} = |F|_{\#}\colon \pi_1(|\mathcal{C}|, *) \longrightarrow \pi_1(BG) = G$$

*be the homomorphism of groups induced by $F$. For each $H \le G$, let $\mathcal{C}_H \subseteq \mathcal{C}$ be the subcategory with the same objects, where for all $\psi \in \mathrm{Mor}(\mathcal{C})$, $\psi \in \mathrm{Mor}(\mathcal{C}_H)$ if and only if $F(\psi) \in H$. Then for each $H \le G$, $|\mathcal{C}_H|$ is homotopy equivalent to the covering space of $|\mathcal{C}|$ whose fundamental group is $\overline{F}^{-1}(H)$.*

*Proof.* This is shown in [OV1, Proposition A.4], and generalizes constructions in [5a2]. We sketch here the proof when $H = 1$.

Let $\widetilde{\mathcal{C}}$ be the pullback category in the following square:

$$
\begin{array}{ccc}
\widetilde{\mathcal{C}} & \longrightarrow & \mathcal{E}(G) \\
\downarrow & & \downarrow{\scriptstyle p_G} \\
\mathcal{C} & \xrightarrow{\ F\ } & \mathcal{B}(G),
\end{array}
$$

where $p_G\colon \mathcal{E}(G) \longrightarrow \mathcal{B}(G)$ is as in Definition 2.4. Thus $\mathrm{Ob}(\widetilde{\mathcal{C}}) = \mathrm{Ob}(\mathcal{C}) \times G$, and

$$\mathrm{Mor}_{\widetilde{\mathcal{C}}}((c, g), (d, h)) = \{\varphi \in \mathrm{Mor}_{\mathcal{C}}(c, d) \mid F(\varphi) = hg^{-1}\}.$$

Identify $\mathcal{C}_1$ with the full subcategory of $\widetilde{\mathcal{C}}$ whose objects are the pairs $(c, 1)$ for $c \in \mathrm{Ob}(\mathcal{C})$.

By assumption, for each $c \in \mathrm{Ob}(\mathcal{C})$ and each $g \in G$, there are $d \in \mathrm{Ob}(\mathcal{C})$ and $\psi \in \mathrm{Iso}_{\mathcal{C}}(c, d)$ such that $F(\psi) = g^{-1}$. Then $\psi \in \mathrm{Iso}_{\widetilde{\mathcal{C}}}((c, g), (d, 1))$. Since $c$ and $g$ were arbitrary, this shows that each object of $\widetilde{\mathcal{C}}$ is isomorphic to an object in the subcategory $\mathcal{C}_1$, and hence that the inclusion $\mathcal{C}_1 \subseteq \widetilde{\mathcal{C}}$ is an equivalence of categories. So by Corollary 2.2(b), it induces a homotopy equivalence $|\mathcal{C}_1| \simeq |\widetilde{\mathcal{C}}|$.

By construction, $G$ acts freely on the sets of objects and morphisms in $\widetilde{\mathcal{C}}$: $a \in G$ sends $(c, g) \in \mathrm{Ob}(\widetilde{\mathcal{C}})$ to $(c, ga^{-1})$ and sends $\varphi \in \mathrm{Mor}_{\widetilde{\mathcal{C}}}((c, g), (d, h))$ to $\varphi \in \mathrm{Mor}_{\widetilde{\mathcal{C}}}((c, ga^{-1}), (d, ha^{-1}))$. The orbit sets of those actions are $\mathrm{Ob}(\mathcal{C})$ and $\mathrm{Mor}(\mathcal{C})$. This free action extends to each of the sets $\mathcal{N}(\widetilde{\mathcal{C}})_n$ which makes up the nerve of $\widetilde{\mathcal{C}}$, and from that to a free action of $G$ on $|\widetilde{\mathcal{C}}|$. The projection functor $\widetilde{\mathcal{C}} \longrightarrow \mathcal{C}$ thus induces a covering space $|\widetilde{\mathcal{C}}| \longrightarrow |\mathcal{C}|$ of geometric realizations. Since $|\widetilde{\mathcal{C}}| \simeq |\mathcal{C}_1|$, this finishes the proof of the proposition.    $\square$

## 2.5. Spaces of maps.

This section plays a relatively minor role in the rest of this survey, and it can easily be skipped. But since some definitions and some of the arguments can be more easily understood or motivated when one knows something about certain spaces of maps between classifying spaces or their $p$-completions, we include here a short discussion, with references, about spaces of maps between a pair of topological spaces. This information will be referred to later, in Section 5.7, when sketching the proofs of some theorems about homotopy classes of maps.

If we want to discuss homotopy groups of the connected components of a space $\mathrm{map}(X, Y)$, we first must explain which topology we choose for the set of all maps. The usual choice, which works when $X$ and $Y$ are CW complexes, is the *compact-open topology* (cf. [Ht, Appendix]). However, what we really need to know, for certain constructions which will be described later (e.g., in Section 5.7) are not the homotopy groups of the space of maps itself, but the groups of homotopy classes of maps $S^k \times X \longrightarrow Y$ relative to a given map on $* \times X$ (for some $* \in S^k$). It is simply more convenient to describe these in terms of homotopy groups of a mapping space $\mathrm{map}(X, Y)$.

In other words, we can replace $\mathrm{map}(X, Y)$ by any topological space $M$, together with a continuous "evaluation map" $\mathrm{ev} \colon M \times X \longrightarrow Y$, such that ev defines a bijection from the path components of $M$ to the homotopy classes of maps $X \longrightarrow Y$, and such that for each $k > 0$ and each $m_0 \in M$, ev defines an isomorphism between $\pi_k(M, m_0)$ and the set of homotopy classes of maps $S^k \times X \longrightarrow Y$ whose restriction to $* \times X$ (for a fixed basepoint $* \in S^k$) is the map $\mathrm{ev}(m_0, -)$. When $X$ and $Y$ are realizations of simplicial sets $K$ and $L$, respectively, and $L$ is a "Kan complex" (cf. [Cu, Definition 1.12] or [GJ, p.12]), then this can be done combinatorially, in terms of simplicial maps (cf. [GJ, §I.5]).

The following is the main result in this subsection. Note that it includes Proposition 1.5 in the case where $X$ is the realization of a simplicial set.

**Proposition 2.10**   *Fix a group $G$, and a simplicial set $K$ such that $|K|$ is connected. Choose a base vertex $x_0 \in K_0$, also regarded as a vertex in $|K|$, and let $* \in BG = |\mathcal{B}(G)|$ be the unique vertex. We identify $\pi_1(BG, *) = G$.*

(a)   *There is a bijection $\chi \colon [|K|, BG] \longrightarrow \mathrm{Rep}(\pi_1(|K|, x_0), G)$ with the property: for any map $f \colon |K| \longrightarrow BG$ such that $f(x_0) = *$, $\chi([f]) = [f_\#]$.*

(b)   *Fix $f \colon |K| \longrightarrow BG$ such that $f(x_0) = *$. Set $H = f_\#(\pi_1(|K|, x_0))$. Then there is an isomorphism $\pi_1(\mathrm{map}(|K|, BG), f) \cong C_G(H)$ defined by restriction to basepoint. Also, $\pi_n(\mathrm{map}(|K|, BG), f) = 0$ for all $n \geq 2$.*

*Proof.* We use here the notation defined before Lemma 2.6 for the vertices and edges of a simplex in $K$.

**Step 1:**   We prove here that each continuous map $f \colon |K| \longrightarrow BG$ is homotopic to a one which is realized by a simplicial map. Moreover, this holds in a relative version: if $L \subseteq K$ is any simplicial subset, $f \colon |K| \longrightarrow BG$ is a continuous map, and $\psi \colon L \longrightarrow \mathcal{N}(\mathcal{B}(G))$ is a simplicial map such that $|\psi| = f|_{|L|}$, then there is a simplicial map $\varphi \colon K \longrightarrow \mathcal{N}(\mathcal{B}(G))$, and a homotopy $\Phi \colon |K| \times I \longrightarrow BG$ from $|\varphi|$ to $f$ which is constant on $|L|$.

This is a special case of the "simplicial extension theorem" in [Cu, Theorem 12.1], where it is proved under the assumption that the target is a Kan complex. That $\mathcal{N}(\mathcal{B}(G))$ is a Kan complex is shown, for example, in [GJ, Lemma 3.5]. Since it is much easier to work with maps to $\mathcal{N}(\mathcal{B}(G))$ than to arbitrary Kan complexes, we sketch the proof here.

For each vertex $v \in K_0$, choose a path $\phi_v \colon I \longrightarrow BG$ such that $\phi_v(0) = *$ and $\phi_v(1) = f(v)$, and such that $\phi_v$ is constant if $f(v) = *$. For each edge $e \in K_1$, we regard $f(e)$ as a path in $BG$ from $f(v_0^*(e))$ to $f(v_1^*(e))$, and set

$$\alpha(e) = [\overline{\phi_{v_1^*(e)}} \cdot f(e) \cdot \phi_{v_0^*(e)}] \in \pi_1(BG, *) = G.$$

Then for each 2-simplex $\sigma \in K_2$,

$$\alpha(e_{12}^*(\sigma))\alpha(e_{01}^*(\sigma)) = [\overline{\phi_{v_2^*(\sigma)}} \cdot f(e_{12}^*(\sigma)) \cdot \phi_{v_1^*(\sigma)}] \cdot [\overline{\phi_{v_1^*(\sigma)}} \cdot f(e_{01}^*(\sigma)) \cdot \phi_{v_0^*(\sigma)}]$$

$$= [\overline{\phi_{v_2^*(\sigma)}} \cdot f(e_{12}^*(\sigma)) \cdot f(e_{01}^*(\sigma)) \cdot \phi_{v_0^*(\sigma)}] = \alpha(e_{02}^*(\sigma)).$$

So by Lemma 2.6, there is a (unique) simplicial map $\varphi \colon K \longrightarrow \mathcal{N}(\mathcal{B}(G))$ such that $\varphi_1 = \alpha$. Furthermore, since $\alpha|_{L_1} = \psi_1$ by construction, $\varphi|_L = \psi$.

Define $X_0 \subseteq X_1 \subseteq X_2 \subseteq \cdots \subseteq X = |K| \times I$, by setting

$$X_n = (|K| \times \{0, 1\}) \cup (|L| \times I) \cup (|K|^{(n)} \times I)$$

for each $n \geq 0$. Here, $|K|^{(n)} \subseteq |K|$ is the union of the realizations of all simplices in $K$ of dimension $\leq n$. Set

$$\Phi_0 = (f \times 0) \cup (|\varphi| \times 1) \cup (f \circ \mathrm{pr}_{|L|}) \cup \{\phi_v\} \colon X_0 \longrightarrow BG.$$

We will construct maps $\Phi_n \colon X_n \longrightarrow BG$ such that $\Phi_n|_{X_{n-1}} = \Phi_{n-1}$ for each $n \geq 0$.

Assume, for some $n \geq 1$, that $\Phi_{n-1}$ has been defined. To define $\Phi_n$, we must define a map $\Phi_\sigma \colon \Delta^n \times I \longrightarrow BG$ for each $\sigma \in K_n$, which extends a given map (the restriction of $\Phi_{n-1}$) on

$$(\Delta^n \times \{0,1\}) \cup (\partial\Delta^n \times I) \cong S^n.$$

When $n = 1$, this is possible since by definition of $\alpha$, the two paths $|\alpha(\sigma)|$ and $\overline{\phi_{v_1^*(\sigma)}} \cdot f(\sigma) \cdot \phi_{v_0^*(\sigma)}$ in $BG$ are homotopic. When $n \geq 2$, this can be done since $\pi_n(BG) \cong \pi_n(EG)$ is the trivial group (so each map $S^n \longrightarrow BG$ extends to $D^{n+1}$). Alternatively, when $n = 2$, since $S^n$ is simply connected, the restriction of $\Phi_{n-1}$ to $(\Delta^n \times \{0,1\}) \cup (\partial\Delta^n \times I)$ lifts to a map to $EG = |\mathcal{E}(G)|$, the lifting extends to a map defined on $\Delta^n \times I$ since $EG$ is contractible, and hence the original map extends to one defined on $\Delta^n \times I$.

In this way, we construct $\Phi_n$ for each $n$. Then $\Phi = \bigcup_{n=0}^{\infty} \Phi_n$ is a homotopy from $|\varphi|$ to $f$ which is constant on $|L|$, and this proves the above claim.

**Step 2:** Choose a tree $T \subseteq K$ such that $T_0 = K_0$. That this always can be done is shown, for example, in [Se2, Proposition 11], applied to the pair $(K_0, K_1)$ regarded as a graph.

Assume $\varphi, \psi \colon K \longrightarrow \mathcal{N}(\mathcal{B}(G))$ are simplicial maps. By Step 1, $|\varphi|$ is homotopic to $|\psi|$ as maps from $|K|$ to $BG$ if and only if $\varphi \simeq \psi$; i.e., homotopic via a simplicial homotopy $K \times \Delta_1 \longrightarrow \mathcal{N}(\mathcal{B}(G))$. Using Lemma 2.6, we see that

$$\varphi \simeq \psi \iff \exists \beta \colon K_0 \longrightarrow G \text{ such that}$$
$$\psi(e) = \beta(v_1^*(e))\varphi(e)\beta(v_0^*(e))^{-1} \; \forall\, e \in K_1. \tag{1}$$

Fix $\varphi \in \mathrm{Mor}_\mathbf{S}(K, \mathcal{N}(\mathcal{B}(G)))$. For each $v \in K_0$, choose a path $\phi_v$ in $|T|$ from $x_0$ to $v$ (unique up to homotopy), and set $\beta(v) = [|\varphi| \circ \phi_v]^{-1} \in \pi_1(BG, *) = G$. By Lemma 2.6, there is $\psi \in \mathrm{Mor}_\mathbf{S}(K, \mathcal{N}(\mathcal{B}(G)))$ such that $\psi(e) = \beta(v_1^*(e))\varphi(e)\beta(v_0^*(e))^{-1}$ for each $e \in K_1$, and $\psi \simeq \varphi$ by (1). Since $\psi(T_1) = 1$ by construction, this proves that each map from $K$ to $BG$ is homotopic to one induced by a simplicial map which sends $|T|$ to the basepoint.

We identify $\mathrm{Mor}_{\mathbf{S}}(K/T, \mathcal{N}(\mathcal{B}(G)))$ with the set of all morphisms $\varphi \in \mathrm{Mor}_{\mathbf{S}}(K, \mathcal{N}(\mathcal{B}(G)))$ such that $\varphi(T_1) = 1$. By Lemma 2.6, and the description of $\pi_1(|K|)$ in Proposition 2.7, there is a bijection

$$\mathrm{Mor}_{\mathbf{S}}(K/T, \mathcal{N}(\mathcal{B}(G))) \longrightarrow \mathrm{Hom}(\pi_1(|K|, x_0), G)$$

which sends $\varphi$ to $|\varphi|_{\#}$. Also, $\varphi, \psi \in \mathrm{Mor}_{\mathbf{S}}(K/T, \mathcal{N}(\mathcal{B}(G)))$ are simplicially homotopic if and only if there is a constant function $\beta$ which satisfies the condition in (1); i.e., there is $g \in G$ such that $|\varphi|_{\#} = c_g \circ |\psi|_{\#}$. Together with Step 1, this finishes the proof that $[|K|, BG] \cong \mathrm{Rep}(\pi_1(|K|, x_0), G)$.

Again assume $\varphi \in \mathrm{Mor}_{\mathbf{S}}(K/T, \mathcal{N}(\mathcal{B}(G)))$. By Step 1, each loop in $\mathrm{map}(|K|, BG)$ based at $|\varphi|$ is homotopic to the realization of a simplicial homotopy from $\varphi$ to itself, which by (1) is determined by an element $g \in G$ which commutes with $\varphi(e)$ for each $e \in K_1$. Thus $g \in C_G(H)$, where $H = \mathrm{Im}(|\varphi|_{\#})$. Two such loops are homotopic (relative to the endpoints) only if they are equal as simplicial homotopies (Lemma 2.6 again), and thus $\pi_1(\mathrm{map}(|K|, BG), |\varphi|) \cong C_G(H)$.

For $n \geq 2$, by Step 1 again, each element of $\pi_n(\mathrm{map}(|K|, BG), |\varphi|)$ is represented by a simplicial map $\Phi \colon K \times \Delta_n \longrightarrow \mathcal{N}(\mathcal{B}(G))$ whose restriction to $K \times \partial\Delta_n$ is $\varphi \circ \mathrm{pr}_K$. Since all 1-simplices of $K \times \Delta_n$ are contained in $K \times \partial\Delta_n$, Lemma 2.6 implies that there is only one such map, and thus that $\pi_n(map(|K|, BG), |\varphi|) = 1$.      $\square$

Thus, in the situation of Proposition 2.10(b), the connected component of $f$ in $\mathrm{map}(|K|, BG)$ has the weak homotopy type of $BC_G(H)$ (i.e., they have the same homotopy groups).

The following, much deeper theorem describes mapping spaces in certain special cases when we replace the target $BG$ by its $p$-completion. This result is in part a special case of Theorem 4.21, whose proof will be sketched in Section 5.7.

**Theorem 2.11**  *Fix a prime $p$, a $p$-group $Q$, and a finite group $G$. Then the space of maps $\mathrm{map}(BP, BG_p^{\wedge})$ has the (weak) homotopy type of the $p$-completion of $\mathrm{map}(BP, BG)$. In particular, $[BP, BG_p^{\wedge}] \cong \mathrm{Rep}(P, G)$, and for each $\rho \in \mathrm{Hom}(Q, G)$,*

$$\pi_1(\mathrm{map}(BP, BG_p^{\wedge}), B\rho) \cong \pi_1\big(BC_G(\rho(P))_p^{\wedge}\big) \cong C_G(\rho(P))/O^p(C_G(\rho(P))).$$

*Proof.* The description of $\mathrm{map}(BP, BG)$ follows as a special case of Proposition 2.10. The description of $\mathrm{map}(BQ, BG_p^{\wedge})$ as the $p$-completion of $\mathrm{map}(BQ, BG)$ is shown in [BL, Proposition 2.1], and also follows as a special case of [BLO2, Theorem 6.3]. The formula for the fundamental group of a component then follows from Proposition 1.11.      $\square$

## 3. Linking systems and classifying spaces of finite groups

The $p$-fusion category of a finite group $G$ was just one of a family of categories defined by Puig [P1], which he called the "$p$-localités". These categories allowed him to give a more uniform treatment of certain results in group theory, and to interpret them in category theoretical language. The close connection between these categories and $p$-completed classifying spaces was discovered much later by Broto, Levi, and Oliver in [BLO1], while attempting to prove the Martino-Priddy conjecture.

### 3.1. The linking category of a finite group.

For any group $G$ and any pair of subgroups $H, K \leq G$, we let $T_G(H, K)$ denote the *transporter set*:

$$T_G(H, K) \overset{\text{def}}{=} \{ g \in G \mid {}^g H \leq K \}.$$

In other words, $T_G(H, K)$ is the set of all elements of $G$ which induce homomorphisms in $\mathrm{Hom}_G(H, K)$. Clearly, $T_G(H, H) = N_G(H)$ is the normalizer of $H$ in $G$.

**Definition 3.1**   Fix a finite group $G$ and a Sylow subgroup $S \in \mathrm{Syl}_p(G)$.

- The *transporter category* of $G$ is the category $\mathcal{T}(G)$ whose objects are the subgroups of $G$, and whose morphism sets are the transporter sets:
$$\mathrm{Mor}_{\mathcal{T}(G)}(H, K) = T_G(H, K).$$
  Let $\mathcal{T}_S(G) \subseteq \mathcal{T}(G)$ be the full subcategory whose objects are the subgroups of $S$. More generally, if $\mathcal{H}$ is any set of subgroups of $G$, then $\mathcal{T}_{\mathcal{H}}(G) \subseteq \mathcal{T}(G)$ denotes the full subcategory with object set $\mathcal{H}$.

- A $p$-subgroup $P \leq G$ is *$p$-centric in $G$* if $Z(P) \in \mathrm{Syl}_p(C_G(P))$; equivalently, if $C_G(P) = Z(P) \times C'_G(P)$ for some (unique) subgroup $C'_G(P)$ of order prime to $p$.

- The *centric linking category* of $G$ over $S$ is the category $\mathcal{L}^c_S(G)$ whose objects are the subgroups of $S$ which are $p$-centric in $G$, and whose morphism sets are given by
$$\mathrm{Mor}_{\mathcal{L}^c_S(G)}(P, Q) = T_G(P, Q)/C'_G(P)$$
  for all $P, Q \leq S$ which are $p$-centric in $G$.

In addition, when needed, we let $\mathcal{T}^c_S(G) \subseteq \mathcal{T}_S(G)$ and $\mathcal{F}^c_S(G) \subseteq \mathcal{F}_S(G)$ be the full subcategories whose objects are the subgroups of $S$ which are $p$-centric in $G$. Thus in general, the superscript "$c$" means restriction to subgroups which are $p$-centric in $G$.

The transporter category $\mathcal{T}_S(G)$ was called by Puig [P1] the 1-*localité* of $G$, and the centric linking category $\mathcal{L}^c_S(G)$ is a full subcategory of his *O-localité*. In the same way, the fusion category $\mathcal{F}_S(G)$ is called the *C-localité* in his terminology, because it is defined by dividing out by centralizers.

There are obvious functors

$$\mathcal{B}(G) \xrightarrow{\ \Phi\ } \mathcal{T}_S(G) \xrightarrow{\ \Psi\ } \mathcal{B}(G),$$

where $\Phi$ sends the object $o_G$ to the trivial subgroup $1$ (and sends $g \in \mathrm{Aut}_{\mathcal{B}(G)}(o_G)$ to $g \in \mathrm{Aut}_{\mathcal{T}_S(G)}(1)$), and $\Psi$ sends each morphism set $T_G(P,Q)$ to $G$ via the inclusion.  Then $\Psi \circ \Phi = \mathrm{Id}_{\mathcal{B}(G)}$, and there is a natural transformation of functors from $\Phi \circ \Psi$ to $\mathrm{Id}_{\mathcal{T}_S(G)}$.  So by Proposition 2.1, $|\mathcal{T}_S(G)| \simeq |\mathcal{B}(G)| = BG$.

Thus the transporter category completely describes the classifying space $BG$.  The problem is that it contains too much information about $G$ itself.  What we need is a category which only contains the information about the $p$-local structure of $G$, and which can be used as a tool for describing the homotopy properties of the space $BG^\wedge_p$.  This is what the linking system $\mathcal{L}^c_S(G)$ does for us, by providing a good intermediate category between the transporter and fusion categories.  It contains enough information to completely determine the $p$-completed homotopy type of $BG$, but contains only that information and no more.

The following theorem, proven in [BLO1], was the first indication of the importance of the linking categories when working with classifying spaces.

**Theorem 3.2**   *For any finite group $G$ and any $S \in \mathrm{Syl}_p(G)$,*

$$BG^\wedge_p \simeq |\mathcal{L}^c_S(G)|^\wedge_p.$$

*Proof.* We outline the proof here; the details are given in [BLO1, Section 1].

Consider the functors

$$\mathcal{L}^c_S(G) \xleftarrow{\ \rho_1\ } \mathcal{T}^c_S(G) \subseteq \mathcal{T}_S(G) \xrightarrow{\ \rho_2\ } \mathcal{B}(G),$$

where $\rho_1$ is the identity on objects and sends $g \in T_G(P,Q)$ to its class modulo $C'_G(P)$; and where $\rho_2$ sends all objects to the unique object $o_G$ of $\mathcal{B}(G)$ and sends a morphism $g \in T_G(P,Q)$ to $g$ as a morphism in $\mathcal{B}(G)$.  Consider the maps between spaces induced by their geometric realizations:

$$|\mathcal{L}^c_S(G)| \xleftarrow{\ |\rho_1|\ } |\mathcal{T}^c_S(G)| \subseteq |\mathcal{T}_S(G)| \xrightarrow{\ |\rho_2|\ } |\mathcal{B}(G)| = BG.$$

Then $|\rho_1|$ is a $p$-equivalence by Lemma 2.3, while $|\rho_2|$ is a homotopy equivalence by the above remarks.  The inclusion of $|\mathcal{T}^c_S(G)|$ into $|\mathcal{T}_S(G)|$ is a $p$-equivalence by a theorem of Dwyer [Dw, Theorem 8.3] (suitably interpreted), and the proof of this is sketched in Lemma 5.30 below.  Thus all of these maps become homotopy equivalences after $p$-completion, and in particular, $|\mathcal{L}_S^c(G)|_p^\wedge \simeq BG_p^\wedge$. $\qquad\square$

## 3.2. Fusion and linking categories of spaces.

Fusion and linking categories can also be defined for spaces. The definition of the fusion category of a space is fairly straightforward, while the definition of its linking category is more subtle. The latter was first defined in [BLO1], while the idea of the fusion category of a space has been around much longer.

Recall that we now define $BG \overset{\text{def}}{=} |\mathcal{B}(G)|$ for each (finite) group $G$. In particular, this allows us to regard $B$ as a functor from groups to spaces: each $\varphi \in \mathrm{Hom}(G, H)$ induces a map $B\varphi\colon BG \longrightarrow BH$. Also, it means that when $P \leq Q$, we regard $BP$ as a subspace of $BQ$.

There are several possible definitions of fusion and linking systems for spaces, which differ in their precise choices of objects. The following definition, taken from [BLO2], is the most convenient for our purposes here.

**Definition 3.3** Fix a space $X$, a $p$-group $S$, and a map $f\colon BS \longrightarrow X$. The *fusion category of* $X$ with respect to $(S, f)$ is the category $\mathcal{F} = \mathcal{F}_{S,f}(X)$, whose objects are the subgroups of $S$, and where

$$\mathrm{Mor}_{\mathcal{F}}(P, Q) = \{\varphi \in \mathrm{Inj}(P, Q) \mid f|_{BQ} \circ B\varphi \simeq f|_{BP}\}.$$

Here, $\mathrm{Inj}(P, Q)$ denotes the set of all injective homomorphsms from $P$ to $Q$.

By comparison, for any space $X$, $\mathcal{F}_p(X)$ was defined in [BLO1] to be the category whose objects are the pairs $(P, f)$, where $P$ is a $p$-group and $f\colon BP \longrightarrow X$ is a "monomorphism" in a certain sense made precise there. Morphisms in $\mathcal{F}_p(X)$ were defined in a way analogous to those of $\mathcal{F}_{S,f}(X)$. It was in part to avoid the technicalities of defining and working with monomorphisms between topological spaces that we decided to use the above definition. In the cases which interest us (when $X = BG$ or $BG_p^\wedge$ for a finite group $G$), the categories $\mathcal{F}_{S,f}(X)$ and $\mathcal{F}_p(X)$ are equivalent.

As an immediate consequence of Proposition 1.5 and Theorem 1.16, we have

**Proposition 3.4** *Fix a finite group $G$ and a Sylow subgroup $S \in \mathrm{Syl}_p(G)$. Let $f\colon BS \longrightarrow BG$ be the inclusion, and let $\widehat{f}\colon BS \longrightarrow BG_p^\wedge$ be its composite with $p$-completion. Then*

$$\mathcal{F}_{S,f}(BG) \cong \mathcal{F}_{S,\widehat{f}}(BG_p^\wedge) \cong \mathcal{F}_S(G).$$

*Proof.* For each subgroup $P \leq S$, $[BP, BG] \cong [BP, BG_p^\wedge] \cong \mathrm{Rep}(P, G)$ by Proposition 2.10(a) and Theorem 2.11. Using this, one sees immediately, for $\varphi \in \mathrm{Inj}(P, Q)$, that $f|_{BQ} \circ B\varphi \simeq f|_{BP}$ if and only if $\varphi \in \mathrm{Hom}_G(P, Q)$; and similarly for $\widehat{f}|_{BQ} \circ B\varphi \simeq \widehat{f}|_{BP}$. $\qquad\square$

The definition of the linking category of a space is more complicated. It is again motivated by the cases $X = BG$ (for a finite group $G$) and $X = BG_p^\wedge$. When $S \in \mathrm{Syl}_p(G)$ and $P, Q \le S$ are $p$-centric in $G$, we need to extend $\mathrm{Mor}_{\mathcal{F}_{S,f}(BG)}(P, Q)$ by $Z(P)$ or by $C_G(P)$. Since $C_G(P) \cong \pi_1(\mathrm{map}(BP, BG)_{\mathrm{incl}})$ by Proposition 2.10(b) and

$$Z(P) = C_G(P)/O^p(C_G(P)) \cong \pi_1(\mathrm{map}(BP, BG_p^\wedge)_{\mathrm{incl}})$$

by Theorem 2.11, this suggests that we add paths in $\mathrm{map}(BP, X)$ to the data which determines a morphism in $\mathcal{F}_{S,f}(X)$.

**Definition 3.5**  Fix a space $X$, a $p$-group $S$, and a map $f \colon BS \longrightarrow X$. The *linking category of $X$* with respect to $(S, f)$ is the category $\mathcal{L}_{S,f}(X)$, whose objects are the subgroups of $S$, and where

$$\mathrm{Mor}_{\mathcal{L}_{S,f}(X)}(P, Q) = \big\{ (\varphi, [H]) \,\big|\, \varphi \in \mathrm{Inj}(P, Q),\ H \colon BP \times I \longrightarrow X,$$
$$H(-, 0) = f|_{BP},\ H(-, 1) = f|_{BQ} \circ B\varphi \big\}.$$

Here, $[H]$ denotes the class of $H$ modulo homotopies which leave fixed $BP \times \{0, 1\}$. The identity morphism of an object $P \le S$ is the pair $(\mathrm{Id}_P, [C_{f|_{BP}}])$, where $C_{f|_{BP}}$ is the constant homotopy at $f|_{BP}$. The composite of two morphisms

$$P \xrightarrow{\ (\varphi, [H])\ } Q \xrightarrow{\ (\psi, [K])\ } R$$

is the morphism $\big(\psi \circ \varphi, [(K \circ (B\varphi \times \mathrm{Id}_I)) \cdot H]\big)$, where $(-) \cdot (-)$ denotes the composite (from right to left) of the two homotopies.

In other words, a morphism in $\mathcal{L}_{S,f}(X)$ from $P$ to $Q$ consists of a pair $(\varphi, [H])$, where $\varphi \in \mathrm{Mor}_{\mathcal{F}_{S,f}(X)}(P, Q)$, $H$ is a path from $f|_{BP}$ to $f|_{BQ} \circ B\varphi$ in the mapping space $\mathrm{map}(BP, X)$, and $[H]$ is its class modulo homotopies which fix the endpoints of the paths. Let

$$\pi_{S,f} \colon \mathcal{L}_{S,f}(X) \longrightarrow \mathcal{F}_{S,f}(X)$$

denote the forgetful functor which sends each object to itself and sends a morphism $(\varphi, [H])$ to $\varphi$. A functor

$$\delta_{S,f} \colon \mathcal{T}_S(S) \longrightarrow \mathcal{L}_{S,f}(X)$$

can also be constructed, induced by the isomorphism $\mathcal{T}_S(S) \cong \mathcal{L}_{S,\mathrm{Id}}(BS)$ of the next proposition, followed by composition with $f$.

For $f \colon BS \longrightarrow X$ as above, $\mathcal{L}_{S,f}^c(X) \subseteq \mathcal{L}_{S,f}(X)$ denotes the full subcategory whose objects are the $\mathcal{F}_{S,f}(X)$-centric subgroups of $S$: those $P \le S$ such that each $P' \le S$ which is isomorphic to $P$ in the fusion category $\mathcal{F}_{S,f}(X)$ is centric in $S$. All of these definitions were motivated by the following proposition.

**Proposition 3.6** ([BLO1, Proposition 2.6])  *Fix a prime $p$, a finite group $G$, and a Sylow subgroup $S \in \mathrm{Syl}_p(G)$. Let $f \colon BS \longrightarrow BG$ be the inclusion, and let $\widehat{f} \colon BS \longrightarrow BG_p^\wedge$ be its composite with $p$-completion. Then*

$$\mathcal{L}_{S,f}(BG) \cong \mathcal{T}_S(G) \qquad \text{and} \qquad \mathcal{L}^c_{S,\widehat{f}}(BG_p^\wedge) \cong \mathcal{L}^c_S(G).$$

*Proof.* We sketch here a slightly different argument from that given in [BLO1].

Let $* \in BG = |\mathcal{B}(G)|$ denote the basepoint: the vertex corresponding to the object $o_G$ in $\mathcal{B}(G)$. Thus $* \in BP$ for all $P \leq S \leq G$. Define a functor

$$\kappa \colon \mathcal{L}_{S,f}(BG) \xrightarrow{\ \cong\ } \mathcal{T}_S(G)$$

by setting $\kappa(P) = P$ for all $P \leq S$, and

$$\kappa(\varphi, [H]) = \big[H|_{* \times I}\big] \in \pi_1(BG, *) = G$$

for all $P, Q \leq S$ and $(\varphi, [H]) \in \mathrm{Mor}_{\mathcal{L}_{S,f}(BG)}(P, Q)$. For each $g \in P$, if we let $\rho_g \colon I \longrightarrow BP$ be the inclusion of the edge corresponding to $g \in \mathrm{Mor}(\mathcal{B}(P))$, then the composite

$$I \times I \xrightarrow{\ \rho_g \times \mathrm{Id}\ } BP \times I \xrightarrow{\ H\ } BG$$

defines a homotopy $(H|_{* \times I}) \cdot (f \circ \rho_g) \simeq (f \circ B\varphi \circ \rho_g) \cdot (H|_{* \times I})$ of loops in $BG$. So if we set $h = [H|_{* \times I}] \in G$, then $hg = \varphi(g)h$ in $\pi_1(BG) = G$, hence $\varphi = c_h \in \mathrm{Hom}(P, Q)$, and $\kappa(\varphi, [H]) = h \in T_G(P, Q)$. This proves that $\kappa$ is a well defined functor.

By definition, $\kappa$ induces a bijection on objects, and commutes with the projections to $\mathcal{F}_{S,f}(BG) = \mathcal{F}_S(G)$. For each pair of objects $P, Q \leq S$, if $(\varphi, [H_1])$ and $(\varphi, [H_2])$ are two morphisms in $\mathcal{L}_{S,f}(BG)$ from $P$ to $Q$ which are both sent to $\varphi \in \mathrm{Mor}_{\mathcal{F}_{S,f}(BG)}(P, Q)$, then $H_1$ and $H_2$ can be regarded as paths in $\mathrm{map}(BP, BG)$ from $f|_{BP}$ to $f|_{BQ} \circ B\varphi$. Hence they differ by the loop $\overline{H_2} \cdot H_1$ based at $f|_{BP}$, which represents an element of $\pi_1(\mathrm{map}(BP, BG)_{\mathrm{incl}})$. In other words, in the following commutative diagram

$$
\begin{array}{ccc}
\mathrm{Mor}_{\mathcal{L}_{S,f}(BG)}(P,Q) & \xrightarrow{\ \ \kappa_{P,Q}\ \ } & \mathrm{Mor}_{\mathcal{T}_S(G)}(P,Q) = T_G(P,Q) \\[2pt]
{\scriptstyle (\pi_{S,f})_{P,Q}}\Big\downarrow & & \Big\downarrow {\scriptstyle g \mapsto c_g} \\[2pt]
\mathrm{Mor}_{\mathcal{F}_{S,f}(BG)}(P,Q) & =\!=\!=\!=\!= & \mathrm{Mor}_{\mathcal{F}_S(G)}(P,Q) = \mathrm{Hom}_G(P,Q)
\end{array}
$$

the two vertical maps are orbit maps for free actions of the two groups $\pi_1(\mathrm{map}(BP, BG)_{\mathrm{incl}})$ and $C_G(P)$, respectively. Since $\kappa_P$ sends $\pi_1(\mathrm{map}(BP, BG)_{\mathrm{incl}})$ isomorphically to $C_G(P)$ by Proposition 2.10(b), this proves that $\kappa_{P,Q}$ is a bijection, and finishes the proof that $\kappa$ is an isomorphism of categories.

It is straightforward to check that the objects of $\mathcal{L}^c_{S,f}(BG)$ and of $\mathcal{L}^c_{S,f}(BG^\wedge_p)$ are precisely the subgroups of $S$ which are $p$-centric in $G$ (cf. [BLO1, Lemma A.5]). Let $\kappa^c$ be the restriction of $\kappa$ to the full subcategories with these objects.

The isomorphism $\mathcal{L}^c_{S,f}(BG^\wedge_p) \cong \mathcal{L}^c_S(G)$ follows upon showing that the natural functor from $\mathcal{L}^c_{S,f}(BG)$ to $\mathcal{L}^c_{S,f}(BG^\wedge_p)$ is surjective on morphisms, and that $\kappa^c$ factors through an isomorphism of categories from $\mathcal{L}^c_{S,f}(BG^\wedge_p)$ to $\mathcal{L}^c_S(G)$. By definition, for each pair of objects $P$ and $Q$, the map $T_G(P,Q) \longrightarrow \mathrm{Mor}_{\mathcal{L}^c_S(P)}(P,Q)$ is the orbit map of a free action of $C'_G(P) = O^p(C_G(P))$ on $T_G(P,Q)$. By comparison, the map

$$\mathrm{Mor}_{\mathcal{L}^c_{S,f}(BG)}(P,Q) \longrightarrow \mathrm{Mor}_{\mathcal{L}^c_{S,f}(BG^\wedge_p)}(P,Q)$$

is the orbit map of a free action of the group

$$\mathrm{Ker}\big[\pi_1(\mathrm{map}(BP,BG),\mathrm{incl}) \longrightarrow \pi_1(\mathrm{map}(BP,BG^\wedge_p),\mathrm{incl})\big].$$

This kernel is also isomorphic to $O^p(C_G(P))$ (see Theorem 2.11), and this finishes the proof that $\mathcal{L}^c_{S,f}(BG^\wedge_p) \cong \mathcal{L}^c_S(G)$. $\qquad\square$


The argument outlined in the proof of Proposition 3.6 shows, in fact, that for all subgroups $P,Q \le S$ ($p$-centric or not), $\mathrm{Mor}_{\mathcal{L}_{S,f}(BG^\wedge_p)}(P,Q) \cong T_G(P,Q)/O^p(C_G(P))$. That helps to motivate this as the definition of morphisms in a category $\mathcal{L}_S(G)$, whose objects consist of all subgroups of $S$. However, from the homotopy theoretic point of view, this category is much less interesting than the centric linking category. For example, when $G = O^p(G)$, the trivial subgroup is an initial object in $\mathcal{L}_S(G)$, and hence $|\mathcal{L}_S(G)|$ is contractible. More generally, when $G$ is an arbitrary finite group and $S \in \mathrm{Syl}_p(G)$, it is not hard to show that $|\mathcal{L}_S(G)| \simeq B(G/O^p(G))$.

There is, however, a category intermediate between $\mathcal{L}^c_S(G)$ and $\mathcal{L}_S(G)$ which has all of the homotopy theoretic properties of the former. Define a $p$-subgroup $P \le G$ to be $p$-quasicentric if $O^p(C_G(P))$ has order prime to $p$. Let $\mathcal{L}^q_S(G) \subseteq \mathcal{L}_S(G)$ be the full subcategory whose objects are the subgroups of $S$ which are $p$-quasicentric in $G$. This is a linking system in the sense of Definition 4.1 (while $\mathcal{L}_S(G)$ is not); and the inclusion $\mathcal{L}^c_S(G) \subseteq \mathcal{L}^q_S(G)$ induces a homotopy equivalence $|\mathcal{L}^q_S(G)| \simeq |\mathcal{L}^c_S(G)|$ by [5a1, Theorem 3.5] (see Theorem 4.20 below).


### 3.3. Linking systems and equivalences of $p$-completed classifying spaces.

If the centric linking categories $\mathcal{L}^c_{S_1}(G_1)$ and $\mathcal{L}^c_{S_2}(G_2)$ of two finite groups are equivalent as categories, then by Corollary 2.2, their geometric realizations are homotopy equivalent. So $BG_1{}^\wedge_p \simeq BG_2{}^\wedge_p$ by Theorem 3.2.

Conversely, if two $p$-completed classifying spaces $BG_1{}^{\wedge}_p$ and $BG_2{}^{\wedge}_p$ are homotopy equivalent, then as noted above, Mislin's theorem (Theorem 1.16) implies that they have isomorphic Sylow $p$-subgroups $S_i \leq G_i$. From this, together with Proposition 3.6, it follows that there are equivalences of categories

$$\mathcal{L}^c_{S_1}(G_1) \cong \mathcal{L}^c_{S_1,f}(BG_1{}^{\wedge}_p) \cong \mathcal{L}^c_{S_2,f'}(BG_2{}^{\wedge}_p) \cong \mathcal{L}^c_{S_2}(G_2).$$

We can thus conclude:

**Theorem 3.7** ([BLO1])   *For any prime $p$ and any pair $G_1, G_2$ of finite groups, $BG_1{}^{\wedge}_p \simeq BG_2{}^{\wedge}_p$ if and only if for some $S_i \in \mathrm{Syl}_p(G_i)$, $\mathcal{L}^c_{S_1}(G_1)$ and $\mathcal{L}^c_{S_2}(G_2)$ are equivalent as categories.*

Hence to prove the Martino-Priddy conjecture for a pair of finite groups $G_1$ and $G_2$, with $S_i \in \mathrm{Syl}_p(G_i)$, it remains to show that any fusion preserving isomorphism $S_1 \longrightarrow S_2$ lifts to an equivalence of categories $\mathcal{L}^c_{S_1}(G_1) \simeq \mathcal{L}^c_{S_2}(G_2)$. There is a fairly straightforward obstruction theory for doing this, analogous to the obstruction theory for the uniqueness of group extensions. This will be described in Section 5.4.

## 4. ABSTRACT FUSION AND LINKING SYSTEMS

So far, we have described the fusion and linking systems $\mathcal{F}_S(G)$ and $\mathcal{L}^c_S(G)$ of a finite group $G$, and their close connection with homotopy properties of the $p$-completed classifying space $BG^{\wedge}_p$. We now turn our attention to abstract versions of these structures. Abstract fusion systems have already been defined in Section I.2, and many of their properties were described throughout the rest of Part I.

Taking as starting point Puig's original ideas about abstract fusion systems, Benson predicted, in [Be3] and in unpublished work, that there should be linking systems and classifying spaces associated to abstract fusion systems. His predictions were quite similar to what was actually constructed much later in [BLO2], and which will be described here.

In this section, we first define abstract linking systems, $p$-local finite groups, and their classifying spaces. We then look at fundamental groups and covering spaces of the geometric realizations of these categories, and their connection with certain types of fusion or linking subsystems. Afterwards, we list some of the other homotopy theoretic properties of the classifying spaces, showing that they keep many of the properties already stated for the spaces $BG^{\wedge}_p$.

## 4.1. Linking systems, centric linking systems and $p$-local finite groups.

We first consider linking systems associated to abstract fusion systems. For a $p$-group $S$ and a set $\mathcal{H}$ of subgroups of $S$, $\mathcal{T}_{\mathcal{H}}(S)$ denotes the full subcategory of the transporter category $\mathcal{T}_S(S)$ (Definition 3.1) whose object set is $\mathcal{H}$.

We recall some definitions from Part I (Definition I.3.2). When $\mathcal{F}$ is a fusion system over a $p$-group $S$ and $P \leq S$, we say $P$ is $\mathcal{F}$-*centric* if $C_S(Q) = Z(Q)$ for each $Q \in P^{\mathcal{F}}$, and $P$ is $\mathcal{F}$-*radical* if $O_p(\mathrm{Out}_{\mathcal{F}}(P)) = 1$. We let $\mathcal{F}^{cr} \subseteq \mathcal{F}^c \subseteq \mathcal{F}$ denote the full subcategories with objects the $\mathcal{F}$-centric-radical subgroups and the $\mathcal{F}$-centric subgroups, respectively. We also write $\mathcal{F}^{cr} \subseteq \mathcal{F}^c$ to denote the sets of $\mathcal{F}$-centric-radical and $\mathcal{F}$-centric subgroups of $S$.

**Definition 4.1** ([BLO2, Definition 1.7], [5a1, Definition 3.3], [O4, Definition 3])　Let $\mathcal{F}$ be a fusion system over a p-group $S$. A *linking system* associated to $\mathcal{F}$ is a finite category $\mathcal{L}$, together with a pair of functors

$$\mathcal{T}_{\mathrm{Ob}(\mathcal{L})}(S) \xrightarrow{\ \delta\ } \mathcal{L} \xrightarrow{\ \pi\ } \mathcal{F},$$

satisfying the following conditions:

(A1)　$\mathrm{Ob}(\mathcal{L})$ is a set of subgroups of $S$ closed under $\mathcal{F}$-conjugacy and over-groups, and contains $\mathcal{F}^{cr}$. Each object in $\mathcal{L}$ is isomorphic (in $\mathcal{L}$) to one which is fully centralized in $\mathcal{F}$.

(A2)　$\delta$ is the identity on objects, and $\pi$ is the inclusion on objects. For each $P, Q \in \mathrm{Ob}(\mathcal{L})$ such that $P$ is fully centralized in $\mathcal{F}$, $C_S(P)$ acts freely on $\mathrm{Mor}_{\mathcal{L}}(P, Q)$ via $\delta_{P,P}$ and right composition, and

$$\pi_{P,Q} \colon \mathrm{Mor}_{\mathcal{L}}(P, Q) \longrightarrow \mathrm{Hom}_{\mathcal{F}}(P, Q)$$

is the orbit map for this action.

(B)　For each $P, Q \in \mathrm{Ob}(\mathcal{L})$ and each $g \in T_S(P, Q)$, $\pi_{P,Q}$ sends $\delta_{P,Q}(g) \in \mathrm{Mor}_{\mathcal{L}}(P, Q)$ to $c_g \in \mathrm{Hom}_{\mathcal{F}}(P, Q)$.

(C)　For all $\psi \in \mathrm{Mor}_{\mathcal{L}}(P, Q)$ and all $g \in P$, the diagram

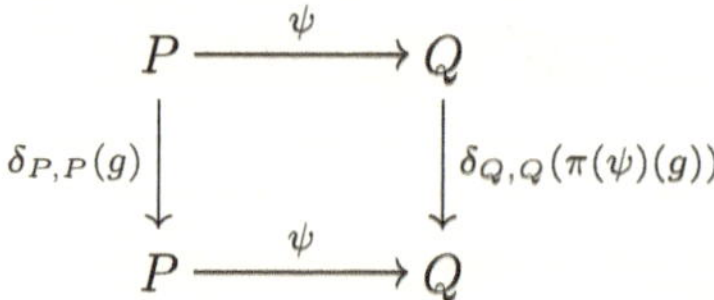

commutes in $\mathcal{L}$.

A *centric linking system* associated to $\mathcal{F}$ is just a linking system $\mathcal{L}$ such that $\mathrm{Ob}(\mathcal{L}) = \mathcal{F}^c$. For centric linking systems, the only difference between the above definition and that in [BLO2] is that we define here $\delta$ as a functor on the transporter category of $S$. That $\delta$ can be defined on $\mathcal{T}_{\mathrm{Ob}(\mathcal{L})}(S)$ follows as a consequence of the earlier definitions (see [BLO2, Proposition 1.11]), and simplifies certain constructions.

For any finite group $G$ with Sylow $p$-subgroup $S$, a subgroup $P \leq S$ is $\mathcal{F}_S(G)$-centric if and only if it is $p$-centric in $G$ (see Definition 3.1). This follows easily from the fact that $P$ is fully centralized in $\mathcal{F}_S(G)$ if and only if $C_S(P) \in \mathrm{Syl}_p(C_G(P))$. The category $\mathcal{L}_S^c(G)$ (Definition 3.1 again) is easily seen to satisfy conditions (A1), (A2), (B), and (C) above, and hence is a centric linking system associated to $\mathcal{F}_S(G)$.

The following proposition describes some of the basic properties of linking systems. Most of these results go back to [5a1, §3], but we refer to [O4] where they are stated and proven more explicitly.

**Proposition 4.2** ([O4, Proposition 4])   *The following hold for any linking system $\mathcal{L}$ associated to a saturated fusion system $\mathcal{F}$ over a p-group $S$.*

(a)   *For each $P, Q \in \mathrm{Ob}(\mathcal{L})$, the subgroup*

$$E(P) \stackrel{\mathrm{def}}{=} \mathrm{Ker}[\mathrm{Aut}_{\mathcal{L}}(P) \longrightarrow \mathrm{Aut}_{\mathcal{F}}(P)]$$

*acts freely on $\mathrm{Mor}_{\mathcal{L}}(P, Q)$ via right composition, and $\pi_{P,Q}$ induces a bijection*

$$\mathrm{Mor}_{\mathcal{L}}(P, Q)/E(P) \stackrel{\cong}{\longrightarrow} \mathrm{Hom}_{\mathcal{F}}(P, Q).$$

(b)   *A morphism $\psi \in \mathrm{Mor}(\mathcal{L})$ is an isomorphism if and only if $\pi(\psi)$ is an isomorphism in $\mathcal{F}$.*

(c)   *If $P \in \mathrm{Ob}(\mathcal{L})$ is fully normalized in $\mathcal{F}$, then*

$$\delta_P(N_S(P)) \in \mathrm{Syl}_p(\mathrm{Aut}_{\mathcal{L}}(P)).$$

(d)   *All morphisms in $\mathcal{L}$ are monomorphisms and epimorphisms in the categorical sense.*

*Proof.* The only part not stated explicitly in [O4, Proposition 4] is (b), and this follows easily from (a).   $\square$

When $\mathcal{L}$ is a linking system associated to a saturated fusion system $\mathcal{F}$ over $S$, then Proposition 4.2(a) implies that the structural functor $\pi\colon \mathcal{L} \longrightarrow \mathcal{F}$ is surjective on morphism sets (between objects of $\mathcal{L}$). Also, by [O4, Proposition 4(c)], the structural functor $\delta$ is injective on all morphism sets.

When $P \leq Q$ are objects in a linking system $\mathcal{L}$, we define $\iota_{P,Q} = \delta_{P,Q}(1) \in \mathrm{Mor}_{\mathcal{L}}(P, Q)$, and regard this as the *inclusion morphism*. This terminology is motivated by axiom (B), which says that $\pi\colon \mathcal{L} \longrightarrow \mathcal{F}$ sends

$\iota_{P,Q}$ to the (ordinary) inclusion of $P$ in $Q$ as a morphism in $\mathcal{F}$. Once inclusions have been defined, it is natural to consider restrictions and extensions of morphisms. Since all morphisms in a linking system are monomorphisms and epimorphisms, restrictions and extensions are unique whenever they exist. The following proposition describes the conditions under which they do exist.

**Proposition 4.3** ([O4, Proposition 4(b,e)])  *The following hold for any linking system $\mathcal{L}$ associated to a saturated fusion system $\mathcal{F}$ over a $p$-group $S$.*

(a)  *For every morphism $\psi \in \mathrm{Mor}_{\mathcal{L}}(P,Q)$, and every $P_0, Q_0 \in \mathrm{Ob}(\mathcal{L})$ such that $P_0 \leq P$, $Q_0 \leq Q$, and $\pi(\psi)(P_0) \leq Q_0$, there is a unique morphism $\psi|_{P_0,Q_0} \in \mathrm{Mor}_{\mathcal{L}}(P_0,Q_0)$ (the "restriction" of $\psi$) such that $\psi \circ \iota_{P_0,P} = \iota_{Q_0,Q} \circ \psi|_{P_0,Q_0}$.*

(b)  *Let $P, Q, \overline{P}, \overline{Q} \in \mathrm{Ob}(\mathcal{L})$ and $\psi \in \mathrm{Mor}_{\mathcal{L}}(P,Q)$ be such that $P \trianglelefteq \overline{P}$, $Q \leq \overline{Q}$, and for each $g \in \overline{P}$ there is $h \in \overline{Q}$ such that $\iota_{Q,\overline{Q}} \circ \psi \circ \delta_P(g) = \delta_{Q,\overline{Q}}(h) \circ \psi$. Then there is a unique morphism $\overline{\psi} \in \mathrm{Mor}_{\mathcal{L}}(\overline{P},\overline{Q})$ such that $\overline{\psi}|_{P,Q} = \psi$.*

We are now ready to define a $p$-local finite group.

**Definition 4.4**  A *$p$-local finite group* is defined to be a triple $(S, \mathcal{F}, \mathcal{L})$, where $S$ is a $p$-group, $\mathcal{F}$ is a saturated fusion system over $S$, and $\mathcal{L}$ is a centric linking system associated to $\mathcal{F}$. The *classifying space* of the triple $(S, \mathcal{F}, \mathcal{L})$ is the $p$-completed nerve $|\mathcal{L}|_p^{\wedge}$.

In particular, for any finite group $G$ with Sylow $p$-subgroup $S \leq G$, $(S, \mathcal{F}_S(G), \mathcal{L}_S^c(G))$ is a $p$-local finite group, with classifying space $|\mathcal{L}_S^c(G)|_p^{\wedge} \simeq BG_p^{\wedge}$ (see Theorem 3.2).

It is not yet known whether every saturated fusion system has an associated linking system, nor whether it is possible for it to have more than one of them. In other words, for all we know now, a saturated fusion system might not give rise to any $p$-local finite group, or it might give rise to several of them. This problem of the existence and uniqueness of associated linking systems will be discussed in detail in Section 5.3.

Linking systems are a special case of a more general type of structure called "transporter systems", which were defined and studied in [OV1]. These categories are modelled on the transporter category of a finite group (Definition 3.1). The most important difference between linking systems and transporter systems is that in a transporter system $\mathcal{T}$ associated to a saturated fusion system $\mathcal{F}$, the condition that the kernel subgroups $E(P) \stackrel{\mathrm{def}}{=} \mathrm{Ker}\big[\mathrm{Aut}_{\mathcal{T}}(P) \longrightarrow \mathrm{Aut}_{\mathcal{F}}(P)\big]$ always be $p$-groups (and equal to

$C_S(P)$ when $P$ is fully centralized) is relaxed. Transporter systems were first defined for use in studying extensions of linking systems with kernel a $p$-group, but they are also useful in many other situations where one wants to weaken the much more rigid conditions imposed on a linking system.

## 4.2. **Quasicentric subgroups and quasicentric linking systems.**

As long as we are working with linking systems associated to one fusion system at a time, it is usually simplest to work with centric linking systems. But as soon as we are working with a pair of linking systems associated to distinct fusion systems (as in Section 4.4), or with a functor between linking systems, it is often necessary to consider linking systems with more objects than just those which are centric. This is why the quasicentric subgroups play an important role: they provide an upper bound for the set of possible objects in a linking system.

**Definition 4.5**  Let $\mathcal{F}$ be a saturated fusion system over a $p$-group $S$. A subgroup $Q \leq S$ is $\mathcal{F}$-*quasicentric* if for each $P \in Q^{\mathcal{F}}$ which is fully centralized in $\mathcal{F}$, the centralizer fusion system $C_{\mathcal{F}}(P)$ is the fusion system of the $p$-group $C_S(P)$. Let $\mathcal{F}^q \subseteq \mathcal{F}$ be the full subcategory whose objects are the $\mathcal{F}$-quasicentric subgroups of $S$. We also let $\mathcal{F}^q = \mathrm{Ob}(\mathcal{F}^q)$ denote the set of $\mathcal{F}$-quasicentric subgroups of $S$.

Note in particular that each $\mathcal{F}$-centric subgroup is $\mathcal{F}$-quasicentric. The following lemma lists some of the important properties of quasicentric subgroups.

**Lemma 4.6**  *Let $\mathcal{F}$ be a saturated fusion system over a $p$-group $S$.*

(a)  *A subgroup $P \leq S$ fully centralized in $\mathcal{F}$ is $\mathcal{F}$-quasicentric if and only if for each $Q \leq P{\cdot}C_S(P)$ containing $P$, and each $\alpha \in \mathrm{Aut}_{\mathcal{F}}(Q)$ such that $\alpha|_P = \mathrm{Id}$, $\alpha$ has $p$-power order.*

(b)  *An arbitrary subgroup of $S$ is $\mathcal{F}$-quasicentric if and only if some fully centralized subgroup in its $\mathcal{F}$-conjugacy class is $\mathcal{F}$-quasicentric.*

(c)  *Assume $Q \leq P \leq S$ are such that $Q \in \mathcal{F}^q$. Let $\varphi, \varphi' \in \mathrm{Hom}_{\mathcal{F}}(P, S)$ be such that $\varphi|_Q = \varphi'|_Q$, and $\varphi(Q) = \varphi'(Q)$ is fully centralized in $\mathcal{F}$. Then there is $x \in C_S(\varphi(Q))$ such that $\varphi = c_x \circ \varphi'$.*

(d)  *If $Q \leq P \leq S$ and $Q \in \mathcal{F}^q$, then $P \in \mathcal{F}^q$. If $Q$ is also fully centralized in $\mathcal{F}$, then so is $P$.*

(e)  *If $\mathcal{F} = \mathcal{F}_S(G)$ for some finite group $G$ with $S \in \mathrm{Syl}_p(G)$, then a subgroup $P \leq S$ is $\mathcal{F}$-quasicentric if and only if $O^p(C_G(P))$ has order prime to $p$.*

*Proof.* **(a)**  If $P$ and $R$ are both fully centralized in $\mathcal{F}$ and $R \in P^{\mathcal{F}}$, then by the extension axiom, there is $\varphi \in \mathrm{Iso}_{\mathcal{F}}(P{\cdot}C_S(P), R{\cdot}C_S(R))$ such that $\varphi(P) = R$. For each $Q_1, Q_2 \leq P{\cdot}C_S(P)$, $\varphi$ induces a bijection from $\mathrm{Hom}_{\mathcal{F}}(Q_1, Q_2)$ to $\mathrm{Hom}_{\mathcal{F}}(\varphi(Q_1), \varphi(Q_2))$ which sends $\psi$ to $(\varphi|_{Q_2})\psi(\varphi|_{Q_1})^{-1}$. In particular, $\varphi$ induces an isomorphism of categories from $C_{\mathcal{F}}(P)$ to $C_{\mathcal{F}}(R)$, and so the first is the fusion system of $C_S(P)$ if and only if the second is the fusion system of $C_S(R)$.

Thus a fully centralized subgroup $P$ is $\mathcal{F}$-quasicentric if and only if $C_{\mathcal{F}}(P)$ is the fusion system of $C_S(P)$. This clearly implies the condition in (a), and the converse follows from Corollary I.3.7.

**(b)** This is immediate from the definition.

**($c_0$)** We first prove this under the additional assumption that $Q \trianglelefteq P$. It was shown in [5a1, Lemma 3.8], but we give here a different, shorter proof. Upon replacing $P$ by $\varphi'(P)$, $Q$ by $\varphi(Q) = \varphi'(Q)$, and $\varphi$ by $\varphi \circ (\varphi')^{-1}$, we can assume that $\varphi' = \mathrm{incl}_P^S$ and $\varphi|_Q = \mathrm{incl}_Q^S$. We are thus reduced to the case where $Q$ is fully centralized and $\varphi|_Q = \mathrm{Id}$, and must show that $\varphi = c_x$ for some $x \in C_S(Q)$.

Set $K = \mathrm{Aut}_P(Q)$. Since $\mathrm{Aut}_{\mathcal{F}}^K(Q) = \mathrm{Aut}_S^K(Q) = K$ and $Q$ is fully centralized, $Q$ is fully $K$-normalized by Proposition I.5.2. Hence by Theorem I.5.5, the normalizer subsystem $N_{\mathcal{F}}^K(Q)$ over $N_S^K(Q) = P{\cdot}C_S(Q)$ is saturated. Also, since $\varphi|_Q = \mathrm{Id}$, $\mathrm{Aut}_{\varphi(P)}(Q) = \mathrm{Aut}_P(Q) = K$. Thus $\varphi(P) \leq N_S^K(Q)$, and $\varphi \in \mathrm{Mor}(N_{\mathcal{F}}^K(Q))$.

It thus suffices to prove that $N_{\mathcal{F}}^K(Q)$ is the fusion system of $N_S^K(Q)$. To show this, it suffices by Corollary I.3.7 to show that all automorphism groups in $N_{\mathcal{F}}^K(Q)$ are $p$-groups. Assume otherwise: fix a subgroup $R \leq N_S^K(Q)$ and $\alpha \in \mathrm{Aut}_{\mathcal{F}}(R)$ such that $Q \leq R$, $\alpha(Q) = Q$, $\alpha|_Q \in K$, and $\alpha \neq \mathrm{Id}_R$ has order prime to $p$. Since $\alpha|_Q \in K$ and $K$ is a $p$-group, $\alpha^{p^k}$ is the identity on $Q$ for some $k$. But then $\alpha^{p^k}|_{C_R(Q)} \in \mathrm{Mor}(C_{\mathcal{F}}(Q))$, $\alpha^{p^k}|_{C_R(Q)}$ has $p$-power order since $C_{\mathcal{F}}(Q)$ is the fusion system of $C_S(Q)$, and thus $\alpha^{p^\ell}|_{Q{\cdot}C_R(Q)} = \mathrm{Id}$ for some $\ell \geq k$. Hence for $g \in R$, $c_{\alpha(g)} = c_g \in \mathrm{Aut}(Q{\cdot}C_R(Q))$, so $g^{-1}\alpha(g) \in C_R(Q{\cdot}C_R(Q)) \leq Q{\cdot}C_R(Q)$, and thus $\alpha$ induces the identity on $R/QC_R(Q)$. So $\alpha = \mathrm{Id}_R$ by Lemma A.2, and this is a contradiction.

**(d)** It suffices to show this when $Q < P$. Under this assumption, since $Q < N_P(Q) \leq P$ (see Lemma A.1), it suffices (by iteration) to prove this when $Q \trianglelefteq P$. By Lemma I.2.6(c), there is a morphism $\chi \in \mathrm{Hom}_{\mathcal{F}}(N_S(Q), S)$ such that $\chi(Q)$ is fully normalized in $\mathcal{F}$. It thus suffices to consider the case where $Q$ is fully centralized (and $Q \in \mathcal{F}^q$).

We first prove that $P$ is fully centralized. Fix $P^* \in P^{\mathcal{F}}$ which is fully centralized in $\mathcal{F}$. Choose $\varphi \in \mathrm{Iso}_{\mathcal{F}}(P^*, P)$, and set $Q^* = \varphi^{-1}(Q)$. Since $Q$ is fully centralized and $\varphi|_{Q^*} \in \mathrm{Iso}_{\mathcal{F}}(Q^*, Q)$ extends to $P^*$, the extension axiom implies there is $\psi \in \mathrm{Hom}_{\mathcal{F}}(P^*{\cdot}C_S(Q^*), S)$ such that $\psi|_{Q^*} = \varphi|_{Q^*}$.

By $(c_0)$, there is $x \in C_S(Q)$ such that $\varphi = c_x \circ \psi|_{P^*}$ in $\mathrm{Iso}_{\mathcal{F}}(P^*, P)$. Set $\psi' = c_x \circ \psi \in \mathrm{Hom}_{\mathcal{F}}(P^* \cdot C_S(Q^*), S)$. Since $\psi'(P^*) = P$, $\psi'(C_S(P^*)) \leq C_S(P)$, and so $P$ is fully centralized in $\mathcal{F}$ since $P^*$ is.

Now fix $R \leq P \cdot C_S(P)$ which contains $P$, and $\alpha \in \mathrm{Aut}_{\mathcal{F}}(R)$ such that $\alpha|_P = \mathrm{Id}_P$. Set $R_0 = R \cap (Q \cdot C_S(Q)) = Q \cdot C_R(Q)$, and $\alpha_0 = \alpha|_{R_0}$. Then $R = PR_0$, $\alpha_0(R_0) = R_0$, and $\alpha_0$ and $\alpha$ have the same order as automorphisms. Since $Q \in \mathcal{F}^q$, $\alpha_0$ has $p$-power order in $\mathrm{Aut}_{\mathcal{F}}(R_0)$, and hence $\alpha$ also has $p$-power order. Thus $P \in \mathcal{F}^q$.

**(c)** Now assume $Q < P \leq S$ are such that $Q \in \mathcal{F}^q$, where $Q$ need not be normal in $P$. Using Lemma A.1 again, we can choose a chain of subgroups $Q = P_0 \trianglelefteq P_1 \trianglelefteq \cdots \trianglelefteq P_k = P$ connecting $Q$ to $P$, each normal in the following one. By (d), the $P_i$ are all $\mathcal{F}$-quasicentric. If $\varphi, \varphi' \in \mathrm{Hom}_{\mathcal{F}}(P, S)$ are such that $\varphi|_Q = \varphi'|_Q$ and $\varphi(Q)$ is fully centralized in $\mathcal{F}$, then each $\varphi(P_i)$ is fully centralized by (d) again. So by $(c_0)$, $\varphi|_{P_1} = c_{x_1} \circ \varphi'|_{P_1}$ for some $x_1 \in C_S(\varphi(Q))$, $\varphi|_{P_2} = c_{x_2} \circ c_{x_1} \circ \varphi'_{P_2}$ for some $x_2 \in C_S(\varphi(P_1))$, etc. Since $C_S(P_i) \leq C_S(Q)$ for each $i$, we conclude that $\varphi = c_x \circ \varphi'$ for some $x \in C_S(Q)$.

**(e)** It suffices to prove this when $P$ is fully centralized in $\mathcal{F} = \mathcal{F}_S(G)$. By (a), $P \in \mathcal{F}^q$ if and only if for each $Q \leq C_S(P)$ and each $\alpha \in \mathrm{Aut}_{C_G(P)}(Q)$, $\alpha$ has $p$-power order. By a theorem of Frobenius (cf. [A4, Theorem 39.4]), this is the case if and only if $C_G(P)$ has a normal $p$-complement; equivalently, $O^p(C_G(P))$ has order prime to $p$. $\qquad\square$

When $G$ is a finite group, we can define a $p$-quasicentric subgroup of $G$ to be a $p$-subgroup $P \leq G$ such that $O^p(C_G(P))$ has order prime to $p$. Thus Lemma 4.6(e) says that if $P \leq S \in \mathrm{Syl}_p(G)$, then $P$ is $p$-quasicentric in $G$ if and only if $P$ is $\mathcal{F}_S(G)$-quasicentric.

We are now ready to describe the role played by quasicentric subgroups as upper bounds to the set of objects in a linking system.

**Proposition 4.7** ([O4, Proposition 4(g)]) *For any linking system $\mathcal{L}$ associated to a saturated fusion system $\mathcal{F}$ over a $p$-group $S$, $\mathrm{Ob}(\mathcal{L}) \subseteq \mathcal{F}^q$.*

*Proof.* We sketch briefly the proof. It suffices to consider $P \in \mathrm{Ob}(\mathcal{L})$ which is fully centralized in $\mathcal{F}$. Fix $Q \leq P \cdot C_S(P)$ which contains $P$, and fix $\alpha \in \mathrm{Aut}_{\mathcal{F}}(Q)$ such that $\alpha|_P = \mathrm{Id}_P$; we must show $\alpha$ has $p$-power order. Choose $\beta \in \mathrm{Aut}_{\mathcal{L}}(Q)$ such that $\pi(\beta) = \alpha$. Then $\beta|_{P,P} \in \mathrm{Aut}_{\mathcal{L}}(P)$ lies in $\mathrm{Ker}(\pi_P) = \delta_P(C_S(P))$, hence has $p$-power order. Thus $\beta^{p^k}|_{P,P} = \mathrm{Id}$ for some $k \geq 0$. But then $\beta^{p^k} = \mathrm{Id}$ in $\mathrm{Aut}_{\mathcal{L}}(Q)$ by Proposition 4.2(d), so $\beta$ and $\alpha$ both have $p$-power order. $\qquad\square$

A *quasicentric linking system* associated to $\mathcal{F}$ is a linking system $\mathcal{L}$ such that $\mathrm{Ob}(\mathcal{L}) = \mathcal{F}^q$. For quasicentric linking systems, the difference between the above definition and that in [5a1, Definition 3.3] is that we define here $\delta$ as a functor on the transporter category of $S$, and drop the axiom $(D)_q$ needed there. Our definition here is easily seen to imply that of [5a1], while the definition in [5a1] implies this one by [5a1, Lemma 3.7] (and since the set of quasicentrics is closed under overgroups by Lemma 4.6(d)).

Fix a finite group $G$, $S \in \mathrm{Syl}_p(G)$. Let $\mathcal{H}$ be a set of subgroups of $S$ which is closed under $G$-conjugacy and overgroups, contains all subgroups which are centric and radical in $\mathcal{F}_S(G)$, and contains only quasicentric subgroups. Let $\mathcal{L}_{\mathcal{H}}(G)$ be the category with object set $\mathcal{H}$, and where

$$\mathrm{Mor}_{\mathcal{L}_{\mathcal{H}}(G)}(P, Q) = T_G(P, Q)/O^p(C_G(P)).$$

Then $\mathcal{L}_{\mathcal{H}}(G)$ is a linking system associated to $\mathcal{F}_S(G)$. The only thing which has to be checked is that this does, in fact, define a category (that composition is well defined on these quotient sets). By Proposition 4.7, when $\mathcal{H}$ is the set of all $\mathcal{F}_S(G)$-quasicentric subgroups, then $\mathcal{L}_S^q(G) \overset{\mathrm{def}}{=} \mathcal{L}_{\mathcal{H}}(G)$ is the largest linking system associated to $\mathcal{F}_S(G)$, in the sense of having the largest possible set of objects.

When $\mathcal{L}_1$ and $\mathcal{L}_2$ are two linking systems associated to the same fusion system $\mathcal{F}$, with structure functors $\delta_i$ and $\pi_i$ and with $\mathrm{Ob}(\mathcal{L}_1) = \mathrm{Ob}(\mathcal{L}_2)$, an *isomorphism of linking systems* from $\mathcal{L}_1$ to $\mathcal{L}_2$ is a functor $\chi \colon \mathcal{L}_1 \longrightarrow \mathcal{L}_2$ such that $\pi_1 = \pi_2 \circ \chi$ and $\delta_2 = \chi \circ \delta_1$. In particular, $\chi$ is the identity on objects. Axiom (A2) implies that any such functor is an isomorphism of categories. The next proposition says, for example, that every centric linking system can be extended to a quasicentric linking system, unique up to isomorphism.

**Proposition 4.8**  *Fix a saturated fusion system $\mathcal{F}$ over a $p$-group $S$. Let $\mathcal{F}^{cr} \subseteq \mathcal{H} \subseteq \widehat{\mathcal{H}} \subseteq \mathcal{F}^q$ be sets of subgroups closed under $\mathcal{F}$-conjugacy and overgroups. Then each linking system $\mathcal{L}$ associated to $\mathcal{F}$ with $\mathrm{Ob}(\mathcal{L}) = \mathcal{H}$ is contained in a linking system $\widehat{\mathcal{L}}$ with $\mathrm{Ob}(\widehat{\mathcal{L}}) = \widehat{\mathcal{H}}$. Also, $\widehat{\mathcal{L}}$ is unique: if $\mathcal{L}^*$ is another linking system associated to $\mathcal{F}$ with $\mathrm{Ob}(\mathcal{L}^*) = \widehat{\mathcal{H}}$, then any isomorphism $\mathcal{L} \cong \mathcal{L}^*|_{\mathcal{H}}$ extends to an isomorphism $\widehat{\mathcal{L}} \cong \mathcal{L}^*$.*

*Proof.* When $\mathcal{H} = \mathcal{F}^c$ and $\widehat{\mathcal{H}} = \mathcal{F}^q$, this was shown in [5a1, Proposition 3.4], and also in [P7, Chapter 20]. The proof in [5a1] is homotopy theoretic — $\widehat{\mathcal{L}}$ is constructed as a full subcategory of the linking system of the space $|\mathcal{L}|_p^\wedge$ (see Definition 3.5) — while the proof in [P7] is algebraic.

We give here a different algebraic proof. Via induction on the number of $\mathcal{F}$-conjugacy classes in $\widehat{\mathcal{H}} \smallsetminus \mathcal{H}$, it suffices to prove the proposition when $\widehat{\mathcal{H}} \smallsetminus \mathcal{H} = \mathcal{P}$ for some $\mathcal{F}$-conjugacy class $\mathcal{P}$. Since $\widehat{\mathcal{H}}$ is closed under overgroups, $Q > P \in \mathcal{P}$ implies $Q \in \mathcal{H}$.

**Step 1:**  Fix a subgroup $P_* \in \mathcal{P}$ which is fully automized and receptive in $\mathcal{F}$. For each $P \in \mathcal{P}$, we fix morphisms

$$\bar{\varphi}_P \in \mathrm{Hom}_{\mathcal{F}}(N_S(P), N_S(P_*)), \quad \psi_P \in \mathrm{Mor}_{\mathcal{L}}(N_S(P), N_S(P_*)),$$

$$\text{and} \quad \varphi_P \in \mathrm{Iso}_{\mathcal{F}}(P, P_*)$$

such that

$$\bar{\varphi}_P = \pi(\psi_P), \qquad \varphi_P = \bar{\varphi}_P|_P, \qquad \text{and} \qquad \psi_{P_*} = \mathrm{Id}_{P_*}.$$

To arrange this, first choose $\bar{\varphi}_P$ such that $\bar{\varphi}_P(P) = P^*$ using Lemma I.2.6(c). Then set $\varphi_P = \bar{\varphi}_P|_P$, and choose any $\psi_P \in \pi^{-1}(\bar{\varphi}_P)$ (a nonempty set by Proposition 4.2(a)).

Set
$$\widehat{P} = \{g \in N_S(P_*) \,|\, c_g \in O_p(\mathrm{Aut}_{\mathcal{F}}(P_*))\}.$$

Since $O_p(\mathrm{Aut}_{\mathcal{F}}(P_*)) \leq \mathrm{Aut}_S(P_*) \in \mathrm{Syl}_p(\mathrm{Aut}_{\mathcal{F}}(P_*))$, we have $\mathrm{Aut}_{\widehat{P}}(P_*) = O_p(\mathrm{Aut}_{\mathcal{F}}(P_*))$. Since $P_* \notin \mathcal{H} \supseteq \mathcal{F}^{cr}$, either $O_p(\mathrm{Aut}_{\mathcal{F}}(P_*)) > \mathrm{Inn}(P_*)$ or $C_S(P_*) \nleq P_*$. In either case, $\widehat{P} > P_*$, and hence $\widehat{P} \in \mathcal{H} = \mathrm{Ob}(\mathcal{L})$. Set

$$\Gamma = \{\gamma \in \mathrm{Aut}_{\mathcal{L}}(\widehat{P}) \,|\, \pi(\gamma)(P_*) = P_*\}.$$

Since $\mathrm{Aut}_{\widehat{P}}(P_*) = O_p(\mathrm{Aut}_{\mathcal{F}}(P_*))$ is normal in $\mathrm{Aut}_{\mathcal{F}}(P_*)$ (and since $P_*$ is receptive),

$$\text{for each } \alpha \in \mathrm{Aut}_{\mathcal{F}}(P_*) \text{ there is } \psi \in \Gamma \text{ such that } \pi(\psi)|_{P_*} = \alpha. \quad (1)$$

Intuitively, to construct $\widehat{\mathcal{L}}$, we will identify $\mathrm{Aut}_{\widehat{\mathcal{L}}}(P_*)$ with $\Gamma$ (each element will be the restriction to $P_*$ of a unique element of $\Gamma$), identify each element of $\mathrm{Iso}_{\widehat{\mathcal{L}}}(P, R)$ (for $P, R \in \mathcal{P}$) with a composite of restrictions of $\psi_P$, $\psi_R$, and an element of $\Gamma$, and then regard each morphism with source group in $\mathcal{P}$ and target in $\mathcal{H}$ as an isomorphism followed by an inclusion. In order to define composition, we must show how to restrict morphisms in $\mathcal{L}$ to isomorphisms between subgroups in $\mathcal{P}$ (when appropriate), and this is what will be done in Step 2.

**Step 2:**  For each $P_1, P_2 \in \mathcal{P}$, set

$$\Psi_{P_1, P_2} = \{(Q_1, \psi, Q_2) \,|\, P_i \trianglelefteq Q_i \in \mathcal{H}, \ \psi \in \mathrm{Mor}_{\mathcal{L}}(Q_1, Q_2), \ \pi(\psi)(P_1) = P_2\}.$$

Let $\tau_{P_1, P_2} \colon \Psi_{P_1, P_2} \longrightarrow \Psi_{P_*, P_*}$ be the "translation map", defined by

$$\tau_{P_1, P_2}(Q_1, \psi, Q_2) = \big(R_1, \psi_{P_2}|_{Q_2, R_2} \circ \psi \circ (\psi_{P_1}|_{Q_1, R_1})^{-1}, R_2\big)$$

$$\text{where } R_i = \bar{\varphi}_{P_i}(Q_i).$$

Set $\Psi = \Psi_{P_*, P_*}$. We claim that for each $(Q_1, \psi, Q_2) \in \Psi$, there is $\widehat{\psi} \in \mathrm{Mor}_{\mathcal{L}}(Q_1\widehat{P}, Q_2\widehat{P})$ such that $\widehat{\psi}|_{Q_1, Q_2} = \psi$. To see this, set $\varphi_0 = \pi(\psi)|_{P_*} \in \mathrm{Aut}_{\mathcal{F}}(P_*)$. Since $P_*$ is receptive (and since $\varphi_0$ extends to $\pi(\psi) \in \mathrm{Hom}_{\mathcal{F}}(Q_1, Q_2)$), $\varphi_0$ extends to some $\varphi_1 \in \mathrm{Hom}_{\mathcal{F}}(Q_1\widehat{P}, S)$. Since $\varphi_1|_{P_*} = \pi(\psi)|_{P_*}$, Lemma 4.6(c) implies there is $x \in C_S(P_*) \leq \widehat{P}$ such that $c_x \circ \varphi_1|_{Q_1} = \pi(\psi)$. Thus $c_x \circ \varphi_1(Q_1) = Q_2$, and $c_x \circ \varphi_1(\widehat{P}) = \widehat{P}$ by definition of

$\widehat{P}$ (and since $c_x \circ \varphi_1(P_*) = P_*$). Choose any $\psi_1 \in \mathrm{Mor}_{\mathcal{L}}(Q_1\widehat{P}, Q_2\widehat{P})$ such that $\pi(\psi_1) = c_x \circ \varphi_1$. By axiom (A2) for the linking system $\mathcal{L}$, and since all subgroups containing $P_*$ are fully centralized in $\mathcal{F}$ by Lemma 4.6(d), $\psi = \psi_1|_{Q_1, Q_2} \circ \delta_{Q_1}(y)$ for some $y \in C_S(Q_1) \leq \widehat{P}$. Thus $\widehat{\psi} \overset{\mathrm{def}}{=} \psi_1 \circ \delta_{Q_1, \widehat{P}}(y)$ lies in $\mathrm{Mor}_{\mathcal{L}}(Q_1\widehat{P}, Q_2\widehat{P})$ and extends $\psi$.

We can thus define a "restriction" map

$$\rho \colon \Psi = \Psi_{P_*, P_*} \longrightarrow \Gamma,$$

by sending $(Q_1, \psi, Q_2) \in \Psi$ to $\widehat{\psi}|_{\widehat{P}, \widehat{P}} \in \Gamma$, where $\widehat{\psi}$ is the extension of $\psi$ constructed above (and is unique by Proposition 4.2(d)). For each $(Q_1, \psi, Q_2)$ and $(Q_2, \chi, Q_3)$ in $\Psi$,

$$\rho(Q_2, \chi, Q_3) \circ \rho(Q_1, \psi, Q_2) = \rho(Q_1, \chi\psi, Q_3)$$

$$\text{and} \quad \pi\big(\rho(Q_1, \psi, Q_2)\big)|_{P_*} = \pi(\psi)|_{P_*}. \quad (2)$$

Upon combining the first equality with the definition of the $\tau_{P,Q}$, we see that for $P_1, P_2, P_3 \in \mathcal{P}$, and any composable pair $(Q_1, \psi, Q_2) \in \Psi_{P_1, P_2}$ and $(Q_2, \chi, Q_3) \in \Psi_{P_2, P_3}$,

$$\rho\tau_{P_2, P_3}(Q_2, \chi, Q_3) \circ \rho\tau_{P_1, P_2}(Q_1, \psi, Q_2) = \rho\tau_{P_1, P_3}(Q_1, \chi\psi, Q_3) \in \Gamma. \quad (3)$$

We claim that

$$P \in \mathcal{P}, \ P \trianglelefteq Q \in \mathcal{H}, \ g \in Q \implies \rho\tau_{P,P}(Q, \delta_Q(g), Q) = \delta_{\widehat{P}}(\bar{\varphi}_P(g)). \quad (4)$$

Set $R = \bar{\varphi}_P(Q)$, so that $P_* \trianglelefteq R$. Then

$$\tau_{P,P}(Q, \delta_Q(g), Q) = \big(R, \psi_P|_{Q,R} \circ \delta_Q(g) \circ (\psi_P|_{Q,R})^{-1}, R\big) = (R, \delta_R(\bar{\varphi}_P(g)), R),$$

where the last equality follows from axiom (C) for the linking system $\mathcal{L}$ (and since $\pi(\psi_P) = \bar{\varphi}_P$). Also, $\delta_R(\bar{\varphi}_P(g))$ extends to $\delta_{R\widehat{P}}(\bar{\varphi}_P(g))$, and hence $\rho(R, \delta_R(\bar{\varphi}_P(g)), R) = \delta_{\widehat{P}}(\bar{\varphi}_P(g))$.

**Step 3:** We are now ready to construct the linking system $\widehat{\mathcal{L}}$ with $\mathrm{Ob}(\widehat{\mathcal{L}}) = \widehat{\mathcal{H}}$. For $P, Q \in \widehat{\mathcal{H}}$, set $\mathrm{Mor}_{\widehat{\mathcal{L}}}(P, Q) = \mathrm{Mor}_{\mathcal{L}}(P, Q)$ if $P, Q \in \mathcal{H}$, and set $\mathrm{Mor}_{\widehat{\mathcal{L}}}(P, Q) = \varnothing$ if $P \in \mathcal{H}$ and $Q \in \mathcal{P}$. When $P \in \mathcal{P}$, set

$$\mathrm{Mor}_{\widehat{\mathcal{L}}}(P, Q) = \big\{\gamma_P^{P_2, Q} \,\big|\, \gamma \in \Gamma, \ Q \geq P_2 \in \mathcal{P}\big\}.$$

When $Q \in \mathcal{P}$ (so $P_2 = Q$), we write $\gamma_P^Q = \gamma_P^{Q,Q} \in \mathrm{Iso}_{\widehat{\mathcal{L}}}(P, Q)$. Intuitively, $\gamma_P^{P_2}$ is the image of $\gamma|_{P_*, P_*} \in \mathrm{Aut}_{\widehat{\mathcal{L}}}(P_*)$ under the bijection $\mathrm{Aut}_{\widehat{\mathcal{L}}}(P_*) \overset{\cong}{\longrightarrow} \mathrm{Iso}_{\widehat{\mathcal{L}}}(P, P_2)$ induced by composition with $(\psi_{P_2}|_{P_2, P_*})^{-1}$ and with $\psi_P|_{P, P_*}$; and $\gamma_P^{P_2, Q}$ is then $\gamma_P^{P_2}$ followed by the inclusion $\iota_{P_2, Q}$.

Define composition in $\widehat{\mathcal{L}}$ as follows. Composition of morphisms in $\mathcal{L}$ is unchanged. For $P_1, P_2, P_3 \in \mathcal{P}$, $P_3 \leq Q_3 \in \widehat{\mathcal{H}}$, $\gamma_{P_1}^{P_2} \in \mathrm{Mor}_{\widehat{\mathcal{L}}}(P_1, P_2)$, and

$\beta^{P_3,Q_3}_{P_2} \in \mathrm{Mor}_{\widehat{\mathcal{L}}}(P_2,Q_3)$, set

$$\beta^{P_3,Q_3}_{P_2} \circ \gamma^{P_2}_{P_1} = (\beta\gamma)^{P_3,Q_3}_{P_1} \in \mathrm{Mor}_{\widehat{\mathcal{L}}}(P_1,Q_3). \tag{5}$$

For $P_1, P_2 \in \mathcal{P}$, $Q_2 > P_2$, $Q_3 \in \mathcal{H}$, $\gamma^{P_2,Q_2}_{P_1} \in \mathrm{Mor}_{\widehat{\mathcal{L}}}(P_1,Q_2)$, and $\psi \in \mathrm{Mor}_{\mathcal{L}}(Q_2,Q_3)$, write $P_3 = \pi(\psi)(P_2) < Q_3$ and $N_i = N_{Q_i}(P_i) > P_i$ $(i = 2,3)$, and set

$$\psi \circ \gamma^{P_2,Q_2}_{P_1} = \left(\rho\tau_{P_2,P_3}(N_2, \psi|_{N_2,N_3}, N_3) \circ \gamma\right)^{P_3,Q_3}_{P_1} \in \mathrm{Mor}_{\widehat{\mathcal{L}}}(P_1,Q_3). \tag{6}$$

The only tricky point when proving that composition is associative comes when taking a composite of three morphisms of which the last two are in $\mathcal{L}$, and in this case, associativity follows from (3). Thus $\widehat{\mathcal{L}}$ is a category.

<u>Definition of $\widehat{\pi}$:</u>  Define $\widehat{\pi}\colon \widehat{\mathcal{L}} \longrightarrow \mathcal{F}$ as follows. When $P, Q \in \mathcal{H}$, set $\widehat{\pi}_{P,Q} = \pi_{P,Q}$. Thus $\widehat{\pi}|_{\mathcal{L}} = \pi$. When $P \in \mathcal{P}$, $Q \in \widehat{\mathcal{H}}$, $Q \geq P_2 \in \mathcal{P}$, and $\gamma \in \Gamma$, set

$$\widehat{\pi}(\gamma^{P_2,Q}_P) = \mathrm{incl}^Q_{P_2} \circ \varphi^{-1}_{P_2} \circ \pi(\gamma)|_{P_*,P_*} \circ \varphi_P \in \mathrm{Hom}_{\mathcal{F}}(P,Q). \tag{7}$$

To see that this defines a functor on $\widehat{\mathcal{L}}$, we must check that it preserves composition as defined in (5) and (6). The first case is clear, while in the situation of (6),

$$\widehat{\pi}(\psi \circ \gamma^{P_2,Q_2}_{P_1}) = \mathrm{incl}^{Q_3}_{P_3} \circ \varphi^{-1}_{P_3} \circ \left(\left(\varphi_{P_3} \circ \pi(\psi)|_{P_2,P_3} \circ \varphi^{-1}_{P_2}\right) \circ \pi(\gamma)|_{P_*,P_*}\right) \circ \varphi_{P_1}$$

$$= \pi_{Q_2,Q_3}(\psi) \circ \mathrm{incl}^{Q_2}_{P_2} \circ \varphi^{-1}_{P_2} \circ \pi(\gamma)|_{P_*,P_*} \circ \varphi_{P_1} = \pi(\psi) \circ \widehat{\pi}(\gamma^{P_2,Q_2}_{P_1}).$$

<u>Definition of $\widehat{\delta}$:</u>  For each $P \in \mathcal{P}$ and $g \in S$, define

$$\gamma(P,g) = \rho\tau_{P,{}^gP}\left(N_S(P), \delta_{N_S(P),N_S({}^gP)}(g), N_S({}^gP)\right) \in \Gamma. \tag{8}$$

Define $\widehat{\delta}\colon \mathcal{T}_{\widehat{\mathcal{H}}}(S) \longrightarrow \widehat{\mathcal{L}}$ by setting $\widehat{\delta}|_{\mathcal{T}_{\mathcal{H}}(S)} = \delta$; and for $P \in \mathcal{P}$, $Q \in \widehat{\mathcal{H}}$, and $g \in S$ such that ${}^gP \leq Q$,

$$\widehat{\delta}_{P,Q}(g) = \gamma(P,g)^{{}^gP,Q}_P.$$

In particular, by (4),

$$P \in \mathcal{P}, \ g \in N_S(P) \implies$$

$$\gamma(P,g) = \delta_{\widehat{P}}(\bar{\varphi}_P(g)) \quad \text{and} \quad \widehat{\delta}_P(g) = \left(\delta_{\widehat{P}}(\bar{\varphi}_P(g))\right)^P_P. \tag{9}$$

We must show $\widehat{\delta}$ is a functor; i.e., it commutes with composition. Fix $P_1, Q_2, Q_3 \leq S$ and $g, h \in S$ such that $P_1 \in \mathcal{P}$, ${}^gP_1 \leq Q_2$, and ${}^hQ_2 \leq Q_3$. Set $P_2 = {}^gP_1$ and $P_3 = {}^hP_2 = {}^{hg}P_1$. If $P_2 = Q_2 \in \mathcal{P}$, then $\gamma(P_2,h) \circ \gamma(P_1,g) = \gamma(P_1,hg)$ by (3) and (8), and so

$$\widehat{\delta}_{P_2,Q_3}(h) \circ \widehat{\delta}_{P_1,P_2}(g) = \gamma(P_2,h)^{P_3,Q_3}_{P_2} \circ \gamma(P_1,g)^{P_2}_{P_1}$$

$$= \gamma(P_1,hg)^{P_3,Q_3}_{P_1} = \widehat{\delta}_{P_1,Q_3}(hg)$$

by (5). If $P_2 < Q_2 \in \mathcal{H}$, set $N_i = N_{Q_i}(P_i)$ $(i = 2, 3)$ and $N_i^* = N_S(P_i)$ $(i = 1, 2, 3)$; then

$$\widehat{\delta}_{Q_2,Q_3}(h) \circ \widehat{\delta}_{P_1,Q_2}(g) = \delta_{Q_2,Q_3}(h) \circ \gamma(P_1, g)_{P_1}^{P_2,Q_2}$$

$$= \left(\rho\tau_{P_2,P_3}(N_2, \delta_{N_2,N_3}(h), N_3) \circ \rho\tau_{P_1,P_2}(N_1^*, \delta_{N_1^*,N_2^*}(g), N_2^*)\right)_{P_1}^{P_3,Q_3}$$

$$\text{(by (6) and (8))}$$

$$= \left(\rho\tau_{P_1,P_3}(N_1^*, \delta_{N_1^*,N_3^*}(hg), N_3^*)\right)_{P_1}^{P_3,Q_3} \qquad \text{(by (3))}$$

$$= \gamma(P_1, hg)_{P_1}^{P_3,Q_3} = \widehat{\delta}_{P_1,Q_3}(hg).$$

<u>Proof of axiom (B):</u> Fix $P, Q \in \widehat{\mathcal{H}}$ and $g \in G$ such that $^g P \leq Q$. If $P \in \mathcal{H}$, then $\widehat{\pi}(\widehat{\delta}_{P,Q}(g)) = c_g|_{P,Q}$ by axiom (B) for $\mathcal{L}$. If $P \in \mathcal{P}$, set $P_2 = {}^g P \leq Q$, and then

$$\widehat{\pi}(\widehat{\delta}_{P,Q}(g)) = \widehat{\pi}(\gamma(P, g)_P^{P_2,Q}) = \text{incl}_{P_2}^Q \circ \varphi_{P_2}^{-1} \circ \pi(\gamma(P, g))|_{P_*,P_*} \circ \varphi_P$$

$$= \text{incl}_{P_2}^Q \circ \varphi_{P_2}^{-1} \circ \left(\varphi_{P_2} \circ \pi(\delta_{N_S(P),N_S(P_2)}(g))|_{P,P_2} \circ \varphi_P^{-1}\right) \circ \varphi_P = c_g|_{P,Q}.$$

<u>Proof of axiom (C):</u> It suffices to consider morphisms $\psi$ in $\widehat{\mathcal{L}}$ of the following three types (since each morphism in $\widehat{\mathcal{L}}$ not in $\mathcal{L}$ is a composite of such morphisms).

- $\psi = \gamma_{P_*}^{P_*} \in \text{Aut}_{\widehat{\mathcal{L}}}(P_*)$ for $\gamma \in \Gamma$. For $g \in P_*$,

$$\psi \circ \widehat{\delta}_{P_*}(g) = \left(\gamma \circ \delta_{\widehat{P}}(g)\right)_{P_*}^{P_*} = \left(\delta_{\widehat{P}}(\pi(\gamma)(g)) \circ \gamma\right)_{P_*}^{P_*} = \widehat{\delta}_{P_*}(\widehat{\pi}(\psi)(g)) \circ \psi,$$

  where the second equality holds by axiom (C) for $\mathcal{L}$, and the first and third each holds by (9) and (5).

- $\psi = 1_P^{P_*} \in \text{Iso}_{\widehat{\mathcal{L}}}(P, P_*)$ for $P \in \mathcal{P}$. For $g \in P$, $\widehat{\delta}_P(g) = \left(\delta_{\widehat{P}}(\varphi_P(g))\right)_P^P$ by (9), and so

$$\psi \circ \widehat{\delta}_P(g) = \left(\delta_{\widehat{P}}(\varphi_P(g))\right)_P^{P_*} = \widehat{\delta}_{P_*}(\varphi_P(g)) \circ \psi$$

  by (5). (Note that $\widehat{\pi}(\psi) = \varphi_P$ in this case.)

- $\psi = 1_P^{P,Q} \in \text{Mor}_{\widehat{\mathcal{L}}}(P, Q)$ for $Q > P \in \mathcal{P}$ (the inclusion of $P$ in $Q$). For $g \in P$,

$$\psi \circ \widehat{\delta}_P(g) = \left(\delta_{\widehat{P}}(\varphi_P(g))\right)_P^{P,Q} = \left(\rho\tau_{P,P}(Q, \delta_Q(g), Q)\right)_P^{P,Q} = \delta_Q(g) \circ \psi,$$

  where the first equality holds by (9) and (5), the second by (4), and the third by (6).

<u>Proof of (A1), (A2):</u> Axiom (A1) follows from the assumptions on $\widehat{\mathcal{H}}$, and the first statement in axiom (A2) holds by construction. Since the rest of (A2) holds for $\mathcal{L}$, it suffices to prove it for morphism sets $\text{Mor}_{\widehat{\mathcal{L}}}(P, Q)$ for $P \in \mathcal{P}$ and $Q \in \widehat{\mathcal{H}}$, where $P$ is fully centralized in $\mathcal{F}$. In this case, $\widehat{\pi}_{P,Q}$

is onto by (1) (and (7)). Also, $\widehat{\delta}_P(C_S(P))$ acts freely on $\operatorname{Mor}_{\widehat{\mathcal{L}}}(P, Q)$ by construction and by (9), and $\widehat{\pi}(\widehat{\delta}_P(C_S(P))) = 1$ by (B).

Now assume $\beta_P^{P_2,Q}$ and $\gamma_P^{P_3,Q}$ are morphisms in $\operatorname{Mor}_{\widehat{\mathcal{L}}}(P, Q)$ such that $\widehat{\pi}(\beta_P^{P_2,Q}) = \widehat{\pi}(\gamma_P^{P_3,Q})$, where $P_2, P_3 \in \mathcal{P}$. Since $P$ is fully centralized in $\mathcal{F}$, $\bar{\varphi}_P(C_S(P)) = C_S(P_*)$. Then $P_2 = \operatorname{Im}(\widehat{\pi}(\beta_P^{P_2,Q})) = \operatorname{Im}(\widehat{\pi}(\gamma_P^{P_3,Q})) = P_3$, and $\pi(\beta)|_{P_*} = \pi(\gamma)|_{P_*} \in \operatorname{Aut}_{\mathcal{F}}(P_*)$ by (7). By Lemma 4.6(c), there is $x \in C_S(P_*) \leq \widehat{P}$ such that $\pi(\beta) = \pi(\gamma) \circ c_x$ in $\operatorname{Aut}_{\mathcal{F}}(\widehat{P})$. By axiom (A2) for the linking system $\mathcal{L}$, there is $y \in C_S(\widehat{P}) = Z(\widehat{P})$ such that $\beta = \gamma \circ \delta_{\widehat{P}}(xy)$ in $\Gamma \leq \operatorname{Aut}_{\widehat{\mathcal{L}}}(\widehat{P})$ ($\widehat{P}$ is fully centralized in $\mathcal{F}$ by Lemma 4.6(d)). Then $xy \in C_S(P_*)$, $xy = \bar{\varphi}_P(z)$ for some $z \in C_S(P)$, and

$$\beta_P^{P_2,Q} = \gamma_P^{P_3,Q} \circ \left(\delta_{\widehat{P}}(xy)\right)_P^P = \gamma_P^{P_3,Q} \circ \widehat{\delta}_P(z)$$

by (5) and (9). This proves that $\widehat{\pi}_{P,Q}$ is the orbit map for the action of $C_S(P)$ on $\operatorname{Mor}_{\widehat{\mathcal{L}}}(P, Q)$, and finishes the proof of axiom (A2).

**Step 4:** Let $\mathcal{L}^*$ be another linking system associated to $\mathcal{F}$ with $\operatorname{Ob}(\mathcal{L}^*) = \widehat{\mathcal{H}}$, and assume there is an isomorphism of linking systems $\chi \colon \mathcal{L} \longrightarrow \mathcal{L}^*|_{\mathcal{H}}$. Let $\delta^*$ and $\pi^*$ be the structure functors of $\mathcal{L}^*$. We claim $\chi$ can be extended to an isomorphism of linking systems $\widehat{\chi} \colon \widehat{\mathcal{L}} \longrightarrow \mathcal{L}^*$.

Let $\widehat{\chi}$ be the identity on objects, and set $\widehat{\chi}(\psi) = \chi(\psi)$ for $\psi \in \operatorname{Mor}(\mathcal{L})$. For each $P \in \mathcal{P}$, $Q \in \widehat{\mathcal{H}}$, and $\gamma_P^{R,Q} \in \operatorname{Mor}_{\widehat{\mathcal{L}}}(P, Q)$, set

$$\widehat{\chi}(\gamma_P^{R,Q}) = \delta_{R,Q}^*(1) \circ \left(\chi(\psi_R)|_{R,P_*}\right)^{-1} \circ \chi(\gamma)|_{P_*,P_*} \circ \chi(\psi_P)|_{P,P_*} \in \operatorname{Mor}_{\mathcal{L}^*}(P, Q).$$

By definition of $\rho$ in Step 2, for each $(R, \psi, T) \in \Psi$, $\psi$ and $\rho(R, \psi, T)$ are both restrictions of some morphism $\widehat{\psi} \in \operatorname{Mor}_{\mathcal{L}}(R\widehat{P}, T\widehat{P})$, and thus

$$\chi(\rho(R, \psi, T))|_{P_*,P_*} = \chi(\psi)|_{P_*,P_*}. \tag{10}$$

Hence for each $P, P_2 \in \mathcal{P}$, $Q_2 > P_2$, $Q_3 \in \mathcal{H}$, $\gamma_P^{P_2,Q_2} \in \operatorname{Mor}_{\widehat{\mathcal{L}}}(P, Q_2)$, and $\psi \in \operatorname{Mor}_{\mathcal{L}}(Q_2, Q_3)$, if we set $P_3 = \pi(\psi)(P_2) < Q_3$ and $N_i = N_{Q_i}(P_i)$ ($i = 2, 3$), then by (6),

$$\widehat{\chi}(\psi \circ \gamma_P^{P_2,Q_2}) = \delta_{P_3,Q_3}^*(1) \circ \left(\chi(\psi_{P_3})|_{P_3,P_*}\right)^{-1}$$

$$\circ \chi\left(\rho_{T P_2,P_3}(N_2, \psi|_{N_2,N_3}, N_3)\right)\big|_{P_*,P_*} \circ \chi(\gamma)|_{P_*,P_*} \circ \chi(\psi_P)|_{P,P_*}$$

$$= \delta_{P_3,Q_3}^*(1) \circ \left(\chi(\psi_{P_3})|_{P_3,P_*}\right)^{-1}$$

$$\circ \left(\chi(\psi_{P_3})|_{P_3,P_*} \circ \chi(\psi)|_{P_2,P_3} \circ \left(\chi(\psi_{P_2})|_{P_2,P_*}\right)^{-1}\right)$$

$$\circ \chi(\gamma)|_{P_*,P_*} \circ \chi(\psi_P)|_{P,P_*}$$

$$= \delta_{P_3,Q_3}^*(1) \circ \chi(\psi)|_{P_2,P_3} \circ \left(\chi(\psi_{P_2})|_{P_2,P_*}\right)^{-1} \circ \chi(\gamma)|_{P_*,P_*} \circ \chi(\psi_P)|_{P,P_*}$$

$$= \chi(\psi) \circ \delta_{P_2,Q_2}^*(1) \circ \left(\chi(\psi_{P_2})|_{P_2,P_*}\right)^{-1} \circ \chi(\gamma)|_{P_*,P_*} \circ \chi(\psi_P)|_{P,P_*}$$

$$= \widehat{\chi}(\psi) \circ \widehat{\chi}(\gamma_P^{P_2,Q_2}).$$

Composition in the situation of (5) is easily handled, and thus $\widehat{\chi}$ is a functor.

Now, $\pi^* \circ \widehat{\chi} = \widehat{\pi}$ and $\delta^* = \widehat{\chi} \circ \widehat{\delta}$ by the definitions of $\widehat{\pi}$ and $\widehat{\delta}$ in Step 3, and by (10) in the latter case. This finishes the proof that $\widehat{\chi}$ is an isomorphism of linking systems.                                                                                        $\square$


## 4.3. Automorphisms of fusion and linking systems.

We next look at certain groups of self-equivalences of fusion and linking systems. We first consider automorphisms of fusion systems. In this case, rather than look at self-equivalences or automorphisms of the category, it turns out to be enough (and much easier) to look at automorphisms of the underlying $p$-group.

Fusion preserving isomorphisms between Sylow subgroups of finite groups were defined in Definition 1.14. When $\mathcal{F}$ is an (abstract) saturated fusion system over a $p$-group $S$, an automorphism $\alpha \in \mathrm{Aut}(S)$ is fusion preserving if $^\alpha\mathcal{F} = \mathcal{F}$ (Definition I.3.4). Set

$$\mathrm{Aut}(S, \mathcal{F}) = \{\alpha \in \mathrm{Aut}(S) \,|\, ^\alpha\mathcal{F} = \mathcal{F}\}$$
$$\mathrm{Out}(S, \mathcal{F}) = \mathrm{Aut}(S, \mathcal{F})/\mathrm{Aut}_{\mathcal{F}}(S).$$


In other words, $\alpha \in \mathrm{Aut}(S)$ is fusion preserving exactly when it induces an automorphism $c_\alpha$ of $\mathcal{F}$, by sending $P$ to $\alpha(P)$ and $\varphi$ to $^\alpha\varphi$. An arbitrary automorphism of the category $\mathcal{F}$ is realized in this way if and only if it sends inclusions to inclusions, and is "isotypical" in the sense defined by Martino and Priddy [MP2]: it commutes up to natural isomorphism of functors with the forgetful functor from $\mathcal{F}$ to groups. Note, however, that there can be nontrivial elements of $\mathrm{Out}(S, \mathcal{F})$ which act trivially on $\mathcal{F}$; for example, when $\mathcal{F}$ is the fusion system of an abelian $p$-group (and $p > 2$).

Before discussing automorphisms of linking systems, we first recall some basic facts about equivalences of categories. A functor $\Phi \colon \mathcal{C} \longrightarrow \mathcal{D}$ is an equivalence if it induces a bijection between the isomorphism classes of objects, and bijections on all morphism sets. This is equivalent to the condition that there be a functor $\Psi$ in the other direction such that $\Phi \circ \Psi$ and $\Psi \circ \Phi$ are both naturally isomorphic to the identity on $\mathcal{C}$ or $\mathcal{D}$. Hence for any small category $\mathcal{C}$, the set of all self equivalences $\mathcal{C} \longrightarrow \mathcal{C}$ up to natural isomorphism is a group under composition, which we denote $\mathrm{Out}(\mathcal{C})$.

Now fix a linking system $\mathcal{L}$ associated to a saturated fusion system $\mathcal{F}$ over a $p$-group $S$. Recall that as part of the structure of the linking system, there is a functor $\delta$ from $\mathcal{T}_{\mathrm{Ob}(\mathcal{L})}(S)$ to $\mathcal{L}$. Thus for each $P \in \mathrm{Ob}(\mathcal{L})$, $\delta_P$ sends $N_G(P)$ to $\mathrm{Aut}_{\mathcal{L}}(P)$. An equivalence of categories $\alpha \colon \mathcal{L} \longrightarrow \mathcal{L}$ is *isotypical* if for each $P \in \mathrm{Ob}(\mathcal{L})$, $\alpha(\delta_P(P)) = \delta_{\alpha(P)}(\alpha(P))$. Let $\mathrm{Out}_{\mathrm{typ}}(\mathcal{L}) \leq \mathrm{Out}(\mathcal{L})$ be the subgroup of all natural isomorphism classes of isotypical self equivalences of $\mathcal{L}$. (A submonoid of the finite group $\mathrm{Out}(\mathcal{L})$ must be a subgroup.)

Finally, define

$$\mathrm{Aut}^I_{\mathrm{typ}}(\mathcal{L}) = \{\alpha \in \mathrm{Aut}_{\mathrm{typ}}(\mathcal{L}) \,|\, \alpha \text{ sends inclusions to inclusions}\}.$$

By analogy with the definition by Martino and Priddy of an isotypical equivalence of fusion systems, an equivalence $\alpha \colon \mathcal{L}_1 \longrightarrow \mathcal{L}_2$ between centric linking systems was defined in [BLO1] to be isotypical whenever $F_1 \cong F_2 \circ \alpha$ (naturally isomorphic as functors), where $F_i \colon \mathcal{L}_i \longrightarrow \mathrm{Gps}$ denotes the forgetful functor. For *centric* linking systems, this definition of an isotypical equivalence was shown in [BLO2, Lemma 8.2] to be equivalent to the one given above. That proof extends easily to the case of isotypical equivalences of linking systems all of whose objects are centric, but the result is definitely not true in general for equivalences of linking systems which contain noncentric objects. Also, Proposition 4.11 below is false if the earlier definition of "isotypical" (in terms of the forgetful functor to groups) is used. This is why the definition in terms of the "distinguished subgroups" $\delta_P(P)$ seems more appropriate.

In general, it is much easier to work with isotypical equivalences of a linking system $\mathcal{L}$ which sends inclusions to inclusions, than to work with arbitrary isotypical equivalences. Some of the reasons for this will be seen shortly in Proposition 4.11. The next lemma explains how $\mathrm{Out}_{\mathrm{typ}}(\mathcal{L})$ is described in terms of $\mathrm{Aut}^I_{\mathrm{typ}}(\mathcal{L})$.

For $\mathcal{L}$ and $\mathcal{F}$ as above, and any $\gamma \in \mathrm{Aut}_{\mathcal{L}}(S)$, we let $c_\gamma \in \mathrm{Aut}^I_{\mathrm{typ}}(\mathcal{L})$ be the automorphism which sends $P \in \mathrm{Ob}(\mathcal{L})$ to $c_\gamma(P) \overset{\mathrm{def}}{=} \pi(\gamma)(P)$, and sends $\psi \in \mathrm{Mor}_{\mathcal{L}}(P,Q)$ to

$$c_\gamma(\psi) \overset{\mathrm{def}}{=} (\gamma|_{Q,c_\gamma(Q)}) \circ \psi \circ (\gamma|_{P,c_\gamma(P)})^{-1} \in \mathrm{Mor}_{\mathcal{L}}(c_\gamma(P), c_\gamma(Q))$$

for $P, Q \in \mathrm{Ob}(\mathcal{L})$. This is clearly isotypical, since for $g \in P \in \mathrm{Ob}(\mathcal{L})$, $c_\gamma(\delta_P(g)) = \delta_{c_\gamma(P)}(\pi(\gamma)(g))$ by axiom (C). It sends inclusions to inclusions since the restrictions of $\gamma$ are defined with respect to the inclusions.

**Lemma 4.9** ([AOV, Lemma 2.7]) *For any saturated fusion system $\mathcal{F}$ over a p-group $S$, and any linking system $\mathcal{L}$ associated to $\mathcal{F}$, the sequence*

$$\mathrm{Aut}_{\mathcal{L}}(S) \xrightarrow{\;\gamma \mapsto c_\gamma\;} \mathrm{Aut}^I_{\mathrm{typ}}(\mathcal{L}) \longrightarrow \mathrm{Out}_{\mathrm{typ}}(\mathcal{L}) \longrightarrow 1$$

*is exact. Each $\alpha \in \mathrm{Aut}^I_{\mathrm{typ}}(\mathcal{L})$ is an automorphism of the category $\mathcal{L}$, and hence $\mathrm{Aut}^I_{\mathrm{typ}}(\mathcal{L})$ (as well as $\mathrm{Out}_{\mathrm{typ}}(\mathcal{L})$) is a group.*

The sequence of Lemma 4.9 is analogous to the following exact sequence involving automorphsms of groups. For any finite group $G$ and any $S \in \mathrm{Syl}_p(G)$, the sequence

$$N_G(S) \xrightarrow{\;g \mapsto c_g\;} \mathrm{Aut}(G, S) \longrightarrow \mathrm{Out}(G) \longrightarrow 1$$

is exact, where $\operatorname{Aut}(G, S) = \{\alpha \in \operatorname{Aut}(G) \mid \alpha(S) = S\}$.

Since one sometimes has to work with two or more linking systems associated to the same fusion system but having different objects, it is useful to know that they have the same outer automorphism groups.

**Lemma 4.10** ([AOV, Lemma 1.17])   *Fix a saturated fusion system $\mathcal{F}$ over a $p$-group $S$. Let $\mathcal{L}_0 \subseteq \mathcal{L}$ be a pair of linking systems associated to $\mathcal{F}$, such that $\operatorname{Ob}(\mathcal{L}_0)$ and $\operatorname{Ob}(\mathcal{L})$ are both $\operatorname{Aut}(S, \mathcal{F})$-invariant. Then restriction defines an isomorphism*

$$\operatorname{Out}_{\mathrm{typ}}(\mathcal{L}) \xrightarrow[\cong]{R} \operatorname{Out}_{\mathrm{typ}}(\mathcal{L}_0).$$

The injectivity of the above map $R$ is an easy consequence of the uniqueness of extensions in a linking system (Proposition 4.3(b)). The surjectivity of $R$ is basically a consequence of the uniqueness statement in Proposition 4.8: each automorphism of the subsystem extends to an automorphism of the larger linking system.

The next proposition describes how an isotypical equivalence of a linking system $\mathcal{L}$ induces a fusion preserving automorphism of the associated fusion system $\mathcal{F}$.

**Proposition 4.11** ([O4, Proposition 6])   *Let $\mathcal{L}$ be a linking system associated to a saturated fusion system $\mathcal{F}$ over a $p$-group $S$. Assume $\alpha \in \operatorname{Aut}^I_{\mathrm{typ}}(\mathcal{L})$, and let $\beta \in \operatorname{Aut}(S)$ be the automorphism such that $\alpha(\delta_S(g)) = \delta_S(\beta(g))$ for each $g \in S$. Then $\beta \in \operatorname{Aut}(S, \mathcal{F})$, and $\alpha$ sends each $P \in \operatorname{Ob}(\mathcal{L})$ to $\beta(P)$. Let $c_\beta \in \operatorname{Aut}(\mathcal{F})$ be the functor which sends $P$ to $\beta(P)$ and $\varphi \in \operatorname{Mor}(\mathcal{F})$ to ${}^\beta\varphi$. Then the following diagram of categories and functors commutes*

$$
\begin{array}{ccccc}
\mathcal{T}_{\mathcal{H}}(S) & \xrightarrow{\ \delta\ } & \mathcal{L} & \xrightarrow{\ \pi\ } & \mathcal{F} \\
\downarrow{\scriptstyle \beta_*} & & \downarrow{\scriptstyle \alpha} & & \downarrow{\scriptstyle c_\beta} \\
\mathcal{T}_{\mathcal{H}}(S) & \xrightarrow{\ \delta\ } & \mathcal{L} & \xrightarrow{\ \pi\ } & \mathcal{F}
\end{array}
$$

*where $\mathcal{H} = \operatorname{Ob}(\mathcal{L})$, and where $\beta_*$ sends $P \in \mathcal{H}$ to $\beta(P)$ and sends $g \in T_S(P, Q)$ to $\beta(g) \in T_S(\beta(P), \beta(Q))$.*

*Proof.* This is all shown in [O4], except for the claim that the left square in the diagram commutes. It commutes on elements of $\operatorname{Aut}_{\mathcal{T}_{\mathcal{H}}(S)}(S) = S$ by definition of $\beta$. It commutes on other objects and morphisms since $\alpha$ sends inclusions to inclusions and hence commutes with restrictions.    $\square$

By Proposition 4.11, for each linking system $\mathcal{L}$ associated to a saturated fusion system $\mathcal{F}$, there is a well defined homomorphism

$$\widetilde{\mu} \colon \operatorname{Aut}^I_{\mathrm{typ}}(\mathcal{L}) \longrightarrow \operatorname{Aut}(S, \mathcal{F}),$$

which sends $\alpha \in \mathrm{Aut}^I_{\mathrm{typ}}(\mathcal{L})$ to the restriction of $\alpha_S$ to $\delta_S(S) \cong S$. By the exact sequence of Lemma 4.9, this factors through a homomorphism

$$\mu_{\mathcal{L}} \colon \mathrm{Out}_{\mathrm{typ}}(\mathcal{L}) \longrightarrow \mathrm{Out}(S, \mathcal{F}).$$

The kernel and cokernel of $\mu_{\mathcal{L}}$ will be described later, in Proposition 5.12. The automorphism group $\mathrm{Out}_{\mathrm{typ}}(\mathcal{L})$ plays an important role when describing self equivalences of the space $|\mathcal{L}_p^\wedge|$ (Theorem 4.22), and also in Section 6.1 (Definition 6.3).

## 4.4. Normal fusion and linking subsystems.

Recall the definition of a weakly normal fusion subsystem in Section I.6. If $\mathcal{F}_0 \subseteq \mathcal{F}$ is a pair of saturated fusion systems over $p$-groups $S_0 \leq S$, then $\mathcal{F}_0$ is weakly normal in $\mathcal{F}$ (denoted $\mathcal{F}_0 \trianglelefteq \mathcal{F}$) if

(i)  $S_0$ is strongly closed in $\mathcal{F}$;

(ii)  (Frattini condition) for each $P \leq S_0$ and each $\varphi \in \mathrm{Hom}_{\mathcal{F}}(P, S_0)$, there are morphisms $\alpha \in \mathrm{Aut}_{\mathcal{F}}(S_0)$ and $\varphi_0 \in \mathrm{Hom}_{\mathcal{F}_0}(P, S_0)$ such that $\varphi = \alpha \circ \varphi_0$; and

(iii)  (Invariance condition) for each $P, Q \leq S_0$, each $\varphi \in \mathrm{Hom}_{\mathcal{F}_0}(P, Q)$, and each $\beta \in \mathrm{Aut}_{\mathcal{F}}(S_0)$, ${}^{\beta}\varphi \in \mathrm{Hom}_{\mathcal{F}_0}(\beta(P), \beta(Q))$.

Given this definition, it is natural to impose the analogous conditions when defining a normal linking subsystem. The only problem is choosing exactly which conditions to put on the sets of objects in the two linking systems. The following definition was used in [O4] and [AOV], where normal linking subsystems played an important role.

**Definition 4.12**  Fix a pair of saturated fusion systems $\mathcal{F}_0 \trianglelefteq \mathcal{F}$ over $p$-groups $S_0 \trianglelefteq S$ such that $\mathcal{F}_0$ is weakly normal in $\mathcal{F}$, and let $\mathcal{L}_0 \subseteq \mathcal{L}$ be associated linking systems. Then $\mathcal{L}_0$ is *normal* in $\mathcal{L}$ ($\mathcal{L}_0 \trianglelefteq \mathcal{L}$) if

(i)  $\mathrm{Ob}(\mathcal{L}) = \{P \leq S \mid P \cap S_0 \in \mathrm{Ob}(\mathcal{L}_0)\}$;

(ii)  for all $P \in \mathrm{Ob}(\mathcal{L}_0)$ and $\psi \in \mathrm{Mor}_{\mathcal{L}}(P, S_0)$, there are morphisms $\gamma \in \mathrm{Aut}_{\mathcal{L}}(S_0)$ and $\psi_0 \in \mathrm{Mor}_{\mathcal{L}_0}(P, S_0)$ such that $\psi = \gamma \circ \psi_0$; and

(iii)  for all $\gamma \in \mathrm{Aut}_{\mathcal{L}}(S_0)$, $P, Q \in \mathrm{Ob}(\mathcal{L}_0)$, and $\psi \in \mathrm{Mor}_{\mathcal{L}_0}(P, Q)$,

$$\gamma|_{Q, \gamma(Q)} \circ \psi \circ \gamma|^{-1}_{P, \gamma(P)} \in \mathrm{Mor}_{\mathcal{L}_0}(\gamma(P), \gamma(Q)).$$

Here, we write $\gamma(P) = \pi(\gamma)(P)$ and $\gamma(Q) = \pi(\gamma)(Q)$ for short. In this situation, we define

$$\mathcal{L}/\mathcal{L}_0 = \mathrm{Aut}_{\mathcal{L}}(S_0)/\mathrm{Aut}_{\mathcal{L}_0}(S_0).$$

In fact, condition (ii) in Definition 4.12 follows from other conditions in that definition, together with the Frattini condition in Definition I.6.1. It is included here only to make the analogy between the two definitions clearer.

The following lemma helps to motivate the notation $\mathcal{L}/\mathcal{L}_0$: it describes a functor from $\mathcal{L}$ to $\mathcal{B}(\mathcal{L}/\mathcal{L}_0)$ with $\mathcal{L}_0$ as a "kernel". By Proposition 2.8, this in turn induces a surjection of $\pi_1(|\mathcal{L}|)$ onto $\mathcal{L}/\mathcal{L}_0$.

**Lemma 4.13**  *For each linking system $\mathcal{L}$ associated to a fusion system $\mathcal{F}$ over $S$, and each normal subsystem $\mathcal{L}_0 \trianglelefteq \mathcal{L}$ associated to $\mathcal{F}_0$ over $S_0$, there is a functor*

$$F\colon \mathcal{L} \longrightarrow \mathcal{B}(\mathcal{L}/\mathcal{L}_0)$$

*which sends $\mathrm{Aut}_{\mathcal{L}}(S_0)$ to $\mathcal{L}/\mathcal{L}_0 = \mathrm{Aut}_{\mathcal{L}}(S_0)/\mathrm{Aut}_{\mathcal{L}_0}(S_0)$ via the projection and sends inclusion morphisms to the identity. Also, $S_0 = \mathrm{Ker}(\Delta_S \circ \delta_S)$; and for $P, Q \in \mathrm{Ob}(\mathcal{L}_0)$ and $\varphi \in \mathrm{Mor}_{\mathcal{L}}(P, Q)$, $\varphi \in \mathrm{Mor}_{\mathcal{L}_0}(P, Q)$ if and only if $F(\varphi) = 1$.*

*Proof.* In general, for $P \in \mathrm{Ob}(\mathcal{L})$, we write $P_0 = P \cap S_0$ for short. Likewise, for $\varphi \in \mathrm{Mor}_{\mathcal{L}}(P, Q)$, set $\varphi_0 = \varphi|_{P_0, Q_0} \in \mathrm{Mor}_{\mathcal{L}}(P_0, Q_0)$. By condition (ii), for any such $\varphi$, we can write $\varphi_0 = \varphi^* \circ \alpha|_{P_0, \alpha(P_0)}$ for some $\alpha \in \mathrm{Aut}_{\mathcal{L}}(S_0)$ and some $\varphi^* \in \mathrm{Mor}_{\mathcal{L}_0}(\alpha(P_0), Q_0)$. Set $F(\varphi) = [\alpha] \in \mathcal{L}/\mathcal{L}_0$: the class of $\alpha$ modulo $\mathrm{Aut}_{\mathcal{L}_0}(S_0)$.

If in addition, $\varphi_0 = \psi^* \circ \beta|_{P_0, \beta(P_0)}$, where $\beta \in \mathrm{Aut}_{\mathcal{L}}(S_0)$ and $\psi^* \in \mathrm{Mor}_{\mathcal{L}_0}(\beta(P_0), Q_0)$, then

$$(\alpha\beta^{-1})|_{\beta(P_0), \alpha(P_0)} = (\varphi^*)^{-1} \circ \psi^* \in \mathrm{Mor}_{\mathcal{L}_0}(\beta(P_0), \alpha(P_0)).$$

Since $(\varphi^*)^{-1} \circ \psi^*$ extends to an automorphism of $S_0$ in $\mathcal{L}$, it satisfies (by axiom (C)) the hypotheses in Proposition 4.3(b) for it to have an extension in the category $\mathcal{L}_0$ with domain $N_{S_0}(\beta(P_0))$. By the same proposition, the extension is unique in $\mathcal{L}$, and hence is again the restriction of $\alpha\beta^{-1}$. Upon repeating this procedure, we see that $(\varphi^*)^{-1} \circ \psi^*$ extends to a unique automorphism of $S_0$ in $\mathcal{L}_0$, and hence that $\alpha\beta^{-1} \in \mathrm{Aut}_{\mathcal{L}_0}(S_0)$. Thus $[\alpha] = [\beta]$ in $\mathcal{L}/\mathcal{L}_0$, and this proves that $F(\varphi)$ is well defined. The other properties of $F$ follow immediately from this construction. $\qquad\square$

A converse to Lemma 4.13 is given in [OV1, Proposition 4.1]. If $\mathcal{L}$ is a linking system associated to $\mathcal{F}$, $\Gamma$ is a group, and $F\colon \mathcal{L} \longrightarrow \mathcal{B}(\Gamma)$ is a functor, then under certain conditions (especially conditions on the objects in $\mathcal{L}$), a kernel category $\mathcal{L}_0$ can be defined which is a normal linking subsystem of $\mathcal{L}$ associated to a saturated fusion subsystem $\mathcal{F}_0 \trianglelefteq \mathcal{F}$. This is stated in [OV1] in terms of transporter systems, but restricts to a result about linking systems as a special case.

The following lemma describes how, under certain conditions, one constructs a normal linking subsystem associated to a given weakly normal fusion subsystem.

**Lemma 4.14** ([AOV, Lemma 1.30])   *Fix a weakly normal pair of fusion systems $\mathcal{F}_0 \trianglelefteq \mathcal{F}$ over $p$-groups $S_0 \trianglelefteq S$. Assume there are sets $\mathcal{H}_0$ and $\mathcal{H}$ of subgroups of $S_0$ and $S$, respectively, both closed under $\mathcal{F}_0$- or $\mathcal{F}$-conjugacy and overgroups, such that $\mathcal{H} = \{P \leq S \mid P \cap S_0 \in \mathcal{H}_0\}$, $\mathcal{H} \subseteq \mathcal{F}^c$, and $\mathcal{H}_0 \supseteq \mathcal{F}_0{}^{cr}$. Assume $\mathcal{F}$ has an associated centric linking system $\mathcal{L}^c$, and let $\mathcal{L} \subseteq \mathcal{L}^c$ be the full subcategory with $\mathrm{Ob}(\mathcal{L}) = \mathcal{H}$. Let $\mathcal{L}_0 \subseteq \mathcal{L}$ be the subcategory with $\mathrm{Ob}(\mathcal{L}_0) = \mathcal{H}_0$, where for $P, Q \in \mathcal{H}_0$,*

$$\mathrm{Mor}_{\mathcal{L}_0}(P, Q) = \{\psi \in \mathrm{Mor}_{\mathcal{L}}(P, Q) \mid \pi(\psi) \in \mathrm{Hom}_{\mathcal{F}_0}(P, Q)\}. \qquad (11)$$

*Then $\mathcal{L}_0 \trianglelefteq \mathcal{L}$ is a normal pair of linking systems associated to $\mathcal{F}_0 \trianglelefteq \mathcal{F}$.*

Once we have defined normal linking subsystems, and the "quotient group" of a normal pair of linking systems, it is natural to ask what conditions are needed to specify an extension of a linking system by a finite group. This is what is described in the following theorem.

**Theorem 4.15** ([O4, Theorem 9])   *Fix a saturated fusion system $\mathcal{F}_0$ over a $p$-group $S_0$, and let $\mathcal{L}_0$ be a linking system associated to $\mathcal{F}_0$. Set $\mathcal{H}_0 = \mathrm{Ob}(\mathcal{L}_0)$, and assume it is closed under overgroups. Set $\Gamma_0 = \mathrm{Aut}_{\mathcal{L}_0}(S_0)$, and regard $S_0$ as a subgroup of $\Gamma_0$ via the inclusion of $\mathcal{T}_{\mathcal{H}_0}(S_0)$ into $\mathcal{L}_0$. Fix a finite group $\Gamma$ such that $\Gamma_0 \trianglelefteq \Gamma$, and a homomorphism $\tau \colon \Gamma \to \mathrm{Aut}^I_{\mathrm{typ}}(\mathcal{L}_0)$ which makes both triangles in the following diagram commute:*

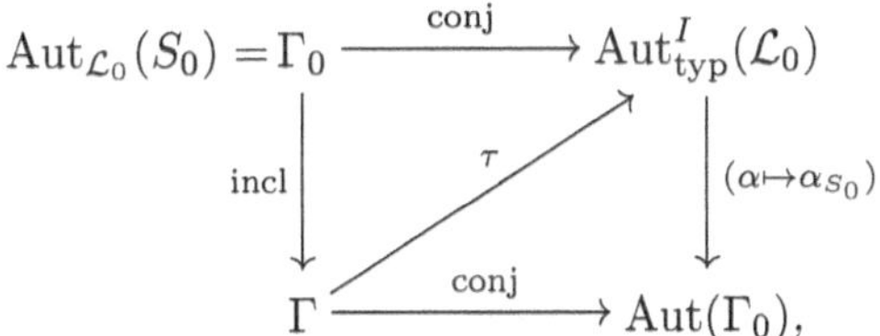

*Let $\mathcal{F}_1$ be the smallest fusion system over $S_0$ (not necessarily saturated) which contains $\mathcal{F}_0$ and $\mathrm{Aut}_\Gamma(S_0)$, where $\Gamma$ acts on $S_0$ via conjugation on $S_0 = O_p(\Gamma_0) \trianglelefteq \Gamma$. Fix $S \in \mathrm{Syl}_p(\Gamma)$. Then there is a saturated fusion system $\mathcal{F}$ over $S$ which contains $\mathcal{F}_1$ as full subcategory, and such that $\mathcal{F}_0 \trianglelefteq \mathcal{F}$.*

*Assume, in addition, that $C_\Gamma(S_0)$ is a $p$-group; and also that either $\Gamma/\Gamma_0$ is a $p$-group or $\mathcal{H}_0 \subseteq \mathcal{F}_0{}^c$. Then $\mathcal{F}$ can be chosen so as to have an associated linking system $\mathcal{L}$ for which $\mathcal{L}_0 \trianglelefteq \mathcal{L}$,*

$$\mathrm{Ob}(\mathcal{L}) = \mathcal{H} \overset{\mathrm{def}}{=} \{P \leq S \mid P \cap S_0 \in \mathcal{H}_0\},$$

*and $\mathrm{Aut}_{\mathcal{L}}(S_0) = \Gamma$ with the given action on $\mathcal{L}_0$. If $\mathcal{L}'$ is another linking system, associated to a saturated fusion system $\mathcal{F}'$ over $S$, such that $\mathcal{L}_0 \trianglelefteq$*

$\mathcal{L}'$, $\mathcal{F}_0 \trianglelefteq \mathcal{F}'$, and $\mathrm{Aut}_{\mathcal{L}'}(S_0) = \Gamma$ *with the given action on* $\mathcal{L}_0$ *and the given inclusion* $S \leq \Gamma$, *then* $\mathcal{F}' = \mathcal{F}$ *and* $\mathcal{L}' \cong \mathcal{L}$.

An early version of Theorem 4.15 was proven in [5a2, Theorem 4.6], and applied to construct extensions containing a given $p$-local finite group with $p$-power index. Later, a different version was proven by Castellana and Libman [CL, Theorem 5.2], and applied to define wreath products of $p$-local finite groups by permutation groups. The result in the above form was needed to prove [AOV, Theorem B] (see Theorem 6.5 below).

### 4.5. Fundamental groups and covering spaces.

Certain constructions made with fusion and linking systems can be explained in terms of the fundamental group and covering spaces of their geometric realizations. By Proposition 2.8, for any fusion system $\mathcal{F}$, $\pi_1(|\mathcal{F}^c|)$ is isomorphic to the free group on elements $[\varphi]$ for $\varphi \in \mathrm{Mor}(\mathcal{F}^c)$, modulo the relations $[\psi \circ \varphi] = [\psi][\varphi]$ and $[\mathrm{incl}_P^S] = 1$. Similarly, for any linking system $\mathcal{L}$, $\pi_1(|\mathcal{L}|)$ is isomorphic to the free group on elements $[\varphi]$ for $\varphi \in \mathrm{Mor}(\mathcal{L})$, modulo the relations $[\psi \circ \varphi] = [\psi][\varphi]$ and $[\iota_{P,S}] = 1$.

In contrast, $\pi_1(|\mathcal{F}|) = 1$ for any fusion system $\mathcal{F}$. In fact, $|\mathcal{F}|$ is contractible since the trivial subgroup is an initial object in $\mathcal{F}$.

The following proposition is an application of Proposition 2.9, and gives a first example of the usefulness of this viewpoint. It is essentially shown in [5a2, Corollary 3.10], although it is stated there with different hypotheses.

**Proposition 4.16** *Fix a linking system* $\mathcal{L}$ *associated to a fusion system* $\mathcal{F}$ *over the $p$-group* $S$. *Let* $\mathcal{L}_0 \trianglelefteq \mathcal{L}$ *be a normal subsystem, associated to* $\mathcal{F}_0 \trianglelefteq \mathcal{F}$ *over* $S_0$. *Then the inclusion of the geometric realization* $|\mathcal{L}_0|$ *into* $|\mathcal{L}|$ *is homotopy equivalent to a covering space over* $|\mathcal{L}|$ *with covering group* $\mathcal{L}/\mathcal{L}_0$.

*Proof.* Let $\mathcal{L}^* \subseteq \mathcal{L}$ be the full subcategory with $\mathrm{Ob}(\mathcal{L}^*) = \mathrm{Ob}(\mathcal{L}_0) = \{P \cap S_0 \mid P \in \mathrm{Ob}(\mathcal{L})\}$, and let $i \colon \mathcal{L}^* \longrightarrow \mathcal{L}$ be the inclusion. Let $r \colon \mathcal{L} \longrightarrow \mathcal{L}^*$ be the functor which sends $P$ to $P \cap S_0$ and sends a morphism to its restriction (Proposition 4.3(a)). Thus $r \circ i = \mathrm{Id}_{\mathcal{L}^*}$. There is a natural morphism of functors from $i \circ r$ to $\mathrm{Id}_{\mathcal{L}}$ which sends an object $P$ to the inclusion $\iota_{P_0,P} \in \mathrm{Mor}_{\mathcal{L}}(P_0, P)$ (where $P_0 = P \cap S_0$), so $|i \circ r| \simeq \mathrm{Id}_{|\mathcal{L}|}$, and $r$ is a deformation retraction. The inclusion $|\mathcal{L}^*| \subseteq |\mathcal{L}|$ is a thus a homotopy equivalence.

Let $F \colon \mathcal{L}^* \longrightarrow \mathcal{B}(\mathcal{L}/\mathcal{L}_0)$ be the functor defined in Lemma 4.13. Then $\mathcal{L}_0$ is the "kernel" of $F$ in the sense of Proposition 2.9: it has the same objects as $\mathcal{L}^*$, and $\varphi \in \mathrm{Mor}(\mathcal{L}^*)$ is in $\mathrm{Mor}(\mathcal{L}_0)$ if and only if $F(\varphi) = 1$ by Lemma 4.13 again. For each $P \in \mathrm{Ob}(\mathcal{L}^*)$ and each $\alpha \in \mathrm{Aut}_{\mathcal{L}}(S_0)$, there is some $P^*$ such that $\alpha$ restricts to an isomorphism $\beta = \alpha|_{P,P^*} \in \mathrm{Iso}_{\mathcal{L}}(P, P^*)$.

The hypotheses of Proposition 2.9 are thus satisfied, and so $|\mathcal{L}_0|$ has the homotopy type of a covering space over $|\mathcal{L}^*|$ (hence over $|\mathcal{L}|$) with covering group $\mathcal{L}/\mathcal{L}_0$. $\qquad\square$

We next look at the properties of $\pi_1(|\mathcal{L}|)$ and $\pi_1(|\mathcal{L}|_p^\wedge)$. For example, it is important that the space $|\mathcal{L}|$ is $p$-good (otherwise we don't expect any reasonable behavior from its $p$-completion). Recall the definition of the hyperfocal subgroup in a fusion system (Definition I.7.1):

$$\mathfrak{hyp}(\mathcal{F}) = \langle g^{-1}\alpha(g) \mid g \in P \leq S,\ \alpha \in O^p(\mathrm{Aut}_\mathcal{F}(P))\rangle \trianglelefteq S.$$

**Theorem 4.17** ([BLO2] & [5a2])   *Let $(S, \mathcal{F}, \mathcal{L})$ be any p-local finite group. Then $|\mathcal{L}|$ is p-good. Let $* \in |\mathcal{L}|$ be the vertex corresponding to $S \in \mathrm{Ob}(\mathcal{L})$. Consider the composite*

$$\Psi\colon S \xrightarrow{\ \delta_S\ } \mathrm{Aut}_\mathcal{L}(S) \xrightarrow{\ \theta_S\ } \pi_1(|\mathcal{L}|, *) \longrightarrow \pi_1(|\mathcal{L}|_p^\wedge, *),$$

*where $\theta_S$ sends an automorphism of $S$ to the corresponding loop in $|\mathcal{L}|$. Then $\Psi$ is surjective, $\mathrm{Ker}(\Psi) = \mathfrak{hyp}(\mathcal{F})$, and thus $\Psi$ induces an isomorphism $\pi_1(|\mathcal{L}|_p^\wedge, *) \cong S/\mathfrak{hyp}(\mathcal{F})$.*

*Proof.* By [BLO2, Proposition 1.11], $|\mathcal{L}|$ is $p$-good, and $S$ surjects onto $\pi_1(|\mathcal{L}|_p^\wedge)$. We sketch here the idea of the proof. Let $H \trianglelefteq \pi_1(|\mathcal{L}|, *)$ be the subgroup generated by all elements of finite order prime to $p$. In particular, $H$ is $p$-perfect. One first shows, using the presentation in Proposition 2.8, that $\pi_1(|\mathcal{L}|, *)$ is generated by $H$ together with $\theta_S(\delta_S(S))$. Thus $H$ has finite index in $\pi_1(|\mathcal{L}|, *)$. It then follows by Proposition 1.11 that $|\mathcal{L}|$ is $p$-good, and that $\pi_1(|\mathcal{L}|_p^\wedge, *) \cong \pi_1(|\mathcal{L}|, *)/H$.

The kernel of $\Psi$ was shown in [5a2, Theorem B] to be equal to $\mathfrak{hyp}(\mathcal{F})$. The hard part, the inclusion $\mathrm{Ker}(\Psi) \leq \mathfrak{hyp}(\mathcal{F})$, was shown by explicitly (and laboriously) constructing a functor from $\mathcal{L}$ to $\mathcal{B}(S/\mathfrak{hyp}(\mathcal{F}))$ which sends $\delta_S(g)$ (for $g \in S$) to the class of $g$ modulo $\mathfrak{hyp}(\mathcal{F})$. $\qquad\square$

Thus the fundamental group $|\mathcal{L}|_p^\wedge$ for a linking system $\mathcal{L}$ depends only on the fusion system to which $\mathcal{L}$ is associated. We will see later that many of the other homotopy theoretic properties of $|\mathcal{L}|_p^\wedge$ also depend only on the fusion system.

In [5a2], we showed that for each $p$-local finite group $(S, \mathcal{F}, \mathcal{L})$, each covering space over $|\mathcal{L}|_p^\wedge$ is the classifying space of another $p$-local finite group, and each covering space over $|\mathcal{F}^c|$ is the geometric realization of the centric fusion system in another $p$-local finite group. This came out of the study of fusion subsystems of $p$-power index and of index prime to $p$, as defined in Part I (Definition I.7.3).

We saw in Theorem I.7.4 that for each saturated fusion system $\mathcal{F}$ over a $p$-group $S$, and each $T \leq S$ which contains $\mathfrak{hyp}(\mathcal{F})$, there is a unique saturated fusion subsystem $\mathcal{F}_T \subseteq \mathcal{F}$ over $T$ of $p$-power index. We now look at associated linking systems in this context.

**Theorem 4.18** ([5a2, Theorem 4.4])  *Fix a saturated fusion system $\mathcal{F}$ over a $p$-group $S$, and a quasicentric linking system $\mathcal{L}$ associated to $\mathcal{F}$. Fix a subgroup $T \leq S$ which contains $\mathfrak{hyp}(\mathcal{F})$, and let $\mathcal{F}_T \subseteq \mathcal{F}$ be the unique saturated fusion subsystem of $p$-power index over $T$. Then there is a quasicentric linking subsystem $\mathcal{L}_T \subseteq \mathcal{L}$ associated to $\mathcal{F}_T$, such that $|\mathcal{L}_T|$ is homotopy equivalent to a covering space over $|\mathcal{L}|$ of degree $[S{:}T]$, and such that $|\mathcal{L}_T|_p^\wedge$ is homotopy equivalent to the covering space over $|\mathcal{L}|_p^\wedge$ with fundamental group $T/\mathfrak{hyp}(\mathcal{F})$. Also, if $T \trianglelefteq S$, then $\mathcal{L}_T \trianglelefteq \mathcal{L}$.*

*Proof.* Except for the last statement ($\mathcal{L}_T \trianglelefteq \mathcal{L}$), this was stated explicitly in [5a2, Theorem 4.4]. The normality of $\mathcal{L}_T$ is shown in [AOV, Propositions 1.31] for $T = \mathfrak{hyp}(\mathcal{F})$ (and for a slightly different choice of objects in the linking systems). The key point is that $\mathcal{L}_T$ can be described as the "kernel" of a map $\lambda\colon \mathrm{Mor}(\mathcal{L}) \longrightarrow S/T$ constructed in [5a2, Proposition 2.4]. The general case follows by a similar argument.  $\square$

Since $\pi_1(|\mathcal{L}|_p^\wedge) \cong S/\mathfrak{hyp}(\mathcal{F})$ by Theorem 4.17, Theorem 4.18 says that every connected covering space of $|\mathcal{L}|_p^\wedge$ is realized (up to homotopy type) as the classifying space of a $p$-local finite subgroup of $p$-power index in $(S, \mathcal{F}, \mathcal{L})$. We thus have a bijective correspondence between subgroups of $\pi_1(|\mathcal{L}|_p^\wedge)$ (equivalently, connected covering spaces of $|\mathcal{L}|_p^\wedge$), and the fusion subsystems of $p$-power index in $\mathcal{F}$. This is analogous to the situation for the classifying space of a finite group $G$: since $\pi_1(BG_p^\wedge) \cong G/O^p(G)$, there is a bijective correspondence between connected covering spaces of $BG_p^\wedge$ and subgroups of $G$ containing $O^p(G)$.

We now look at subsystems of index prime to $p$, and their connection with $\pi_1(|\mathcal{F}^c|)$ which was first shown by Aschbacher.

**Theorem 4.19** ([5a2, Theorem 5.4])  *Fix a saturated fusion system $\mathcal{F}$ over a $p$-group $S$. Let $* \in |\mathcal{F}^c|$ be the vertex corresponding to $S \in \mathrm{Ob}(\mathcal{F})$, and let*

$$\theta\colon \mathrm{Mor}(\mathcal{F}^c) \longrightarrow \pi_1(|\mathcal{F}^c|, *)$$

*be the function of Proposition 2.8 which sends $\varphi \in \mathrm{Hom}_{\mathcal{F}}(P, Q)$ to the class of the loop formed by the edges in $|\mathcal{F}^c|$ corresponding to $\varphi$, $\mathrm{incl}_P^S$, and $\mathrm{incl}_Q^S$ in $\mathcal{F}^c$.*

(a)  *For each $H \leq \pi_1(|\mathcal{F}^c|, *)$, there is a unique saturated fusion subsystem $\mathcal{F}_H \subseteq \mathcal{F}$ of index prime to $p$ such that $\mathrm{Mor}(\mathcal{F}_H{}^c) = \theta^{-1}(H)$ and $\mathrm{Ob}(\mathcal{F}_H{}^c) = \mathrm{Ob}(\mathcal{F}^c)$. Conversely, for each saturated fusion subsystem $\mathcal{F}_* \subseteq \mathcal{F}$ of index prime to $p$, $\mathcal{F}_* = \mathcal{F}_H$ for some $H$.*

(b)  *For each $H \leq \pi_1(|\mathcal{F}^c|, *)$, $|\mathcal{F}_H{}^c|$ has the homotopy type of the connected covering space of $|\mathcal{F}^c|$ with fundamental group $H$.*

(c)   *The restriction of $\theta$ to $\mathrm{Aut}_{\mathcal{F}}(S)$ induces an isomorphism*

$$\pi_1(|\mathcal{F}^c|,*) \cong \mathrm{Aut}_{\mathcal{F}}(S)/\mathrm{Aut}_{O^{p'}(\mathcal{F})}(S),$$

*where $O^{p'}(\mathcal{F}) = \mathcal{F}_1$. In particular, $\pi_1(|\mathcal{F}^c|,*)$ is finite of order prime to $p$.*

*Proof.* Let $\theta' \colon \mathrm{Mor}(\mathcal{F}^c) \longrightarrow \Gamma_{\mathcal{F}}$ be the map and group defined in Theorem I.7.7. Thus $\theta'$ sends composites of morphisms to products in $\Gamma_{\mathcal{F}}$, $\theta'(O^{p'}(\mathrm{Aut}_{\mathcal{F}}(P))) = 1$ for each $P \in \mathcal{F}^c$, and $\theta'$ is universal among all maps from $\mathrm{Mor}(\mathcal{F}^c)$ to groups having these properties. By Theorem I.7.7 and Lemma I.7.6, points (a) and (c) hold if $\pi_1(|\mathcal{F}^c|,*)$ and $\theta$ are replaced by $\Gamma_{\mathcal{F}}$ and $\theta'$. Also, (b) follows from (a) and Proposition 2.9.

By Proposition 2.8, $\theta$ is universal among all maps from $\mathrm{Mor}(\mathcal{F}^c)$ to a group which send composites of morphisms to products, and such that $\theta(\mathrm{incl}_P^S) = 1$ for each $P \in \mathcal{F}^c$. Hence there is a surjective homomorphism $\lambda \colon \pi_1(|\mathcal{F}^c|,*) \xrightarrow{\cong} \Gamma_{\mathcal{F}}$ such that $\lambda \circ \theta = \theta'$, and we must show that $\lambda$ is an isomorphism. Set $\Gamma = \pi_1(|\mathcal{F}^c|,*)$ for short.

For $g \in S$, we write $\theta(c_g) \in \Gamma$ to denote the image of $c_g \in \mathrm{Inn}(S)$. For each $P \in \mathcal{F}^c$ and $c_g|_P \in \mathrm{Hom}_{\mathcal{F}}(P, {}^gP)$, $\theta(c_g|_P) = \theta(c_g)$ since $\theta(\mathrm{incl}_P^S) = \theta(\mathrm{incl}_{{}^gP}^S) = 1$. If $P, Q \in \mathcal{F}^c$, $\varphi \in \mathrm{Iso}_{\mathcal{F}}(P,Q)$, and $g \in P$, then $\theta(c_g|_P)$ and $\theta(c_{\varphi(g)}|_Q)$ are conjugate in $\Gamma$ by $\theta(\varphi)$, and thus $\theta(c_g)$ is $\Gamma$-conjugate to $\theta(c_{\varphi(g)})$. Hence by Alperin's fusion theorem (Theorem I.3.6), if $g, h \in S$ are $\mathcal{F}$-conjugate, then $\theta(c_g)$ is $\Gamma$-conjugate to $\theta(c_h)$.

For each $g \in S$ such that $\langle g \rangle$ is fully centralized, $C_S(g) \in \mathcal{F}^c$, and $c_g|_{C_S(g)} = \mathrm{Id} \in \mathrm{Aut}_{\mathcal{F}}(C_S(g))$. Hence $\theta(c_g|_{C_S(g)}) = 1$, and so $\theta(c_g) = 1$ in this case. Since every element of $S$ is $\mathcal{F}$-conjugate to such an element $g$, this proves that $\theta(\mathrm{Inn}(S)) = 1$. So by restriction, $\theta(\mathrm{Aut}_S(P)) = 1$ for all $P$. Thus $\theta(O^{p'}(\mathrm{Aut}_{\mathcal{F}}(P))) = 1$ when $P$ is fully normalized in $\mathcal{F}$, since $O^{p'}(\mathrm{Aut}_{\mathcal{F}}(P))$ is the normal closure of $\mathrm{Aut}_S(P) \in \mathrm{Syl}_p(\mathrm{Aut}_{\mathcal{F}}(P))$ in this case; and $\theta(O^{p'}(\mathrm{Aut}_{\mathcal{F}}(P))) = 1$ for all $P$ since this clearly depends only on the $\mathcal{F}$-isomorphism class of $P$. This finishes the proof that $\lambda$ sends $\Gamma$ isomorphically to $\Gamma_{\mathcal{F}}$.  $\square$

When $\mathcal{L}$ is a centric linking system associated to $\mathcal{F}$, and $\mathcal{F}_* \subseteq \mathcal{F}$ is a saturated fusion subsystem of index prime to $p$, then by Lemma 4.14, a centric linking subsystem $\mathcal{L}_* \subseteq \mathcal{L}$ associated to $\mathcal{F}_*$ can be defined simply as the pullback of $\mathcal{L}$ and $\mathcal{F}_*$ over $\mathcal{F}$. See [5a2, Theorem 5.5] for more details.

### 4.6. Homotopy properties of classifying spaces.

We now list some of the other basic results on classifying spaces of $p$-local finite groups, which show that they have many of the nice homotopy theoretic properties of the $p$-completed classifying spaces $BG_p^\wedge$. We begin with the following result, which shows that the homotopy type of $|\mathcal{L}|$ is independent of the choice of objects.

**Theorem 4.20** ([5a1, Theorem 3.5])   *Fix a saturated fusion system $\mathcal{F}$ over a $p$-group $S$. Let $\mathcal{L}_0 \subseteq \mathcal{L}$ be two linking systems associated to $\mathcal{F}$ with different sets of objects. Then the inclusion induces a homotopy equivalence of spaces $|\mathcal{L}_0| \simeq |\mathcal{L}|$.*

*Proof.* This is stated in [5a1] under the assumption that $\mathcal{L}$ is a quasicentric linking system, and $\mathcal{L}_0 \subseteq \mathcal{L}$ is any full subcategory which contains all subgroups which are $\mathcal{F}$-centric and $\mathcal{F}$-radical. Axiom (A1) in Definition 4.1 says that $\mathrm{Ob}(\mathcal{L}_0) \supseteq \mathcal{F}^{cr}$, and axiom (A2) implies that $\mathcal{L}_0$ is a full subcategory. Thus the only thing assumed in [5a2] and not assumed here is that $\mathrm{Ob}(\mathcal{L}) = \mathcal{F}^q$. But in fact, the proof of [5a1, Proposition 3.11] (which is the main induction step when proving the theorem and is based on Quillen's Theorem A) does not use this assumption at all, and so the result holds as stated above. $\qquad\square$

Theorem 4.20 helps to motivate using the homotopy type of $|\mathcal{L}|$ and $|\mathcal{L}|_p^\wedge$, and their homotopy properties, as important invariants of the $p$-local finite group. It also allows a certain flexibility when working with geometric realizations of linking systems which is crucial, for example, when comparing linking systems associated to a pair $\mathcal{F}_0 \subseteq \mathcal{F}$ of fusion systems.

We next look at mapping spaces. The following result is the version of Theorem 1.16 for classifying spaces of $p$-local finite groups. When $Q$ is a $p$-group and $\mathcal{F}$ is a saturated fusion system over a $p$-group $S$, we set

$$\mathrm{Rep}(Q, \mathcal{F}) = \mathrm{Hom}(Q, S)/\sim,$$

where for $\rho, \sigma \in \mathrm{Hom}(Q, S)$, $\rho \sim \sigma$ if there is $\alpha \in \mathrm{Iso}_{\mathcal{F}}(\rho(Q), \sigma(Q))$ such that $\alpha \circ \rho = \sigma$.

**Theorem 4.21** ([BLO2, Corollary 4.5])   *For any $p$-group $Q$ and any $p$-local finite group $(S, \mathcal{F}, \mathcal{L})$, the map*

$$\mathrm{Rep}(Q, \mathcal{F}) \xrightarrow{\ \cong\ } [BQ, |\mathcal{L}|_p^\wedge],$$

*defined by sending the class of $\rho\colon Q \longrightarrow S$ to the class of the composite*

$$BQ \xrightarrow{\ B\rho\ } BS \xrightarrow{\ \mathrm{incl}\ } |\mathcal{L}| \xrightarrow{\ \phi_{|\mathcal{L}|}\ } |\mathcal{L}|_p^\wedge$$

*is a bijection.*

The components of the mapping space $\mathrm{map}(BQ, |\mathcal{L}|_p^\wedge)$ can also be described. By [BLO2, Theorem 6.3], for $\rho \in \mathrm{Hom}(Q, S)$ such that $\rho(Q)$ is fully centralized in $\mathcal{F}$, the component of the above composite induced by $\rho$ has the homotopy type of $|C_{\mathcal{L}}(\rho(Q))|_p^\wedge$, where $C_{\mathcal{L}}(\rho(Q))$ is a certain linking system associated to the centralizer fusion system $C_{\mathcal{F}}(\rho(Q))$.

For any topological space $X$, we define

$$\mathrm{Out}(X) = \{\text{homotopy equivalences } f\colon X \xrightarrow{\simeq} X\}/(\text{homotopy}):$$

the group of homotopy classes of self homotopy equivalences. One other property which the classifying spaces $|\mathcal{L}|_p^\wedge$ have is a very nice combinatorial description of $\mathrm{Out}(|\mathcal{L}|_p^\wedge)$ in terms of self equivalences of the category $\mathcal{L}$. Recall the definitions of isotypical equivalences of a linking system, and of the group $\mathrm{Out}_{\mathrm{typ}}(\mathcal{L})$, in Section 4.3. The next theorem helps to explain the importance of $\mathrm{Out}_{\mathrm{typ}}(\mathcal{L})$ as an automorphism group of the $p$-local finite group $(S, \mathcal{F}, \mathcal{L})$.

**Theorem 4.22** ([BLO2, Theorem 8.1])  *For any $p$-local finite group $(S, \mathcal{F}, \mathcal{L})$, there is an isomorphism of groups*

$$\mathrm{Out}_{\mathrm{typ}}(\mathcal{L}) \xrightarrow{\;\cong\;} \mathrm{Out}(|\mathcal{L}|_p^\wedge)$$

*which sends the class of $\alpha \in \mathrm{Aut}_{\mathrm{typ}}^I(\mathcal{L})$ to the class of $|\alpha|_p^\wedge \colon |\mathcal{L}|_p^\wedge \longrightarrow |\mathcal{L}|_p^\wedge$.*

This isomorphism $\mathrm{Out}_{\mathrm{typ}}(\mathcal{L}) \cong \mathrm{Out}(|\mathcal{L}|_p^\wedge)$ was proven in [BLO2] only for *centric* linking systems. However, with the help of Lemma 4.10 and Theorem 4.20 (and Proposition 4.8), it can be shown to hold for an arbitrary linking system whose set of objects is $\mathrm{Aut}(S, \mathcal{F})$-invariant.

For any fusion system $\mathcal{F}$ over a $p$-group $S$, let $H^*(\mathcal{F}; \mathbb{F}_p)$ be the subring of $H^*(BS; \mathbb{F}_p)$ consisting of those elements which are stable under all fusion in $\mathcal{F}$; i.e.,

$$H^*(\mathcal{F}; \mathbb{F}_p) = \left\{ x \in H^*(BS; \mathbb{F}_p) \,\middle|\, \alpha^*(x) = x|_{BP}, \text{ all } \alpha \in \mathrm{Hom}_{\mathcal{F}}(P, S) \right\}$$
$$\cong \varprojlim_{\mathcal{F}} H^*(-; \mathbb{F}_p).$$

**Theorem 4.23** ([BLO2, Theorem B])  *For any $p$-local finite group $(S, \mathcal{F}, \mathcal{L})$, the natural homomorphism*

$$H^*(|\mathcal{L}|_p^\wedge; \mathbb{F}_p) \xrightarrow{\;\cong\;} H^*(\mathcal{F}; \mathbb{F}_p),$$

*induced by the inclusion of $BS$ in $|\mathcal{L}|$, is an isomorphism. Furthermore, the ring $H^*(|\mathcal{L}|_p^\wedge; \mathbb{F}_p)$ is noetherian.*

The proof of Theorem 4.23 depends on the existence of a characteristic biset for $\mathcal{F}$ (Definition I.8.3).

An isomorphism $(S_1, \mathcal{F}_1, \mathcal{L}_1) \longrightarrow (S_2, \mathcal{F}_2, \mathcal{L}_2)$ of $p$-local finite groups consists of a triple

$$S_1 \xrightarrow[\cong]{\alpha} S_2, \qquad \mathcal{F}_1 \xrightarrow[\cong]{\alpha_{\mathcal{F}}} \mathcal{F}_2, \qquad \text{and} \qquad \mathcal{L}_1 \xrightarrow[\cong]{\alpha_{\mathcal{L}}} \mathcal{L}_2$$

of isomorphisms of groups and categories, such that $\alpha_{\mathcal{F}}(P) = \alpha(P)$ for all $P \leq S_1$, $\alpha_{\mathcal{L}}(P) = \alpha(P)$ for all $P \in \mathrm{Ob}(\mathcal{L}_1)$, and such that they commute in the obvious way with the structural functors $\mathcal{L}_i \xrightarrow{\pi_i} \mathcal{F}_i$ and $\mathcal{T}_{\mathrm{Ob}(\mathcal{L}_i)}(S_i) \xrightarrow{\delta_i} \mathcal{L}_i$.

This is slightly stronger than the definition in [BLO2, §7] in that in [BLO2], $\alpha$ and $\alpha_{\mathcal{L}}$ were not required to commute with $\delta$, but only with the restricted homomorphisms $\delta_P|_P \colon P \longrightarrow \mathrm{Aut}_{\mathcal{L}}(P)$. So we check here that the two definitions are equivalent. In fact, we can show the following:

**Lemma 4.24** *For $i = 1, 2$, let $\mathcal{F}_i$ be a saturated fusion system over the $p$-group $S_i$, and let $\mathcal{L}_i$ be a linking system associated to $\mathcal{F}_i$. Let*

$$S_1 \xrightarrow[\cong]{\alpha} S_2, \qquad \mathcal{F}_1 \xrightarrow[\cong]{\alpha_{\mathcal{F}}} \mathcal{F}_2, \qquad \text{and} \qquad \mathcal{L}_1 \xrightarrow[\cong]{\alpha_{\mathcal{L}}} \mathcal{L}_2$$

*be isomorphisms of groups and categories such that $\pi_2 \circ \alpha_{\mathcal{L}} = \alpha_{\mathcal{F}} \circ \pi_1$, $\alpha_{\mathcal{F}}(P) = \alpha(P)$ for each $P \le S_1$ and $\alpha_{\mathcal{L}}(\delta_{1 S_1}(g)) = \delta_{2 S_2}(\alpha(g))$ for each $g \in S_1$. Then there is $\alpha'_{\mathcal{L}} \in \mathrm{Iso}(\mathcal{L}_1, \mathcal{L}_2)$ such that $(\alpha, \alpha_{\mathcal{F}}, \alpha'_{\mathcal{L}})$ is an isomorphism of $p$-local finite groups.*

*Proof.* For each $P \in \mathrm{Ob}(\mathcal{L}_1)$, let

$$\iota_P \in \mathrm{Mor}_{\mathcal{L}_1}(P, S_1) \qquad \text{and} \qquad \iota_{\alpha(P)} \in \mathrm{Mor}_{\mathcal{L}_2}(\alpha(P), S_2)$$

be the two inclusions. Since $\pi_2(\alpha_{\mathcal{L}}(\iota_P)) = \pi_2(\iota_{\alpha(P)})$, there is by Proposition 4.2(a) a unique element $\zeta_P \in \mathrm{Aut}_{\mathcal{L}_1}(P)$ such that $\pi_1(\zeta_P) = \mathrm{Id}_P$ and $\alpha_{\mathcal{L}}(\iota_P) = \iota_{\alpha(P)} \circ \zeta_P$. In particular, $\zeta_{S_1} = \mathrm{Id}$. Define $\alpha'_{\mathcal{L}}$ by setting $\alpha'_{\mathcal{L}}(P) = \alpha(P)$ for each $P \in \mathrm{Ob}(\mathcal{L}_1)$; and for each $\varphi \in \mathrm{Mor}_{\mathcal{L}_1}(P, Q)$,

$$\alpha'_{\mathcal{L}}(\varphi) = \zeta_Q \circ \alpha_{\mathcal{L}}(\varphi) \circ \zeta_P^{-1}.$$

Then $\alpha'_{\mathcal{L}}$ sends inclusions to inclusions, $\pi_2 \circ \alpha'_{\mathcal{L}} = \pi_2 \circ \alpha_{\mathcal{L}} = \alpha_{\mathcal{F}} \circ \pi_1$, and $\alpha'_{\mathcal{L}}(\varphi) = \alpha_{\mathcal{L}}(\varphi)$ for all $\varphi \in \mathrm{Aut}_{\mathcal{L}_1}(S_1)$. Since the $\delta_i$ are determined uniquely by the inclusions and $\delta_{i S_i}$, $\alpha'_{\mathcal{L}} \circ \delta_1 = \delta_2 \circ \mathcal{T}(\alpha)$. Thus $(\alpha, \alpha_{\mathcal{F}}, \alpha'_{\mathcal{L}})$ is an isomorphism of $p$-local finite groups. $\square$

The next theorem says that the isomorphism type of $(S_i, \mathcal{F}_i, \mathcal{L}_i)$ is completely determined by the homotopy type of the space $|\mathcal{L}_i|_p^{\wedge}$.

**Theorem 4.25** *If $(S_1, \mathcal{F}_1, \mathcal{L}_1)$ and $(S_2, \mathcal{F}_2, \mathcal{L}_2)$ are $p$-local finite groups, then any homotopy equivalence $|\mathcal{L}_1|_p^{\wedge} \xrightarrow{\simeq} |\mathcal{L}_2|_p^{\wedge}$ induces an isomorphism $(S_1, \mathcal{F}_1, \mathcal{L}_1) \xrightarrow{\cong} (S_2, \mathcal{F}_2, \mathcal{L}_2)$ of $p$-local finite groups.*

*Proof.* See [BLO2, Theorem 7.4]. The basic idea of the proof is to show that $\mathcal{F}_i$ and $\mathcal{L}_i$ are isomorphic to the fusion and linking categories of the space $|\mathcal{L}_i|_p^{\wedge}$ in a natural way which commutes with the structural functors.

More precisely, by [BLO2, Proposition 7.3], if $(S, \mathcal{F}, \mathcal{L})$ is a $p$-local finite group, and $f \colon BS \longrightarrow |\mathcal{L}|_p^{\wedge}$ is the natural inclusion, then there are equivalences of categories

$$\mathcal{F} \cong \mathcal{F}_{S,f}(|\mathcal{L}|_p^{\wedge}) \qquad \text{and} \qquad \mathcal{L} \cong \mathcal{L}_{S,f}^c(|\mathcal{L}|_p^{\wedge}).$$

This implies the theorem, since by Theorem 4.21, for any homotopy equivalence $|\mathcal{L}_1|_p^\wedge \xrightarrow[\simeq]{\psi} |\mathcal{L}_2|_p^\wedge$, there is an isomorphism $S_1 \xrightarrow[\cong]{\alpha} S_2$ such that the following square commutes up to homotopy:

$$
\begin{array}{ccc}
BS_1 & \xrightarrow{\ f_1\ } & |\mathcal{L}_1|_p^\wedge \\
\Big\downarrow{\scriptstyle B\alpha} & & \Big\downarrow{\scriptstyle \psi} \\
BS_2 & \xrightarrow{\ f_2\ } & |\mathcal{L}_2|_p^\wedge
\end{array}
$$

where $f_1$ and $f_2$ are the natural inclusions. $\qquad\square$

### 4.7. Classifying spectra of fusion systems.

This subsection is mainly intended for algebraic topologists who already know what a spectrum is. But to keep it from being completely inaccessible to others, we begin with a quick summary of some basic definitions.

For any space $X$, the *suspension* of $X$ is the space

$$\Sigma X = (X \times I)/\sim \qquad \text{where} \qquad (x, s) \sim (y, t) \text{ if } s = t \in \{0, 1\}.$$

In other words, each of the subspaces $X \times \{0\}$ and $X \times \{1\}$ in $X \times [0, 1]$ is identified to a point. One important property of the suspension is that it increases the degree of all (reduced) homology groups: $\widetilde{H}_q(\Sigma X) \cong \widetilde{H}_{q-1}(X)$ for all $q$, where $\widetilde{H}_q(X) = H_q(X)/H_q(\mathrm{pt})$. A *spectrum* consists of a sequence $\mathbf{X} = \{X_n\}$ of spaces, together with maps $\Sigma X_n \longrightarrow X_{n+1}$ for each $n$. These maps induce homomorphisms $H_q(X_n) \longrightarrow H_{q+1}(X_{n+1})$ for each $q$ and $n$; and the spectrum homology of $\mathbf{X} = \{X_n\}$ is defined by setting $H_q(\mathbf{X}) = \mathrm{colim}_n \widetilde{H}_{q+n}(X_n)$ and similarly for cohomology. The *suspension spectrum* of a space $X$ is the sequence $\Sigma^\infty X = \{\Sigma^n X\}_{n \geq 0}$ of iterated suspensions of $X$, and $H_q(\Sigma^\infty X) \cong \widetilde{H}_q(X)$.

To avoid technical details, we omit here the general definition of maps (and homotopy classes of maps) between spectra, and refer to [Ad, §III.2] and [Sw, 8.12–15]. In general, for any pair of spectra $\mathbf{X}$ and $\mathbf{Y}$, $[\mathbf{X}, \mathbf{Y}]$ has the structure of an abelian group (defined roughly by adding along the suspension coordinates), and the structure of a ring (with multiplication defined by composition) when $\mathbf{X} = \mathbf{Y}$. When $X$ and $Y$ are CW complexes, one writes $\{X, Y\} = [\Sigma^\infty X, \Sigma^\infty Y]$ for short; this is equal to $\mathrm{colim}_n[\Sigma^n X, \Sigma^n Y]$ when $X$ is a finite complex.

In [BLO2, p. 815], a procedure was very loosely described for constructing a classifying spectrum for a saturated fusion system $\mathcal{F}$, whether or not a classifying space (equivalently, a linking system) exists. This construction, which takes as starting point a characteristic biset for $\mathcal{F}$ as described in Lemma I.8.3, was made more precise by Kári Ragnarsson [Rg]. He showed how to assign, to each saturated fusion system $\mathcal{F}$ over a $p$-group $S$, a canonical idempotent $e_{\mathcal{F}} \in \{BS, BS\}$, a "classifying spectrum" $\mathbb{B}\mathcal{F}$, together with a structure map $\sigma_{\mathcal{F}} \in [\Sigma^\infty BS, \mathbb{B}\mathcal{F}]$ (projection onto a direct factor), such that

$$H^*(\mathbb{B}\mathcal{F}; \mathbb{F}_p) \cong H^*(e_{\mathcal{F}}) \cdot H^*(BS; \mathbb{F}_p) \cong \varprojlim_{\mathcal{F}} H^*(-; \mathbb{F}_p).$$

Hence by Theorem 4.23, for any centric linking system $\mathcal{L}$ associated to $\mathcal{F}$, $H^*(\mathbb{B}\mathcal{F}; \mathbb{F}_p) \cong H^*(|\mathcal{L}|_p^\wedge; \mathbb{F}_p)$. Using this, he then showed that $\mathbb{B}\mathcal{F}$ is the suspension spectrum of $|\mathcal{L}|_p^\wedge$. In particular, if $\mathcal{F} = \mathcal{F}_S(G)$ for some finite group $G$ with $S \in \mathrm{Syl}_p(G)$, then $\mathbb{B}\mathcal{F}$ is the suspension spectrum of $BG_p^\wedge$, and $\sigma_{\mathcal{F}}$ is the suspension of the canonical map $BS \longrightarrow BG_p^\wedge$ induced by the inclusion $S \leq G$.

Thus whether or not classifying spaces exist or are unique, a unique classifying spectrum can be associated to every saturated fusion system, a spectrum which does have some of the properties (such as homology) which the classifying spaces would have if they existed.

In [Rg, Theorem A], Ragnarsson proves that for any $\mathcal{F}$ and $S$, and any pair of subgroups $P, Q \leq S$, $\mathrm{Hom}_{\mathcal{F}}(P, Q)$ is precisely the set of all homomorphisms $\varphi \in \mathrm{Hom}(P, Q)$ such that $\sigma_{\mathcal{F}}|_{\Sigma^\infty BP}$ and $\sigma_{\mathcal{F}}|_{\Sigma^\infty BQ} \circ \Sigma^\infty(B\varphi)$ are equal in $[\Sigma^\infty BP, \mathbb{B}\mathcal{F}]$. In other words, the fusion system $\mathcal{F}$ can be recovered from the classifying spectrum $\mathbb{B}\mathcal{F}$ together with the structure map $\sigma_{\mathcal{F}}$, in a way very similar to that in which $\mathcal{F}$ can be recovered from $|\mathcal{L}|_p^\wedge$ when there is an associated linking system $\mathcal{L}$ (see Theorem 4.25 and its proof). It is, however, important here that $\sigma_{\mathcal{F}}$ is part of the structure: Martino and Priddy [MP1, Example 5.2] gave examples of pairs of finite groups $G_1$ and $G_2$ whose fusion systems at $p$ are not isomorphic, but such that $\Sigma^\infty(BG_{1p}^\wedge) \simeq \Sigma^\infty(BG_{2p}^\wedge)$.

In addition, Ragnarsson described the group of homotopy classes of stable maps $[\mathbb{B}\mathcal{F}_1, \mathbb{B}\mathcal{F}_2]$ between the classifying spectra of two saturated fusion systems by giving an explicit basis [Rg, Theorem B]. This was then used in [Rg, Theorem C] to show that these classifying spectra are functorial: that $\mathbb{B}$ defines a functor from the category of saturated fusion systems (for fixed $p$), with fusion preserving homomorphisms as its morphisms, to the category of $p$-local spectra and homotopy classes of stable maps between them.

More recently, Ragnarsson and Stancu [RSt] have shown that saturated fusion systems over a $p$-group $S$ are in bijective correspondence with a certain family of idempotents in the ring $\{BS, BS\}$ (which can be described purely algebraically as the "$p$-adic double Burnside ring" $A(S, S)_p^\wedge$ for $S$). The conditions on the idempotents [RSt, Theorem D] are that they not be in the Nishida ideal (a certain ideal in $A(S, S)_p^\wedge$), and that they satisfy a Frobenius reciprocity condition. This result provides a completely new way to characterize saturated fusion systems, this time using stable homotopy theory.

## 4.8. An infinite version: $p$-local compact groups.

A *discrete $p$-toral group* is a group $S$ which contains a normal subgroup $S_0 \trianglelefteq S$ such that $S/S_0$ is a (finite) $p$-group, and $S_0 \cong (\mathbb{Z}/p^\infty)^r$ for some $r$. Here, $\mathbb{Z}/p^\infty \cong \mathbb{Z}[\frac{1}{p}]/\mathbb{Z}$ is the increasing union of the $\mathbb{Z}/p^n$ for all $n$.

In [BLO3], a saturated fusion system over a discrete $p$-toral group $S$ is defined to be a category $\mathcal{F}$ whose objects are the subgroups of $S$, which satisfies the same conditions as those used to define a fusion system over a $p$-group (Definitions I.2.1 and I.2.2), and which in addition satisfies a "continuity" axiom: if $P \leq S$ is the increasing union of subgroups $P_n$ for $n \geq 1$, and $\varphi \in \mathrm{Hom}(P, S)$ is such that $\varphi|_{P_n} \in \mathrm{Hom}_{\mathcal{F}}(P_n, S)$ for all $n$, then $\varphi \in \mathrm{Hom}_{\mathcal{F}}(P, S)$. A centric linking system associated to $\mathcal{F}$ is then defined exactly as in the finite case (Definition 4.1).

A *$p$-local compact group* is a triple $(S, \mathcal{F}, \mathcal{L})$, where $S$ is a discrete $p$-toral group, $\mathcal{F}$ is a saturated fusion system over $S$, and $\mathcal{L}$ is a linking system associated to $\mathcal{F}$. As in the finite case, the classifying space of the triple $(S, \mathcal{F}, \mathcal{L})$ is defined to be the $p$-completed space $|\mathcal{L}|_p^\wedge$. By results in [BLO3, Sections 6–7], classifying spaces of $p$-local compact groups satisfy homotopy theoretic properties similar to those just described in Section 4.6.

By [BLO3, Theorem 9.10], for every compact Lie group $G$, there is a maximal discrete $p$-toral subgroup $S \leq G$ which is unique up to conjugacy, a saturated fusion system $\mathcal{F}_S(G)$ over $S$, and an associated linking system $\mathcal{L}_S^c(G)$ with the property that $|\mathcal{L}_S^c(G)|_p^\wedge \simeq BG_p^\wedge$. The definition of $\mathcal{F}_S(G)$ is essentially identical to that in the finite case, but the definition of $\mathcal{L}_S^c(G)$ turns out to be quite a bit trickier.

Other examples of $p$-local compact groups studied in [BLO3] include those coming from $p$-compact groups, as defined by Dwyer and Wilkerson, and those coming from torsion linear groups (subgroups of $GL_n(K)$ for $\mathrm{char}(K) \neq p$, all of whose elements have finite order). We do not know any examples for which we can prove that they do not come from one of these three sources, but we have constructed many for which this is unknown. So while $p$-local compact groups were originally defined in order to construct a "compact Lie group version" of $p$-local finite groups, it seems to be a much more varied class of objects, and we are far from understanding what the class really contains.

## 5. THE ORBIT CATEGORY AND ITS APPLICATIONS

The orbit category $\mathcal{O}(G)$ of a group $G$ is usually defined to be the category with one object $G/H = \{gH \mid g \in G\}$ for each subgroup $H \leq G$ (the "orbits" of $G$), and with morphism sets defined by

$$\mathrm{Mor}_{\mathcal{O}(G)}(G/H, G/K) = \mathrm{map}_G(G/H, G/K).$$

However, by analogy with the notation for fusion and linking systems, it will be more convenient for our purposes to regard the objects of $\mathcal{O}(G)$ as subgroups rather than as orbits of $G$, and to define morphisms from $H$ to $K$ in terms of the transporter set $T_G(H, K)$ (defined in Section 3.1).

**Definition 5.1**  For any group $G$, the *orbit category of $G$* is the category $\mathcal{O}(G)$ whose objects are the subgroups of $G$, and where for each $H, K \leq G$,

$$\mathrm{Mor}_{\mathcal{O}(G)}(H, K) = K \backslash T_G(H, K).$$

When $G$ is finite and $S \in \mathrm{Syl}_p(G)$, then $\mathcal{O}_S(G) \subseteq \mathcal{O}(G)$ denotes the full subcategory whose objects are the subgroups of $S$. More generally, for any set $\mathcal{H}$ of subgroups of $G$, $\mathcal{O}_{\mathcal{H}}(G) \subseteq \mathcal{O}(G)$ is the full subcategory with object set $\mathcal{H}$.

The equivalence of these two definitions comes from identifying $f \in \mathrm{map}_G(G/H, G/K)$ with $Kx^{-1} \in K \backslash T_G(H, K)$, when $x \in G$ is such that $f(H) = xK$. Note that for $h \in H$, $xK = f(hH) = hxK$, so $H^x \leq K$, and thus $x^{-1} \in T_G(H, K)$. Conversely, for $y \in T_G(H, K)$, the coset $Ky$ corresponds to the map of orbits $(gH \mapsto gy^{-1}K)$.

Thus $\mathcal{O}(G)$ can be thought of as a quotient category of the transporter category $\mathcal{T}(G)$, where we divide out by the action of the target group on the morphism sets. By analogy, the "orbit category" of a fusion system is defined by replacing the homomorphisms in the fusion system by conjugacy classes of homomorphisms.

**Definition 5.2**  Let $\mathcal{F}$ be a saturated fusion system over a $p$-group $S$. The *orbit category of $\mathcal{F}$* is the category $\mathcal{O}(\mathcal{F})$ where $\mathrm{Ob}(\mathcal{O}(\mathcal{F})) = \mathrm{Ob}(\mathcal{F})$, and where

$$\mathrm{Mor}_{\mathcal{O}(\mathcal{F})}(P, Q) = \mathrm{Rep}_{\mathcal{F}}(P, Q) \stackrel{\mathrm{def}}{=} \mathrm{Inn}(Q) \backslash \mathrm{Hom}_{\mathcal{F}}(P, Q).$$

For any full subcategory $\mathcal{F}_0 \subseteq \mathcal{F}$, $\mathcal{O}(\mathcal{F}_0) \subseteq \mathcal{O}(\mathcal{F})$ denotes the full subcategory with $\mathrm{Ob}(\mathcal{O}(\mathcal{F}_0)) = \mathrm{Ob}(\mathcal{F}_0)$. In particular, the *centric orbit category* $\mathcal{O}(\mathcal{F}^c)$ of $\mathcal{F}$ is the full subcategory of $\mathcal{O}(\mathcal{F})$ with $\mathrm{Ob}(\mathcal{O}(\mathcal{F}^c)) = \mathcal{F}^c$.

When $\mathcal{F} = \mathcal{F}_S(G)$ for a finite group $G$ with $S \in \mathrm{Syl}_p(G)$, then for arbitrary subgroups $P, Q \leq S$,

$$\mathrm{Mor}_{\mathcal{O}(G)}(P, Q) \cong Q \backslash T_G(P, Q)$$

while

$$\mathrm{Mor}_{\mathcal{O}(\mathcal{F})}(P, Q) \cong \mathrm{Inn}(Q)\backslash\mathrm{Hom}_G(P, Q) \cong Q\backslash T_G(P, Q)/C_G(P).$$

Thus there is a natural functor

$$\mathcal{O}_S(G) \longrightarrow \mathcal{O}(\mathcal{F}_S(G)),$$

which is the identity on objects and surjective on morphism sets, but where the morphism sets can be very different. For example, the trivial subgroup $1 \leq G$ is an initial object in $\mathcal{O}(\mathcal{F}_S(G))$, while $\mathrm{Aut}_{\mathcal{O}(G)}(1) \cong G$.

However, when $P, Q \leq S$ are both $p$-centric in $G$ (equivalently, $\mathcal{F}_S(G)$-centric), then since $C_G(P) = Z(P) \times C'_G(P)$,

$$\mathrm{Mor}_{\mathcal{O}(\mathcal{F}^c)}(P, Q) \cong Q\backslash T_G(P, Q)/C_G(P) \cong Q\backslash T_G(P, Q)/C'_G(P).$$

In other words, by comparison with $\mathrm{Mor}_{\mathcal{O}(G)}(P, Q)$, we are dividing out by the group $C'_G(P)$ of order prime to $p$. The restricted functor

$$\mathcal{O}^c_S(G) \longrightarrow \mathcal{O}(\mathcal{F}^c_S(G))$$

thus satisfies the hypotheses of Lemma 2.3, and hence induces a $p$ equivalence from $|\mathcal{O}^c_S(G)|$ to $|\mathcal{O}(\mathcal{F}^c_S(G))|$ and a homotopy equivalence $|\mathcal{O}^c_S(G)|^{\wedge}_p \simeq |\mathcal{O}(\mathcal{F}^c_S(G))|^{\wedge}_p$. This helps to explain why the centric orbit category $\mathcal{O}(\mathcal{F}^c)$ of a fusion system $\mathcal{F}$ in practice seems to be more useful than the full orbit category $\mathcal{O}(\mathcal{F})$.

These orbit categories play a central role when proving many of the results listed in earlier sections. For example, when $(S, \mathcal{F}, \mathcal{L})$ is a $p$-local finite group, then $\mathcal{O}(\mathcal{F}^c)$ is the indexing category of one of the important decompositions of $|\mathcal{L}|$ as a "homotopy colimit" of the classifying spaces of its $p$-subgroups (Section 5.6). Orbit categories are also used when defining many of the obstruction groups which arise in connection with fusion and linking systems, such as those to the existence and uniqueness of linking systems (Section 5.3).

The original motivation in [BLO2] for defining abstract linking systems associated to fusion systems was as a tool for constructing a classifying space associated to a given fusion system $\mathcal{F}$. We now sketch briefly a different, and equivalent, way of doing this (details will be given in Section 5.6). For given $\mathcal{F}$, one would like to define a functor from $\mathcal{O}(\mathcal{F}^c)$ to spaces by sending $P \leq S$ to the classifying space $BP$, but on morphisms this is defined only up to homotopy. Proposition 5.31 describes a bijective correspondence between centric linking systems $\mathcal{L}$ associated to $\mathcal{F}$ and "rigidifications" (actual functors) $\widetilde{B} \colon \mathcal{O}(\mathcal{F}^c) \longrightarrow$ Top of the "homotopy functor" $P \mapsto BP$. Under this correspondence, the homotopy type of the space $|\mathcal{L}|$ can be recovered as a direct limit (or homotopy version of a direct limit) of the functor $\widetilde{B}$. This correspondence provides a justification (at least to a topologist!) for why centric linking systems are natural objects to choose as a means of "linking" fusion systems to associated classifying spaces.

We begin this section by describing the bar resolution for higher derived functors of inverse limits. This does not seem to be very useful for making computations, at least not directly, but it is the key to showing how these groups appear as obstruction groups for certain lifting problems. Next, as a first application of the bar resolution for functors defined on a group, we prove that constrained fusion systems (see Definition I.4.8) are always (uniquely) realizable (as announced in Theorem I.4.9). Afterwards, we describe concretely the obstructions to the existence and uniqueness of linking systems, and also the obstructions to existence and uniqueness when lifting automorphisms of a fusion system to automorphisms of an associated linking system.

We then explain what a homotopy decomposition is in general, and describe in particular the subgroup decomposition of $|\mathcal{L}|$ for a $p$-local finite group $(S, \mathcal{F}, \mathcal{L})$. On the way, we also describe some of the techniques which can be used when computing in these obstruction groups. We end the section with an outline of the proofs of Theorems 4.21 and 4.22 (two results describing certain sets of maps between classifying spaces), to illustrate the role which homotopy decompositions and the orbit category play in those proofs.

## 5.1. Higher limits of functors and the bar resolution.

Let $\mathcal{C}$ be a small category. For any functor $F\colon \mathcal{C}^{\mathrm{op}} \longrightarrow \mathrm{Ab}$, the *inverse limit of $F$* is defined by setting

$$\varprojlim_{\mathcal{C}}(F) = \big\{(x_c)_{c \in \mathrm{Ob}(\mathcal{C})} \,\big|\, x_c \in F(c) \text{ for all } c \in \mathrm{Ob}(\mathcal{C}),$$

$$x_c = F(\alpha)(x_d) \text{ for all } \alpha \in \mathrm{Mor}_{\mathcal{C}}(c, d)\big\}.$$

This is characterized by the universal property: there are homomorphisms $\phi_c\colon \varprojlim(F) \longrightarrow F(c)$ which make the obvious triangles (one for each morphism in $\mathcal{C}$) commute. Of course, all of this applies equally well to covariant functors $\mathcal{C} \longrightarrow \mathrm{Ab}$, but to simplify the discussion (and since the applications we will describe all involve contravariant functors), we restrict to that case.

Inverse limits are left exact, and hence have (right) derived functors $\varprojlim^i(F)$ for $i \geq 0$. When describing these, it is useful to work in the category $\mathcal{C}\text{-mod}$ whose objects are the functors $\mathcal{C}^{\mathrm{op}} \longrightarrow \mathrm{Ab}$ (contravariant functors from $\mathcal{C}$ to abelian groups), and whose morphisms are the natural transformations of functors. A sequence of functors $F' \to F \to F''$ in $\mathcal{C}\text{-mod}$ is *exact* if $F'(c) \to F(c) \to F''(c)$ is exact for each $c \in \mathrm{Ob}(\mathcal{C})$.

Injective and surjective morphisms in $\mathcal{C}$-mod are defined similarly. A functor $I \colon \mathcal{C}^{\mathrm{op}} \longrightarrow \mathbf{Ab}$ is *injective* if for each injective morphism $F \longrightarrow F'$, every morphism $\varphi \colon F \longrightarrow I$ can be extended to some $\psi \colon F' \longrightarrow I$.

An *injective resolution* of $F \colon \mathcal{C}^{\mathrm{op}} \longrightarrow \mathbf{Ab}$ is an exact sequence

$$0 \longrightarrow F \xrightarrow{\ d_0\ } I_0 \xrightarrow{\ d_1\ } I_1 \xrightarrow{\ d_2\ } \cdots$$

in $\mathcal{C}$-mod such that each $I_n$ is injective. Such a resolution exists for any functor $F$; see, e.g., [Wei, Example 2.3.13, p. 43] or [BK, p.305] for a proof. For any such injective resolution, the (right) derived functors of $\varprojlim(F)$ are defined by setting, for each $i \geq 0$,

$$\varprojlim_{\mathcal{C}}{}^{i}(F) = H^{i}\Big(\varprojlim_{\mathcal{C}}(I_*), d_*\Big).$$

Since $\varprojlim(-)$ is left exact, $\varprojlim(F) \cong \varprojlim^{0}(F)$. For a proof that these groups are independent (up to isomorphism) of the choice of resolution, we refer to [McL, Theorem IX.4.3] or [Wei, Lemma 2.4.1]. (Those result are stated and proven for projective resolutions and left derived functors, but our situation can be reduced to that case upon replacing $\mathbf{Ab}$ by its opposite category $\mathbf{Ab}^{\mathrm{op}}$.)

We now want to present a more concrete formula for describing these higher limit functors. For any functor $F \colon \mathcal{C}^{\mathrm{op}} \longrightarrow \mathbf{Ab}$, let $C^*(\mathcal{C}; F)$ denote the chain complex induced by the "bar resolution" for $\mathcal{C}$:

$$C^n(\mathcal{C}; F) - \prod_{c_0 \to \cdots \to c_n} F(c_0),$$

where the product is taken over all composable sequences of $n$ morphisms in $\mathcal{C}$. We regard an element $\xi \in C^n(\mathcal{C}; F)$ as a function which sends each $(c_0 \to \cdots \to c_n) \in \mathcal{N}(\mathcal{C})_n$ to an element of $F(c_0)$. Coboundary maps

$$d \colon C^n(\mathcal{C}; F) \longrightarrow C^{n+1}(\mathcal{C}; F),$$

are defined by setting

$$d\xi(c_0 \xrightarrow{\ \alpha\ } c_1 \to \cdots \to c_{n+1}) = F(\alpha)\big(\xi(c_1 \to \cdots \to c_{n+1})\big)$$

$$+ \sum_{i=1}^{n+1} (-1)^i \xi(c_0 \to \cdots \to c_{i-1} \to c_{i+1} \to \cdots) \in F(c_0).$$

This is easily checked to define a chain complex $(C^*(\mathcal{C}; F), d)$. There is also a subcomplex $(\overline{C}^*(\mathcal{C}; F), d)$ of *normalized chains*, where $\overline{C}^n(\mathcal{C}, F)$ is the group of those $\xi \in C^n(\mathcal{C}; F)$ such that $\xi(\eta) = 0$ for each $\eta \in \mathcal{N}(\mathcal{C})_n$ which contains an identity morphism.

When $\mathcal{C} = \mathcal{B}(G)$ for a group $G$, this is just the chain complex induced by the classical *bar resolution* for $G$ (cf. [McL, §IV.5]). So we think of the above complexes as being induced by a "bar resolution" for functors over categories. This bar resolution will be constructed explicitly in the proof of the following proposition.

**Proposition 5.3** *For any small category $\mathcal{C}$ and any functor $F\colon \mathcal{C}^{\mathrm{op}} \to$* Ab,

$$\varprojlim_{\mathcal{C}}{}^{i}(F) \cong H^{i}\big(C^{*}(\mathcal{C};F),d\big) \cong H^{i}\big(\overline{C}^{*}(\mathcal{C};F),d\big).$$

*Proof.* The first isomorphism is shown in [GZ, Appendix II, Proposition 3.3], and the second follows upon modifying the proof of the first. They also follow from [BK, Proposition XI.6.2]. A different, more explicit proof is given in [O1, Lemma 2]. We sketch that proof here, for the normalized chain complex $(\overline{C}^{*}(\mathcal{C};F),d)$.

Let $\underline{\mathbb{Z}}\colon \mathcal{C}^{\mathrm{op}} \longrightarrow$ Ab be the constant functor which sends each object to $\mathbb{Z}$ and each morphism to the identity. From the above definition of inverse limits, it follows immediately that for any $F$,

$$\varprojlim_{\mathcal{C}}(F) \cong \mathrm{Hom}_{\mathcal{C}\text{-mod}}(\underline{\mathbb{Z}},F).$$

We can thus identify the derived functors of $\varprojlim(F)$ with the derived functors of $\mathrm{Hom}_{\mathcal{C}\text{-mod}}(\underline{\mathbb{Z}},F)$; i.e., with the groups $\mathrm{Ext}^{i}_{\mathcal{C}\text{-mod}}(\underline{\mathbb{Z}},F)$. As usual, these groups can be defined not only via a $\mathcal{C}$-mod-injective resolution of $F$, but also using a $\mathcal{C}$-mod-projective resolution of $\underline{\mathbb{Z}}$ (cf. [Wei, Theorem 2.7.6]).

To construct a projective resolution of $\underline{\mathbb{Z}}$, define functors

$$\overline{C}_{n}(\mathcal{C})\colon \mathcal{C}^{\mathrm{op}} \longrightarrow \text{Ab}$$

(for each $n \geq 0$) as follows. For $c \in \mathrm{Ob}(\mathcal{C})$, let $\mathrm{ch}_{n}(\mathcal{C},c)$ be the set of all chains $c \to c_{0} \to \cdots \to c_{n}$ of morphisms in $\mathcal{C}$. Let $\mathrm{ch}^{0}_{n}(\mathcal{C},c) \subseteq \mathrm{ch}_{n}(\mathcal{C},c)$ be the subset of those chains which are "degenerate": those where at least one of the morphisms $c_{i} \to c_{i+1}$ (for $0 \leq i \leq n-1$) is an identity morphism. Let $C_{n}(\mathcal{C})(c)$ be the free abelian group with basis $\mathrm{ch}_{n}(\mathcal{C},c)$, and set $\overline{C}_{n}(\mathcal{C})(c) = C_{n}(\mathcal{C})(c)/\langle \mathrm{ch}^{0}_{n}(\mathcal{C},c)\rangle$. A morphism $c \to d$ in $\mathcal{C}$ induces a map from $\mathrm{ch}_{n}(d)$ to $\mathrm{ch}_{n}(c)$ by composition with the first map in the chain, and hence homomorphisms from $C_{n}(\mathcal{C})(d)$ to $C_{n}(\mathcal{C})(c)$ and from $\overline{C}_{n}(\mathcal{C})(d)$ to $\overline{C}_{n}(\mathcal{C})(c)$.

For each $n$ and each $F\colon \mathcal{C}^{\mathrm{op}} \longrightarrow$ Ab, there is a natural isomorphism

$$\mathrm{Hom}_{\mathcal{C}\text{-mod}}(\overline{C}_{n}(\mathcal{C}),F) \cong \overline{C}^{n}(\mathcal{C};F).$$

Since $\overline{C}^{n}(\mathcal{C};-)$ sends surjections in $\mathcal{C}$-mod to surjections of groups, this shows that $\overline{C}_{n}(\mathcal{C})$ is projective in $\mathcal{C}$-mod.

For each $n$ and $c$, define $\partial\colon \bar{C}_n(\mathcal{C})(c) \longrightarrow \bar{C}_{n-1}(\mathcal{C})(c)$ by setting

$$\partial(c \to c_0 \to \cdots c_n) = \sum_{k=0}^{n} (-1)^k (c \to c_0 \to \cdots \widehat{c_k} \cdots \to c_n).$$

Here, $\widehat{c_k}$ means that that term is removed from the chain. It is straightforward to check that these are well defined, and define a chain complex

$$\cdots \xrightarrow{\partial} \bar{C}_2(\mathcal{C}) \xrightarrow{\partial} \bar{C}_1(\mathcal{C}) \xrightarrow{\partial} \bar{C}_0(\mathcal{C}) \xrightarrow{\varepsilon} \underline{\mathbb{Z}} \longrightarrow 0 \qquad (1)$$

of functors on $\mathcal{C}$, where $\underline{\mathbb{Z}}$ denotes the constant functor with value $\mathbb{Z}$, and $\varepsilon(c \to c_0) = 1$ for each basis element $(c \to c_0)$ in $\bar{C}_0(\mathcal{C})(c) = C_0(\mathcal{C})(c)$. Furthermore, for each $c \in \mathrm{Ob}(\mathcal{C})$, the functions $T_n\colon \bar{C}_n(\mathcal{C})(c) \longrightarrow \bar{C}_{n+1}(\mathcal{C})(c)$ which send a chain $(c \to c_0 \to \cdots \to c_n)$ to $(c \xrightarrow{\mathrm{Id}} c \to c_0 \to \cdots \to c_n)$ (or send $1 \in \mathbb{Z}$ to $(c \xrightarrow{\mathrm{Id}} c)$) define a splitting of the chain complex $(\bar{C}_*(\mathcal{C})(c), \partial)$ for each $c \in \mathrm{Ob}(\mathcal{C})$. Thus (1) is a projective resolution of $\underline{\mathbb{Z}}$, $(\mathrm{Hom}_{\mathcal{C}\text{-mod}}(\bar{C}_*(\mathcal{C}), F), \partial^*) \cong (\bar{C}^*(\mathcal{C}; F), d)$, and its homology groups are the higher derived functors of $\mathrm{Hom}_{\mathcal{C}\text{-mod}}(\underline{\mathbb{Z}}, F) \cong \varprojlim_{\mathcal{C}}(F)$.   $\square$

The following is an immediate consequence of Proposition 5.3, together with the description in Section 2.1 of the cohomology of the realization of a simplicial set.

**Corollary 5.4**   *Fix a small category $\mathcal{C}$ and an abelian group $A$, and let $\underline{A}\colon \mathcal{C}^{\mathrm{op}} \longrightarrow \mathrm{Ab}$ be the constant functor: $\underline{A}$ sends each object to $A$ and sends each morphism to $\mathrm{Id}_A$. Then $H^*(|\mathcal{C}|; A) \cong \varprojlim_{\mathcal{C}}^{*}(\underline{A})$.*

When $\mathcal{C}$ is a finite category, the bar resolution provides an easy way to prove finite generation of higher limits of certain functors on $\mathcal{C}$.

**Proposition 5.5**   *Assume $\mathcal{C}$ has finitely many objects and finitely many morphisms. Then the following hold for any commutative noetherian ring $R$ and any functor $F\colon \mathcal{C}^{\mathrm{op}} \to R\text{-mod}$.*

(a)   *If $F(c)$ is finitely generated as an $R$-module for each $c \in \mathrm{Ob}(\mathcal{C})$, then $\varprojlim^{i}(F)$ is finitely generated as an $R$-module for each $i$.*

(b)   *In general, for each $i$, $\varprojlim^{i}(F)$ is the direct limit of the groups $\varprojlim^{i}(F_0)$, taken over all subfunctors $F_0 \subseteq F$ for which $F_0(c)$ is finitely generated for each $c$.*

*Proof.* For convenience, we say $F$ is finitely generated if $F(c)$ is finitely generated for each $c$. When $F$ is finitely generated, then $C^i(\mathcal{C}; F)$ is finitely generated for each $i$ since the set $\mathcal{N}(\mathcal{C})_i$ is finite. Since $R$ is noetherian, this implies the cohomology groups $\varprojlim^{i}(F)$ are all finitely generated.

Now let $F$ be arbitrary. For each $c \in \mathrm{Ob}(\mathcal{C})$ and each $x \in F(c)$, let $F_x \subseteq F$ be the subfunctor where $F_x(d) = \langle \varphi^*(x) \, | \, \varphi \in \mathrm{Mor}_{\mathcal{C}}(d, c) \rangle$ for each $d \in \mathrm{Ob}(\mathcal{C})$. This is a finitely generated subfunctor, and $x \in F_x(c)$. Thus $F$ is the direct limit (or union) of its finitely generated subfunctors, so $C^*(\mathcal{C}; F)$ is the direct limit of the subcomplexes $C^*(\mathcal{C}; F_0)$ for $F_0 \subseteq F$ finitely generated. Since (filtered) direct limits preserve exact sequences, we conclude that for each $i$, $\varprojlim^i(F)$ is the direct limit of the groups $\varprojlim^i(F_0)$ for $F_0 \subseteq F$ finitely generated. $\qquad\square$

When $\mathcal{C}$ and $\mathcal{D}$ are small categories, and $\Phi \colon \mathcal{C} \to \mathcal{D}$ and $F \colon \mathcal{D}^{\mathrm{op}} \to \mathrm{Ab}$ are functors, composition with $\Phi$ defines a homomorphism from $C^*(\mathcal{D}; F)$ to $C^*(\mathcal{C}; F \circ \Phi^{\mathrm{op}})$, and hence a homomorphism

$$\Phi^{\#} \colon \varprojlim_{\mathcal{D}}{}^{\!*}(F) \longrightarrow \varprojlim_{\mathcal{C}}{}^{\!*}(F \circ \Phi^{\mathrm{op}}).$$

Of course, this can also be defined directly using injective resolutions of these functors. By the analogy with cohomology of spaces, it is natural to expect this to be invariant under natural isomorphisms of functors, interpreted correctly. This is true, but we will not prove it here. Instead, we just note the following special case.

**Proposition 5.6** *Let $\Phi \colon \mathcal{C} \longrightarrow \mathcal{D}$ be an equivalence of small categories. Then for any functor $F \colon \mathcal{D}^{\mathrm{op}} \longrightarrow \mathrm{Ab}$, the induced homomorphism*

$$\Phi^{\#} \colon \varprojlim_{\mathcal{D}}{}^{\!*}(F) \overset{\cong}{\longrightarrow} \varprojlim_{\mathcal{C}}{}^{\!*}(F \circ \Phi^{\mathrm{op}})$$

*is an isomorphism. In particular, restricting a functor to an equivalent subcategory doesn't change its higher limits.*

*Proof.* Since $\Phi$ is an equivalence of categories, one easily checks that for any injective functor $I \colon \mathcal{C}^{\mathrm{op}} \longrightarrow \mathrm{Ab}$, the composite $I \circ \Phi$ is also injective. Hence for any injective resolution

$$0 \longrightarrow F \longrightarrow I_0 \longrightarrow I_1 \longrightarrow I_2 \longrightarrow \cdots$$

of $F$, composition with $\Phi$ defines an injective resolution

$$0 \longrightarrow F \circ \Phi \longrightarrow I_0 \circ \Phi \longrightarrow I_1 \circ \Phi \longrightarrow I_2 \circ \Phi \longrightarrow \cdots$$

of $F \circ \Phi$. Also,

$$\varprojlim_{\mathcal{D}}(F) \cong \varprojlim_{\mathcal{C}}(F \circ \Phi) \qquad \text{and} \qquad \varprojlim_{\mathcal{D}}(I_k) \cong \varprojlim_{\mathcal{C}}(I_k \circ \Phi),$$

and the proposition follows immediately. $\qquad\square$

As usual when working with cohomology, we write $\overline{Z}^n(\mathcal{C}; F)$ to denote the group of normalized $n$-cocycles; i.e., the kernel of the coboundary map from $\overline{C}^n(\mathcal{C}; F)$ to $\overline{C}^{n+1}(\mathcal{C}; F)$. Similarly, $\overline{B}^n(\mathcal{C}; F)$ denotes the group of normalized $n$-coboundaries; i.e., the image of the coboundary map from $\overline{C}^{n-1}(\mathcal{C}; F)$ to $\overline{C}^n(\mathcal{C}; F)$. Thus

$$\varprojlim_{\mathcal{C}}{}^n(F) \cong \overline{Z}^n(\mathcal{C}; F)/\overline{B}^n(\mathcal{C}; F).$$

## 5.2. Constrained fusion systems.

Recall that a saturated fusion system $\mathcal{F}$ over a $p$-group $S$ is constrained if it contains a normal centric $p$-subgroup; i.e., if $O_p(\mathcal{F}) \in \mathcal{F}^c$. In this section, we prove the "model theorem" for constrained fusion systems (Theorem I.4.9): if $\mathcal{F}$ is a constrained fusion system over $S$, and $Q \trianglelefteq \mathcal{F}$ is normal and centric, then there is a finite group $G$, unique up to isomorphism, such that $S \in \mathrm{Syl}_p(G)$, $Q \trianglelefteq G$, $C_G(Q) \leq Q$, and $\mathcal{F} \cong \mathcal{F}_S(G)$. The existence of such $G$, and a weak form of the uniqueness statement, were proven in [5a1, Proposition 4.3], by first constructing a linking system $\mathcal{L}$ associated to $\mathcal{F}$, using the obstruction theory described in Section 5.3, and then showing that we can set $G = \mathrm{Aut}_{\mathcal{L}}(Q)$. We give a more direct proof of this result here; one which does not involve linking systems.

The first lemma describes the connection between certain group automorphisms and cohomology. In its proof, and also in the proof of Proposition 5.8 which follows, we describe elements in group cohomology in terms of the classical bar resolution.

When $G$ is a group and $M$ is a $\mathbb{Z}[G]$-module, we can regard $M$ as a functor from $\mathcal{B}(G)$ to abelian groups, and $H^n(G; M)$ is just the $n$-th derived functor of the inverse limit of $M$. So we can describe elements in the cohomology using the bar resolution of Section 5.1. Thus $H^*(G; M)$ is the homology of the cochain complex $(\overline{C}^*(G; M), d)$, where $\overline{C}^n(G; M)$ is the group of all $\theta\colon G^n \longrightarrow M$ such that $\theta(g_1, \ldots, g_n) = 0$ if $g_i = 1$ for some $i$. However, we use here the more traditional definition of the boundary map (cf. [Br, p. 59] or [McL, p. 116]): for $\xi \in \overline{C}^n(G; M)$,

$$d\xi(g_1, \ldots, g_{n+1}) = g_1\big(\xi(g_2, \ldots, g_{n+1})\big)$$

$$+ \sum_{i=1}^{n}(-1)^i \xi(g_1, \ldots, g_i g_{i+1}, \ldots, g_{n+1}) + (-1)^{n+1}\xi(g_1, \ldots, g_n).$$

**Lemma 5.7** *Fix a group $G$ and a normal subgroup $H \trianglelefteq G$ such that $C_G(H) \leq H$. Set $\mathcal{A}(G) = \{\alpha \in \mathrm{Aut}(G) \,|\, \alpha|_H = \mathrm{Id}\}$. Then there is an isomorphism*

$$\psi\colon \mathcal{A}(G) \xrightarrow{\ \cong\ } \overline{Z}^1(G/H; Z(H)),$$

*where $\psi(\alpha)(gH) = \alpha(g)g^{-1} \in Z(H)$ for all $g \in G$ and $\alpha \in \mathcal{A}(G)$. Also, $\overline{B}^1(G/H; Z(H)) = \psi(\mathrm{Aut}_{Z(H)}(G))$, and thus*

$$\mathcal{A}/\mathrm{Aut}_{Z(H)}(G) \cong H^1(G/H; Z(H)).$$

*Proof.* Fix $\alpha \in \mathcal{A}(G)$. For each $g \in G$ and each $x \in H$, $\alpha(g)x = \alpha(gx) = gx$, so $\alpha(g) \equiv g \pmod{C_G(H) = Z(H)}$, and $\alpha(g)g^{-1} \in Z(H)$. Also, $\alpha(e) = e$, and $\alpha(g)g^{-1} = \alpha(g')(g')^{-1}$ if $g \equiv g' \pmod H$. Thus $\psi(\alpha)$ as defined above is a normalized 1-cochain in $\bar{C}^1(G/H; Z(H))$. For all $g, h \in G$,

$$\psi(\alpha)(gh) = \alpha(gh)(gh)^{-1} = \alpha(g)\psi(\alpha)(h)g^{-1} = \psi(\alpha)(g)\cdot{}^g\psi(\alpha)(h)g^{-1},$$

and so $\psi(\alpha) \in \bar{Z}^1(G/H; Z(H))$.

Conversely, for any normalized cochain $\xi \in \bar{Z}^1(G/H; Z(H))$, one easily checks that the coboundary condition implies that the map $\alpha\colon G \longrightarrow G$, defined by setting $\alpha(g) = \xi(gH)\cdot g$, is an automorphism. This proves that $\psi$ is a bijection; and this bijection is easily checked to be an isomorphism of groups.

The last statement follows since for each $a \in Z(H)$, regarded as a 0-chain (as an element in $C^0(G/H; Z(H))$), $\psi^{-1}(da) \in \mathcal{A}(G)$ sends $g \in G$ to $({}^ga\cdot a^{-1})g = (a^{-1}\cdot{}^ga)g = {}^{a^{-1}}g$, and thus $\psi^{-1}(da) = c_a^{-1} \in \mathrm{Aut}_{Z(H)}(G)$.   $\square$

In the following proposition, we collect those results involving models for constrained fusion systems which can be proven using group cohomology. We are *not* yet claiming that the group $G$ constructed there is a model for the fusion system $\mathcal{F}$.

Recall, for any fusion system $\mathcal{F}$ over a $p$-group $S$, that $\mathrm{Aut}(S, \mathcal{F})$ denotes the group of fusion preserving automorphisms of $S$: the group of all $\alpha \in \mathrm{Aut}(S)$ such that ${}^\alpha\mathcal{F} = \mathcal{F}$.

**Proposition 5.8**  *Let $\mathcal{F}$ be a constrained saturated fusion system over a $p$-group $S$. Fix $Q \in \mathcal{F}^c$ such that $Q \trianglelefteq \mathcal{F}$.*

(a)  *There is a finite group $G$ containing $S$ as a Sylow $p$-subgroup such that $Q \trianglelefteq G$, $C_G(Q) \le Q$, and $\mathrm{Aut}_G(Q) = \mathrm{Aut}_{\mathcal{F}}(Q)$.*

(b)  *If $G_1$ and $G_2$ are two groups as in (a), then there is an isomorphism $\psi\colon G_1 \xrightarrow{\cong} G_2$ such that $\psi|_Q = \mathrm{Id}_Q$.*

(c)  *If $G$ is as in (a), then for any $\alpha \in \mathrm{Aut}(S, \mathcal{F}_S(G))$ such that $\alpha|_Q = \mathrm{Id}_Q$, there is $\beta \in \mathrm{Aut}(G)$ such that $\beta|_S = \alpha$. If $\beta, \beta' \in \mathrm{Aut}(G)$ are two extensions of $\alpha$, then there is $z \in Z(S)$ such that $\beta' = \beta \circ c_z$.*

*Proof.* Set $\bar{G} = \mathrm{Out}_{\mathcal{F}}(Q)$. In general, for subgroups $P \le S$ containing $Q$, we write $\bar{P} = \mathrm{Out}_P(Q) \le \bar{G}$; thus $\bar{P} \cong P/Q$. Also, $\bar{S} \in \mathrm{Syl}_p(\bar{G})$. We regard $Z(Q)$ as a $\mathbb{Z}[\bar{G}]$-module.

Consider, for $n > 0$, the homomorphism

$$\mathrm{Rs}_{\bar{G}, \bar{S}}\colon H^n(\bar{G}; Z(Q)) \longrightarrow H^n(\bar{S}; Z(Q))$$

induced by restriction. By [CE, Theorem XII.10.1] or [Br, Theorem III.10.3],

(i)   $\mathrm{Rs}_{\bar{G},\bar{S}}$ is injective; and

(ii)   $\mathrm{Im}(\mathrm{Rs}_{\bar{G},\bar{S}})$ is the subgroup of "stable elements" in $H^n(\bar{S}; Z(Q))$: the group of all $\xi \in H^n(\bar{S}; Z(Q))$ such that for each $\bar{P}, \bar{R} \leq \bar{S}$ and each $x \in \bar{G}$ such that $^x\bar{P} = \bar{R}$, $c_x(\mathrm{Rs}_{\bar{S},\bar{P}}(\xi)) = \mathrm{Rs}_{\bar{S},\bar{R}}(\xi)$.

Here, when $^x\bar{P} = \bar{R}$, $c_x \colon H^n(\bar{P}; Z(Q)) \longrightarrow H^n(\bar{R}; Z(Q))$ sends the class of a cocycle $\theta \colon \bar{P}^n \longrightarrow Z(Q)$ to the class of the cocycle $c_x(\theta)$, where

$$c_x(\theta)(g_1, \ldots, g_n) = {}^x(\theta(g_1{}^x, \ldots, g_n{}^x))$$

(see [CE, § XII.8] or [Br, § III.8]).

(a)   Let $\mathfrak{X}$ be the set of all isomorphism classes of extensions

$$1 \longrightarrow Q \longrightarrow G \longrightarrow \bar{G} \longrightarrow 1$$

where the conjugation action of $\bar{G} = \mathrm{Out}_{\mathcal{F}}(Q)$ on $Q$ is the canonical one. For each $T \leq S$ containing $Q$, let $\mathfrak{X}_T$ be the set of isomorphism classes of extensions of $Q$ by $\bar{T} = \mathrm{Out}_T(Q)$, and let $[T] \in \mathfrak{X}_T$ be the class of the extension $T$. There are obvious maps $R_T \colon \mathfrak{X} \longrightarrow \mathfrak{X}_T$ induced by restriction. We must show there is $[G] \in \mathfrak{X}$ such that $R_S([G]) = [S]$.

By [McL, Theorem IV.8.7], the obstruction to $\mathfrak{X}$ being nonempty lies in $H^3(\bar{G}; Z(Q))$. This obstruction vanishes by (i), since its restriction to $H^3(\bar{S}; Z(Q))$ vanishes ($[S] \in \mathfrak{X}_S \neq \varnothing$). Fix an extension $G^*$ of $Q$ by $\bar{G}$ such that $[G^*] \in \mathfrak{X}$. For each $Q \leq T \leq S$, let $T^* \leq G^*$ be the subextension of $Q$ by $\bar{T}$. Thus $[T^*] = R_T([G^*]) \in \mathfrak{X}_T$.

For $P, R \leq S$ containing $Q$, each $\gamma \in T_{\bar{G}}(\bar{P}, \bar{R})$ (the transporter set) induces a map $\gamma^* \colon \mathfrak{X}_R \longrightarrow \mathfrak{X}_P$ by taking a pullback over $c_\gamma \colon \bar{P} \longrightarrow \bar{R}$ and twisting by $\gamma \in \bar{G} = \mathrm{Out}_{\mathcal{F}}(Q)$. This is uniquely defined, since twisting by an element of $\mathrm{Inn}(Q)$ (and the identity on $\bar{P}$) induces the identity on $\mathfrak{X}_P$. If $\gamma = c_g \in \mathrm{Out}_{G^*}(Q)$ for $g \in G^*$, then $c_g \in \mathrm{Hom}(P^*, R^*)$ induces a monomorphism of extensions which implies $\varphi^*([R^*]) = [P^*]$. Since $Q$ is receptive in $\mathcal{F}$, any $\tilde{\gamma} \in \mathrm{Aut}_{\mathcal{F}}(Q)$ in the class of $\gamma \in \mathrm{Out}_{\mathcal{F}}(Q)$ extends to some $\bar{\gamma} \in \mathrm{Hom}_{\mathcal{F}}(P, R)$, and this monomorphism of extensions shows that $\gamma^*([R]) = [P]$.

By [McL, Theorem IV.8.8], there is a natural action of $H^2(\bar{G}; Z(Q))$ on $\mathfrak{X}$ which is free and transitive. Similarly, for each $Q \leq P \leq S$, the group $H^2(\bar{P}; Z(Q))$ acts freely and transitively on $\mathfrak{X}_P$. Let $\xi_S \in H^2(\bar{S}; Z(Q))$ be such that $\xi_S([S^*]) = [S]$. Since the elements $[P^*] \in \mathfrak{X}_P$ are stable under fusion in $\mathcal{F}_{\bar{S}}(\bar{G})$, and similarly for the elements $[P] \in \mathfrak{X}_P$, $\xi_S$ is also stable under such fusion. Hence by (ii) above, $\xi_S$ is the restriction of an element $\xi \in H^2(\bar{G}; Z(Q))$. Set $[G] = \xi([G^*]) \in \mathfrak{X}$; then $R_S([G]) = [S]$.

**(b)**  Now let $G_1$ and $G_2$ be two groups such that $S \in \mathrm{Syl}_p(G_i)$, $Q \trianglelefteq G_i$, $C_{G_i}(Q) \leq Q$, and $\mathrm{Aut}_{G_i}(Q) = \mathrm{Aut}_{\mathcal{F}}(Q)$ for $i = 1, 2$. Then $[G_1], [G_2] \in \mathfrak{X}$, and $R_S([G_i]) = [S]$. Since the restriction map from $H^2(\overline{G}; Z(Q))$ to $H^2(\overline{S}; Z(Q))$ is injective by (i) again, $R_S$ is also injective, and so $[G_1] = [G_2]$. Thus there is an isomorphism $\psi \colon G_1 \overset{\cong}{\longrightarrow} G_2$ such that $\psi|_Q = \mathrm{Id}_Q$.

**(c)**  Fix $G$ as in (a), and identify $\overline{G} = G/Q$. Set $\bar{g} = gQ \in G/Q$ for all $g \in G$.

Fix $\alpha \in \mathrm{Aut}(S, \mathcal{F}_S(G))$ such that $\alpha|_Q = \mathrm{Id}_Q$. Let $\theta \in \overline{Z}^1(\overline{S}; Z(Q))$ be the 1-cocycle of Lemma 5.7: $\theta(\bar{g}) = \alpha(g)g^{-1}$ for all $g \in S$. We must show that $\theta$ extends to a 1-cocycle on $\overline{G}$. By (ii), it suffices to show that $[\theta] \in H^1(\overline{S}; Z(Q))$ is a stable element in the sense defined above.

Fix $\bar{x} \in \overline{G}$ and $\overline{P} = P/Q \leq \overline{S}$ such that ${}^{\bar{x}}\overline{P} \leq \overline{S}$. Thus $\bar{x} = c_x|_Q$ for some $x \in G$ such that ${}^x P \leq S$ and hence $c_x \in \mathrm{Hom}_{\mathcal{F}_S(G)}(P, S)$. Since $\alpha(P) = P$, $\alpha(S) = S$, and $\alpha \in \mathrm{Aut}(S, \mathcal{F}_S(G))$ ($\alpha$ is fusion preserving), there is $y \in G$ such that $\alpha \circ c_x \circ (\alpha|_P)^{-1} = c_y$ as homomorphisms from $P$ to $S$. Since $\alpha|_Q = \mathrm{Id}_Q$, this means that $c_x|_Q = c_y|_Q$, and thus that $x^{-1}y \in C_G(Q) = Z(Q)$. Set $z = x^{-1}y$, so $y = xz$.

For all $g \in P$,

$$\alpha(xgx^{-1}) = y\alpha(g)y^{-1} \quad \Longrightarrow \quad \theta(\overline{xgx^{-1}})\cdot xgx^{-1} = xz(\theta(\bar{g})g)z^{-1}x^{-1};$$

and hence (since $[z, \theta(\bar{g})] \in [Z(Q), Z(Q)] = 1$)

$$x^{-1}\theta(\overline{{}^{\bar{x}}g}) = z\theta(\bar{g})gz^{-1}g^{-1} = \theta(\bar{g})\cdot zgz^{-1}g^{-1} = \theta(\bar{g})\cdot(dz(g))^{-1}.$$

Here, we regard $z \in Z(Q)$ as a 0-cochain, so that $dz \in \overline{B}^1(\overline{P}; Z(Q))$ (and $dz(g) = {}^g z \cdot z^{-1}$). Hence if we define $\theta_x \in \overline{Z}^1(\overline{P}; Z(Q))$ by setting $\theta_x(g) = x^{-1}\theta(\overline{{}^{\bar{x}}g})$, then $[\theta_x] = [\theta|_{\overline{P}}]$ in $H^1(\overline{P}; Z(Q))$.

By definition, the conjugation homomorphism

$$c_x \colon H^1(P; Z(Q)) \longrightarrow H^1({}^x P; Z(Q))$$

sends the class of $\theta_x$ as defined above to the class of $\theta|_{\overline{{}^x P}}$. So what we have shown is that for each $P \leq S$ containing $Q$ and each $x \in G$ with ${}^x P \leq S$, $c_x([\theta|_{\overline{P}}]) = [\theta|_{\overline{{}^x P}}]$. In other words, $[\theta] \in H^1(\overline{S}; Z(Q))$ is a stable element with respect to the inclusion $\overline{S} \leq \overline{G}$, and so by (ii), it is the restriction of an element $[\eta] \in H^1(\overline{G}; Z(Q))$. Since $\overline{C}^0(\overline{G}; Z(Q)) = \overline{C}^0(\overline{S}; Z(Q)) = Z(Q)$, this means that there is $\eta \in \overline{Z}^1(\overline{G}; Z(Q))$ such that $\eta|_{\overline{S}} = \theta$.

Define $\beta \colon G \longrightarrow G$ by setting $\beta(g) = \eta(\bar{g})g$. Then $\beta \in \mathrm{Aut}(G)$ by Lemma 5.7, and $\beta|_S = \alpha$.

Assume $\beta' \in \mathrm{Aut}(G)$ is such that $\beta'|_S = \beta|_S$. Set $\eta' = \psi(\beta')$ in the notation of Lemma 5.7; thus $[\eta], [\eta'] \in H^1(\overline{G}; Z(Q))$ and $\eta|_{\overline{S}} = \eta'|_{\overline{S}}$. So

$[\eta] = [\eta']$ by (i) again, and $\eta' = \eta \cdot dz$ for some $z \in Z(Q)$. In other words, for each $g \in G$, $\eta'(\bar{g}) = \eta(\bar{g})gzg^{-1}z^{-1}$, so

$$\beta'(g) = \eta'(\bar{g})g = \eta(\bar{g}) \cdot gzg^{-1} \cdot z^{-1} \cdot g = \eta(\bar{g}) \cdot z^{-1} \cdot gzg^{-1} \cdot g = \beta(z^{-1}gz).$$

Thus $\beta' = \beta \circ c_z^{-1}$ for some $z \in Z(Q)$, and $z \in Z(S)$ since $\beta'|_S = \beta|_S$.  $\square$

The proof of the model theorem given here, like the proof in [5a1], depends on the vanishing of a certain group of higher limits over the orbit category of $\mathrm{Out}_{\mathcal{F}}(Q)$. This vanishing result follows as a special case of a much more general theorem of Jackowski and McClure.

**Lemma 5.9** *Fix a finite group $G$, a Sylow subgroup $S \in \mathrm{Syl}_p(G)$, and a $\mathbb{Z}[G]$-module $M$. For $i \geq 0$, let $H^i M \colon \mathcal{O}_S(G)^{\mathrm{op}} \longrightarrow \mathbf{Ab}$ be the functor $(H^i M)(P) = H^i(P; M)$, where a morphism $[g] \in Q \backslash T_G(P, Q)$ in $\mathcal{O}_S(G)$ induces a homomorphism between the cohomology groups via composition with $c_g \in \mathrm{Hom}(P, Q)$ and the action of $g^{-1}$ on $M$. Then $\underset{\mathcal{O}_S(G)}{\varprojlim}{}^{k}(H^i M) = 0$ for all $k \geq 1$.*

*Proof.* Via the transfer homomorphism for group cohomology (see [CE, § XII.8] or [Br, § III.9]), $H^i M$ becomes a Mackey functor in the sense of Dress [Dr] and Jackowski and McClure [JM]. Hence $\varprojlim^{k}(H^i M) = 0$ for $k > 0$ by [JM, Proposition 5.14], which says that higher limits vanish for all Mackey functors over $\mathcal{O}_S(G)$.  $\square$

We are now ready to prove Theorem I.4.9: all constrained fusion systems have models, which are unique in a very strong sense. What makes this difficult is that when $\mathcal{F}$, $S$, $Q$, and $G$ are as in Proposition 5.8(a), then $\mathcal{F}$ need not be *equal* to $\mathcal{F}_S(G)$ as fusion systems over $S$. Instead, as we will see, there is always $\beta \in \mathrm{Aut}(S)$ such that $\beta|_Q = \mathrm{Id}_Q$ and $\mathcal{F}_S(G) = {}^{\beta}\mathcal{F}$. In other words, $\mathcal{F}$ and $\mathcal{F}_S(G)$ are always isomorphic, but need not be equal. When choosing the automorphism $\beta$ as just described, we must take all morphisms in the fusion system into account, and this is why Lemma 5.9 is needed.

**Theorem 5.10** (Model theorem for constrained fusion systems) *Let $\mathcal{F}$ be a constrained, saturated fusion system over a $p$-group $S$. Fix $Q \in \mathcal{F}^c$ such that $Q \trianglelefteq \mathcal{F}$. Then the following hold.*

(a) *There is a model for $\mathcal{F}$: a finite group $G$ with $S \in \mathrm{Syl}_p(G)$ such that $Q \trianglelefteq G$, $C_G(Q) \leq Q$, and $\mathcal{F}_S(G) = \mathcal{F}$.*

(b) *For any finite group $G$ such that $S \in \mathrm{Syl}_p(G)$, $Q \trianglelefteq G$, $C_G(Q) \leq Q$, and $\mathrm{Aut}_G(Q) = \mathrm{Aut}_{\mathcal{F}}(Q)$, there is $\beta \in \mathrm{Aut}(S)$ such that $\beta|_Q = \mathrm{Id}_Q$ and $\mathcal{F}_S(G) = {}^{\beta}\mathcal{F}$.*

(c) *The model $G$ is unique in the following strong sense: if $G_1, G_2$ are two finite groups such that $S \in \mathrm{Syl}_p(G_i)$, $Q \trianglelefteq G_i$, $\mathcal{F}_S(G_i) = \mathcal{F}$, and $C_{G_i}(Q) \leq Q$ for $i = 1, 2$, then there is an isomorphism $\psi \colon G_1 \xrightarrow{\cong} G_2$ such that $\psi|_S = \mathrm{Id}_S$. If $\psi$ and $\psi'$ are two such isomorphisms, then $\psi' = \psi \circ c_z$ for some $z \in Z(S)$.*

*Proof.* Points (a) and (b), and a weaker version of (c), were shown in [5a1, Proposition C]. We give a different proof here.

**(a,b)** Let $G$ be as in Proposition 5.8(a): $S \in \mathrm{Syl}_p(G)$, $Q \trianglelefteq G$, $C_G(Q) \leq Q$, and $\mathrm{Aut}_G(Q) = \mathrm{Aut}_{\mathcal{F}}(Q)$. We must show that $\mathcal{F}_S(G) = {}^{\beta}\mathcal{F}$ for some $\beta \in \mathrm{Aut}(S)$ such that $\beta|_Q = \mathrm{Id}_Q$. Set $\mathcal{F}_1 = \mathcal{F}$ and $\mathcal{F}_2 = \mathcal{F}_S(G)$.

Let $\mathcal{O} = \mathcal{O}_{\bar{S}}(\bar{G})$ be the orbit category of the group $\bar{G}$ with objects the subgroups of $\bar{S}$. Thus $\mathrm{Mor}_{\mathcal{O}}(\bar{P}, \bar{R}) = \bar{R}\backslash T_{\bar{G}}(\bar{P}, \bar{R})$. Consider the functor

$$F \colon \mathcal{O}^{\mathrm{op}} \longrightarrow p\text{-}\mathrm{Gps}$$

which sends $\bar{P}$ to $H^1(\bar{P}; Z(Q))$. A morphism $\varphi = [\gamma] \in \mathrm{Mor}_{\mathcal{O}}(\bar{P}, \bar{R})$ is sent to the homomorphism $\gamma^*$ from $H^1(\bar{R}; Z(Q))$ to $H^1(\bar{P}; Z(Q))$ induced by $c_\gamma \in \mathrm{Hom}(\bar{P}, \bar{R})$ and by $(\gamma|_{Z(Q)})^{-1}$.

For each $P \leq S$ containing $Q$, set $\mathcal{A}(P) = \{\alpha \in \mathrm{Aut}(P) \,|\, \alpha|_Q = \mathrm{Id}\}$. For each $\bar{P} \leq \bar{S}$, consider the isomorphism

$$\psi_P \colon \mathcal{A}(P)/\mathrm{Aut}_{Z(Q)}(P) \xrightarrow{\cong} H^1(\bar{P}; Z(Q)) = F(\bar{P})$$

of Lemma 5.7, which sends the class of $\alpha \in \mathcal{A}(P)$ to the class of the 1-cocycle $\widehat{\psi}_P(\alpha)$ defined by $\widehat{\psi}_P(\alpha)(\bar{g}) = \alpha(g) \cdot g^{-1}$. Here, for $g \in P$, $\bar{g} = [c_g|_Q] \in \bar{P} = \mathrm{Out}_P(Q)$.

Fix $P, R \leq S$ containing $Q$ and $\gamma \in T_{\bar{G}}(\bar{P}, \bar{R})$, and choose $\tilde{\gamma} \in \mathrm{Aut}_{\mathcal{F}}(Q)$ in the class of $\gamma \in \bar{G} = \mathrm{Out}_{\mathcal{F}}(Q)$. Since $Q$ is receptive in $\mathcal{F}_1$ and $\mathcal{F}_2$, $\tilde{\gamma}$ extends to morphisms $\gamma_i \in \mathrm{Hom}_{\mathcal{F}_i}(P, R)$ ($i = 1, 2$). We claim that

$$\gamma^* \circ \psi_R = \psi_P \circ c_{\gamma_i}^{-1} \colon \mathcal{A}(R)/\mathrm{Aut}_{Z(Q)}(R) \longrightarrow H^1(\bar{P}; Z(Q)) \qquad (2)$$

for $i = 1, 2$. Here, $c_{\gamma_i}^{-1}$ sends the class of $\alpha \in \mathcal{A}(R)$ to the class of $\gamma_i^{-1}\alpha\gamma_i|_P \in \mathcal{A}(P)$, which is defined since $\alpha \equiv \mathrm{Id} \pmod{Q}$. For $g \in P$, since $\bar{g} = [c_g|_Q] \in \bar{P}$, we have $\overline{\gamma_i(g)} = [c_{\gamma_i(g)}|_Q] = \gamma[c_g|_Q]\gamma^{-1} = c_\gamma(\bar{g}) \in \bar{R}$. Thus for each $\alpha \in \mathcal{A}(R)$ and each $g \in P$,

$$(\gamma^*(\widehat{\psi}_R(\alpha)))(\bar{g}) = \gamma^{-1}(\widehat{\psi}_R(\alpha)(c_\gamma(\bar{g}))) = \gamma^{-1}(\widehat{\psi}_R(\alpha)(\overline{\gamma_i(g)}))$$

$$= \gamma^{-1}(\alpha(\gamma_i(g)) \cdot \gamma_i(g)^{-1}) = \gamma_i^{-1}(\alpha(\gamma_i(g))) \cdot g^{-1}$$

$$= c_{\gamma_i}^{-1}(\alpha)(g) \cdot g^{-1} = \widehat{\psi}_P(c_{\gamma_i}^{-1}(\alpha))(\bar{g}),$$

and this proves (2).

Again fix $\varphi = [\gamma] \in \mathrm{Mor}_{\mathcal{O}}(\overline{P}, \overline{R})$, choose $\widetilde{\gamma} \in \mathrm{Aut}_{\mathcal{F}}(Q)$ in the class of $\gamma \in \overline{G} = \mathrm{Out}_{\mathcal{F}}(Q)$, and let $\gamma_i \in \mathrm{Hom}_{\mathcal{F}_i}(P, R)$ $(i = 1, 2)$ be extensions of $\widetilde{\gamma}$. Then $\mathrm{Im}(\gamma_1) = \mathrm{Im}(\gamma_2)$, so there is some $\theta_\gamma \in \mathcal{A}(P)$ such that $\gamma_2 = \gamma_1 \circ \theta_\gamma$. By Lemma I.5.6 or 4.6(c), for a given choice of $\widetilde{\gamma}$, the extensions $\gamma_1$ and $\gamma_2$ (and hence $\theta_\gamma$) are unique modulo $\mathrm{Aut}_{Z(Q)}(P)$. Also, if $\widetilde{\gamma}$ is replaced by $c_h \circ \widetilde{\gamma}$ for some $h \in R$, then each $\gamma_i$ can be replaced by $c_h \circ \gamma_i$ without changing $\theta_\gamma$. Thus $[\theta_\gamma] \in \mathcal{A}(P)/\mathrm{Aut}_{Z(Q)}(P)$ depends only on the morphism $\varphi$ in the orbit category, and not on $\gamma$ or $\widetilde{\gamma}$ itself.

Now set $t(\varphi) = \psi_P([\theta_\gamma]) \in F(\overline{P})$ for $\varphi = [\gamma] \in \mathrm{Mor}_{\mathcal{O}}(\overline{P}, \overline{R})$ as above. This defines an element $t \in \overline{C}^1(\mathcal{O}; F)$. We claim $t$ is a cocycle. To see this, fix a pair of morphisms

$$P \xrightarrow{\ \varphi = [\gamma]\ } R \xrightarrow{\ \psi = [\nu]\ } T$$

in $\mathcal{O}$, choose representatives $\widetilde{\gamma}, \widetilde{\nu} \in \mathrm{Aut}_{\mathcal{F}}(Q)$ of $\gamma, \nu \in \overline{G} = \mathrm{Out}_{\mathcal{F}}(Q)$, and choose extensions $\gamma_i \in \mathrm{Hom}_{\mathcal{F}_i}(P, R)$ of $\widetilde{\gamma}$ and $\nu_i \in \mathrm{Hom}_{\mathcal{F}_i}(R, T)$ of $\widetilde{\nu}$. Let $\theta_\gamma, \theta_{\nu\gamma} \in \mathcal{A}(P)$ and $\theta_\nu \in \mathcal{A}(Q)$ be such that $\gamma_2 = \gamma_1 \circ \theta_\gamma$, $\nu_2 = \nu_1 \circ \theta_\nu$, and $\nu_2 \gamma_2 = \nu_1 \gamma_1 \theta_{\nu\gamma}$. Thus $\theta_\nu \circ \gamma_1 \circ \theta_\gamma = \gamma_1 \circ \theta_{\nu\gamma}$, since all of these are injective group homomorphisms, and hence

$$t(\psi\varphi) = \psi_P([\theta_{\nu\gamma}]) = \psi_P([c_{\gamma_1}^{-1}(\theta_\nu) \circ \theta_\gamma])$$
$$= \gamma^*(\psi_R([\theta_\nu])) \cdot \psi_P([\theta_\gamma]) = \gamma^*(t(\psi)) \cdot t(\varphi).$$

Thus $dt(\varphi, \psi) = 1$ for each pair $(\varphi, \psi)$, and so $t \in \overline{Z}^1(\mathcal{O}; F)$.

Now, $\varprojlim^1(F) = 0$ by Lemma 5.9. Hence there is $u \in C^0(\mathcal{O}; F)$ such that $t = du$. In other words, for each $\varphi = [\gamma] \in \mathrm{Mor}_{\mathcal{O}}(\overline{P}, \overline{R})$,

$$t(\varphi) = \gamma^*(u(\overline{R})) \cdot u(\overline{P})^{-1} \in H^1(\overline{P}; Z(Q)). \tag{3}$$

Since all inclusions into $S$ are in both categories $\mathcal{F}$ and $\mathcal{F}_S(G)$, $t(\mathrm{incl}_P^S) = 0$ for each $\overline{P} \leq \overline{S}$, and so $u(\overline{P})$ is the restriction of $u(\overline{S})$. Thus $u$ is determined by $u(\overline{S}) \in H^1(\overline{S}; Z(Q))$.

Choose any $\beta \in \mathcal{A}(S)$ such that $\psi_S([\beta]) = u(\overline{S})$. For each morphism $\varphi = [\gamma] \in \mathrm{Mor}_{\mathcal{O}}(\overline{P}, \overline{R})$, (3) and (2) imply

$$\psi_P([\theta_\gamma]) = t(\varphi) = \gamma^*(u(\overline{R})) \cdot u(\overline{P})^{-1}$$
$$= \gamma^*(\psi_R([\beta|_R])) \cdot (\psi_P([\beta|_P]))^{-1} = \psi_P\big([c_{\gamma_1}^{-1}(\beta|_R) \circ (\beta|_P)^{-1}]\big)$$

and hence $\theta_\gamma \equiv (\gamma_1^{-1}\beta\gamma_1)|_P \circ \beta^{-1}|_P \pmod{\mathrm{Aut}_{Z(Q)}(P)}$. Thus

$$\gamma_2 = \gamma_1\theta_\gamma \equiv (\beta|_R)\gamma_1(\beta|_P)^{-1} \pmod{\mathrm{Aut}_{Z(Q)}(R)},$$

where $\gamma_i \in \mathrm{Hom}_{\mathcal{F}_i}(P, R)$ are extensions defined as before.

This proves that $\mathcal{F}_2$ and $^\beta\mathcal{F}_1$ have the same morphisms between subgroups of $S$ which contain $Q$. By Proposition I.4.5, for each $i = 1, 2$, all subgroups which are $\mathcal{F}_i$-essential contain $Q$. Hence $\mathcal{F}_2 = {}^{\beta}\mathcal{F}_1$ by Alperin's fusion theorem (Theorem I.3.5).

**(c)** Assume $G_1$ and $G_2$ are two models for $\mathcal{F}$. By Proposition 5.8(b), there is an isomorphism $\psi_0 \colon G_1 \xrightarrow{\cong} G_2$ such that $\psi_0|_Q = \mathrm{Id}_Q$. We must find an isomorphism which is the identity on $S$.

Set $\alpha = \psi_0|_S \in \mathrm{Aut}(S)$. Thus $\alpha|_Q = \mathrm{Id}_Q$. Also, $\alpha c_x \alpha^{-1} = c_{\alpha(x)}$ for $x \in G_1$, and so ${}^{\alpha}\mathcal{F} = {}^{\alpha}(\mathcal{F}_S(G_1)) = \mathcal{F}_S(G_2) = \mathcal{F}$. Thus $\alpha \in \mathrm{Aut}(S, \mathcal{F})$, and by Proposition 5.8(c), $\alpha$ extends to an automorphism $\beta \in \mathrm{Aut}(G_1)$. Set $\psi = \psi_0 \circ \beta^{-1}$; then $\psi \in \mathrm{Iso}(G_1, G_2)$ and $\psi|_S = \mathrm{Id}_S$.

If $\psi' \in \mathrm{Iso}(G_1, G_2)$ is another isomorphism such that $\psi'|_S = \mathrm{Id}_S$, then $\psi^{-1}\psi' \in \mathrm{Aut}_{Z(S)}(G_1)$ by the last statement in Proposition 5.8(c), and so $\psi' = \psi \circ c_z$ for some $z \in Z(S)$. $\qquad\square$

### 5.3. Existence, uniqueness, and automorphisms of linking systems.

Consider the functor

$$\mathcal{Z}_{\mathcal{F}} \colon \mathcal{O}(\mathcal{F}^c)^{\mathrm{op}} \longrightarrow \mathbf{Ab},$$

defined for any fusion system $\mathcal{F}$ by setting $\mathcal{Z}_{\mathcal{F}}(P) = Z(P)$ and

$$\mathcal{Z}_{\mathcal{F}}(P \xrightarrow{\varphi} Q) = \left( Z(Q) \xrightarrow{\mathrm{incl}} Z(\varphi(P)) \xrightarrow{\varphi^{-1}} Z(P) \right).$$

(Note that $Z(Q) \leq Z(\varphi(P))$ since $Q \in \mathcal{F}^c$.) The higher derived functors of the inverse limit of $\mathcal{Z}_{\mathcal{F}}$ turn out to be obstruction groups to various problems involving fusion and linking systems.

Two linking systems $\mathcal{L}_1$ and $\mathcal{L}_2$ associated to the same fusion system $\mathcal{F}$ over $S$, with $\mathrm{Ob}(\mathcal{L}_1) = \mathrm{Ob}(\mathcal{L}_2) = \mathcal{H}$, are isomorphic if there is an isomorphism of categories from $\mathcal{L}_1$ to $\mathcal{L}_2$ which commutes with the structural functors to $\mathcal{F}$ and from $\mathcal{T}_{\mathcal{H}}(S)$. By Lemma 4.24, applied with $\alpha = \mathrm{Id}_S$ and $\alpha_{\mathcal{F}} = \mathrm{Id}_{\mathcal{F}}$, this is equivalent to the definition used in [BLO2] for an isomorphism of centric linking systems associated to $\mathcal{F}$.

**Proposition 5.11** ([BLO2, Proposition 3.1]) *Fix a saturated fusion system $\mathcal{F}$ over the p-group $S$. The obstruction to the existence of a centric linking system associated to $\mathcal{F}$ lies in $\varprojlim\limits_{\mathcal{O}(\mathcal{F}^c)}^{3}(\mathcal{Z}_{\mathcal{F}})$. The group $\varprojlim\limits_{\mathcal{O}(\mathcal{F}^c)}^{2}(\mathcal{Z}_{\mathcal{F}})$ acts freely and transitively on the set of isomorphism classes of all centric linking systems if the set is nonempty.*

*Brief sketch of proof.* Assume first $\mathcal{L}_1$ and $\mathcal{L}_2$ are two centric linking systems associated to $\mathcal{F}$, with structural functors $\delta_i$ and $\pi_i$ ($i = 1, 2$). Choose maps

$$\Phi_{P,Q} \colon \mathrm{Mor}_{\mathcal{L}_1}(P, Q) \longrightarrow \mathrm{Mor}_{\mathcal{L}_2}(P, Q),$$

for each pair of objects $P, Q$, which commute with the projections $\pi_1$ and $\pi_2$ to $\mathrm{Hom}_{\mathcal{F}}(P, Q)$. By first defining them on orbit representatives for the free actions of $Q$ (defined via $\delta_{i,Q}$ and left composition), this can be done so that $\Phi_{P,Q}$ is $Q$-equivariant for each $P$ and $Q$. Thus $\Phi_{P,Q}(\delta_{1,Q}(g) \circ \varphi) = \delta_{2,Q}(g) \circ \Phi_{P,Q}(\varphi)$ for all $g \in Q$ and $\varphi \in \mathrm{Mor}_{\mathcal{L}_1}(P, Q)$.

For each composable pair of morphisms $P \xrightarrow{\varphi} Q \xrightarrow{\psi} R$ in $\mathcal{L}_1$, $\Phi_{P,R}(\psi \circ \varphi)$ differs from $\Phi_{Q,R}(\psi) \circ \Phi_{P,Q}(\varphi)$ by $\delta_{2,P}(t(\varphi, \psi))$ for some unique $t(\varphi, \psi) \in Z(P)$ (axiom (A2)). This depends only on the classes of $\varphi$ and $\psi$ in $\mathcal{O}(\mathcal{F}^c)$; and together the elements $t(\varphi, \psi)$ define a normalized 2-cocycle $t \in \bar{Z}^2(\mathcal{O}(\mathcal{F}^c); \mathcal{Z}_{\mathcal{F}})$. If $t = du$ for some 1-cochain $u \in \bar{C}^1(\mathcal{O}(\mathcal{F}^c); \mathcal{Z}_{\mathcal{F}})$, then if we set $\Phi'_{P,Q}(\varphi) = \Phi_{P,Q}(\varphi) \circ \delta_P(u([\varphi]))$ for each $\varphi \in \mathrm{Mor}_{\mathcal{L}_1}(P, Q)$, we obtain a new set of maps $\{\Phi'_{P,Q}\}$ which commute with composition, and hence define an isomorphism of categories $\Phi'$ from $\mathcal{L}_1$ to $\mathcal{L}_2$.

Thus the choice of maps $\Phi_{P,Q}$ determines an element $[t] \in \varprojlim^2(\mathcal{Z}_{\mathcal{F}})$, and $\mathcal{L}_1 \cong \mathcal{L}_2$ if $[t] = 0$. If $\{\Psi_{P,Q}\}_{P,Q \in \mathrm{Ob}(\mathcal{L})}$ is another choice of maps, defining a 2-cocycle $u$, one can show that $t - u \in \bar{B}^2(\mathcal{O}(\mathcal{F}^c); \mathcal{Z}_{\mathcal{F}})$, and hence that $[u] = [t]$. In other words, there is a unique element of $\varprojlim^2(\mathcal{Z}_{\mathcal{F}})$ which measures the "difference" between any pair of linking systems associated to $\mathcal{F}$. More precisely, the group $\varprojlim^2(\mathcal{Z}_{\mathcal{F}})$ acts freely and transitively on the set of isomorphism classes of associated centric linking systems, and the above procedure describes how to construct the unique element that sends $\mathcal{L}_1$ to $\mathcal{L}_2$.

The existence obstruction is constructed in a similar way. One defines a "precategory" $\mathcal{L}$ by setting $\mathrm{Ob}(\mathcal{L}) = \mathrm{Ob}(\mathcal{F}^c)$, and $\mathrm{Mor}_{\mathcal{L}}(P, Q) = Q \times \mathrm{Mor}_{\mathcal{O}(\mathcal{F}^c)}(P, Q)$ for each pair of objects $P$ and $Q$. One then defines a projection of each $\mathrm{Mor}_{\mathcal{L}}(P, Q)$ to $\mathrm{Hom}_{\mathcal{F}}(P, Q)$, and composition maps for each triple $P, Q, R$ of objects. If this is done with sufficient care (see the proof of [BLO2, Proposition 3.1] for details), then the error in associativity defines a function which sends each composable triple of morphisms with source $P$ to $Z(P)$, and which depends only on the classes of those morphisms in $\mathcal{O}(\mathcal{F}^c)$. In this way, we get a 3-cocycle, whose class in $\varprojlim^3(\mathcal{Z}_{\mathcal{F}})$ vanishes if and only if there exists some linking system associated to $\mathcal{F}$. $\qquad\square$

These obstructions are analogous to the obstructions to the existence and uniqueness of group extensions. For any pair of groups $G$ and $K$, with an outer action $\chi\colon G \longrightarrow \mathrm{Out}(K)$ of $G$ on $K$, the obstruction to the existence of an extension

$$1 \longrightarrow K \longrightarrow \Gamma \longrightarrow G \longrightarrow 1$$

for which the conjugation action of $G$ on $K$ equals $\chi$ lies in $H^3(G; Z(K))$; while the obstruction to its uniqueness lies in $H^2(G; Z(K))$. More precisely, the group $H^2(G; Z(K))$ acts freely and transitively on the set of isomorphism classes of all extensions if it is nonempty. See, e.g., [McL, Theorems IV.8.7–8] for a detailed description of this theory. By comparison, we think of a linking system $\mathcal{L}$ associated to $\mathcal{F}$ as an extension

$$1 \longrightarrow \{P\}_{P \in \mathrm{Ob}(\mathcal{F}^c)} \longrightarrow \mathcal{L} \overset{\tilde{\pi}}{\longrightarrow} \mathcal{O}(\mathcal{F}^c) \longrightarrow 1.$$

We know of no examples of saturated fusion systems for which either of these obstruction groups is nonvanishing, and it is quite possible that there is a unique linking system associated to each saturated fusion system. This question is discussed in more detail later in the section. But we first look at the (very closely related) obstruction groups which appear when comparing automorphisms of fusion and linking systems.

Fix a saturated fusion system $\mathcal{F}$ over a $p$-group $S$, and an associated (centric) linking system $\mathcal{L}$. We refer to Section 4.3 for the definitions of outer automorphism groups $\mathrm{Out}(S, \mathcal{F})$ and $\mathrm{Out}_{\mathrm{typ}}(\mathcal{L})$. By Theorem 4.22, $\mathrm{Out}(|\mathcal{L}|_p^\wedge) \cong \mathrm{Out}_{\mathrm{typ}}(\mathcal{L})$ for any $p$-local finite group $(S, \mathcal{F}, \mathcal{L})$, and the group $\mathrm{Out}(|\mathcal{L}|_p^\wedge)$ of homotopy classes of self equivalences of the classifying space is thus described combinatorially in terms of automorphisms of the finite category $\mathcal{L}$. For this and other reasons (see, e.g., Section 6.1), $\mathrm{Out}_{\mathrm{typ}}(\mathcal{L})$ seems to be the most important automorphism group of $(S, \mathcal{F}, \mathcal{L})$. However, in practice, the group $\mathrm{Out}(S, \mathcal{F})$ of fusion preserving (outer) automorphisms of $\mathcal{F}$ is much easier to compute. Hence the importance of being able to compare these two groups.

Recall the formula of Lemma 4.9: for any linking system $\mathcal{L}$ associated to $\mathcal{F}$ over $S$,

$$\mathrm{Out}_{\mathrm{typ}}(\mathcal{L}) \cong \mathrm{Aut}^I_{\mathrm{typ}}(\mathcal{L})/\{c_\gamma \mid \gamma \in \mathrm{Aut}_{\mathcal{L}}(S)\}.$$

Here, $\mathrm{Aut}^I_{\mathrm{typ}}(\mathcal{L})$ is the group of isotypical automorphisms of $\mathcal{L}$ which send inclusions to inclusions.

**Proposition 5.12** *For each $p$-local finite group $(S, \mathcal{F}, \mathcal{L})$, there is an exact sequence*

$$1 \longrightarrow \varprojlim_{\mathcal{O}(\mathcal{F}^c)}{}^1(\mathcal{Z}_{\mathcal{F}}) \overset{\lambda_{\mathcal{L}}}{\longrightarrow} \mathrm{Out}_{\mathrm{typ}}(\mathcal{L}) \overset{\mu_{\mathcal{L}}}{\longrightarrow} \mathrm{Out}(S, \mathcal{F}) \overset{\omega_{\mathcal{L}}}{\longrightarrow} \varprojlim_{\mathcal{O}(\mathcal{F}^c)}{}^2(\mathcal{Z}_{\mathcal{F}}).$$

*Here, $\lambda_{\mathcal{L}}$ and $\mu_{\mathcal{L}}$ are homomorphisms, and $\mu_{\mathcal{L}}([\alpha]) = [\beta]$ for $\alpha \in \mathrm{Aut}^I_{\mathrm{typ}}(\mathcal{L})$ and $\beta \in \mathrm{Aut}(S, \mathcal{F})$ such that $\delta_S(\beta(g)) = \alpha_S(\delta_S(g))$ for $g \in S$.*

*Proof.* When $(S, \mathcal{F}, \mathcal{L})$ is realized by a finite group $G$ with $S \in \mathrm{Syl}_p(G)$, this was shown in [BL, Theorem 1.6] with $\mathrm{Out}(BG_p^\wedge)$ instead of $\mathrm{Out}_{\mathrm{typ}}(\mathcal{L}_S(G))$, and in [BLO1, Theorem 6.2] in the above form. The general case (for arbitrary linking and fusion systems) follows by essentially the same proof as that of [BLO1]. Here, we just give part of the proof, enough to illustrate how the bar resolution is used to identify these obstruction groups.

For each $\varphi \in \mathrm{Mor}(\mathcal{L})$, we write $[\varphi] = \tilde{\pi}(\varphi) \in \mathrm{Mor}(\mathcal{O}(\mathcal{F}^c))$ for short. A (normalized) 1-cochain $t \in \bar{C}^1(\mathcal{O}(\mathcal{F}^c); \mathcal{Z}_{\mathcal{F}})$ is a map from $\mathrm{Mor}(\mathcal{O}(\mathcal{F}^c))$ to $S$ which sends $\mathrm{Mor}_{\mathcal{L}}(P, Q)$ to $Z(P)$ for each $P, Q \in \mathrm{Ob}(\mathcal{L})$, and such that $t(\mathrm{Id}_P) = 1$ for each $P$. Such a 1-cochain is a 1-cocycle (and hence represents an element in $\varprojlim^1(\mathcal{Z}_{\mathcal{F}})$) if for each $(P \xrightarrow{\varphi} Q \xrightarrow{\psi} R)$ in $\mathcal{L}$,

$$dt\big(P \xrightarrow{\ \varphi\ } Q \xrightarrow{\ \psi\ } R\big) = \mathcal{Z}_{\mathcal{F}}(\varphi)\big(t([\psi])\big) \cdot t([\psi\varphi])^{-1} \cdot t([\varphi]) = 1 \in Z(P). \quad (4)$$

Note that $t([\delta_P(g)]) = t(\mathrm{Id}_P) = 1$ for all $g \in P$, since $[\delta_P(g)] = \mathrm{Id}_P$ in $\mathcal{O}(\mathcal{F}^c)$.

For each such 1-cocycle $t$, set $\widehat{\lambda}(t) = \Psi_t$, where $\Psi_t \in \mathrm{Aut}_{\mathrm{typ}}(\mathcal{L})$ is the identity on objects, and

$$\Psi_t(\varphi) = \varphi \circ \delta_P(t([\varphi])) \qquad \text{for all } \varphi \in \mathrm{Mor}_{\mathcal{L}}(P, Q).$$

Then $\Psi_t$ is a functor, since for each sequence $(P \xrightarrow{\ \varphi\ } Q \xrightarrow{\ \psi\ } R)$ of morphisms in $\mathcal{L}$,

$$\begin{aligned}
\Psi_t(\psi \circ \varphi) &= \psi \circ \varphi \circ \delta_P(t([\psi\varphi])) \\
&= \psi \circ \varphi \circ \delta_P(\pi(\varphi)^{-1}(t([\psi]))) \circ \delta_P(t([\varphi])) && \text{(by (4))} \\
&= \psi \circ \delta_Q(t([\psi])) \circ \varphi \circ \delta_P(t([\varphi])) && \text{(by axiom (C))} \\
&= \Psi_t(\psi) \circ \Psi_t(\varphi).
\end{aligned}$$

In fact, this computation shows that for all $t \in \bar{C}^1(\mathcal{O}(\mathcal{F}^c); \mathcal{Z}_{\mathcal{F}})$, $\Psi_t$ is a functor if and only if $dt = 0$. Clearly, $\Psi_t$ sends $\delta_\Gamma(P)$ to itself for each $P \in \mathrm{Ob}(\mathcal{L})$, and hence is isotypical.

Let $\widehat{Z}^1(\mathcal{O}(\mathcal{F}^c); \mathcal{Z}_{\mathcal{F}})$ be the subgroup of all $t \in \bar{Z}^1(\mathcal{O}(\mathcal{F}^c); \mathcal{Z}_{\mathcal{F}})$ such that $t([\iota_{P,S}]) = 1$ for each $P \in \mathrm{Ob}(\mathcal{L})$. By definition, $\Psi_t \in \mathrm{Aut}^I_{\mathrm{typ}}(\mathcal{L})$ if and only if $t \in \widehat{Z}^1(\mathcal{O}(\mathcal{F}^c); \mathcal{Z}_{\mathcal{F}})$. Fix $t \in \bar{Z}^1(\mathcal{O}(\mathcal{F}^c); \mathcal{Z}_{\mathcal{F}})$, and let $u \in \bar{C}^0(\mathcal{O}(\mathcal{F}^c); \mathcal{Z}_{\mathcal{F}})$ be the 0-cochain where $u(P) = t([\iota_{P,S}])$ for all $P \in \mathrm{Ob}(\mathcal{L})$. Then

$$du([\iota_{P,S}]) = u(S) \cdot u(P)^{-1} = t([\iota_{P,S}])^{-1},$$

and hence $t + du \in \widehat{Z}^1(\mathcal{O}(\mathcal{F}^c); \mathcal{Z}_{\mathcal{F}})$. Thus every element in $\varprojlim^1(\mathcal{Z}_{\mathcal{F}})$ is represented by a cocycle in $\widehat{Z}^1(\mathcal{O}(\mathcal{F}^c); \mathcal{Z}_{\mathcal{F}})$. Also, the same computation shows that for $u \in \bar{C}^0(\mathcal{O}(\mathcal{F}^c); \mathcal{Z}_{\mathcal{F}})$, $du \in \widehat{Z}^1(\mathcal{O}(\mathcal{F}^c); \mathcal{Z}_{\mathcal{F}})$ if and only if $u$ is a constant cochain; i.e., there is some $g \in Z(S)$ such that $u(P) = g$ for all $P$.

Consider the following sequence

$$1 \longrightarrow \widehat{Z}^1(\mathcal{O}(\mathcal{F}^c); \mathcal{Z}_{\mathcal{F}}) \xrightarrow{\ \widehat{\lambda}\ } \mathrm{Aut}^I_{\mathrm{typ}}(\mathcal{L}) \xrightarrow{\ \widehat{\mu}\ } \mathrm{Aut}(S, \mathcal{F}). \quad (5)$$

Here, $\widehat{\mu}$ sends $\alpha \in \mathrm{Aut}^I_{\mathrm{typ}}(\mathcal{L})$ to $\delta_S^{-1} \circ \alpha_S \circ \delta_S \in \mathrm{Aut}(S)$, and $\widehat{\mu}(\alpha)$ is fusion preserving by Proposition 4.11(b). For each $t \in \widehat{Z}^1(\mathcal{O}(\mathcal{F}^c); \mathcal{Z}_{\mathcal{F}})$ and each $g \in =S$, $[\delta_S(g)]$ is the identity in $\mathcal{O}(\mathcal{F}^c)$, so $t([\delta_S(g)]) = 1$, and $\Psi_t(\delta_S(g)) =

$\delta_S(g)$ by definition of $\Psi_t$. In other words, $\widehat{\mu} \circ \widehat{\lambda} = 0$. If $\Psi \in \mathrm{Ker}(\widehat{\mu})$, then by Proposition 4.11(b) again, $\pi \circ \Psi = \pi$ where $\pi \colon \mathcal{L} \longrightarrow \mathcal{F}$ is the canonical functor. By axiom (A2), for each morphism $\varphi \in \mathrm{Mor}_{\mathcal{L}}(P, Q)$ in $\mathcal{L}$, $\Psi(\varphi) = \varphi \circ \delta_P(t(\varphi))$ for some unique $t(\varphi) \in Z(P)$, this depends only on the class of $\varphi$ in $\mathrm{Mor}_{\mathcal{O}(\mathcal{F}^c)}(P, Q)$ since $\Psi$ is the identity on $\delta_S(S)$ and on inclusions, and thus $\Psi = \Psi_t$, and $[\Psi] = \widehat{\lambda}([t])$, for $t \in \widehat{Z}^1(\mathcal{O}(\mathcal{F}^c); \mathcal{Z}_{\mathcal{F}})$. Since $\widehat{\lambda}$ is clearly injective, this proves the exactness of (5).

Now consider the following diagram:

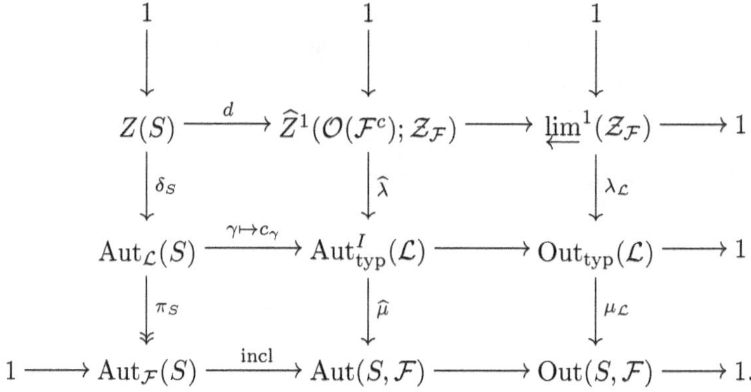

Here, $d$ sends $g \in Z(S)$ to $d(u_g)$, where $u_g$ is the constant 0-cochain with value $g$. The diagram commutes by construction. We have just shown the top row and the second column are exact. The second row is exact by Lemma 4.9, and the third by definition of $\mathrm{Out}(S, \mathcal{F})$. The first column is exact by axiom (A2) for a linking system. By an easy diagram chase, one now sees that the third column in the diagram is exact.

The definition of the map $\omega_{\mathcal{L}}$ and the exactness of the sequence at $\mathrm{Out}(S, \mathcal{F})$ are really a consequence of Proposition 5.11. Very roughly, for each $\beta \in \mathrm{Aut}(S, \mathcal{F})$, let $\mathcal{L}_\beta$ be the linking system whose underlying category is $\mathcal{L}$, and with structural functors

$$\mathcal{T}_{\mathrm{Ob}(\mathcal{L})}(S) \xrightarrow{\ \delta \circ \mathcal{T}(\beta^{-1})\ } \mathcal{L}_\beta \xrightarrow{\ c_\beta \circ \pi\ } \mathcal{F}$$

(and where the identification of objects in $\mathcal{L}$ with subgroups of $S$ is changed accordingly). Then $\omega_{\mathcal{L}}([\beta]) \in \varprojlim^2(\mathcal{Z}_{\mathcal{F}})$ is defined to be the obstruction to the existence of an isomorphism $\mathcal{L} \cong \mathcal{L}_\beta$. In particular, $[\beta] \in \mathrm{Im}(\mu_{\mathcal{L}})$ if and only if this obstruction vanishes. $\qquad\square$

Proposition 5.12 thus gives a very precise description of the kernel of the natural map $\mu_{\mathcal{L}}$ from $\mathrm{Out}_{\mathrm{typ}}(\mathcal{L})$ to $\mathrm{Out}(S, \mathcal{F})$. A very different method for doing this, which in practice seems to give an easy way to make this computation in many cases without using the machinery for computing higher limits, is given in [AOV, Proposition 4.2].

We now sketch what is known about these obstruction groups $\varprojlim^{i}(\mathcal{C}_{\mathcal{F}})$, beginning with the case where $\mathcal{F}$ is realizable.

**Theorem 5.13** *If $\mathcal{F} = \mathcal{F}_S(G)$ for some finite group $G$ and some $S \in \mathrm{Syl}_p(G)$, then*

$$\varprojlim_{\mathcal{O}(\mathcal{F}^c)}^{i}(\mathcal{Z}_{\mathcal{F}}) = 0$$

*for all $i \geq 2$, and for all $i \geq 1$ if $p$ is odd.*

*Proof.* This is shown as [O2, Theorems A & B] when $p$ is odd, and in [O3, Theorems A & B] when $p = 2$. In both cases, the proof consisted first of a reduction to a question involving simple groups, and then a case-by-case check using the classification theorem. $\square$

When $p = 2$, examples do occur of (realizable) fusion systems $\mathcal{F}$ for which $\varprojlim^{1}(\mathcal{Z}_{\mathcal{F}}) \neq 0$. For example, when $\mathcal{F} = \mathcal{F}_S(G)$ and $G = PSL_2(q)$ for an odd prime power $q \equiv \pm 1 \pmod 8$, then $\varprojlim^{1}(\mathcal{Z}_{\mathcal{F}}) \cong \mathbb{Z}/2$ by [BL, Theorem 1.7(3)] when $q = 3^{2^s}$ for some $s \geq 1$, and the same argument applies in the general case. The proof in [BL] is based on a comparison of the automorphisms for $PSL_2(q)$ and $SL_2(q)$; a more direct algebraic proof (using some of the techniques described in Section 5.4) is given in [O3, Proposition 1.6]. Similarly, $\varprojlim^{1}(\mathcal{Z}_{\mathcal{F}}) \cong \mathbb{Z}/2$ when $\mathcal{F} = \mathcal{F}_S(G)$ and $G \cong A_n$ (the alternating group) for $n \equiv 2, 3 \pmod 4$ and $n \geq 6$: this is shown indirectly (via the exact sequence of Proposition 5.12) in [AOV, proof of Proposition 4.8], and a more direct proof is sketched in [O3, Chapter 10].

Upon combining Theorem 5.13 with Propositions 5.11 and 5.12 and Theorem 4.22, we get the following result:

**Corollary 5.14** *For any finite group $G$, and any Sylow $p$-subgroup $S \leq G$, $\mathcal{L}_S^c(G)$ is the unique centric linking system associated to $\mathcal{F}_S(G)$. If $p$ is odd, then*

$$\mathrm{Out}(BG_p^\wedge) \cong \mathrm{Out}_{\mathrm{typ}}(\mathcal{L}_S^c(G)) \cong \mathrm{Out}(S, \mathcal{F}_S(G)).$$

If $G_1$ and $G_2$ are two finite groups, and $S_i \in \mathrm{Syl}_p(G_i)$ are such that there is a fusion preserving isomorphism $S_1 \xrightarrow{\cong} S_2$, then $\mathcal{L}_{S_1}^c(G_1) \cong \mathcal{L}_{S_2}^c(G_2)$ by Corollary 5.14, and hence $BG_{1p}^\wedge \simeq BG_{2p}^\wedge$ by Theorem 3.7. In other words, the Martino-Priddy conjecture (the "if" part of Theorem 1.17) follows as a special case of Theorem 5.13.

Computations of higher limits over these functors will be discussed in more detail in Section 5.4. For now, we just note a few additional cases where we know the existence and/or uniqueness of linking systems.

**Proposition 5.15** ([BLO2, Corollary 3.5]) *Let $\mathcal{F}$ be a saturated fusion system over a $p$-group $S$. If $\mathrm{rk}(S) < p^3$ (if $(C_p)^{p^3} \not\leq S$), then there is a* *centric linking system associated to $\mathcal{F}$. If $\mathrm{rk}(S) < p^2$ (if $(C_p)^{p^2} \not\leq S$), then there is a unique centric linking system associated to $\mathcal{F}$.*

*Proof.* This follows from Proposition 5.11, together with Proposition 5.26 below. $\qquad\square$

When $p$ is odd, the general question of whether $\varprojlim^i(\mathcal{Z}_\mathcal{F}) = 0$ for all $\mathcal{F}$ and all $i > 0$ was reduced in [O2] to a purely group-theoretic question about $p$-groups. Another definition is needed to describe this. For any group $G$ and any $H, K \leq G$, we write $[H, K; n]$ for the $n$-fold iterated commutator subgroup: thus $[H, K; 1] = [H, K]$, $[H, K; 2] = [[H, K], K]$, and $[H, K; n] = [[H, K; n-1], K]$.

**Definition 5.16** ([O2, Definition 3.1])   Fix a prime $p$ and a $p$-group $S$. For any $p$-group $S$, let $\mathfrak{X}(S)$ denote the unique largest subgroup of $S$ for which there is a sequence

$$1 = Q_0 \leq Q_1 \leq \cdots \leq Q_n = \mathfrak{X}(S) \leq S$$

of subgroups, all normal in $S$, such that

$$[\Omega_1(C_S(Q_{i-1})), Q_i; p-1] = 1 \tag{6}$$

for each $i = 1, \ldots, n$.

To see that $\mathfrak{X}(S)$ is well defined (that there is a unique such largest subgroup), note that any two chains which satisfy (6) and terminate in $Q_n$ and $Q'_m$ can be combined to give a chain ending in $Q_n Q'_m$. It follows immediately from the definition that $\mathfrak{X}(S) = S$ if $[S, S; p-1] = 1$; i.e., if $S$ has nilpotence class at most $p - 1$.

Recall (Definition A.15) that the *Thompson subgroup $J(S)$* of a $p$-group $S$ is be the subgroup generated by all elementary abelian subgroups of maximal rank in $S$. The following is one of the main results in [O2].

**Proposition 5.17** ([O2, Proposition 3.5 & Corollary 3.8])   *Assume $p$ is odd, and let $\mathcal{F}$ be a saturated fusion system over a $p$-group $S$.*

(a)   *Assume there is a subgroup $P \leq \mathfrak{X}(S)$ which is $\mathcal{F}$-centric and weakly closed in $\mathcal{F}$. Then $\varprojlim^i(\mathcal{Z}_\mathcal{F}) = 0$ for all $i > 0$.*

(b)   *Let $1 = T_0 \leq T_1 \leq \cdots \leq T_k = S$ be any sequence of subgroups which are all strongly $\mathcal{F}$-closed in $S$. Assume, for all $1 \leq i \leq k$, that $\mathfrak{X}(T_i/T_{i-1}) \geq J(T_i/T_{i-1})$. Then $\varprojlim^i(\mathcal{Z}_\mathcal{F}) = 0$ for all $i > 0$.*

We have stated these two versions of the proposition for simplicity, but there is also a more complicated version, a consequence of [O2, Proposition 3.5], which includes points (a) and (b) as special cases. Note that in point (a), for $P$ to be weakly closed in $\mathcal{F}$, it suffices that it be characteristic in each subgroup of $S$ which contains it. In particular, we can take $P = J(S)$ if it is contained in $\mathfrak{X}(S)$.

Proposition 5.17 motivates the following conjecture, which if true would imply the existence and uniqueness of centric linking systems associated to any saturated fusion system over a $p$-group for $p$ odd.

**Conjecture 5.18** ([O2, Conjecture 3.9])  *For each odd prime $p$ and each $p$-group $S$, $\mathfrak{X}(S) \geq J(S)$.*

In fact, by Proposition 5.17, in order to prove the existence and uniqueness of centric linking systems associated to saturated fusion systems over $p$-groups ($p$-odd), it suffices to prove Conjecture 5.18 for all $p$-groups $S$ for which there is a saturated fusion system over $S$ with no proper strongly closed subgroups of $S$. In principle, this restricts greatly the list of $p$-groups which must be considered, but it is difficult to see how to use this restriction in practice.

This conjecture has recently been reformulated in terms of representation theory by Green, Héthelyi, and Lilienthal [GHL]. Using that reformulation, they show among other things that Conjecture 5.18 holds whenever $S/\mathfrak{X}(S)$ has nilpotency class at most two [GHL, Theorem 1.1]. Building on this, Green, Héthelyi, and Mazza [GHM] then showed that Conjecture 5.18 holds whenever $S/\mathfrak{X}(S)$ has nilpotency class $\leq 4$, or is metabelian, or is of maximal class, or has rank $\leq p$.

### 5.4. Some computational techniques for higher limits over orbit categories.

The bar resolution is very useful for identifying certain obstruction groups as higher derived functors of limits over orbit categories. However, it does not seem to be very useful in practice for making actual computations. For that, other techniques are needed. We describe some of those here, with examples to illustrate how they are applied.

One technique, which works very well in practice when computing higher limits of a functor $F$ over an orbit category, is to filter $F$ by a sequence of subfunctors, $0 = F_0 \subseteq F_1 \subseteq \cdots \subseteq F_n = F$, in such a way that each quotient functor $F_i/F_{i-1}$ vanishes except on one isomorphism class of objects. Surprisingly, when a functor $F'$ vanishes except on the isomorphism class of one object $P$, then $\varprojlim^*(F')$ depends only on the automorphism group of $P$ in the orbit category and its action on $F'(P)$. The following definition helps us to formulate this precisely.

**Definition 5.19**  Fix a finite group $\Gamma$ and a $\mathbb{Z}[\Gamma]$-module $M$. Let $\mathcal{O}_p(\Gamma) \subseteq \mathcal{O}(\Gamma)$ be the full subcategory whose objects are the $p$-subgroups of $\Gamma$. Define

$F_M \colon \mathcal{O}_p(\Gamma)^{\mathrm{op}} \longrightarrow \mathbf{Ab}$ by setting

$$F_M(P) = \begin{cases} M & \text{if } P = 1 \\ 0 & \text{if } P \neq 1. \end{cases}$$

Here, $F_M(1) = M$ has the given action of $\mathrm{Aut}_{\mathcal{O}_p(\Gamma)}(1) = \Gamma$. Set

$$\Lambda^*(\Gamma; M) = \varprojlim_{\mathcal{O}_p(\Gamma)}{}^{*}(F_M).$$

**Proposition 5.20**    (a)  *Fix a finite group $G$. Let*

$$F \colon \mathcal{O}_p(G)^{\mathrm{op}} \longrightarrow \mathbb{Z}_{(p)}\text{-mod}$$

*be any functor which vanishes except on the isomorphism class of some
$p$-subgroup $P \leq G$. Then*

$$\varprojlim_{\mathcal{O}_p(G)}{}^{*}(F) \cong \Lambda^*(N_G(P)/P; F(P)).$$

(b)  *Let $\mathcal{F}$ be a saturated fusion system over a $p$-group $S$. Let*

$$F \colon \mathcal{O}(\mathcal{F}^c)^{\mathrm{op}} \longrightarrow \mathbb{Z}_{(p)}\text{-mod}$$

*be any functor which vanishes except on the isomorphism class of some
subgroup $Q \in \mathcal{F}^c$. Then*

$$\varprojlim_{\mathcal{O}(\mathcal{F}^c)}{}^{*}(F) \cong \Lambda^*(\mathrm{Out}_{\mathcal{F}}(Q); F(Q)).$$

*Proof.* Part (a) is shown in [JMO, Proposition 5.4], and part (b) in [BLO2,
Proposition 3.2]. A more general versiom of both of these results is shown
in [BLO3, Proposition 5.3].

The proofs in [BLO2] and [BLO3] and the first proof in [JMO] are purely
algebraic, based on a series of comparisons of higher limits of different
functors over different categories. There is also a geometric proof of (a)
in [JMO], based on interpreting higher limits of functors over $\mathcal{O}_p(G)$ as
equivariant cohomology over a certain $G$-space associated to this category.
$\qquad\square$

The idea now, for an arbitrary functor $F \colon \mathcal{O}(\mathcal{F}^c)^{\mathrm{op}} \longrightarrow \mathbb{Z}_{(p)}\text{-mod}$, is to
reduce the computation of $\varprojlim{}^{*}(F)$ to that of the graded groups $\Lambda^*(G; M)$
for appropriate $G$ and $M$. The way of doing this is illustrated by the
following corollary to Proposition 5.20.

**Corollary 5.21**    *Let $\mathcal{C}$ be one of the following categories: either $\mathcal{C} =
\mathcal{O}_p(G)$ for some finite group $G$, or $\mathcal{C} = \mathcal{O}(\mathcal{F}^c)$ for some saturated fusion
system $\mathcal{F}$ over a $p$-group $S$. Fix a functor $F \colon \mathcal{C}^{\mathrm{op}} \longrightarrow \mathbb{Z}_{(p)}\text{-mod}$.*

(a)  *Assume, for some $i \geq 0$, that $\Lambda^i(\mathrm{Aut}_{\mathcal{C}}(P); F(P)) = 0$ for each $P \in
\mathrm{Ob}(\mathcal{C})$. Then $\varprojlim{}^{i}(F) = 0$.*

(b)  *Assume, for some $Q \in \mathrm{Ob}(\mathcal{C})$, that $\Lambda^i(\mathrm{Aut}_{\mathcal{C}}(P); F(P)) = 0$ for each $P \in \mathrm{Ob}(\mathcal{C})$ not $\mathcal{C}$-isomorphic to $Q$. Then*

$$\varprojlim{}^*(F) \cong \Lambda^*(\mathrm{Aut}_{\mathcal{C}}(Q); F(Q)).$$

*Proof.* Let $\mathcal{P}_1, \ldots, \mathcal{P}_m$ be the set of $\mathcal{C}$-isomorphism classes of objects in $\mathcal{C}$, arranged from smallest to largest. Thus $P \in \mathcal{P}_i$, $Q \in \mathcal{P}_j$, and $i \leq j$ imply $|P| \leq |Q|$. For each $i = 0, \ldots, m$, let $F_i \subseteq F$ be the subfunctor

$$F_i(P) = \begin{cases} F(P) & \text{if } P \in \mathcal{P}_1 \cup \cdots \cup \mathcal{P}_i \\ 0 & \text{otherwise.} \end{cases}$$

This defines a sequence of subfunctors

$$0 = F_0 \subseteq F_1 \subseteq F_2 \subseteq \cdots \subseteq F_m = F,$$

such that for each $1 \leq i \leq m$,

$$(F_i/F_{i-1})(P) = \begin{cases} F(P) & \text{for } P \in \mathcal{P}_i \\ 0 & \text{for } P \notin \mathcal{P}_i. \end{cases}$$

By Proposition 5.20, for each $j = 1, \ldots, m$,

$$\varprojlim{}^*(F_j/F_{j-1}) \cong \Lambda^*(\mathrm{Aut}_{\mathcal{C}}(P_j); F(P_j))$$

for all $P_j \in \mathcal{P}_j$. Hence under the assumption of (a), $\varprojlim{}^i(F_j/F_{j-1}) = 0$ for each $j$. For each triple of indices $0 \leq j < k < \ell \leq m$, there is an exact sequence

$$\varprojlim{}^i(F_k/F_j) \longrightarrow \varprojlim{}^i(F_\ell/F_j) \longrightarrow \varprojlim{}^i(F_\ell/F_k),$$

and so $\varprojlim{}^i(F) = \varprojlim{}^i(F_m/F_0) = 0$ by induction.

Now assume we are in the situation of (b), and let $j$ be such that $Q \in \mathcal{P}_j$. Then $\varprojlim{}^*(F_i/F_{i-1}) = 0$ for all $i \neq j$. Using the exact sequences for an extension of functors, we now see that $\varprojlim{}^*(F) \cong \varprojlim{}^*(F_j/F_{j-1}) \cong \Lambda^*(\mathrm{Aut}_{\mathcal{C}}(Q); F(Q))$. $\qquad\square$

A similar technique can also be used to compute $\varprojlim{}^*(F)$ in many other cases. As long as most of the groups $\varprojlim{}^*(F_j/F_{j-1})$ vanish, the exact sequences for higher limits of an extension of functors can be used to compute $\varprojlim{}^*(F)$ once one knows the higher limits of all subquotient functors. This procedure works surprisingly well in practice, mostly because the functors $\Lambda^*$ vanish in many cases, as will be seen in Proposition 5.24.

The computation of $\varprojlim{}^*(F)$ using such a filtration can also be organized as a spectral sequence, as was done explicitly by Grodal [Gr, Theorem 1.3]. This spectral sequence was used by Ziemiański [Zi] when making the computations needed to construct maps between 2-completed classifying spaces of certain compact Lie groups.

We now need some tools which allow us to compute $\Lambda^*(G; M)$ in favorable cases. There are two very different ways of doing this, based on the next two propositions. The first is due to Grodal.

**Proposition 5.22** ([Gr])    *Fix a prime $p$, a finite group $G$ such that $p \,\big|\, |G|$, and a $\mathbb{Z}_{(p)}[G]$-module $M$. Let $\mathcal{S}_p(G)$ be the poset of $p$-subgroups $1 \neq P \leq G$, and let $|\mathcal{S}_p(G)|$ be its geometric realization when regarded as a category. Let $\overline{C}_*(|\mathcal{S}_p(G)|)$ be the reduced cellular chain complex for $|\mathcal{S}_p(G)|$; i.e., the chain complex which has $\mathbb{Z}$ in degree $-1$, and which in degree $k$ is the free abelian group with basis the set of all chains $P_0 < \cdots < P_k$ of length $k$ in $\mathcal{S}_p(G)$. Then for all $i > 0$,*

$$\Lambda^i(G; M) \cong H^{i-1}\big(\mathrm{Hom}_{\mathbb{Z}[G]}(\overline{C}_*(|\mathcal{S}_p(G)|), M)\big),$$

*where $G$ acts on $\overline{C}_*(|\mathcal{S}_p(G)|)$ via its conjugation action on $\mathcal{S}_p(G)$.*

*Proof.* This is a special case of [Gr, Theorem 1.2]. The proof is based on a comparison of the orbit and poset categories $\mathcal{O}_p(G)$ and $\mathcal{S}_p(G)$. A similar result, which also involves comparing higher limits over a poset and an associated category but in a more abstract setting, was shown by Słomińska ([JS, Theorem 6.6]).      $\square$

We refer to [Gr, §5] for several examples of how Proposition 5.22 can be applied. Many of these are based on a theorem of Webb [Wb1], which says that the $p$-adic "Steinberg complex" $\overline{C}_*(|\mathcal{S}_p(G)|; \mathbb{Z}_p)$ splits as a sum of an acyclic complex and a complex of projective modules, together with a very careful analysis of which indecomposable projective $\mathbb{Z}_p[G]$-modules can occur in the complex.

The next proposition provides a very different tool for making such computations. Here, for any $\mathbb{Z}[G]$-module $M$ and any $H \leq G$, we write $M^H$ for the subgroup of elements of $M$ fixed by $H$. Note that part of this proposition is a special case of Lemma 5.9.

**Proposition 5.23** ([JMO])    *Fix a finite group and a $\mathbb{Z}_{(p)}[G]$-module $M$. For each $H \leq G$, set $\mathfrak{N}_H = \sum_{h \in H} h \in \mathbb{Z}[G]$. Let*

$$H^0 M,\ \mathfrak{N}M : \mathcal{O}_p(G) \longrightarrow \mathbb{Z}_{(p)}\text{-mod}$$

*be the functors $H^0 M(P) = H^0(P; M) = M^P = C_M(P)$ and $\mathfrak{N}M(P) = \mathfrak{N}_P \cdot M$. Then*

$$\varprojlim_{\mathcal{O}_p(G)}{}^i (H^0 M) = \begin{cases} M^G = C_M(G) & \text{if } i = 0 \\ 0 & \text{if } i > 0 \end{cases}$$

*and*

$$\varprojlim_{\mathcal{O}_p(G)}{}^i (\mathfrak{N}M) = \begin{cases} \mathfrak{N}_G \cdot M & \text{if } i = 0 \\ 0 & \text{if } i > 0. \end{cases}$$

*Proof.* This is shown in [JMO, Proposition 5.2] for the functors $H^0M$ and $H^0M/\mathfrak{N}M$ (note that $(H^0M/\mathfrak{N}M)(P) = \widehat{H}^0(P; M)$). This, in turn, is a special case of a general result of Jackowski and McClure [JM, Proposition 5.14].  $\qquad\square$

In practice, Proposition 5.22 seems to be most useful for getting information about the groups $\Lambda^*(-;-)$ in general situations, while a procedure using Proposition 5.23 often works better for specific computations. Proposition 5.24 was shown in [JMO] mostly using repeated applications of Proposition 5.23 (together with exact sequences for higher limits of a pair of functors), but most or all of its points can also be proven using Proposition 5.22.

**Proposition 5.24** *The following hold for each finite group $G$ and each $\mathbb{Z}_{(p)}[G]$-module $M$.*

(a) *If $(p, |G|) = 1$, then*

$$\Lambda^i(G; M) = \begin{cases} M^G & \text{if } i = 0 \\ 0 & \text{if } i > 0. \end{cases}$$

*If $H \trianglelefteq G$ is a normal subgroup which acts trivially on $M$, and $(p, |H|) = 1$, then $\Lambda^*(G; M) \cong \Lambda^*(G/H; M)$.*

(b) *If $p\,\big|\,|G|$, then $\Lambda^0(G; M) = 0$. If $H \trianglelefteq G$ is a normal subgroup which acts trivially on $M$, and $p\,\big|\,|H|$, then $\Lambda^*(G; M) = 0$. If $O_p(G) \neq 1$, then $\Lambda^*(G; M) = 0$.*

(c) *Assume $p\,\big|\,|G|$, fix $S \in \mathrm{Syl}_p(G)$, and let $\sim$ be the equivalence relation among Sylow $p$-subgroups of $\Gamma$ generated by nontrivial intersection. Set*

$$H = \{g \in G \,|\, {}^gS \sim S\}.$$

*Thus $H < G$ is the minimal strongly $p$-embedded subgroup in $G$ containing $S$ if there is one (see Definition A.6 and Proposition A.7), and $H = G$ otherwise. Then $\Lambda^1(G; M) \cong M^H/M^G$, and $\Lambda^i(G; M) \cong \Lambda^i(H; M)$ for $i \geq 2$.*

(d) *A short exact sequence $0 \longrightarrow M' \longrightarrow M \longrightarrow M'' \longrightarrow 0$ of $\mathbb{Z}_{(p)}[G]$-modules induces a long exact sequence*

$$\longrightarrow \Lambda^i(G; M') \longrightarrow \Lambda^i(G; M) \longrightarrow \Lambda^i(G; M'') \longrightarrow \Lambda^{i+1}(G; M') \longrightarrow .$$

*Proof.* Except for the last statement in (c) ($\Lambda^i(G; M) \cong \Lambda^i(H; M)$ for $i \geq 2$), this is shown in [JMO, Propositions 6.1 and 6.2]. So we prove that here, using Proposition 5.23. We refer to Definition A.6 and Proposition A.7 for the definition and properties of strongly $p$-embedded subgroups.

Let $F_M \subseteq H^0 M \colon \mathcal{O}_p(G) \longrightarrow \mathbb{Z}_{(p)}$-mod be the functors $F_M(1) = M$, $F_M(P) = 0$ for $P \neq 1$, and $H^0(M)(P) = M^P = C_M(P)$ for all $p$-subgroups $P \leq G$. We also use the same names for the restrictions of these functors to $\mathcal{O}_p(H)$. We claim that for all $i \geq 2$,

$$\Lambda^i(G; M) \cong \varprojlim_{\mathcal{O}_p(G)}{}^{i-1}(H^0 M / F_M) \cong \varprojlim_{\mathcal{O}_s(G)}{}^{i-1}(H^0 M / F_M)$$

$$\cong \varprojlim_{\mathcal{O}_s(H)}{}^{i-1}(H^0 M / F_M) \cong \varprojlim_{\mathcal{O}_p(H)}{}^{i-1}(H^0 M / F_M) \cong \Lambda^i(H; M).$$

The first and last isomorphisms follow from the exact sequence of higher limits for the pair $F_M \subseteq H^0 M$, and since $\varprojlim{}^i(H^0 M) = 0$ for $i > 0$ by Proposition 5.23. The second and fourth isomorphisms follow from Corollary 5.6. The third isomorphism follows from Proposition 5.3 ($\varprojlim{}^*(-)$ is the cohomology of the bar resolution): the two chain complexes are isomorphic since $\mathrm{Mor}_{\mathcal{O}_s(G)}(P, Q) = \mathrm{Mor}_{\mathcal{O}_s(H)}(P, Q)$ for all $1 \neq P, Q \leq S$, and since $(H^0 M / F_M)(1) = 0$. This proves the last statement in (c). $\quad\square$

The next proposition provides a simple example of how Propositions 5.22 and 5.23 can both be applied in practice.

**Proposition 5.25** ([BLO2, Corollary 3.4])  *Let $\mathcal{F}$ be a saturated fusion system over a $p$-group $S$ of order $p^m$. Then for each $F \colon \mathcal{O}(\mathcal{F}^c)^{\mathrm{op}} \longrightarrow \mathbb{Z}_{(p)}$-mod, $\varprojlim{}^i(F) = 0$ for all $i > m$.*

*Proof.* By Lemma 5.21(a), it suffices to prove, for each finite group $G$, each $\mathbb{Z}_{(p)}[G]$-module $M$, and each $k \geq 0$,

$$G = p^k r \text{ where } p \nmid r \qquad \text{implies} \qquad \Lambda^i(G; M) = 0 \quad \text{for all } i > k. \qquad (7)$$

This follows immediately from Proposition 5.22, since the geometric realization of the poset $\mathcal{S}_p(G)$ has dimension at most $k - 1$.

We now give an inductive proof of (7) using Proposition 5.23. When $k = 0$, it follows from Proposition 5.23 ($F_M = H^0 M$ as functors on $\mathcal{O}_p(G)$), so assume $k > 0$. Let $F_M \colon \mathcal{O}_p(G) \longrightarrow \mathbb{Z}_{(p)}$-mod be the functor $F_M(1) = M$ and $F_M(P) = 0$ for $p$-subgroups $1 \neq P \leq G$, regarded as a subfunctor of $H^0 M$. Thus $(H^0 M / F_M)(P) = M^P$ for $1 \neq P \leq G$, and $(H^0 M / F_M)(1) = 0$. By the induction hypothesis, $\Lambda^i(N_G(P)/P; M^P) = 0$ for each $1 \neq P \leq G$ and each $i \geq k$; and hence $\varprojlim{}^i(H^0 M / F_M) = 0$ for all $i \geq k$ by Lemma 5.21(a). Since $\varprojlim{}^i(H^0 M) = 0$ for all $i > 0$ by Proposition 5.23, (7) now follows from the long exact sequence of the pair $F_M \subseteq H^0 M$. $\quad\square$

Here is another vanishing result. In fact, it includes Proposition 5.25 as a special case, but its proof is less elementary. As usual, for a $p$-group $P$,

$\mathrm{Fr}(P)$ denotes its Fratini subgroup: the subgroup generated by all commutators and $p$-th powers in $P$. Thus $\mathrm{Fr}(P)$ is the smallest normal subgroup such that $P/\mathrm{Fr}(P)$ is elementary abelian.

**Proposition 5.26** ([BLO2, Corollary 3.5])    *Let $\mathcal{F}$ be a saturated fusion system over a $p$-group $S$, and let $F\colon \mathcal{O}(\mathcal{F}^c)^{\mathrm{op}} \longrightarrow \mathbb{Z}_{(p)}\text{-}\mathrm{mod}$ be a functor. Assume, for some $k > 0$, that either $\mathrm{rk}(P/\mathrm{Fr}(P)) < k$ for each $P \leq S$, or that $\mathrm{rk}_{\mathbb{F}_p}(M/pM) < p^k$ for each $P \in \mathrm{Ob}(\mathcal{F}^c)$ and each $M \subseteq F(P)$. Then $\varprojlim^i(F) = 0$ for all $i \geq k$.*

*Proof.* By Lemma 5.21(a), it suffices to prove that $\Lambda^*(G; M) = 0$ for each finite group $G$ and each $\mathbb{Z}_{(p)}[G]$-module $M$ such that $\mathrm{rk}_p(G) < k$ (i.e., $G$ contains no subgroup $\cong C_p^k$), or $\mathrm{rk}_{\mathbb{F}_p}(M_0/pM_0) < p^k$ for all $M_0 \subseteq M$.

By [BLO1, Proposition 6.3], for each finite group $G$, each $\mathbb{F}_p[G]$-module $M$, and each $k \geq 1$, $\Lambda^k(G; M) = 0$ if $\mathrm{rk}(M) < p^k$ or if $\mathrm{rk}_p(G) < k$ (i.e., $G$ contains no subgroup $C_p^k$). This is proven in [BLO1] using Grodal's theorem (Theorem 5.22), together with a decomposition of the Steinberg complex due to Peter Webb [Wb1, Theorem 2.7.1]. A second argument, using Proposition 5.23, is given in the proof of Lemma 5.27 below.

The general case ($M$ not necessarily of exponent $p$) now follows by standard manipulations of the groups $\Lambda^*(G; -)$, using Propositions 5.5 and 5.24(f). If $M$ is a finitely generated $\mathbb{Z}_{(p)}[G]$-module, then $\Lambda^k(G; M)$ is finitely generated by Proposition 5.5(a) (and since $\mathcal{O}_p(G)$ is a finite category). If, in addition, $\mathrm{rk}(M/pM) < p^k$ or $\mathrm{rk}_p(G) < k$, then $\mathrm{rk}(M_0/pM_0) < p^k$ for each $\mathbb{Z}_{(p)}[G]$-submodule $M_0 \subseteq M$, and hence $\Lambda^k(G; M_0/pM_0) = 0$ for each such $M_0$ by the above remarks. Via the exact sequences of Proposition 5.24(f), this implies that each inclusion $p^{r+1}M \subseteq p^r M$ induces a surjection from $\Lambda^k(G; p^{r+1}M)$ onto $\Lambda^k(G; p^r M)$. Since $M$ is finitely generated, $p^r M$ is torsion free for $r$ large enough, hence $\Lambda^k(G; p^r M)$ is $p$-divisible, and $\Lambda^k(G; p^r M) = 0$ since it is also finitely generated. We just saw that $\Lambda^k(G; p^i M/p^{i+1}M) = 0$ for each $i$, and hence $\Lambda^k(G; M) = 0$ by the exact sequences of Proposition 5.24(f) again.

If $M$ is infinitely generated, then by Proposition 5.5(b), $\Lambda^k(G; M)$ is the direct limit of the groups $\Lambda^k(G; M_0)$ for $M_0 \subseteq M$ finitely generated. Thus $\Lambda^k(G; M) = 0$ if $\mathrm{rk}_p(G) < k$ or $\mathrm{rk}(M_0/pM_0) < p^k$ for all $M_0 \subseteq M$.    $\square$

We now state a more precise version of Proposition 5.26, and show how it can be proven using Proposition 5.23. It is a special case of the more technical result [O3, Proposition 3.5], and the proof given here is a much shortened version of that given in [O3].

For use in the following lemma only, for any set $P_1, \dots, P_m \leq G$ of subgroups, we set $\mathbf{N}_G(P_1, \dots, P_m) = \bigcap_{i=1}^m N_G(P_i)$. A *radical $p$-chain of* *length $n$ in $G$ is a sequence*

$$O_p(G) = P_0 < P_1 < \cdots < P_n$$

of distinct $p$-subgroups of $G$ such that $P_i = O_p(\mathbf{N}_G(P_0, \ldots, P_i))$ for all $i$, and such that $P_n \in \mathrm{Syl}_p(\mathbf{N}_G(P_0, \ldots, P_{n-1}))$. Note that the first condition implies that $P_i \trianglelefteq P_j$ for $i < j$, and that $P_i/P_{i-1}$ is a radical $p$-subgroup of $\mathbf{N}_G(P_0, \ldots, P_{i-1})/P_{i-1}$ for all $i$.

**Lemma 5.27** *Fix a finite group $G$, and a $\mathbb{Z}_{(p)}[G]$-module $M$. Assume, for some $n \geq 1$, that $\Lambda^n(G; M) \neq 0$. Then there is a radical $p$-chain*

$$1 = P_0 < P_1 < P_2 < \cdots < P_n$$

*of length $n$ in $G$ such that $\mathfrak{N}_{\mathbf{N}_G(P_1,\ldots,P_n)} \cdot M \neq 0$. If $M$ is finitely generated, then $M/pM$ (when regarded as an $\mathbb{F}_p[P_n]$-module) contains a copy of the free module $\mathbb{F}_p[P_n]$. In particular,*

$$\mathrm{rk}(M/pM) \geq |P_n| \geq p^n.$$

*Proof.* We prove this by induction on $n \geq 1$. By assumption, $\Lambda^n(G; M) = \varprojlim^n(F_M) \neq 0$, where $F_M(1) = M$, and $F_M(P) = 0$ for all $p$-subgroups $1 \neq P \leq G$. Regard $F_M$ as a subfunctor of the acyclic functor $\mathfrak{N}M$ of Proposition 5.23: the functor on $\mathcal{O}_p(G)$ which sends $P$ to $\mathfrak{N}_P \cdot M$. Thus $\varprojlim^n(\mathfrak{N}M) = 0$, and hence $\varprojlim^{n-1}(\mathfrak{N}M/F_M) \neq 0$.

By Corollary 5.21(a), there is a $p$-subgroup $1 \neq P_1 \leq G$ such that

$$\Lambda^{n-1}(N_G(P_1)/P_1; \mathfrak{N}_{P_1} \cdot M) \cong \Lambda^{n-1}(N_G(P_1)/P_1; (\mathfrak{N}M/F_M)(P_1)) \neq 0.$$

By Proposition 5.24(b), $O_p(N_G(P_1)) = P_1$. If $n = 1$, then $N_G(P_1)/P_1$ has order prime to $p$ by Proposition 5.24(b); so $P_1 \in \mathrm{Syl}_p(G)$ (Lemma A.1). Thus $(1 < P_1)$ is a radical $p$-chain of length one in this case, and $\mathfrak{N}_{N_G(P_1)} \cdot M = (\mathfrak{N}_{P_1} \cdot M)^{N_G(P_1)/P_1} \neq 0$.

If $n > 1$, then set $G_1 = N_G(P_1)/P_1$ and $M_1 = \mathfrak{N}_{P_1} \cdot M$. By the induction hypothesis applied to $\Lambda^{n-1}(G_1; M_1) \neq 0$, there is a radical $p$-chain $(1 < P_2/P_1 < \cdots < P_n/P_1)$ in $G_1$ such that

$$\mathfrak{N}_{\mathbf{N}_{G_1}(P_2/P_1,\ldots,P_n/P_1)} \cdot M_1 = \mathfrak{N}_{\mathbf{N}_G(P_1,\ldots,P_n)} \cdot M \neq 0.$$

This proves the first statement.

Assume $pM = 0$, so that $M$ is an $\mathbb{F}_p[G]$-module. Fix $x \in M$ such that $\mathfrak{N}_{P_n} \cdot x \neq 0$, and consider the $\mathbb{F}_p[P_n]$-linear homomorphism $\varphi \colon \mathbb{F}_p[P_n] \to M$ which sends $1$ to $x$. Every nonzero submodule of $\mathbb{F}_p[P_n]$ contains fixed elements and hence contains $\mathfrak{N}_{P_n}$, so $\mathrm{Ker}(\varphi) = 0$, and $\varphi$ embeds $\mathbb{F}_p[P_n]$ as a submodule of $M$. The general case $(pM \neq 0)$ follows by the same arguments as those used in the proof of Proposition 5.26. $\square$

All of these vanishing results suggest that if there is a fusion system $\mathcal{F}$ for which $\varprojlim^i(\mathcal{Z}_\mathcal{F}) \neq 0$ for some $i > 1$, then this must be due to some very complicated combination of conditions being fulfilled. On the other hand, most of the general techniques we have for proving vanishing of higher limits are designed to prove that $\varprojlim^i(-) = 0$ for all $i > 0$, and hence cannot be applied to this situation since we know examples where $\varprojlim^1(\mathcal{Z}_{\mathcal{F}}) \neq 0$. This might help to explain why the question of the existence and uniqueness of linking systems seems to be so difficult.

## 5.5. **Homotopy colimits and homotopy decompositions.**

By a homotopy decomposition of a space $X$ is meant a homotopy equivalence of $X$ with the homotopy colimit of a functor defined on some small category. So before discussing homotopy decompositions of $|\mathcal{L}|$ and $|\mathcal{L}|_p^\wedge$, we first explain what a homotopy colimit is.

Let $\mathcal{C}$ be a small category, and let $F \colon \mathcal{C} \longrightarrow \mathsf{Top}$ be a (covariant) functor to the category of topological spaces. The ordinary colimit (or direct limit) of $F$ is defined by setting

$$\operatorname*{colim}_{\mathcal{C}}(F) = \left( \coprod_{c \in \mathrm{Ob}(\mathcal{C})} F(c) \right) \Big/ \sim,$$

where $\sim$ is the equivalence relation generated by identifying $x \in F(c)$ with $f_*(x) \in F(d)$ for each morphism $f \in \mathrm{Mor}_{\mathcal{C}}(c, d)$ in $\mathcal{C}$. The problem with this construction is that the homotopy types of the spaces $F(c)$ and the maps between them do not, in general, in any way determine the homotopy type of the colimit. This will be explained more concretely in some of the examples given below.

The *homotopy colimit* of $F \colon \mathcal{C} \longrightarrow \mathsf{Top}$ is defined by setting

$$\operatorname*{hocolim}_{\mathcal{C}}(F) = \left( \coprod_{n \geq 0} \coprod_{c_0 \to \cdots \to c_n} F(c_0) \times \Delta^n \right) \Big/ \sim.$$

Here, each face or degeneracy map between the sequences $c_0 \to \cdots \to c_n$ gives rise to an obvious identification between the corresponding spaces. In other words, we begin by taking the disjoint union of the spaces $F(c)$, for all $c \in \mathrm{Ob}(\mathcal{C})$. Then, for each nonidentity morphism $\varphi \colon c \to d$ in $\mathcal{C}$, we attach a copy of $F(c) \times I$, by identifying $F(c) \times 0$ with $F(c)$ and attaching $F(c) \times 1$ to $F(d)$ via $F(\varphi) \colon F(c) \longrightarrow F(d)$. This can be thought of as the "1-skeleton" of the homotopy colimit. Afterwards, one attaches higher dimensional simplices corresponding to each sequence of two or more composable morphisms in $\mathcal{C}$.

More precisely, let $\mathcal{C}_F$ be the *simplicial space* $\Delta^{\mathrm{op}} \longrightarrow \mathsf{Top}$ defined by sending the object $[n]$ to the disjoint union of the $F(c_0)$, taken over all sequences $c_0 \to \cdots \to c_n$ in $\mathcal{N}(\mathcal{C})_n$ (see Section 2.2). A morphism $\varphi \in \mathrm{Mor}_{\Delta}([n], [m])$ sends $F(c_0)_\xi$, where $\xi = (c_0 \to \cdots \to c_m) \in \mathcal{N}(\mathcal{C})_m$, to

$F(c_{\varphi(0)})_{\varphi^*(\xi)}$ via the map $F(c_0 \to c_{\varphi(0)})$. We then set

$$\operatorname*{hocolim}_{\mathcal{C}}(F) = |\mathcal{C}_F| = \left( \coprod_{n=0}^{\infty} \big( \mathcal{C}_F([n]) \times \Delta^n \big) \right) \Big/ \sim$$

with the quotient topology, where just as for the geometric realization of a simplicial *set*, $(x, \varphi_*(y)) \sim (\varphi^*(x), y)$ for all $x \in \mathcal{C}_F([m])$, $y \in \Delta^n$, and $\varphi \in \operatorname{Mor}_\Delta([n], [m])$.

If $F_1, F_2 \colon \mathcal{C} \longrightarrow \operatorname{Top}$ are two functors, and $\varphi \colon F_1 \longrightarrow F_2$ is a natural transformation of functors, then $\varphi$ induces a map of spaces $\operatorname{hocolim}(\varphi)$ from $\operatorname{hocolim}(F_1)$ to $\operatorname{hocolim}(F_2)$. If $\varphi$ induces a homotopy equivalence $\varphi(c) \colon F_1(c) \xrightarrow{\;\simeq\;} F_2(c)$ for each $c \in \operatorname{Ob}(\mathcal{C})$, then $\operatorname{hocolim}(\varphi)$ is a homotopy equivalence: at least if $F_1$, $F_2$, and $\varphi$ take values in the category of simplicial sets and maps (cf. [GJ, Proposition IV.1.7]). Similarly, if $\varphi(c)$ induces an isomorphism in cohomology (with any given group of coefficients) for each $c \in \operatorname{Ob}(\mathcal{C})$, then so does $\operatorname{hocolim}(\varphi)$. Thus, if we think of a functor $F \colon \mathcal{C} \longrightarrow \operatorname{Top}$ as a diagram of spaces and maps between them, then the the homotopy type or homology of the homotopy colimit depends only on the "homotopy type" or "homology" of the diagram in a very weak sense. This is definitely not true for induced maps between ordinary colimits of diagrams of spaces, and explains why this construction is called a "homotopy" colimit.

Among some simple examples of homotopy colimits, we note the following:

- Let $\mathcal{C}$ be the "pushout category": $\mathcal{C} = (c_1 \xleftarrow{\;f_1\;} c_0 \xrightarrow{\;f_2\;} c_2)$. The homotopy colimit of a functor $F \colon \mathcal{C} \longrightarrow \operatorname{Top}$ can be identified with the double mapping cylinder of the pair of maps $F(c_1) \xleftarrow{\;F(f_1)\;} F(c_0) \xrightarrow{\;F(f_2)\;} F(c_2)$; i.e., with the space

  $$\operatorname*{hocolim}_{\mathcal{C}}(F) = \big( F(c_1) \amalg (F(c_0) \times I) \amalg F(c_2) \big) \big/ \sim,$$

  where for each $x \in F(c_0)$, $(x, 0) \sim f_1(x)$ and $(x, 1) \sim f_2(x)$. In contrast,

  $$\operatorname*{colim}_{\mathcal{C}}(F) \cong \big( F(c_1) \amalg F(c_2) \big) \big/ \sim,$$

  where $f_1(x) \sim f_2(x)$ for each $x \in F(c_0)$. Thus, for example, if $F(c_1)$ and $F(c_2)$ are both points, then $\operatorname{colim}(F)$ is a point, and the structure of $F(c_0)$ is lost. However, $\operatorname{hocolim}(F)$ is the suspension of $F(c_0)$ in this case; and the cohomology (at least) of $F(c_0)$ can be recovered from the cohomology of the homotopy colimit.

- Fix a discrete group $G$, and a space $X$ upon which $G$ acts. We can regard $X$ as a functor from $\mathcal{B}(G)$ to spaces: the functor which sends the unique object to $X$ and which sends a morphism $g \in G$ to the action of $g$ on $X$ (see Definition 2.4). The colimit of this functor is the orbit space $X/G$, while the homotopy colimit of this functor is the Borel construction $EG \times_G X$. Any equivariant map of $G$-spaces $f\colon X \longrightarrow Y$ which is also an (ordinary) homotopy equivalence induces a homotopy equivalence between the Borel constructions, but not in general between the orbit spaces.

- Let $\mathcal{C}$ be any category, and let $*\colon \mathcal{C} \longrightarrow \mathsf{Top}$ be the functor which sends each object to a point. Then $\mathrm{hocolim}_{\mathcal{C}}(*) \cong |\mathcal{C}|$. Thus for arbitrary $F\colon \mathcal{C} \longrightarrow \mathsf{Top}$, the natural morphism of functors from $F$ to $*$ induces a map of spaces from $\mathrm{hocolim}_{\mathcal{C}}(F)$ to $|\mathcal{C}|$ whose fibers (point inverses) all have the form $F(-)$.

There is a spectral sequence which links the cohomology of a homotopy colimit to that of its pieces.

**Proposition 5.28** ([BK, XII.4.5]) *For any small category $\mathcal{C}$, any functor $F\colon \mathcal{C} \longrightarrow \mathsf{Top}$, and any coefficient ring $A$, there is a spectral sequence*

$$E_2^{nk} = \varprojlim_{\mathcal{C}}{}^n\big(H^k(F(-);A)\big) \implies H^{n+k}(\mathrm{hocolim}_{\mathcal{C}}(F);A).$$

*Proof.* This is seen most simply by filtering the homotopy colimit by its "skeleta". For each $n \geq 0$, let $\mathrm{hocolim}^{(n)}(F)$ be the union of the cells $F(c_0) \times \Delta^m$, taken over all $c_0 \to \cdots \to c_m$ and all $m \leq n$. We can identify

$$E_1^{nk} = H^{n+k}\big(\mathrm{hocolim}^{(n)}(F), \mathrm{hocolim}^{(n-1)}(F); A\big)$$

$$\cong \prod_{c_0 \to \cdots \to c_n} H^{n+k}\big(F(c_0) \times \Delta^n, F(c_0) \times \partial\Delta^n; A\big)$$

$$\cong \prod_{c_0 \to \cdots \to c_n} H^k(F(c_0); A) \cong \overline{C}^n\big(H^k(F(-); A)\big),$$

where both products are taken over nondegenerate simplices only. Hence the spectral sequence of this filtration has $E_2$-term

$$E_2^{nk} \cong H^n\big(\overline{C}^*(H^k(F(-); A)), d\big) \cong \varprojlim{}^n\big(H^k(F(-); A)\big),$$

and converges to $H^*(\mathrm{hocolim}(F); A)$. A different argument is given in [BK, XII.4.5]. $\qquad\square$

Note that in the first example (over the pushout category), this spectral sequence reduces to the usual Meyer-Vietoris exact sequence for a union of two spaces. In the second example, it is just the Serre spectral sequence of the fibration $X \longrightarrow EG \times_G X \longrightarrow BG$.

### 5.6. **The subgroup decomposition of $|\mathcal{L}|$.**

Fix a $p$-local finite group $(S, \mathcal{F}, \mathcal{L})$, and let $\mathcal{L} \xrightarrow{\tilde{\pi}} \mathcal{O}(\mathcal{F}^c)$ be the projection. We define a functor

$$\widetilde{B} \colon \mathcal{O}(\mathcal{F}^c) \longrightarrow \mathsf{Top}$$

as follows. For $P \in \mathrm{Ob}(\mathcal{O}(\mathcal{F}^c)) = \mathrm{Ob}(\mathcal{L})$, let $\tilde{\pi} \downarrow P$ be the overcategory whose objects are the pairs $(Q, \alpha)$ for $Q \in \mathrm{Ob}(\mathcal{L})$ and $\alpha \in \mathrm{Rep}_{\mathcal{F}}(Q, P) = \mathrm{Mor}_{\mathcal{O}(\mathcal{F}^c)}(Q, P)$, and where

$$\mathrm{Mor}_{\tilde{\pi} \downarrow P}\big((Q, \alpha), (R, \beta)\big) = \big\{ \psi \in \mathrm{Mor}_{\mathcal{L}}(Q, R) \,\big|\, \beta \circ \tilde{\pi}(\psi) = \alpha \big\}.$$

Define $\widetilde{B}$ on objects by setting

$$\widetilde{B}(P) = \operatorname*{hocolim}_{\tilde{\pi} \downarrow P}(*) = |\tilde{\pi} \downarrow P|.$$

For each morphism $\varphi \in \mathrm{Rep}_{\mathcal{F}}(P_1, P_2)$, set

$$\widetilde{B}(\varphi) = |\tilde{\pi} \downarrow \varphi| \colon |\tilde{\pi} \downarrow P_1| \longrightarrow |\tilde{\pi} \downarrow P_2|,$$

where $\tilde{\pi} \downarrow \varphi$ sends $(Q, \alpha)$ in $\tilde{\pi} \downarrow P_1$ to $(Q, \varphi \circ \alpha)$ in $\tilde{\pi} \downarrow P_2$. Thus $\widetilde{B}$ is the *left homotopy Kan extension* over $\tilde{\pi}$ of the constant functor $* \colon \mathcal{L} \longrightarrow \mathsf{Top}$.

**Proposition 5.29**   *Fix a saturated fusion system $\mathcal{F}$ and an associated centric linking system $\mathcal{L}$. Let $\tilde{\pi} \colon \mathcal{L} \longrightarrow \mathcal{O}(\mathcal{F}^c)$ be the projection functor, and let*

$$\widetilde{B} \colon \mathcal{O}(\mathcal{F}^c) \longrightarrow \mathsf{Top}$$

*be as defined above. Then the following hold:*

(a)   $|\mathcal{L}| \simeq \operatorname{hocolim}_{\mathcal{O}(\mathcal{F}^c)}(\widetilde{B}).$

(b)   *For each $P \in \mathrm{Ob}(\mathcal{L})$, the functor $e_P \colon \mathcal{B}(P) \longrightarrow \tilde{\pi} \downarrow P$, which sends the unique object in $\mathcal{B}(P)$ to $(P, \mathrm{Id})$ and sends $g \in P$ to $\delta_P(g)$, induces a homotopy equivalence $\varepsilon_P \colon BP \longrightarrow \widetilde{B}(P)$. For each $\varphi \in \mathrm{Hom}_{\mathcal{F}}(P, Q)$,*

$$\varepsilon_Q \circ B\varphi \simeq \widetilde{B}([\varphi]) \circ \varepsilon_P,$$

*where $[\varphi] \in \mathrm{Rep}_{\mathcal{F}}(P, Q)$ denotes the class of $\varphi$ modulo $\mathrm{Inn}(Q)$.*

*Proof.* See [BLO2, Proposition 2.2]. Point (a) is a very general fact: if $\Psi \colon \mathcal{C} \longrightarrow \mathcal{D}$ and $F \colon \mathcal{C} \longrightarrow \mathsf{Top}$ are two functors, then $\operatorname{hocolim}_{\mathcal{C}}(F)$ is homotopy equivalent to the homotopy colimit over $\mathcal{D}$ of the left homotopy Kan extension over $\Psi$ of $F$ [HV, Theorem 5.5]. In this case, that reduces to the equivalence

$$\operatorname*{hocolim}_{\mathcal{O}(\mathcal{F}^c)}(\widetilde{B}) \simeq \operatorname*{hocolim}_{\mathcal{L}}(*) = |\mathcal{L}|. \qquad \square$$

To help motivate this property of Kan extensions, we note here as an exercise the analogous (but easier) algebraic result. Let $\Psi\colon \mathcal{C} \longrightarrow \mathcal{D}$ and $T\colon \mathcal{C} \longrightarrow \mathrm{Ab}$ be a pair of functors. Let $\widehat{T}$ be the left Kan extension over $\Psi$ of $T$: $\widehat{T}(d) = \mathrm{colim}_{\Psi\downarrow d}(T)$ for $d \in \mathrm{Ob}(\mathcal{D})$, with induced morphisms defined in the obvious way. Then it is not hard to see that

$$\operatorname*{colim}_{\mathcal{D}}(\widehat{T}) \cong \operatorname*{colim}_{\mathcal{C}}(T).$$

Point (b) in Proposition 5.29 is motivated by regarding $\mathcal{L}$ as an extension of the form

$$1 \longrightarrow \{P\}_{P\in\mathrm{Ob}(\mathcal{F}^c)} \longrightarrow \mathcal{L} \xrightarrow{\ \widetilde{\pi}\ } \mathcal{O}(\mathcal{F}^c) \longrightarrow 1.$$

The actual proof is by straightforward manupulation of categories.

The decomposition of Proposition 5.29 is known as the "subgroup decomposition" of $|\mathcal{L}|$, because it identifies this space with a homotopy colimit of spaces having the homotopy type of classifying spaces of $p$-subgroups of $S$. We will see in Section 5.7 the important role played by this decomposition when studying certain mapping spaces involving $|\mathcal{L}|_p^{\wedge}$.

Now fix a finite group $G$ and $S \in \mathrm{Syl}_p(G)$. If we apply the above procedure to the projection functor $\widetilde{\pi}\colon \mathcal{T}_S(G) \longrightarrow \mathcal{O}_S(G)$, we obtain a functor $\widetilde{B}\colon \mathcal{O}_S(G) \longrightarrow \mathrm{Top}$, where $\widetilde{B}(P) = |\widetilde{\pi}\downarrow P|$ for each $P \le S$, and $\widetilde{B}(P) \simeq BP$. Hence by [HV, Theorem 5.5] again, for each set $\mathcal{H}$ of subgroups of $S$,

$$|\mathcal{T}_{\mathcal{H}}(G)| = \operatorname*{hocolim}_{\mathcal{T}_{\mathcal{H}}(G)}(*) \simeq \operatorname*{hocolim}_{\mathcal{O}_{\mathcal{H}}(G)}(\widetilde{B}|_{\mathcal{H}}).$$

Using this, we prove the following lemma, which is essentially due to Dwyer [Dw, Theorem 8.3], and which was used in Section 3.1 when showing that $|\mathcal{L}_S^c(G)|_p^{\wedge} \simeq |\mathcal{T}_S(G)|_p^{\wedge}$ (and hence that $|\mathcal{L}_S^c(G)|_p^{\wedge} \simeq BG_p^{\wedge}$).

**Lemma 5.30** *Fix a finite group $G$ and $S \in \mathrm{Syl}_p(G)$. Then the inclusion of $|\mathcal{T}_S^c(G)|$ into $|\mathcal{T}_S(G)|$ induces an isomorphism of mod $p$ cohomology.*

*Proof.* We just saw that

$$|\mathcal{T}_S(G)| \simeq \operatorname*{hocolim}_{\mathcal{O}_S(G)}(\widetilde{B}) \quad \text{and} \quad |\mathcal{T}_S^c(G)| \simeq \operatorname*{hocolim}_{\mathcal{O}_S^c(G)}(\widetilde{B}^c),$$

where $\widetilde{B}^c$ is the restriction of $\widetilde{B}$ to $\mathcal{O}_S^c(G)$. By Proposition 5.28, there is a spectral sequence of the form

$$E_2^{ij} = \operatorname*{\underleftarrow{\lim}}_{\mathcal{O}_S(G)}{}^{i}(\Phi^j) \Longrightarrow H^*(|\mathcal{T}_S(G)|, |\mathcal{T}_S^c(G)|; A)$$

where for each $P \le S$,

$$\Phi^j(P) = H^j(\widetilde{B}(P), \widetilde{B}^c(P); A) \cong \begin{cases} 0 & \text{if } P \text{ is } p\text{-centric in } G \\ H^j(BP; A) & \text{otherwise.} \end{cases}$$

When $P$ is not $p$-centric, $PC_G(P)/P$ acts trivially on $H^j(BP; A)$ under the action of $N_G(P)/P$, and $p\,\big|\,|PC_G(P)/P|$ since $Z(P) \notin \mathrm{Syl}_p(C_G(P))$.

Thus by Proposition 5.24(b), $\Lambda^*(N_G(P)/P; \Phi^j(P)) = 0$ for each $P \leq S$ and each $j$. Hence by Corollary 5.21(a), $\varprojlim^i(\Phi^j) = 0$ for each $i$ and $j$, and so $H^*(|\mathcal{T}_S(G)|, |\mathcal{T}_S^c(G)|; A) = 0$ by the above spectral sequence. $\qquad\square$

The functor $\widetilde{B}$ defined in Proposition 5.29 is an example of a *rigidification* of a homotopy functor. To make precise what this means, let $\mathsf{hoTop}$ be the category whose objects are topological spaces and whose morphisms are homotopy classes of continuous maps (the "homotopy category"), and let $h\colon \mathsf{Top} \longrightarrow \mathsf{hoTop}$ be the quotient functor. Thus $h$ is the identity on objects, and sends a map to its homotopy class. A *homotopy functor* is a functor from any small category $\mathcal{C}$ to $\mathsf{hoTop}$. For any homotopy functor $F\colon \mathcal{C} \longrightarrow \mathsf{hoTop}$, a *rigidification* of $F$ is a functor $\widetilde{F}\colon \mathcal{C} \longrightarrow \mathsf{Top}$, together with a natural isomorphism of functors from $h \circ \widetilde{F}$ to $F$.

In general, the homotopy colimit of a homotopy functor is *not* defined, not even up to homotopy equivalence. Hence one must first replace it by a rigidification. The rigidification of homotopy functors was studied by Dwyer and Kan [DK1, DK2], who developed an obstruction theory for the existence and uniqueness of rigidifications.

When $\mathcal{F}$ is a saturated fusion system over a $p$-group $S$, there is a natural homotopy functor $B\colon \mathcal{O}(\mathcal{F}^c) \longrightarrow \mathsf{hoTop}$ which sends $P \leq S$ to $BP$ and sends the class of $\varphi \in \mathrm{Mor}(\mathcal{F})$ to $B\varphi$. This is not a functor to $\mathsf{Top}$, since inner automorphisms of $P$ do not induce the identity on $BP$. Proposition 5.29 says that any linking system $\mathcal{L}$ associated to $\mathcal{F}$ determines a rigidification of $B$, and also that the homotopy colimit of that realization has the homotopy type of $|\mathcal{L}|$. The next proposition says that this in fact defines a bijective correspondence between linking systems associated to $\mathcal{F}$ and rigidifications of $B$ up to homotopy.

The obstructions of Dwyer and Kan to the existence and uniqueness of rigidifications of $B\colon \mathcal{O}(\mathcal{F}^c) \longrightarrow \mathsf{hoTop}$, in the form defined in [DK2], are in fact exactly the same groups as the obstructions of Proposition 5.11 to the existence and uniqueness of centric linking systems associated to $\mathcal{F}$. This was the observation which first motivated us to prove the next proposition. We also note that it was in [DK2] that Dwyer and Kan introduced the word "centric", but as a condition on a continuous map rather than on a subgroup.

**Proposition 5.31**  *Fix a saturated fusion system $\mathcal{F}$ over a p-group $S$. Then there are mutually inverse bijections*

$$
\left\{
\begin{array}{c}
\text{centric linking systems} \\
\text{associated to } \mathcal{F} \\
\text{up to isomorphism}
\end{array}
\right\}
\underset{\mathrm{LS}}{\overset{\mathrm{KE}}{\rightleftarrows}}
\left\{
\begin{array}{c}
\text{rigidifications } \mathcal{O}(\mathcal{F}^c) \to \mathsf{Top} \\
\text{of } (P \mapsto BP) \text{ up to} \\
\text{natural homotopy equiv.}
\end{array}
\right\}
$$

*where* KE *sends the class of a centric linking system $\mathcal{L}$ to the class of the rigidification*

$$
\mathbf{ke}(\mathcal{L})\colon \mathcal{O}(\mathcal{F}^c) \longrightarrow \mathsf{Top},
$$

*defined to be the left homotopy Kan extension of the constant functor $\mathcal{L} \overset{*}{\longrightarrow} \mathsf{Top}$ along the projection $\widetilde{\pi}\colon \mathcal{L} \longrightarrow \mathcal{O}(\mathcal{F}^c)$. Furthermore, for each linking system $\mathcal{L}$,*

$$
|\mathcal{L}| \simeq \underset{\mathcal{O}(\mathcal{F}^c)}{\mathrm{hocolim}}(\mathbf{ke}(\mathcal{L})).
$$

*Proof.* A weaker version of this is stated in [BLO2, Proposition 2.3]. The full proposition (in a more general context) is shown in [BLO3, Proposition 4.6]. Part of it, of course, follows from Proposition 5.29.

One way to construct the linking system $\mathcal{L} = \mathbf{ls}(\widetilde{B})$ associated to a rigidification $\widetilde{B}$ is as the linking category of the space $\mathrm{hocolim}_{\mathcal{O}(\mathcal{F}^c)}(\widetilde{B})$ in the sense of Definition 3.5 (taken with respect to $S$ and the obvious inclusion of $BS$). The difficulty when doing it this way is to show that the linking category of the homotopy colimit actually is a linking system associated to $\mathcal{F}$.

We sketch here another, more direct construction of $\mathbf{ls}(\widetilde{B})$. It is carried out in detail in the proof of [BLO3, Proposition 4.6]. Fix $\widetilde{B}\colon \mathcal{O}(\mathcal{F}^c) \to \mathsf{Top}$. We are given, as part of the rigidification data, homotopy equivalences $BP \xrightarrow{\varepsilon_P} \widetilde{B}(P)$, for all $P \in \mathrm{Ob}(\mathcal{F}^c)$, which make $\{[\varepsilon_P]\}_{P \in \mathrm{Ob}(\mathcal{F}^c)}$ into a natural isomorphism from $B$ to $h \circ \widetilde{B}$. For each $P \in \mathrm{Ob}(\mathcal{F}^c)$, let $*_P \in \widetilde{B}(P)$ be the image under $\varepsilon_P$ of the base point of $BP$, and let

$$
\gamma_P\colon P \xrightarrow{\ \cong\ } \pi_1(\widetilde{B}(P), *_P)
$$

be the isomorphism induced by $\varepsilon_P$ on fundamental groups.

The category $\mathcal{L} = \mathbf{ls}(\widetilde{B})$ is defined by setting $\mathrm{Ob}(\mathcal{L}) = \mathrm{Ob}(\mathcal{F}^c)$, and

$$
\mathrm{Mor}_{\mathcal{L}}(P, Q) = \big\{ (\varphi, u) \,\big|\, \varphi \in \mathrm{Rep}_{\mathcal{F}}(P, Q),\ u \in \pi_1(\widetilde{B}(Q); \widetilde{B}\varphi(*_P), *_Q) \big\}.
$$

Here, $\pi_1(X; x_0, x_1)$ means the set of homotopy classes of paths in $X$ from $x_0$ to $x_1$. Composition in $\mathcal{L}$ is defined by setting

$$
(\psi, v) \circ (\varphi, u) = (\psi\varphi, v \cdot (\widetilde{B}\psi \circ u)),
$$

where paths are composed from right to left. Also, $\pi\colon \mathcal{L} \longrightarrow \mathcal{F}^c$ is the functor which is the identity on objects, and which sends $(\varphi, u) \in \mathrm{Mor}_{\mathcal{L}}(P, Q)$

to the composite

$$P \xrightarrow[\cong]{\gamma_P} \pi_1(\widetilde{B}(P), *_P) \xrightarrow{\widetilde{B}\varphi_\#} \pi_1(\widetilde{B}(Q), \widetilde{B}\varphi(*_P))$$

$$\xrightarrow{c_u} \pi_1(\widetilde{B}(Q), *_Q) \xrightarrow[\cong]{\gamma_Q^{-1}} Q.$$

It remains only to check that $\mathcal{L}$ and $\pi$, together with a functor $\delta$ constructed in an appropriate way, form a centric linking system associated to $\mathcal{F}$. $\qquad\square$

Recall that by Theorem 4.25, a $p$-local finite group $(S, \mathcal{F}, \mathcal{L})$ is determined up to isomorphism by the homotopy type of its classifying space $|\mathcal{L}|_p^\wedge$. So Proposition 5.31 also says that there is a bijective correspondence between classifying spaces of $\mathcal{F}$ (up to homotopy type) and rigidifications of the homotopy functor $B$, where a rigidification corresponds to its homotopy colimit.

## 5.7. An outline of the proofs of Theorems 4.21 and 4.22.

To illustrate the role played by the subgroup decomposition, we now sketch the proofs of two of the results stated earlier about maps to the space $|\mathcal{L}|_p^\wedge$, with emphasis on those parts which most directly involve the decomposition. To be carried out completely, both proofs require more information on the actual components of certain mapping spaces; but we skip over most of those details in this presentation. In fact, both of these theorems in their original form in [BLO2] contain results about the homotopy types of the full mapping spaces — the space of all maps from $BQ$ to $|\mathcal{L}|_p^\wedge$ and the space of self homotopy equivalences of $|\mathcal{L}|_p^\wedge$ — but we restrict attention here as far as possible to the set (or group) of homotopy classes of maps.

Theorem 4.21 says that for any $p$-group $Q$ and any $p$-local finite group $(S, \mathcal{F}, \mathcal{L})$, the natural map from $\mathrm{Rep}(Q, \mathcal{F})$ to $[BQ, |\mathcal{L}|_p^\wedge]$ is a bijection. The starting point for proving this is the following proposition, due originally to Broto and Kitchloo. We say that a category $\mathcal{C}$ has *bounded limits at* $p$ if there is some $N$ such that for any functor $\Phi\colon \mathcal{C}^{\mathrm{op}} \longrightarrow \mathbb{Z}_{(p)}\text{-mod}$, $\varprojlim^i(\Phi) = 0$ for all $i > N$.

**Proposition 5.32** ([BLO2, Proposition 4.2]) *Fix a prime $p$ and a $p$-group $Q$. Let $\mathcal{C}$ be a finite category with bounded limits at $p$, and let $F\colon \mathcal{C} \longrightarrow \mathrm{Top}$ be a functor such that for each $c \in \mathrm{Ob}(\mathcal{C})$ and each $Q_0 \leq Q$, $\mathrm{map}(BQ_0, F(c))$ is $p$-complete and has finite mod $p$ cohomology in each degree. Then the natural map*

$$\Big(\operatorname*{hocolim}_{\mathcal{C}}(\mathrm{map}(BQ, F(-)))\Big)_p^\wedge \longrightarrow \mathrm{map}\Big(BQ, \big(\operatorname*{hocolim}_{\mathcal{C}}(F)\big)_p^\wedge\Big)$$

*is a homotopy equivalence.*

When $Q$ is elementary abelian, Proposition 5.32 follows as a consequence of the description by Lannes [La] of mapping spaces $\mathrm{map}(BQ, -)$ and their cohomology. The general case then follows by induction on $|Q|$.

By Proposition 5.25, the orbit category $\mathcal{O}(\mathcal{F}^c)$ of a saturated fusion system over a $p$-group $S$ has bounded limits at $p$: if $|S| = p^k$, then $\varprojlim^i(F) = 0$ for every $i > k$ and every contravariant functor $F$ from $\mathcal{O}(\mathcal{F}^c)$ to $\mathbb{Z}_{(p)}$-modules. When $Q$ and $P$ are two $p$-groups, $[BQ, BP] \cong \mathrm{Rep}(Q, P)$ by Corollary 1.6. The component of the mapping space containing $B\rho$ (for $\rho \in \mathrm{Hom}(Q, P)$) is homotopy equivalent to $BC_P(\rho(Q))$ by Proposition 2.10(b), and hence $\mathrm{map}(BQ, BP)$ is $p$-complete (Proposition 1.10) and has finite mod $p$ cohomology in each degree. So by Proposition 5.32, for any $p$-group $Q$ and any $p$-local finite group $(S, \mathcal{F}, \mathcal{L})$, the space of maps from $BQ$ to $|\mathcal{L}|_p^\wedge$ has the homotopy type of

$$\left( \operatornamewithlimits{hocolim}_{\mathcal{O}(\mathcal{F}^c)} \left( \mathrm{map}(BQ, \widetilde{B}(-)) \right) \right)_p^\wedge,$$

where $\widetilde{B}\colon \mathcal{O}(\mathcal{F}^c) \longrightarrow \mathbf{Top}$ is as in Proposition 5.29. Using the above description of the homotopy type of $\mathrm{map}(BQ, \widetilde{B}(P)) \simeq \mathrm{map}(BQ, BP)$, and some straightforward manipulations of the homotopy colimit, we are led to the following result:

**Proposition 5.33** ([BLO2, Proposition 4.3])  *Fix a $p$-local finite group $(S, \mathcal{F}, \mathcal{L})$ and a $p$-group $Q$. Let $\mathcal{L}_Q$ be the category whose objects are the pairs $(P, \alpha)$ for $P \in \mathrm{Ob}(\mathcal{L})$ and $\alpha \in \mathrm{Hom}(Q, P)$, and where*

$$\mathrm{Mor}_{\mathcal{L}_Q}\big((P, \alpha), (R, \beta)\big) = \big\{ \varphi \in \mathrm{Mor}_{\mathcal{L}}(P, R) \,\big|\, \beta = \pi(\varphi) \circ \alpha \in \mathrm{Hom}(Q, R) \big\}.$$

*Let $\Phi\colon \mathcal{L}_Q \times \mathcal{B}(Q) \longrightarrow \mathcal{L}$ be the functor defined by setting*

$$\Phi\big((P, \alpha), o_Q\big) = P \qquad \text{and} \qquad \Phi\big((P, \alpha) \xrightarrow{\varphi} (R, \beta)\,, \, x\big) = \varphi \circ \delta_P(\alpha(x)).$$

*Then the map*

$$|\Phi|'\colon |\mathcal{L}_Q|_p^\wedge \longrightarrow \mathrm{map}(BQ, |\mathcal{L}|_p^\wedge)$$

*adjoint to $|\Phi|$ is a homotopy equivalence.*

Thus the set $[BQ, |\mathcal{L}|_p^\wedge]$ is in bijective correspondence with the set of connected components of $|\mathcal{L}_Q|_p^\wedge$ and hence of $|\mathcal{L}_Q|$, and this is in bijective correspondence with $\mathrm{Rep}(Q, \mathcal{F})$. This finishes the proof of Theorem 4.21.

The proof of Theorem 4.22 — the description of $\mathrm{Out}(|\mathcal{L}|_p^\wedge)$ in terms of automorphisms of the category $\mathcal{L}$ — is based on a very different use of the subgroup decomposition of $|\mathcal{L}|$. One wants to understand maps from $|\mathcal{L}|_p^\wedge$ to itself using the subgroup decomposition of $|\mathcal{L}|$, and this is a special case of the following theorem of Wojtkowiak.

**Proposition 5.34** ([Wo])  *Fix a small category $\mathcal{C}$, a functor $F\colon \mathcal{C} \longrightarrow \mathrm{Top}$, and a space $Y$. Consider the map*

$$R\colon \big[\mathop{\mathrm{hocolim}}_{\mathcal{C}}(F), Y\big] \longrightarrow \varprojlim_{\mathcal{C}}([F(-), Y])$$

*which sends the homotopy class $[f]$ of a map $f$ to the family $\{[f|_{F(c)}]\}_{c \in \mathrm{Ob}(\mathcal{C})}$ of homotopy classes of its restrictions. Fix maps $f_c\colon F(c) \longrightarrow Y$ for each object $c$, such that $\mathbf{f} \overset{\mathrm{def}}{=} \{[f_c]\}_{c \in \mathrm{Ob}(\mathcal{C})}$ lies in $\varprojlim([F(-), Y])$. Then the obstructions to the existence of an element in $R^{-1}(\mathbf{f})$ lie in the groups $\varprojlim^{i+1}(\pi_i(\mathrm{map}(F(-), Y), f_c))$ for $i \geq 1$, and the obstructions to the uniqueness of such an element lie in $\varprojlim^{i}(\pi_i(\mathrm{map}(F(-), Y), f_c))$ for $i \geq 1$.*

More precisely, there is a sequence of obstructions to the existence of an element in $R^{-1}(\{[f_c]\})$, where one obstruction is defined only if the preceeding one vanishes. So it is more precise (but also more restrictive) to say that if all of the groups $\varprojlim^{i+1}(\pi_i(\mathrm{map}(F(-), Y), f_c))$ vanish, then there is a map $f$ such that $f|_{F(c)} \simeq f_c$ for each $c \in \mathrm{Ob}(\mathcal{C})$ (and similarly for uniqueness). This is what will be used here.

Consider the following diagram:

$$
\begin{array}{ccccccccc}
1 & \longrightarrow & \varprojlim^1(\mathcal{Z}) & \overset{\lambda_{\mathcal{L}}}{\longrightarrow} & \mathrm{Out}_{\mathrm{typ}}(\mathcal{L}) & \overset{\mu_{\mathcal{L}}}{\longrightarrow} & \mathrm{Out}(S, \mathcal{F}) & \overset{\omega_{\mathcal{L}}}{\longrightarrow} & \varprojlim^2(\mathcal{Z}) \\
 & & \| & & {\scriptstyle |-|_p^{\wedge}}\downarrow & & {\scriptstyle \widehat{r}}\downarrow\,{\scriptstyle \cong} & & \| \\
1 & \longrightarrow & \varprojlim^1(\mathcal{Z}) & \overset{\widehat{\lambda}}{\longrightarrow} & \mathrm{Out}(|\mathcal{L}|_p^{\wedge}) & \overset{\widehat{\mu}}{\longrightarrow} & \varprojlim \mathrm{IRep}(-, \mathcal{F}) & \overset{\widehat{\omega}}{\longrightarrow} & \varprojlim^2(\mathcal{Z}),
\end{array}
\tag{8}
$$

where all limits are taken over $\mathcal{O}(\mathcal{F}^c)$. Here, $\mathcal{Z} = \mathcal{Z}_{\mathcal{F}}$, as defined in Section 5.3. The top row is defined and exact by Proposition 5.12. To prove that $|-|_p^{\wedge}$ is an isomorphism, it suffices to construct the rest of the diagram and prove that it commutes, show the bottom row is exact, and that $\widehat{r}$ is an isomorphism.

Let $\mathrm{IRep}(P, \mathcal{F}) \subseteq \mathrm{Rep}(P, \mathcal{F})$ (for $P \leq S$) denote the set of $\mathcal{F}$-conjugacy classes of injective homomorphisms from $P$ to $S$. Thus the set $\mathrm{IRep}(S, \mathcal{F}) = \mathrm{Aut}(S)/\mathrm{Aut}_{\mathcal{F}}(S)$ contains $\mathrm{Out}(S, \mathcal{F}) = \mathrm{Aut}(S, \mathcal{F})/\mathrm{Aut}_{\mathcal{F}}(S)$ as a subgroup. We can also consider $\varprojlim \mathrm{IRep}(-, \mathcal{F})$ as a subset of $\mathrm{IRep}(S, \mathcal{F})$, and this is equal to $\mathrm{Out}(S, \mathcal{F})$ by definition of fusion preserving. In other words, there is a bijection $\widehat{r}$ from $\mathrm{Out}(S, \mathcal{F})$ to $\varprojlim \mathrm{IRep}(-, \mathcal{F})$, defined by setting $\widehat{r}([\alpha]) = \{[\alpha|_P]\}_{P \in \mathrm{Ob}(\mathcal{F}^c)}$ for $\alpha \in \mathrm{Aut}(S, \mathcal{F})$.

Let $\widehat{\mu}$ be the homomorphism defined by restriction:

$$\mathrm{Out}(|\mathcal{L}|_p^{\wedge}) \subseteq [|\mathcal{L}|_p^{\wedge}, |\mathcal{L}|_p^{\wedge}] \longrightarrow \varprojlim_{\mathcal{O}(\mathcal{F}^c)} [B(-), |\mathcal{L}|_p^{\wedge}] \cong \varprojlim_{\mathcal{O}(\mathcal{F}^c)} \mathrm{Rep}(-, \mathcal{F}).$$

Each self equivalence of $|\mathcal{L}|_p^{\wedge}$ restricts to an automorphism of $S$ (well defined up to $\mathcal{F}$-conjugacy), and hence $\mathrm{Im}(\widehat{\mu})$ is contained in $\varprojlim \mathrm{IRep}(-, \mathcal{F})$. This is exactly the type of problem — describing the point inverses of this map — which Wojtkowiak's theorem is designed to solve.

Let $\mathrm{map}(BP, |\mathcal{L}|_p^\wedge)_{\mathrm{incl}}$ be the space of maps homotopic to the inclusion. By [BLO2, Theorem 4.4(c)], when $P \in \mathcal{F}^c$, $\mathrm{map}(BP, |\mathcal{L}|_p^\wedge)_{\mathrm{incl}} \simeq BZ(P)$; and thus its only nonvanishing homotopy group is $\pi_1\big(\mathrm{map}(BP, |\mathcal{L}|_p^\wedge)_{\mathrm{incl}}\big) \cong Z(P)$. Furthermore, as a functor on $\mathcal{O}(\mathcal{F}^c)$, $\pi_1(\mathrm{map}(B(-), |\mathcal{L}|_p^\wedge)) \cong \mathcal{Z}_\mathcal{F}$. The only obstruction to lifting an element in $\varprojlim \mathrm{IRep}(-, \mathcal{F})$ to $\mathrm{Out}(|\mathcal{L}|_p^\wedge)$ thus lies in $\varprojlim^2(\mathcal{Z}_\mathcal{F})$, and the only obstruction to the uniqueness of such a lifting lies in $\varprojlim^1(\mathcal{Z}_\mathcal{F})$. The definition and exactness of the bottom row of (8) are thus special cases of Proposition 5.34.

The second square in (8) clearly commutes. The proofs that the other two squares commute are based on a closer examination of the obstruction maps, and we refer to Step 2 in the proof of [BLO3, Theorem 7.1] for details.

### 5.8. The centralizer and normalizer decompositions of $|\mathcal{L}|$.

In [Dw], Dwyer studied homotopy decompositions of $BG$ and $BG_p^\wedge$, when $G$ is a finite group, and described three general families of such decompositions. A *subgroup decomposition* identifies $BG_p^\wedge$ as the homotopy colimit of a functor defined on some full subcategory of $\mathcal{O}(G)$, which sends each object $P$ to a space with the homotopy type of $BP$. A *centralizer decomposition* describes $BG_p^\wedge$ as the homotopy colimit of a functor on a category of $p$-subgroups of $G$, which sends each object $P \leq G$ to a space with the homotopy type of $BC_G(P)_p^\wedge$. A *normalizer decomposition* of $BG_p^\wedge$ is the homotopy colimit of a functor defined on a certain category of conjugacy classes of chains of subgroups, and sends each chain to a space with the homotopy type of the classifying space of a certain automorphism group of the chain.

Each of these families of decompositions can be realized as a decomposition of the classifying space of a $p$-local finite group $(S, \mathcal{F}, \mathcal{L})$. In fact, in some cases, this gives us a decomposition of $|\mathcal{L}|$ without $p$-completion. The subgroup decomposition of $\mathcal{L}$, and some of its applications, was discussed in Sections 5.6 and 5.7.

We now describe centralizer and normalizer decompositions of $|\mathcal{L}|$, beginning with the centralizer decomposition. For any $p$-local finite group $(S, \mathcal{F}, \mathcal{L})$, let $\mathcal{F}^e$ denote the full subcategory of $\mathcal{F}$ whose objects are those nontrivial elementary abelian $p$-subgroups of $S$ which are fully centralized in $\mathcal{F}$. For each $E \in \mathrm{Ob}(\mathcal{F}^e)$, let $\overline{C}_\mathcal{L}(E)$ be the category whose objects are the pairs $(P, \alpha)$ for $P$ in $\mathcal{L}$ and $\alpha \in \mathrm{Hom}_\mathcal{F}(E, Z(P))$, and where

$$\mathrm{Mor}_{\overline{C}_\mathcal{L}(E)}((P, \alpha), (Q, \beta)) = \{\varphi \in \mathrm{Mor}_\mathcal{L}(P, Q) \,|\, \pi(\varphi) \circ \alpha = \beta\}.$$

The following decomposition of $|\mathcal{L}|$ played an important role in the proof in [BLO2] of Theorem 4.23 (the computation of $H^*(|\mathcal{L}|; \mathbb{F}_p)$).

**Theorem 5.35**  *For any p-local finite group* $(S, \mathcal{F}, \mathcal{L})$, *the natural map*

$$\operatorname*{hocolim}_{E \in (\mathcal{F}^e)^{op}} |\overline{C}_{\mathcal{L}}(E)| \longrightarrow |\mathcal{L}|,$$

*induced by the forgetful functors* $(P, \alpha) \mapsto P$, *is a homotopy equivalence. Also, for each* $E \in \operatorname{Ob}(\mathcal{F}^e)$, *the functor* $P \mapsto (P, \operatorname{incl})$ *induces a homotopy equivalence* $|C_{\mathcal{L}}(E)| \longrightarrow |\overline{C}_{\mathcal{L}}(E)|$.

Normalizer decompositions of $|\mathcal{L}|$ have been constructed by Assaf Libman [Lb]. We describe here just one version of his decomposition: that based on $\mathcal{F}$-centric subgroups. Fix a $p$-local finite group $(S, \mathcal{F}, \mathcal{L})$. Let $\operatorname{sd}(\mathcal{F}^c)$ be the poset of all chains $P_0 < P_1 < \cdots < P_n$ of elements of $\mathcal{F}^c$, ordered by inclusions of subchains. Let $\overline{\operatorname{sd}}(\mathcal{F}^c)$ be the quotient poset of $\mathcal{F}$-conjugacy classes of such chains. We regard $\overline{\operatorname{sd}}(\mathcal{F}^c)$ as a category in the usual way: there is a unique morphism from one conjugacy class of chains to another if some representative of the first class is a subchain of a representative of the second.

For each $\mathbf{P} = (P_0 < \cdots < P_n) \in \operatorname{sd}(\mathcal{F}^c)$, let $\operatorname{Aut}_{\mathcal{L}}(\mathbf{P}) \leq \prod_{i=0}^n \operatorname{Aut}_{\mathcal{L}}(P_i)$ be the subgroup of all $n$-tuples $(\alpha_0, \ldots, \alpha_n)$ such that for each $1 \leq i \leq n$, $\alpha_i(P_{i-1}) = P_{i-1}$ and $\alpha_i|_{P_{i-1}, P_{i-1}} = \alpha_{i-1}$. This is identified with a subgroup of $\operatorname{Aut}_{\mathcal{L}}(P_0)$ in the obvious way (using the uniqueness of extensions of morphisms in $\mathcal{L}$), and thus $\mathcal{B}(\operatorname{Aut}_{\mathcal{L}}(\mathbf{P}))$ is identified as a subcategory of $\mathcal{L}$.

**Theorem 5.36** ([Lb, Theorem A])  *Fix a p-local finite group* $(S, \mathcal{F}, \mathcal{L})$, *and let* $\overline{\operatorname{sd}}(\mathcal{F}^c)$ *be defined as above. Then there is a functor* $\delta \colon \overline{\operatorname{sd}}(\mathcal{F}^c) \longrightarrow \operatorname{Top}$ *such that the following hold.*

(a)  *There is a homotopy equivalence*

$$\operatorname*{hocolim}_{\overline{\operatorname{sd}}(\mathcal{F}^c)}(\delta) \xrightarrow{\ \simeq\ } |\mathcal{L}|.$$

(b)  *For each* $\mathbf{P} \in \operatorname{sd}(\mathcal{F}^c)$, *there is a natural homotopy equivalence*

$$B\operatorname{Aut}_{\mathcal{L}}(\mathbf{P}) \xrightarrow{\ \simeq\ } \delta([\mathbf{P}]).$$

(c)  *For each* $\mathbf{P} \in \operatorname{sd}(\mathcal{F}^c)$, *the map from* $B\operatorname{Aut}_{\mathcal{L}}(\mathbf{P})$ *to* $|\mathcal{L}|$ *induced by the equivalences in (a) and (b) is that induced by the inclusion of categories from* $\mathcal{B}(\operatorname{Aut}_{\mathcal{L}}(\mathbf{P}))$ *to* $\mathcal{L}$.

This decomposition has been applied by Libman and Viruel [LV], to give precise descriptions of the homotopy type of $|\mathcal{L}|$ in many cases.

## 6. EXAMPLES OF EXOTIC FUSION SYSTEMS

We say that a saturated fusion system is *realizable* if it is the fusion system of a finite group, and is *exotic* otherwise. There are several reasons for looking for exotic saturated fusion systems. From the homotopy theory point of view, their classifying spaces provide new spaces which are not $p$-completed classifying spaces of finite groups (Theorem 4.25), but which have many of the very nice homotopy theoretic properties of such spaces. From the group theory point of view, some exotic fusion systems have properties which are similar to those of certain sporadic simple groups. Also, one can hope that the search for exotic fusion systems, at least at the prime 2, could give some more insight into the proof of the classification theorem for finite simple groups. More generally, we look for exotic fusion system simply because we want to understand better how they arise, and how frequently.

We emphasise that when we say that a saturated fusion system over a $p$-group $S$ is "exotic", we mean it is not isomorphic to $\mathcal{F}_S(G)$ for any *finite* group $G$ which contains $S$ *as a Sylow $p$-subgroup*. Ian Leary and Radu Stancu [LS] and Geoff Robinson [Ro3] have given two very different constructions which show that every saturated fusion system over a $p$-group $S$ is at least realized as the fusion system of some *infinite* group $G$ which contains $S$ as Sylow subgroup, in the sense that every finite $p$-subgroup of $G$ is contained in a subgroup conjugate to $S$. Also, Sejong Park [Pa2] has shown that every saturated fusion system over a $p$-group $S$ is isomorphic to $\mathcal{F}_S(G)$ for some finite $G$ which contains $S$ — but not necessarily as Sylow subgroup.

In almost all cases, fusion systems are shown to be exotic using the classification of finite simple groups, together with the following lemma. As usual, a finite group $G$ is *almost simple* if for some nonabelian simple group $H$, $G$ is isomorphic to a subgroup of $\mathrm{Aut}(H)$ which contains $\mathrm{Inn}(H) \cong H$. Equivalently, $G$ contains a nonabelian simple subgroup $H \trianglelefteq G$ such that $C_G(H) = 1$.

**Lemma 6.1** ([DRV])  *Let $\mathcal{F}$ be a saturated fusion system over a $p$-group $S$. Assume, for each strongly closed subgroup $1 \neq P \trianglelefteq S$ in $\mathcal{F}$, that $P \in \mathcal{F}^c$, $P$ is not elementary abelian, and does not factor as a product of two or more subgroups which are permuted transitively by $\mathrm{Aut}_{\mathcal{F}}(P)$. Then if $\mathcal{F}$ is realizable, it is the fusion system of a finite almost simple group.*

*Proof.* This is shown in [DRV, Proposition 2.19], and we outline their argument here. Assume $\mathcal{F}$ is realized by $G$, where $|G|$ is the smallest possible. Then $O_{p'}(G) = 1$, since $G/O_{p'}(G)$ also realizes $\mathcal{F}$. Let $1 \neq H \trianglelefteq G$ be a minimal normal subgroup, and set $P = S \cap H$. By Lemma A.5, either $H$

is elementary abelian, or it is a product of simple groups which are permuted transitively under conjugation in $G$. Since $P$ is strongly closed in $\mathcal{F}$, the hypotheses of the lemma imply $H$ must be nonabelian and simple. Since $C_G(H) \trianglelefteq G$, this implies that $C_G(H) \cap H = 1$. Also, $C_G(H)$ has order prime to $p$ since $P \in \mathrm{Syl}_p(H)$ is $\mathcal{F}$-centric, and $C_G(H) = 1$ since $O_{p'}(G) = 1$. Thus $G$ is almost simple. $\qquad\square$

Throughout the rest of this section, we attempt to at least list all known examples of exotic fusion systems, giving more discussion in cases where it seems appropriate.

### 6.1. Reduced fusion systems and tame fusion systems.

When searching systematically for exotic fusion systems, it is convenient to have a way of restricting the class of saturated fusion systems one has to examine, without risking to overlook some exotic cases. Also, one would like to focus on exotic fusion systems which are minimal in some sense, without including those which can easily be derived from smaller exotic fusion systems. The class of simple fusion systems doesn't seem appropriate for doing this, since saturated fusion systems in general do not seem to be built up via extensions of simple fusion systems.

This is the question addressed in [AOV], where we settled on the following definition of "reduced fusion systems". Recall the definition of a normal subgroup in a fusion system (Definition I.4.1), and those of fusion subsystems of $p$-power index or of index prime to $p$ (Definition I.7.3).

**Definition 6.2** ([AOV])　A *reduced fusion system* is a saturated fusion system $\mathcal{F}$ such that

- $\mathcal{F}$ has no nontrivial normal $p$-subgroups,

- $\mathcal{F}$ has no proper subsystem of $p$-power index, and

- $\mathcal{F}$ has no proper subsystem of index prime to $p$.

For any fusion system $\mathcal{F}$, the *reduction* of $\mathcal{F}$ is the fusion system $\mathfrak{red}(\mathcal{F})$ obtained by first setting $\mathcal{F}_0 = C_{\mathcal{F}}(O_p(\mathcal{F}))/Z(O_p(\mathcal{F}))$ (see Definition I.5.3 for the definition of the centralizer fusion system), and then alternately taking $O^p(-)$ and $O^{p'}(-)$. More explicitly, let

$$\mathcal{F}_0 \supseteq \mathcal{F}_1 \supseteq \mathcal{F}_2 \supseteq \cdots \supseteq \mathcal{F}_m$$

be such that $\mathcal{F}_i = O^p(\mathcal{F}_{i-1})$ if $i$ is odd, $\mathcal{F}_i = O^{p'}(\mathcal{F}_{i-1})$ if $i$ is even, and $O^p(\mathcal{F}_m) = O^{p'}(\mathcal{F}_m) = \mathcal{F}_m$. Then $\mathfrak{red}(\mathcal{F}) = \mathcal{F}_m$.

By [AOV, Proposition 2.2], the reduction of any saturated fusion system is reduced. (The only condition which is not obvious is that $O_p(\mathfrak{red}(\mathcal{F})) =$

1.) By [AOV, Proposition 2.4], $\mathfrak{red}(\mathcal{F}) = 1$ (the fusion system over the trivial group) if and only if $\mathcal{F}$ is constrained (i.e., $O_p(\mathcal{F}) \in \mathcal{F}^c$).

We do not know an example of an exotic fusion system whose reduction is realizable, but it seems likely that this could happen.

The following definition was formulated to provide a criterion for testing whether or not a given reduced fusion system can be the reduction of an exotic one. Recall the outer automorphism group $\mathrm{Out}_{\mathrm{typ}}(\mathcal{L})$ of a centric linking system $\mathcal{L}$ which was defined in Section 4.3, and its homotopy theoretic interpretation $\mathrm{Out}_{\mathrm{typ}}(\mathcal{L}) \cong \mathrm{Out}(|\mathcal{L}|_p^{\wedge})$ in Theorem 4.22.

**Definition 6.3**  A saturated fusion system $\mathcal{F}$ over $S$ is *tame* if there is a finite group $G$ which satisfies:

- $S \in \mathrm{Syl}_p(G)$ and $\mathcal{F} \cong \mathcal{F}_S(G)$; and

- the natural map

$$\kappa_G \colon \mathrm{Out}(G) \longrightarrow \mathrm{Out}_{\mathrm{typ}}(\mathcal{L}_S^c(G)) \cong \mathrm{Out}(BG_p^{\wedge})$$

  is split surjective.

The following theorem describes how the problem of finding exotic fusion systems is reduced to one involving only reduced fusion systems.

**Theorem 6.4** ([AOV, Theorem A])  *For any saturated fusion system $\mathcal{F}$ over a p-group $S$, if $\mathfrak{red}(\mathcal{F})$ is tame, then $\mathcal{F}$ is also tame, and hence realizable.*

So far, we know of no examples of realizable fusion systems which are not tame, and hence no examples of exotic fusion systems whose reduction is realizable. By contrast, Ruiz [Rz] gave examples of realizable (in fact, tame) fusion systems $\mathcal{F} = \mathcal{F}_S(G)$ for which $\mathfrak{red}(\mathcal{F})$ is exotic (see Section 6.5). However, the following theorem provides a different type of converse to Theorem 6.4.

**Theorem 6.5** ([AOV, Theorem B])  *Let $\mathcal{F}$ be a reduced fusion system which is not tame. Then there is an exotic fusion system $\widetilde{\mathcal{F}}$ such that $\mathfrak{red}(\widetilde{\mathcal{F}}) \cong \mathcal{F}$.*

It is Theorem 6.5 which motivates the split surjectivity condition in the definition of tameness. Assume $\mathcal{F}$ is a saturated fusion system over the $p$-group $S$ which is realizable but not tame. In particular, it has an associated linking system $\mathcal{L}$. Let $A$ be any elementary abelian $p$-group upon which $\mathrm{Out}_{\mathrm{typ}}(\mathcal{L})$ acts faithfully. Consider the $p$-local finite group $(S \times A, \mathcal{F} \times A, \mathcal{L} \times A)$, defined so that if $(S, \mathcal{F}, \mathcal{L})$ is realized by $G$, then $(S \times A, \mathcal{F} \times A, \mathcal{L} \times A)$ is realized by $G \times A$. Using Theorem 4.15, one can construct a fusion system $\widehat{\mathcal{F}}$, with an associated linking system $\widehat{\mathcal{L}}$ which is extension of $\mathcal{L} \times A$ by $\mathrm{Out}_{\mathrm{typ}}(\mathcal{L})$, where $\mathrm{Out}_{\mathrm{typ}}(\mathcal{L})$ acts on $\mathcal{L}$ in the canonical way and on $A$ via the given faithful action. One now checks that $\mathfrak{red}(\widehat{\mathcal{F}}) \cong \mathcal{F}$. If $\widehat{\mathcal{F}}$ is realizable, then it is realized by a group $\widehat{G}$ such that $A \trianglelefteq \widehat{G}$ and $\widehat{G}/C_{\widehat{G}}(A) \cong \mathrm{Out}_{\mathrm{typ}}(\mathcal{L})$, and also such that $G^* \overset{\mathrm{def}}{=} C_{\widehat{G}}(A)/A$ realizes $\mathcal{F}$. Then conjugation induces a homomorphism from $\mathrm{Out}_{\mathrm{typ}}(\mathcal{L})$ to $\mathrm{Out}(G^*)$ which splits $\kappa_{G^*}$, contradicting the original assumption that $\mathcal{F}$ is not tame.

Thus Theorems 6.4 and 6.5 show that a reduced fusion system is tame if and only if it is not the reduction of any exotic fusion system.

As mentioned above, reduced fusion systems are very far from being simple in any sense. For example, any product of reduced fusion systems is again reduced. (See Definition I.3.4(c) for the definition of a product of fusion systems.) However, the following theorem says that when searching for reduced fusion systems which are not tame, it suffices to look for those which are indecomposable: those which do not factor as products of proper fusion subsystems.

**Theorem 6.6** ([AOV, Theorem C])  *Each reduced fusion system $\mathcal{F}$ has a unique factorisation $\mathcal{F} = \mathcal{F}_1 \times \cdots \times \mathcal{F}_m$ as a product of indecomposable fusion subsystems. If $\mathcal{F}_i$ is tame for each $i$, then $\mathcal{F}$ is tame.*

### 6.2. The Ruiz-Viruel examples.

Probably the simplest examples of exotic fusion systems are those constructed by Ruiz and Viruel [RV]. They classified all saturated fusion systems over extraspecial $p$-groups of order $p^3$ and exponent $p$ (for odd $p$), of which three turned out to be exotic (all when $p = 7$). We give a quick sketch here of their results.

Let $S$ be extraspecial of order $p^3$ and exponent $p$. Then $S$ contains $p+1$ subgroups $V_0, \ldots, V_p$ of order $p^2$, each of which is isomorphic to $C_p^2$. The $V_i$ are the only proper subgroups which are centric in $S$. So by Theorem I.3.5, each saturated fusion system $\mathcal{F}$ over $S$ is generated by (hence determined by) $\mathrm{Out}_{\mathcal{F}}(S)$, together with the groups $\mathrm{Aut}_{\mathcal{F}}(V_i)$ for those $V_i$ which are $\mathcal{F}$-radical. If $V_i$ is $\mathcal{F}$-radical, then $\mathrm{Aut}_{\mathcal{F}}(V_i) \leq \mathrm{Aut}(V_i) \cong GL_2(p)$ contains at least two subgroups of order $p$, and hence must contain $SL_2(p)$.

This, together with the saturation axioms, lead to the following conditions which relate these different groups:

- $\mathrm{Out}_{\mathcal{F}}(S)$ has order prime to $p$, and is a subgroup of $\mathrm{Out}(S) \cong GL_2(p)$.

- For each $i$, and each $\alpha \in \mathrm{Aut}_{\mathcal{F}}(S)$ such that $\alpha(V_i) = V_i$, $\alpha|_{V_i} \in \mathrm{Aut}_{\mathcal{F}}(V_i)$.

- If $V_i$ is $\mathcal{F}$-radical, then $SL_2(p) \leq \mathrm{Aut}_{\mathcal{F}}(V_i) \leq GL_2(p)$. If $V_i$ is not $\mathcal{F}$-radical, then $\mathrm{Aut}_{\mathcal{F}}(V_i)$ contains only the restrictions of automorphisms in $\mathrm{Aut}_{\mathcal{F}}(S)$ which leave $V_i$ invariant.

- If $V_i$ is $\mathcal{F}$-radical, then for each $\beta \in \mathrm{Aut}_{\mathcal{F}}(V_i)$ such that $\beta(Z(S)) = Z(S)$, $\beta$ extends to an element of $\mathrm{Aut}_{\mathcal{F}}(S)$.

Conversely, any fusion system over $S$ which satisfies these conditions, and for which all other morphisms are composites of restrictions of automorphisms of these subgroups, is saturated by Theorem I.3.10.

Using these criteria, Ruiz and Viruel made a complete list of all saturated fusion systems over $S$ (for any odd prime $p$). To simplify the statement of their result, we list in Table 6.1 only those fusion systems which are reduced. In this situation, $\mathcal{F}$ is reduced if and only if there are at least two $\mathcal{F}$-radical subgroups among the $V_i$ and $\mathcal{F}$ has no fusion subsystems of index prime to $p$.

| $p$ | $\mathrm{Out}_{\mathcal{F}}(S)$ | nr. $V_i$ rad. | $\mathrm{Aut}_{\mathcal{F}}(V_i)$ | group |
|---|---|---|---|---|
| $3{\nmid}(p{-}1)$ | $C_{p-1} \times C_{p-1}$ | $1+1$ | $GL_2(p)$ | $PSL_3(p)$ |
| $3{\mid}(p{-}1)$ | $C_{p-1} \times C_{(p-1)/3}$ | $1+1$ | $SL_2(p) \rtimes C_{(p-1)/3}$ | $PSL_3(p)$ |
| $3$ | $D_8$ | $2+2$ | $GL_2(p)$ | ${}^2F_4(2)'$ |
| $5$ | $4 S_4$ | $6$ | $GL_2(p)$ | $F_3 = \mathrm{Th}$ |
| $7$ | $S_3 \times C_3$ | $3$ | $SL_2(p)$ | He |
| $7$ | $S_3 \times C_6$ | $3+3$ | $SL_2(p) \rtimes C_2$ | $\mathrm{Fi}'_{24}$ |
| $7$ | $(C_6 \times C_6) \rtimes C_2$ | $6+2$ | $SL_2(p) \rtimes C_2,\ GL_2(p)$ | — |
| $7$ | $D_8 \times C_3$ | $2+2$ | $SL_2(p) \rtimes C_2$ | O'N |
| $7$ | $D_{16} \times C_3$ | $4+4$ | $SL_2(p) \rtimes C_2$ | — |
| $13$ | $C_3 \times 4 S_4$ | $6$ | $SL_2(p) \rtimes C_4$ | $F_1 = \mathrm{M}$ |

TABLE 6.1

The third column in the table gives the numbers of subgroups $V_i$ in each $\mathcal{F}$-conjugacy class of radical subgroups. Thus "6" means there are six $\mathcal{F}$-radical subgroups among the $V_i$, all of which are $\mathcal{F}$-conjugate, while "$1 + 1$" means there are two $\mathcal{F}$-conjugacy classes containing one subgroup each. The fusion systems listed in the seventh and ninth lines of the table were shown to be exotic using a version of Lemma 6.1. (The third exotic fusion system found by Ruiz and Viruel is not reduced.)

## 6.3. Saturated fusion systems over 2-groups.

The following was proven in [LO] (with a correction in [LO2]).

**Theorem 6.7** *Let $q$ be an odd prime power, and fix $S \in \mathrm{Syl}_2(\mathrm{Spin}_7(q))$. Let $z \in Z(\mathrm{Spin}_7(q))$ be the central element of order 2. Then there is a saturated fusion system $\mathcal{F} = \mathcal{F}_{\mathrm{Sol}}(q)$ which satisfies the following conditions:*

(a) *$C_{\mathcal{F}}(z) = \mathcal{F}_S(\mathrm{Spin}_7(q))$ as fusion systems over $S$.*

(b) *All involutions of $S$ are $\mathcal{F}$-conjugate.*

*Furthermore, there is a unique centric linking system $\mathcal{L} = \mathcal{L}_{\mathrm{Sol}}^c(q)$ associated to $\mathcal{F}$.*

The existence of these fusion systems, and certain homotopical properties of their classifying spaces, were predicted by Benson [Be3]. This, in turn, was motivated by a theorem of Solomon [So, Theorem 3.2], which said that when $q \equiv \pm 3 \pmod 8$, there is no finite group $G$ whose 2-fusion system satisfies the above condtions. Thus $\mathcal{F}_{\mathrm{Sol}}(q)$ is exotic in these cases by Solomon's theorem. In fact, Solomon showed that this is true for each prime power $q$, and the remaining details of his argument were written down in the proof of [LO, Proposition 3.4].

We sketch here the construction of $\mathcal{F}_{\mathrm{Sol}}(q)$ given in [LO, §2]. Set $G = \mathrm{Spin}_7(q)$ for short, and fix $S \in \mathrm{Syl}_2(G)$. Let $z \in Z(G) = Z(S)$ be the central involution, and choose $U \trianglelefteq S$ such that $z \in U \cong C_2^2$. The centralizer of $U$ in $\mathrm{Spin}_7(\bar{\mathbb{F}}_q)$ is isomorphic to $SL_2(\bar{\mathbb{F}}_q)^3/\langle(-I,-I,-I)\rangle$, and $C_G(U)$ contains $SL_2(q)^3/\langle(-I,-I,-I)\rangle$ with index 2. Thus $S_3$ acts on $C_G(U)$ and on $C_S(U)$ in a natural way. (Note, however, that the action depends on the choice of identification of $C_G(U)$ as a subgroup of $SL_2(\bar{\mathbb{F}}_q)^3/\langle(-I,-I,-I)\rangle$; this was the source of the error which made a correction necessary in [LO2].) The fusion system $\mathcal{F}_{\mathrm{Sol}}(q)$ was defined to be the fusion system over $S$ generated by $\mathcal{F}_S(G)$, together with this action of $S_3$ on $C_S(U)$ (and its restrictions).

A different construction of these fusion and linking systems was given by Aschbacher and Chermak in [AC]. Again set $G = \mathrm{Spin}_7(q)$ and fix $U \trianglelefteq S \in \mathrm{Syl}_2(G)$. Set $K = N_G(U)$ and $K_0 = C_G(U)$. There is a group $H \geq K$ such that $K_0 \trianglelefteq H$ and $H/K_0 \cong S_3$ acts on $K_0 = C_G(U)$ via the same action as that described in the last paragraph. Aschbacher and Chermak showed that $\mathcal{F}_{\mathrm{Sol}}(q)$ is the fusion system of the amalgamated free product $G \underset{K}{*} H$, and gave a new proof that this fusion system is saturated. They also constructed an associated linking system $\mathcal{L}_{\mathrm{Sol}}^c(q)$ as the quotient of the transporter system of $G \underset{K}{*} H$ by a certain "signaliser functor".

In contrast, the linking system $\mathcal{L}^c_{\mathrm{Sol}}(q)$ was not constructed explicitly in [LO], but rather shown to exist because the obstruction groups of Proposition 5.11 to its existence and uniqueness vanish. In fact, the existence of a linking system follows from Proposition 5.15, since the Sylow 2-subgroups of $\mathrm{Spin}_7(q)$ have rank 4, but its uniquenss requires some more calculations.

The smallest of the Solomon fusion systems is over a group of order $2^{10}$. Work in progress by Andersen, Oliver, and Ventura seems to show that there are no exotic fusion systems over 2-groups of smaller order. Using a computer search for 2-groups which meet certain necessary conditions, we explicitly listed all indecomposable, reduced fusion systems over 2-groups of order $\leq 2^9$, and showed that they are all tame.

### 6.4. **Mixing related fusion systems.**

Fix a prime $p \geq 5$. Let $q$ be a prime power such that $p|(q-1)$ but $p^2 \nmid (q-1)$. Set $G = PSL_p(q)$, let $T \cong C_{q-1}^{p-2} \times C_{(q-1)/p}$ be the subgroup of classes of diagonal matrices, and set $N = N_G(T) \cong T \rtimes S_p$. Fix $S \in \mathrm{Syl}_p(N)$; then $S \cong C_p^{p-2} \rtimes C_p$ is also a Sylow $p$-subgroup of $G$.

Let $V_1,\ldots,V_p$ be conjugacy class representatives for those subgroups $V \leq S$ such that $V \cong C_p^2$ and $V \nleq T$. (All of these lift to extraspecial subgroups in $SL_p(q)$ of order $p^3$ and exponent $p$, and they are conjugate to each other in $PGL_p(q)$.) The fusion systems $\mathcal{F}_S(N) \subseteq \mathcal{F}_S(G)$ differ only in that all of the $V_i$ are radical in $\mathcal{F}_S(G)$ (with $\mathrm{Aut}_G(V_i) \cong SL_2(p)$), while none of them are radical in $\mathcal{F}_S(N)$. For each subset $I \subseteq \{1,\ldots,p\}$, let $\mathcal{F}_I$ be the fusion system such that $\mathcal{F}_S(N) \subseteq \mathcal{F}_I \subseteq \mathcal{F}_S(G)$, and such that $\mathrm{Aut}_{\mathcal{F}_I}(V_i) = \mathrm{Aut}_G(V_i)$ when $i \in I$, and $\mathrm{Aut}_{\mathcal{F}_I}(V_i) = \mathrm{Aut}_N(V_i)$ when $i \notin I$. Thus, for example, $\mathcal{F}_{\mathrm{all}} = \mathcal{F}_S(G)$ and $\mathcal{F}_\varnothing = \mathcal{F}_S(N)$.

Using the fact that $\mathcal{F}_S(G)$ and $\mathcal{F}_S(N)$ are both saturated, it is not hard to see that these "mixed" fusion systems $\mathcal{F}_I$ are also saturated. Using Lemma 6.1, they are all shown to be exotic when $1 \leq |I| < p$ (see [BLO2, Example 9.3] for details).

### 6.5. **Other examples.**

We end with a brief survey of all other examples we know of exotic fusion systems.

We first describe some examples constructed by Ruiz [Rz]. Fix a prime $p \geq 5$, and a prime power $q$ such that $p \nmid q$. Let $e$ be the multiplicative order of $q$ in $\mathbb{F}_p^\times$, and assume $e > 1$ ($p \nmid (q-1)$). Fix any $n \geq ep$, set $G_{n,q} = GL_n(q)$ (a simple group), fix $S_{n,q} \in \mathrm{Syl}_p(G)$, and consider the fusion system $\mathcal{F}_{n,q} = \mathcal{F}_{S_{n,q}}(G)$. By [Rz, Theorem B], $O^{p'}(\mathcal{F})$ has index $e$ in $\mathcal{F}_{n,q}$, and thus $\mathcal{F}_{n,q}$ is not reduced. Furthermore, for each divisor $d|e$ with $d > 2$, the normal fusion subsystem of index $d$ in $\mathcal{F}_{n,q}$ is exotic.

In [DRV, Theorem 1.1], Díaz, Ruiz, and Viruel classify all saturated fusion systems over $p$-groups of rank 2 when $p$ is odd. In addition to the exotic fusion systems over extraspecial groups of order $7^3$ found in [RV], they showed in [DRV, Theorem 5.10] that there are several infinite families of exotic fusion systems over 3-groups of rank 2.

Several families of exotic saturated fusion systems were constructed by Clelland and Parker [CP], and by Broto, Levi, and Oliver [BLO4], as the fusion systems of the amalgamated free products associated to certain trees of finite groups.

Another source of exotic fusion systems comes from nonabelian $p$-groups with abelian subgroup of index $p$. When $p = 2$, each reduced fusion system over a 2-group $S$ of this type is the fusion system of one of the simple groups $PSL_2(q)$ or $PSL_3(q)$ for $q$ odd, and $S$ is dihedral, semidihedral, or a wreath product $C_{2^n} \wr C_2$ (to appear in a future paper by Andersen, Oliver, and Ventura). But when $p$ is odd, this author has shown in unpublished work that there are many families of exotic (reduced) fusion systems over groups of this form. For example, one can give a complete list of all reduced fusion systems $\mathcal{F}$ over $p$-groups $S$ ($p$ odd) which contain a unique abelian subgroup $A \trianglelefteq S$ of index $p$, and such that $A$ is not $\mathcal{F}$-essential — and almost all of these (with a few exceptions when $p = 3$) are exotic.

In addition, a few other examples of exotic fusion systems are given in [BLO2, Example 9.4].

## 7. OPEN PROBLEMS

We finish this survey with a list of important problems in the subject which as far as we know are still open.

1. **Prove the existence and uniqueness of centric linking systems (equivalently, of classifying spaces) associated to any given saturated fusion system, or find a counterexample.** This was discussed in detail in Section 5.3.

   *Added in proof:* This problem has now been solved by Andy Chermak. He proves this by direct constructions of linking systems and isomorphisms between them, but his proof can also be formulated in terms of the obstructions described in Section 5.3.

2. **Define morphisms between $p$-local finite groups.** Once $p$-local finite groups have been defined, it is natural to want to construct a category with them as objects. But it is not clear how morphisms should be defined in such a category. The simplest way to define a morphism from $(S_1, \mathcal{F}_1, \mathcal{L}_1)$ to $(S_2, \mathcal{F}_2, \mathcal{L}_2)$ is as a basepoint preserving map (or homotopy class of maps) from $|\mathcal{L}_1|_p^\wedge$ to $|\mathcal{L}_2|_p^\wedge$. However, one

would like to find a more combinatorial definition; for example, one induced by a functor between finite categories. A map between the classifying spaces as above does induce a functor $\Phi\colon \mathcal{F}_1 \longrightarrow \mathcal{F}_2$. But it need not induce a functor between the linking systems (at least not in an obvious way) since $\Phi$ need not send $\mathcal{F}_1{}^c$ into $\mathcal{F}_2{}^c$ (not even when it is induced by a homomorphism of groups).

A related problem was studied by Castellana and Libman [CL]: that of extending a homomorphism between Sylow subgroups to a map between classifying spaces. First, they construct wreath products in this setting: for each $p$-local finite group $(S, \mathcal{F}, \mathcal{L})$, each $n > 1$, and each $K \le S_n$ (the symmetric group on $n$ letters), they define a $p$-local finite group $(S \wr K, \mathcal{F} \wr K, \mathcal{L} \wr K)$ such that $|\mathcal{L} \wr K|$ has the homotopy type of the wreath product of spaces $|\mathcal{L}| \wr K = EK \times_K (|\mathcal{L}|^n)$. They then show [CL, Theorem B] that for any pair of $p$-local finite groups $(S_i, \mathcal{F}_i, \mathcal{L}_i)$ ($i = 1, 2$), and any fusion preserving homomorphism $\varphi\colon S_1 \longrightarrow S_2$ (whether or not it defines a functor between the fusion systems), there is a map of spaces from $|\mathcal{L}_1|_p^\wedge$ to $|\mathcal{L}_2 \wr S_{p^k}|_p^\wedge$, for $k$ large enough, which realizes $\varphi$ in a certain explicit sense.

3. **Find more exotic fusion systems at the prime 2, or prove there are none.** As noted in Section 6.3, there is only one known family of exotic fusion systems over 2-groups. Motivated by the proof of the classification of finite simple groups, Ron Solomon has conjectured that these are the only ones which are simple. Whether or not the fusion systems $\mathcal{F}_{\mathrm{Sol}}(q)$ are the only exotic fusion systems (aside from the obvious constructions taking them as starting point), we currently have little idea how to search for other examples.

Here is a slightly more general question. Does there exist an exotic, reduced fusion system over a finite 2-group which does not contain a normal subsystem isomorphic to a product of Solomon fusion systems $\mathcal{F}_{\mathrm{Sol}}(q)$ (Section 6.3)? Note that this includes as a special case the question of whether there exist any realizable fusion systems over finite 2-groups which are not tame.

One project which could lead to such examples is that of Aschbacher, to repeat some of the steps in the proof of the classification theorem in the context of fusion systems. This is discussed in much more detail in Part II (Sections II.14–15), and also in [A5, A6, A7]. Another (less ambitious) project is the attempt by Andersen, Oliver, and Ventura to carry out a systematic computer search for exotic examples.

4. **Try to better understand how exotic fusion systems arise at odd primes; or (more realistically) look for patterns which explain how certain large families of them arise.** There seem

to be many different examples of exotic fusion systems at odd primes, and they are constructed using a large range of techniques. Is there any way to categorise certain types? Is there an algorithm for systematically constructing large families of exotic examples, other than under certain very restrictive conditions?

5.   **Find criteria, especially at odd primes, which at least in some cases allow us to prove that a saturated fusion system is exotic without using the classification theorem.** Currently, there is no known easily checked property of fusion systems which holds for all realizable fusion systems, but which fails for at least some exotic ones. In fact, there is *no* fusion system over a $p$-group for $p$ odd which we can prove is exotic without using the classification of finite simple groups. This is a surprising and unsatisfactory situation. Any property which allows us to distinguish (even in some cases) between realizable and exotic fusion systems would be of great interest.

6.   **Find an example of a saturated fusion system which is realizable but not tame, or prove that there are none.** As described in Theorem 6.5, finding such an example would provide a new way of constructing exotic fusion system as extensions of realizable fusion systems.

7.   **Prove that each $p$-local finite group $(S, \mathcal{F}, \mathcal{L})$ can be realized by some (possibly infinite) group, and some choice of "signaliser functor" in the sense of Aschbacher and Chermak.** As noted earlier, Leary and Stancu [LS], and Robinson [Ro3] have shown via separate constructions that one can always find an infinite group $G$ with Sylow $p$-subgroup $S$ such that $\mathcal{F} \cong \mathcal{F}_S(G)$. Can $G$ always be chosen so that $\mathcal{L}$ can be constructed as a quotient category of the centric transporter category $\mathcal{T}_S^c(G)$? (See Definition 3.1.) This would mean choosing subgroups $C_G^*(P) \leq C_G(P)$ for each $P \in \mathcal{F}^c$ (the "signaliser functor"), such that $\mathrm{Mor}_{\mathcal{L}}(P, Q) \cong T_G(P, Q)/C_G^*(P)$ for each $P$ and $Q$.

*Added in proof:*   This has now been shown by Libman and Seeliger [LbS]. They prove that there always exists a group $G$ and a signalizer functor which determines $\mathcal{L}$, although the signalizer functor is not constructed explicitly.

8.   **Fundamental groups of linking systems.** What, in general, can be said about $\pi_1(|\mathcal{L}|)$, when $\mathcal{L}$ is a linking system, especially the linking system of a finite group $G$? For example, in [COS], it was shown that $\pi_1(|\mathcal{L}_{\mathrm{Sol}}^c(q)|) = 1$ when $\mathcal{L}_{\mathrm{Sol}}^c(q)$ is the centric linking system associated to the Solomon fusion system $\mathcal{F}_{\mathrm{Sol}}(q)$ (Section 6.3). In unpublished work by Grodal and Oliver, they construct some examples

where $\pi_1(|\mathcal{L}_S^c(G)|)$ is isomorphic to $G$, others where it is a group of Lie type over an infinite ring, and yet others where it is infinite without any obvious group to identify it with. Also, together with Shpectorov, they show that when $G$ is the sporadic simple group of Lyons, $\pi_1(|\mathcal{L}_S^c(G)|) = 1$. So far, these are relatively isolated examples, and the computations in most cases depend on earlier results about geometries for $G$. Is there anything more general one can say about these groups? What, if any, significance might they have for the properties of $G$, its fusion system, and its $p$-completed classifying space?

9. **Get a more precise understanding of $H^*(|\mathcal{L}|; \mathbb{F}_p)$ for a $p$-local finite group $(S, \mathcal{F}, \mathcal{L})$, and find ways to compute $H^*(|\mathcal{L}|; M)$ when $M$ is a finite abelian $p$-group with action of $\pi_1(|\mathcal{L}|)$.** For example, for a $p$-local finite group $(S, \mathcal{F}, \mathcal{L})$, it would be useful to know whether the functors $H^i(-; \mathbb{F}_p)$ (as functors on $\mathcal{O}(\mathcal{F}^c)$) all have vanishing higher limits. If so, this would give a new proof of Theorem 4.23, by showing that the cohomology spectral sequence for the subgroup decomposition of $\mathcal{L}$ (see Proposition 5.28) collapses (thus this would show that the decomposition is "sharp" in the terminology of Dwyer).

More generally, when $M$ is any finite abelian $p$-group with action of $\pi_1(|\mathcal{L}|)$, we know examples where $H^*(|\mathcal{L}|; M)$ is not the inverse limit over $\mathcal{O}(\mathcal{F}^c)$ of the cohomology groups $H^*(-; M)$, and we know examples where it is the inverse limit (including all cases where the action is trivial). But we do not yet have a good understanding of what properties of the action distinguish these two cases. Some results along these lines have been obtained by Levi and Ragnarsson [LR].

10. **Find other classes of fusion systems, or fusion systems over classes of groups other than $p$-groups or discrete $p$-toral groups, which could be of interest in group theory or homotopy theory.** The generalization to fusion systems over discrete $p$-toral groups is described in Section 4.8. In work which has not yet appeared, Stancu and Symonds develop a theory of fusion systems over pro-$p$-groups. In [ABC, § X.4.1], the authors discuss the possibility of defining fusion systems over $p$-unipotent groups of finite Morley rank.

Are there any other generalizations which arise naturally?

# Part IV. Fusion and Representation theory

## Radha Kessar

Let $G$ be a finite group, $p$ a prime number dividing the order of $G$ and $k$ a field of characteristic $p$. In the 1930s Richard Brauer, partly in joint work with Nesbitt, initiated the systematic study of the representations of $G$ over $k$, nearly four decades after the first paper on representations of $G$ over $\mathbb{C}$ by Frobenius. In contrast to the complex group algebra $\mathbb{C}G$ of $G$, the modular group algebra $kG$ is not a direct product of simple algebras; the indecomposable factors of $kG$, called blocks, have a very rich representation theory.

Between 1942 and 1946, Brauer introduced two fundamental notions. The first, which is now known as the Brauer homomorphism, is a $k$-algebra homomorphism from the center $Z(kG)$ of $kG$ to $Z(kH)$, for $H$ a local subgroup of $G$, that is $H$ is a subgroup of $G$ such that $C_G(Q) \leq H \leq N_G(Q)$, for some $p$-subgroup $Q$ of $G$. To each block $B$ of $kG$ the Brauer homomorphism associates a set of blocks of $kH$ called the Brauer correspondents of $B$. The second notion is that of the defect groups of a block algebra — a certain conjugacy class of $p$-subgroups of $G$. The two notions are intimately connected, for instance Brauer's First Main Theorem gives a bijection, via the Brauer homomorphism, between the set of blocks of $kG$ with a given defect group $P$ and the set of blocks of $kN_G(P)$ with defect group $P$. The representation theory of $B$ is deeply influenced by the nature of the defect groups of $B$, the local subgroups of $G$ and the represention theory of Brauer correspondents of $B$.

Building on the ideas behind Brauer's First Main Theorem, in the late 1970s Alperin and Broué introduced the $G$-poset of Brauer pairs associated to a block $B$ of $kG$. The elements of this poset are pairs $(Q, e)$ where $Q$ is a $p$-subgroup of $G$ and $e$ is a block of $kC_G(Q)$ in Brauer correspondence with $B$; the inclusion relationship is defined through the Brauer homomorphism, and the first component of any maximal pair is a defect group of $B$. A certain sub-poset of these pairs (which in the language of this part of the book corresponds to the poset of centric Brauer pairs) had been studied earlier by Brauer. The $G$-poset of Brauer pairs associated to $B$ has many similarities with the $G$-poset of p-subgroups of $G$. It was these similarities that led Puig to distill the essence of "fusion" through abstractly defined categories, and the results of Alperin and Broué may be reformulated as the statement that for any maximal Brauer pair $(P, e)$ associated to $B$, the category $\mathcal{F}$ whose objects are subgroups of $P$ and whose morphism sets

$\operatorname{Hom}_{\mathcal{F}}(Q, R)$ consist of morphisms induced by conjugations in the $G$-sub poset of Brauer pairs contained in $(P, e)$ is a saturated fusion system on $P$.

The first three sections of this part of the book discuss Brauer pairs and associated fusion systems. The theory of Brauer pairs as developed by Alperin and Broué and later by Broué and Puig applies more generally to $p$-permutation $G$-algebras and we work in this setting. In particular, to every $p$-permutation $G$-algebra $A$ over $k$ and to every primitive idempotent $b$ of the subalgebra of $A$ of $G$-fixed points is associated an isomorphism class of fusion system through the fusion of Brauer pairs. These fusion systems are not always saturated. Section 3 gives a sufficient condition for saturation. The special case of block fusion systems is recovered in subsections 3.4 and 3.5.

Let $B$ be a block of $kG$ and let $\mathcal{F}$ be the saturated fusion system arising from some maximal Brauer pair $(P, e)$ associated to $B$. Many results and conjectures in modular representation theory are statements relating the representation theory of $B$ with $\mathcal{F}$. Section 4 contains some background results and terminology from group representation theory. Section 5 is a selection of topics which highlight the connections between the representation theory of $B$ and $\mathcal{F}$. This section is not a comprehensive survey, but is rather intended to give a flavour of the subject. For instance, the nilpotent block theorem of Puig states that if $\mathcal{F} = \mathcal{F}_P(P)$ then the module category of $B$ is equivalent to that of the algebra $kP$. As another example, by a result of Rickard, building on the Brauer-Thompson-Dade-Green-Janush theory of cyclic blocks, if $P$ is cyclic then the derived module category $B$ is equivalent to the derived module category of the algebra $k(P \rtimes \operatorname{Out}_{\mathcal{F}}(P))$. Broué's abelian defect group conjecture asserts that an analogue of the above result holds whenever $P$ is abelian, with $k(P \rtimes \operatorname{Out}_{\mathcal{F}}(P))$ replaced by a twisted group algebra $k_{\alpha}(P \rtimes \operatorname{Out}_{\mathcal{F}}(P))$, where $\alpha$ is a certain element in the second cohomogy group $H^2(\operatorname{Out}_{\mathcal{F}}(P), k^*)$. Alperin's weight conjecture is a formula which expresses the number of isomorphism classes of simple $B$-modules in terms of the corresponding saturated fusion system $\mathcal{F}$ and certain elements in $H^2(\operatorname{Out}_{\mathcal{F}}(Q), k^*)$, where $Q$ runs over the centric-radical objects of $\mathcal{F}$.

A saturated fusion system is called block realisable if it is the fusion system of a block of a finite group, and is called block-exotic otherwise. By Brauer's Third Main Theorem, any realisable saturated fusion system is block realisable, but the converse is open. A possible approach is to reduce the problem to blocks of finite simple or quasi-simple groups. For this it is important to understand the relationship between the fusion sytem of the blocks of $kG$ and those of $kN$ for $N$ a normal subgroup of $G$. This is discussed in Section 6.

Sections 2 and 3 contain detailed proofs — the intention is that a reader reasonably familiar with the background material of Section 1 will be able to get a detailed picture of how fusion systems arise in $p$-permutation $G$-algebras and in particular, in block algebras. By contrast, Sections 4, 5, and 6 should be read as a "guided tour", and proofs are either omitted or only sketchily outlined.

I would like to thank Kasper Andersen, Fei Xu, Ellen Henke and Matthew Gelvin for their careful reading of this part of the manuscript, and their many corrections and suggestions.

**Notation.** Group actions and maps are written on the left. If a group $X$ acts on a set $\mathcal{B}$, denote by $Orb_X(\mathcal{B})$ the set of orbits of $X$ on $\mathcal{B}$, by $\mathcal{B}^X$ the subset of $X$-fixed points in $\mathcal{B}$ and by $C_X(a)$ the stabilizer in $X$ of an element $a$ of $\mathcal{B}$. If $\mathcal{B}$ is a $X$-poset, $Y$ is a subgroup of $X$ and $\mathcal{C}$ is a subset of $\mathcal{B}$ such that ${}^y\mathcal{C} \subseteq \mathcal{C}$, for all $y \in Y$, then $\mathcal{C}$ is considered as a $Y$-poset through the structure inherited from $\mathcal{B}$, and $\mathcal{C}$ is said to be a $Y$-subposet of $\mathcal{B}$. If $(A, +)$ is an abelian group, then for any finite subset $C$ of $A$, denote by $C^+$ the sum of the elements of $C$ in $A$.

## 1. Algebras and $G$-Algebras

This section is intended as a quick introduction to some basic notions and terminology for finite dimensional $G$-algebras over fields. Throughout this section, $k$ will denote an algebraically closed field, and $A$ will denote a ring with identity $1_A \neq 0$. We will denote by $A^*$ the group of invertible elements of $A$. If $a \in A$ and $u \in A^*$, we will denote by ${}^u a$ the conjugate $uau^{-1}$ of $a$ by $u$. By an ideal of $A$, we will always mean a two-sided ideal. If $A$ is a $k$-algebra, we will always assume that $A$ is finite dimensional as vector space over $k$.

All $A$-modules will be left $A$-modules, and will be assumed to be finitely generated over $A$, so in particular, if $A$ is a $k$-algebra, then all $A$-modules considered will be finite dimensional as $k$-vector spaces. An $A$-module $M$ is *simple* if $M \neq 0$ and $M$ does not properly contain any non-zero $A$-module, $M$ is *indecomposable* if $M \not\cong M_1 \oplus M_2$ for any non-zero $A$-modules $M_1$, $M_2$, and $M$ is *projective* if $M \neq 0$ and there exists a positive integer $n$ such that for some $A$-module $N$, $A^n \cong M \oplus N$, where $A^n$ denotes the direct sum of $n$ copies of $A$.

### 1.1. **Ideals and Idempotents.**

This subsection recalls some background results from ring theory. This is standard material and can be found in many representation theory texts, for example, in the books of Benson [Be1], Curtis and Reiner [CuR1], [CuR2], [CuR3], Külshammer [Ku] and Nagao and Tsushima [NT]. The treatment below is not comprehensive and only results which are used in the sequel are mentioned here.

An element $a$ of $A$ is called *nilpotent* if $a^n = 0$ for some natural number $n$. Similarly, an ideal $I$ of $A$ is nilpotent if $I^n = 0$ for some natural number $n$. Clearly, every element of a nilpotent ideal of $A$ is nilpotent. The *Jacobson radical* $J(A)$ of $A$ is the intersection of the maximal left ideals of $A$.

**Proposition 1.1**  *Suppose that $A$ is a $k$-algebra.*

(a)  *$A$ has finitely many maximal two-sided ideals and $J(A)$ is the intersection of these. In particular, $J(A)$ is an ideal of $A$.*

(b)  *$J(A)$ is nilpotent and $J(A)$ contains every nilpotent ideal of $A$.*

(c)  *$A/J(A)$ is isomorphic to a direct product*

$$A/J(A) = \prod_{i=1}^{r} \mathrm{Mat}_{n_i}(k)$$

*of matrix algebras over $k$. Further, $J(A)$ is contained in any ideal $I$ of $A$ such that $A/I$ is a direct product of matrix algebras over $k$.*

(d)  *An element $a$ of $A$ is in $J(A)$ if and only if $aV = 0$ for every simple $A$-module $V$.*

An *idempotent* of $A$ is an element $e$ of $A$ such that $0 \neq e = e^2$. In particular, $1_A$ is an idempotent of $A$. Two idempotents $e, f$ of $A$ are *orthogonal* if $ef = fe = 0$. An idempotent $e$ of $A$ is *primitive* in $A$ if $e$ cannot be decomposed as a direct sum $e = e_1 + e_2$, with $e_1, e_2$ idempotent and orthogonal. For an idempotent $e$ of $A$, a (primitive) decomposition of $e$ in $A$ is a finite set of (primitive) idempotents of $A$, which are pairwise orthogonal and which sum to $e$. If $X$ is a decomposition of $e$, we often abusively say that $e = \sum_{f \in X} f$ is an idempotent decomposition of $e$.

**Proposition 1.2**  *Suppose that $A$ is a $k$-algebra and let $e \in A$ be an idempotent. Then $e$ has a primitive decomposition. Further, this primitive decomposition is unique up to conjugation in $A$. Precisely, if $I$ and $I'$ are two primitive decompositions of $e$, then there exists a bijection $i \to i'$ from $I$ to $I'$ and $u \in A^*$ such that $i' = {}^{u}i$ for all $i \in I$.*

**Proposition 1.3**  *Suppose that $A$ is a $k$-algebra and let $e$ and $f$ be primitive idempotents of $A$. Then $e$ and $f$ are conjugate in $A$ if and only if $eAf$ is not contained in $J(A)$.*

**Proposition 1.4**  *(Idempotent Lifting)  Let $\pi : A \to B$ be a surjective homomorphism of $k$-algebras.*

(a)  *Let $e \in A$ be an idempotent. If $e$ is primitive, then $\pi(e)$ is either $0$ or a primitive idempotent of $B$. If $\pi(e)$ is primitive, then there exists an* *orthogonal idempotent decomposition $e = e' + e''$, where $e'$ is primitive in $A$, and $\pi(e'') = 0$, so that $\pi(e') = \pi(e)$.*

(b)  *If $\mathrm{Ker}(\pi)$ is nilpotent, then $\pi(e)$ is a primitive idempotent of $B$ for any primitive idempotent $e$ of $A$.*

(c)  *Let $\{j_1, \cdots, j_t\}$ be a primitive decomposition of an idempotent $f$ in $B$. Then there exists an idempotent $e \in A$ and a primitive decomposition $\{i_1, \cdots, i_t\}$ of $e$ in $A$ such that $\pi(e) = f$ and $\pi(i_s) = j_s$ for all $s$, $1 \le s \le t$. In particular, any primitive idempotent $f$ of $B$ lifts to a primitive idempotent of $A$ via $\pi$.*

**Lemma 1.5**  *(Rosenberg's Lemma)  Suppose that $A$ is a $k$-algebra. Let $e$ be a primitive idempotent of $A$, and let $\mathcal{X}$ be a family of ideals of $A$. If $e \in \sum_{I \in \mathcal{X}} I$, then $e \in I$ for some $I \in \mathcal{X}$.*

**Definition 1.6**  Let $e$ and $f$ be idempotents of $A$.  We say that $f$ is contained in $e$ and write $f \le e$ if

$$ fe = f = ef. $$

It is immediate from the definition that the relation $\le$ is transitive on the set of idempotents of $A$. Also, it is easy to see that $f \le e$ if and only if $f = efe$. If $f \le e$ and $f \ne e$, then $\{f, e - f\}$ is an idempotent decomposition of $e$. In particular, if $A$ is a $k$-algebra and if $f$ is primitive in $A$, then $f \le e$ if and only if there is a primitive decomposition of $e$ containing $f$.

If $e$ is an idempotent of $A$, then $eAe$ is a subring of $A$ with identity element $1_{eAe} = e$. An element $a \in A$ belongs to $eAe$ if and only if $ea = a = ae$. In particular, $e$ is a primitive idempotent of $A$ if and only if $e$ is the only idempotent of $eAe$.

The ring $A$ is *local* if the set of its non-invertible elements forms an ideal of $A$.

**Proposition 1.7**  *Suppose that $A$ is a $k$-algebra and let $e$ be an idempotent of $A$. Then $J(eAe) = eJ(A)e$. Further, the following are equivalent.*

(a)  *$eAe$ is local.*

(b)  *$e$ is a primitive idempotent of $A$.*

(c)  *$e$ is a primitive idempotent of $eAe$.*

(d)  *$eJ(A)e$ is the unique maximal ideal of $eAe$.*

(e)  *Every proper ideal of $eAe$ is nilpotent.*

(f)  *$eAe/eJ(A)e \cong k$ as $k$-algebras.*

If $e$ is an idempotent of $A$, then $Ae$ and $A(1_A - e)$ are $A$-modules via left multiplication by elements of $A$, and $A = Ae \oplus A(1_A - e)$ is a direct sum of $A$-modules. In particular, $Ae$ is a projective $A$-module.

**Proposition 1.8** *Suppose that $A$ is a $k$-algebra and let $I$ be a set of representatives of the $A$-conjugacy classes of primitive idempotents of $A$.*

(a) *The set $\{Ai : i \in I\}$ is a set of representatives of the projective indecomposable $A$-modules.*

(b) *The set $\{Ai/J(A)i : i \in I\}$ is a set of representatives of the simple $A$-modules.*

A *central idempotent* of $A$ is an idempotent of the center $Z(A)$ of $A$. A primitive idempotent of $Z(A)$ is called a *block* of $A$. If $e$ is a central idempotent of $A$, then for any $a \in A$, $ae = ea = eae$. By applying Proposition 1.2 to the commutative $k$-algebra $Z(A)$, we get the following.

**Proposition 1.9** *Suppose that $A$ is a $k$-algebra.*

(a) *Any central idempotent of $A$ has a unique primitive decomposition in $Z(A)$ and this decomposition is a subset of the unique primitive decomposition of $1_A$ in $Z(A)$. In particular, the blocks of $A$ all belong to the unique primitive decomposition of $1_A$ in $Z(A)$ and any two distinct blocks of $A$ are orthogonal.*

(b) *If $b$ is a block of $A$ and $d$ is a central idempotent of $A$, then $b \leq d$ if and only if $bd \neq 0$.*

(c) *If $e$ is a primitive idempotent of $A$, then there is a unique block, $b$ of $A$ such that $be \neq 0$ and for this $b$, $e \leq b$.*

(d) *Let*

$$1_A = b_1 + \cdots + b_r$$

*be the primitive decomposition of $1_A$ in $Z(A)$. Then for each $i$, $1 \leq i \leq r$, $Ab_i = b_i Ab_i$ is a $k$-algebra, with identity $b_i$; $Ab_i$ is an indecomposable ring and*

$$A = Ab_1 \times \cdots \times Ab_r$$

*is the unique decomposition of $A$ into a product of indecomposable factors. Further, for each $i$, $1 \leq i \leq r$, $J(Ab_i) = J(A)b_i$ and $Ab_i/J(A)b_i$ is isomorphic to $\mathrm{Mat}_{n_i}(k)$ for some $n_i \geq 1$.*

The decomposition of $1_A$ (respectively $A$) in (d) is called the *block decomposition* of $1_A$ (respectively $A$).

### 1.2. *G*-algebras.

Thevenaz's book [Th] is a good source for the material in this and the following subsections. The material here is less standard than in the previous subsection, and proofs are provided for many of the statements. Throughout, $G$ will denote a finite group.

**Definition 1.10**   A $G$-algebra (over $k$) is a pair $(A, \psi)$ where $A$ is a $k$-algebra and $\psi : G \to \mathrm{Aut}(A)$ is a group homomorphism.

Thus, a $G$-algebra is a $k$-algebra $A$ endowed with an action of $G$ by algebra automorphisms. If $(A, \psi)$ is a $G$-algebra, we suppress the notation $\psi$ by writing ${}^{g}a$ for $\psi(g)(a)$ for $g \in G$ and $a \in A$.

**Definition 1.11**   An *interior $G$-algebra* over $k$ is a pair $(A, \varphi)$, such that $A$ is a $k$-algebra and $\varphi : G \to A^{*}$ is a group homomorphism.

An interior $G$-algebra $(A, \varphi)$ is a $G$-algebra via ${}^{g}a = {}^{\varphi(g)}a$, for $g \in G$, $a \in A$. The $G$-algebra structure on an interior $G$-algebra will always be the inherited one unless stated otherwise. The map $\varphi$ will usually be clear from context and will be suppressed. Note that if $(A, \varphi)$ is an interior $G$-algebra over $k$ then $A$ is an $k[G \times G]$-module through the map $(g, h).a = \varphi(g)a\varphi(h)^{-1}$, for $a \in A$ and $(g, h) \in G \times H$.

We list some examples of $G$-algebras and interior $G$-algebras - the main example is the first one.

**Example 1.12**   (a)   As a $k$-vector space the group algebra $kG$ has a basis consisting of the elements of $G$, multiplication in $kG$ is the unique $k$-linear extension to $kG$ of the group multiplication in $G$ and the $k$-algebra structure is through the map

$$\lambda \to \lambda 1_{G}, \quad \lambda \in k.$$

The natural inclusion of $G$ in $kG$ gives $kG$ an interior $G$-algebra structure; the corresponding $G$-algebra structure on $kG$ is through conjugation, that is

$$ {}^{g}a = gag^{-1}, \quad g \in G, \ a \in kG.$$

(b)   If $N$ is a normal subgroup of $G$, then $kN$ is a $G$-algebra via the restriction of the $G$-action on $kG$ to the subalgebra $kN$. The $G$-algebra structure on $kN$ may not be the restriction of an interior $G$-algebra structure on $kN$.

(c)   Let $V$ be a left $kG$-module. For each $g \in G$ let $\varphi_{g} \in \mathrm{End}_{k}(V)$ be defined by $\varphi_{g}(v) = gv$, $v \in V$. Clearly, $\varphi_{g}$ is invertible and the map $g \to \varphi_{g}$, $g \in G$ endows $\mathrm{End}_{k}(V)$ with an interior $G$-algebra structure. The inherited $G$-algebra structure satisfies ${}^{g}(\phi)(v) = g\phi(g^{-1}v)$ for $g \in G$, $\phi \in \mathrm{End}_{k}(V)$ and $v \in V$.

(d)  Let $A$ be an (interior) $G$-algebra and let $H \leq G$. Then, $A$ is an (interior) $H$-algebra via restriction.

(e)  Let $A$ be a $G$-algebra and let $e$ be an idempotent of $A$ such that $^g e = e$ for all $g \in G$. Then $eAe$ is a $G$-algebra through the restriction of the $G$ action on $A$ to $eAe$; if in addition, $A$ is an interior $G$-algebra, then $eAe$ is an interior $G$-algebra through the map which sends an element $g$ of $G$ to the element $ege$ of $eAe$.

### 1.3. Relative trace maps and Brauer homomorphisms.

If $A$ is a $G$-algebra, and $H$ is a subgroup of $G$, then the $H$-fixed point set $A^H = \{a \in A : {}^h a = a \ \ \forall h \in H\}$ of $A$ is a $k$-subalgebra of $A$ and if $L \leq H$, then $A^H \subseteq A^L$. Further, if $g \in G$, then $^g(A^H) = A^{{}^g H}$.

**Definition 1.13**   Let $A$ be a $G$-algebra and let $L \leq H \leq G$.

(a)  Denote by $\operatorname{Res}_L^H : A^H \to A^L$ the inclusion map.

(b)  The *relative trace map* from $H$ to $L$ is the map $\operatorname{Tr}_L^H : A^L \to A^H$, defined by $\operatorname{Tr}_L^H(a) := \sum_{[h \in H/L]} {}^h a$, for $a \in A^L$. We denote by $A_L^H$ the image of $\operatorname{Tr}_L^H$.

(c)  For $g \in G$, denote by $c_{g,H} : A^H \to A^{{}^g H}$ the conjugation map defined by $a \to {}^g a$ for $a \in A^H$.

The maps $\operatorname{Res}_{\{-\}}^{\{-\}}$ are inclusions. So, for subgroups $L \leq II$ of $G$, a map $f$ with domain $A^L$ and a subset $X$ of $A^H$, we will simply write $f(X)$ instead of $f(\operatorname{Res}_L^H(X))$.

We record some of the basic properties of the above maps. The proofs are an easy exercise and can be found in [Th, Proposition 11.4].

**Proposition 1.14**   *Let $A$ be a $G$-algebra.*

(a)  $\operatorname{Tr}_L^H$, $\operatorname{Res}_L^H$ *and* $c_{g,H}$ *are $k$-linear and* $\operatorname{Res}_L^H$ *and* $c_{g,H}$ *are unitary $k$-algebra homomorphisms for all* $L \leq H \leq G$ *and all* $g \in G$.

(b)  $c_{xg,H} = c_{x,\,{}^g H} c_{g,H}$ *for all* $g, x \in G$ *and all* $H \leq G$.

(c)  $\operatorname{Tr}_H^H$, $\operatorname{Res}_H^H$ *and* $c_{h,H}$ *are identity maps for all* $H \leq G$ *and all* $h \in H$.

(d)  $\operatorname{Tr}_L^H \operatorname{Tr}_K^L = \operatorname{Tr}_K^H$ *and* $\operatorname{Res}_K^L \operatorname{Res}_L^H = \operatorname{Res}_K^H$ *for all* $K \leq L \leq H \leq G$.

(e)  $\operatorname{Res}_{{}^g L}^{{}^g H} c_{g,H} = c_{g,L} \operatorname{Res}_L^H$ *and* $\operatorname{Tr}_{{}^g L}^{{}^g H} c_{g,L} = c_{g,H} \operatorname{Tr}_L^H$ *for all* $L \leq H \leq G$ *and all* $g \in G$.

(f)  *(Mackey formula)* $\operatorname{Res}_L^H \operatorname{Tr}_K^H = \sum_{x \in [L \backslash H / K]} \operatorname{Tr}_{L \cap {}^x K}^L c_{x,\,{}^{x^{-1}} L \cap K} \operatorname{Res}_{{}^{x^{-1}} L \cap K}^K$ *for all* $L, K \leq H \leq G$. *In other words,* $\operatorname{Tr}_K^H(a) = \sum_{x \in [L \backslash H / K]} \operatorname{Tr}_{L \cap {}^x K}^L({}^x a)$ *for all* $L, K \leq H \leq G$ *and all* $a \in A^K$.

(g) *(Frobenius relations)* $\mathrm{Tr}_L^H(ea) = e\mathrm{Tr}_L^H(a)$ *and* $\mathrm{Tr}_L^H(ae) = \mathrm{Tr}_L^H(a)e$ *for all* $L \le H \le G$ *and all* $a \in A^L$, $e \in A^H$. *In particular, if $I$ is an ideal of $A^L$, then $\mathrm{Tr}_L^H(I)$ is an ideal of $A^H$.*

(h) $\mathrm{Tr}_L^H(a) = |H{:}L|{\cdot}a$ *for all* $L \le H \le G$ *and all* $a \in A^H$.

The above properties can be summed up by saying that the datum

$$(A^H, \mathrm{Res}_L^H, \mathrm{Tr}_L^H, c_{g,H})_{L \le H \le G, g \in G}$$

defines a cohomological Green functor for $G$ over $k$ (see [Th, Examples 53.1, 53.2]).

**Definition 1.15**　Let $A$ be a $G$-algebra. For $H \le G$, let $A_{<H}^H$ be the sum of all relative traces $A_L^H$ of proper subgroups $L$ of $H$, and let $A(H)$ denote the quotient $A^H/A_{<H}^H$. The Brauer homomorphism $\mathrm{Br}_H^A$ is the canonical surjection

$$\mathrm{Br}_H^A : A^H \to A(H).$$

By Proposition 1.14 (g), for $H \le G$, $A_{<H}^H$ is an ideal of $A^H$, hence $\mathrm{Br}_H^A$ is a homomorphism of $k$-algebras. If $g \in G$, then ${}^g(A^H) = A^{{}^gH}$ and by Proposition 1.14(e) ${}^g\mathrm{Tr}_L^H(a) = \mathrm{Tr}_{{}^gL}^{{}^gH}({}^ga)$ for $L \le H \le G$, and $a \in A^L$. So, ${}^g\mathrm{Ker}(\mathrm{Br}_H^A) = \mathrm{Ker}(\mathrm{Br}_{{}^gH}^A)$ and conjugation by $g$ induces an algebra isomorphism between $A(H)$ and $A({}^gH)$. In particular, if $A(H) \ne 0$, then $A(H)$ inherits a natural structure of $N_G(H)$-algebra. Further, if $K \le N_G(H)$, then the image of $A(H)^K$ under the isomorphism between $A(H)$ and $A({}^gH)$ induced by $g$ is $A({}^gH)^{{}^gK}$ and the image of $\mathrm{Ker}(\mathrm{Br}_K^{A(H)})$ is $\mathrm{Ker}(\mathrm{Br}_{{}^gK}^{{}^gA(H)})$, hence $g$ induces an isomorphism between $A(H)(K)$ and $A({}^gH)({}^gK)$.

**Definition 1.16**　For $H \le G$, $a \in A^H$ and $g \in G$, define ${}^g(\mathrm{Br}_H^A(a)) := \mathrm{Br}_{{}^gH}^A({}^ga)$ and set ${}^gA(H) = A({}^gH)$; for $K \le N_G(H)$ and $\alpha \in A(H)^K$, denote by ${}^g\mathrm{Br}_K^{A(H)}(\alpha)$ the element $\mathrm{Br}_{{}^gH}^{A({}^gH)}({}^g\alpha)$ of $A({}^gH)({}^gK)$ and set ${}^gA(H)(K) = A({}^gH)({}^gK)$.

If $k$ has characteristic $p$ and $S$ is a Sylow $p$-subgroup of $H$, then by Proposition 1.14 (h), the $k$-linearity of trace maps, and the fact that the index of $S$ in $H$ is invertible in $k$,

$$1_A = \frac{1}{|H{:}S|}\mathrm{Tr}_S^H(1_A) = \mathrm{Tr}_S^H\left(\frac{1}{|H{:}S|}1_A\right).$$

So, if $S$ is proper in $H$, then $1_A \in A_{<H}^H$ and $A(H) = 0$. Thus we get the following.

**Proposition 1.17**　*Suppose that $\mathrm{char}(k) = p$ and let $A$ be a $G$-algebra. Then $A(H)$ is non-zero only if $H$ is a $p$-group.*

The next few results sum up the interaction between relative trace maps, Brauer homomorphisms and primitive idempotents when $k$ is of characteristic $p$. They will be crucial in the next sections. The first is a consequence of the Mackey formula.

**Lemma 1.18**  *Suppose that $char(k) = p$. Let $A$ be a $G$-algebra and let $P, Q$ be subgroups of $G$. Then,*

(a)  *For all $a \in A^P$,*

$$\mathrm{Br}_Q^A(\mathrm{Tr}_P^G(a)) = \sum_{\substack{x \in [Q \backslash G / P], \\ Q \leq {}^x P}} \mathrm{Br}_Q^A({}^x a).$$

(b)  $\mathrm{Br}_Q^A(\mathrm{Tr}_Q^G(A^Q)) = \mathrm{Tr}_Q^{N_G(Q)}(A(Q)).$

*Proof.* The statement of (b) makes sense as $A(Q)$ is an $N_G(Q)$-algebra (see discussion before Definition 1.16). Let $a \in A^P$. By the Mackey formula (Lemma 1.14 (f)), $\mathrm{Tr}_P^G(a) = \sum_{x \in [Q \backslash G / P]} \mathrm{Tr}_{Q \cap {}^x P}^Q({}^x a)$, hence $\mathrm{Br}_Q^A(\mathrm{Tr}_P^G(a)) = \sum_{x \in [Q \backslash G / P]} \mathrm{Br}_Q^A(\mathrm{Tr}_{Q \cap {}^x P}^Q({}^x a))$. If $Q \cap {}^x P$ is a proper subgroup of $Q$, then $\mathrm{Br}_Q^A(\mathrm{Tr}_{Q \cap {}^x P}^Q({}^x a)) = 0$. On the other hand, if $Q \leq {}^x P$, then $\mathrm{Tr}_{Q \cap {}^x P}^Q({}^x a) = {}^x a$. This proves (a). If $P = Q$, the sum in (a) runs over a set of coset representatives of $Q$ in $N_G(Q)$. So,

$$\mathrm{Br}_Q^A(\mathrm{Tr}_Q^G(a)) = \sum_{x \in [N_G(Q)/Q]} \mathrm{Br}_Q^A({}^x a) = \sum_{x \in [N_G(Q)/Q]} {}^x \mathrm{Br}_Q^A(a)$$

$$= \mathrm{Tr}_Q^{N_G(Q)}(\mathrm{Br}_Q^A(a)),$$

for all $a \in A^P$ and this proves (b). $\qquad\qquad\qquad\square$

In the next sections, we will often be dealing with idempotents in $A^H$ for various subgroups $H$ of $G$. For instance, if $A = kG$, then $A^G = Z(kG)$, and hence the idempotents of $A^G$ are the central idempotents of $kG$ and the primitive idempotents of $A^G$ are the blocks of $kG$. Note that if $L \leq H \leq G$, and if $i$ is a primitive idempotent in $A^H$, then since $A^H \subseteq A^L$, $i$ is also an idempotent of $A^L$, but since $A^L$ may be strictly larger than $A^H$, $i$ may no longer be primitive in $A^L$.

**Lemma 1.19**  *Suppose that $char(k) = p$. Let $A$ be a $G$-algebra, $b$ a primitive idempotent of $A^G$ and $P$ a $p$-subgroup of $G$.*

(a)  *If $\mathrm{Br}_P^A(b) \neq 0$, then there exists a primitive idempotent $i$ of $A^P$ such that $i \leq b$ and $\mathrm{Br}_P^A(i) \neq 0$.*

(b)  *If $b \in \mathrm{Tr}_P^G(A^P)$, then there exists a primitive idempotent $i$ of $A^P$ such that $i \leq b$ and $b \in \mathrm{Tr}_P^G(A^P i A^P)$. Here $A^P i A^P$ denotes the smallest ideal of $A^P$ containing $i$.*

*Proof.* Let $X$ be primitive idempotent decomposition of $b$ in $A^P$. Note that this makes sense by the remark above the statement of the lemma. Suppose that $\operatorname{Br}_P^A(b) \neq 0$. Applying $\operatorname{Br}_P^A$ to both sides of the equation $b = \sum_{i \in X} i$ gives

$$0 \neq \operatorname{Br}_P^A(b) = \sum_{i \in X} \operatorname{Br}_P^A(i).$$

Thus, $\operatorname{Br}_P^A(i) \neq 0$ for some $i \in X$, proving (a).

Now suppose that $b \in \operatorname{Tr}_P^G(A^P)$, say $b = \operatorname{Tr}_P^G(a)$, $a \in A^P$. So,

$$b = b^2 = b\operatorname{Tr}_P^G(a) = \operatorname{Tr}_P^G(ba) = \operatorname{Tr}_P^G(\sum_{i \in X} ia) = \sum_{i \in X} \operatorname{Tr}_P^G(ia).$$

Here, the third equality follows from Proposition 1.14 (g) and the fifth by Proposition 1.14 (a). Now $\operatorname{Tr}_P^G(ia) \in \operatorname{Tr}_P^G(A^P i A^P)$, for all $i \in X$. So, the above displayed equation gives that $b \in \sum_{i \in X} \operatorname{Tr}_P^G(A^P i A^P)$. By Proposition 1.14 (g), for each $i \in X$, $\operatorname{Tr}_P^G(A^P i A^P)$ is an ideal of $A^G$, and $b$ is a primitive idempotent of $A^G$. Thus, by Rosenberg's Lemma (Lemma 1.5), $b \in \operatorname{Tr}_P^G(A^P i A^P)$ for some $i \in X$, proving (b). $\qquad\square$

**Lemma 1.20**  *Suppose that $char(k) = p$. Let $A$ be a $G$-algebra, and let $P$ and $Q$ be subgroups of $G$. Let $b$ be a primitive idempotent of $A^G$, $i$ a primitive idempotent of $A^P$ and $j$ a primitive idempotent of $A^Q$ such that $i \leq b$ and $j \leq b$. Suppose that $\operatorname{Br}_Q^A(j) \neq 0$ and $b \in \operatorname{Tr}_P^G(A^P i A^P)$. Then, there exists $x \in G$ and $u \in A^{Q^*}$ such that $Q \leq {}^x P$ and ${}^u j \leq {}^x i$.*

*Proof.* Write $b = \operatorname{Tr}_P^G(a)$, $a \in A^P i A^P$. By Lemma 1.18 (a)

$$\operatorname{Br}_Q^A(b) = \sum_{\substack{x \in [Q \backslash G / P], \\ Q \leq {}^x P}} \operatorname{Br}_Q^A({}^x a)$$

and by hypothesis, $j = jbj$. So,

$$\operatorname{Br}_Q^A(j) = \sum_{\substack{x \in [Q \backslash G / P], \\ Q \leq {}^x P}} \operatorname{Br}_Q^A(j\,{}^x aj).$$

By hypothesis $\operatorname{Br}_Q^A(j) \neq 0$ and $\operatorname{Br}_Q^A$ is a surjective homomorphism of $k$-algebras. So, by idempotent lifting, $\operatorname{Br}_Q^A(j)$ is a primitive idempotent of $A(Q)$. By Proposition 1.7, it follows that $\operatorname{Br}_Q^A(j)A(Q)\operatorname{Br}_Q^A(j)$ is a local algebra, with identity $\operatorname{Br}_Q^A(j)$. Since by definition, the set of non-invertible elements of a local algebra forms a proper ideal of the algebra, it follows from the above equation that there exists some $x \in G$ such that $Q \leq {}^x P$ and such that $\operatorname{Br}_Q^A(j\,{}^x aj) = \operatorname{Br}_Q^A(j)\operatorname{Br}_Q^A({}^x a)\operatorname{Br}_Q^A(j)$ is an invertible element of $\operatorname{Br}_Q^A(j)A(Q)\operatorname{Br}_Q^A(j)$. Consequently, $\operatorname{Br}_Q^A(j\,{}^x aj)$ and hence $j\,{}^x aj \in A^Q$ is not nilpotent. Since $J(A^Q)$ contains only nilpotent elements, this means that $j\,{}^x aj \notin J(A^Q)$. By hypothesis, $a \in A^P i A^P$, hence $j\,{}^x aj$ is a finite sum of elements of the form $jd\,{}^xicj$, where $d, c \in A^{{}^xP} \subseteq A^Q$. Further, $j \in A^Q$ and $J(A^Q)$ is an ideal of $A^Q$. So, $j\,{}^xaj \notin J(A^Q)$ implies that there exists $d \in A^Q$ such that $jd\,{}^xi \notin J(A^Q)$. Let $I$ be a primitive idempotent decomposition of ${}^xi$ in $A^Q$, so $jd\,{}^xi = jd\sum_{i' \in I} i'$. Again, the fact that $J(A^Q)$ is an ideal of $A^Q$ implies that $jdi' \notin J(A^Q)$ for some $i' \in I$. So, by Proposition 1.3, applied to the primitive idempotents $j$ and $i'$ of the algebra $A^Q$, $i' = {}^uj$ for some $u \in A^{Q*}$. This proves the result as clearly, $i' \leq {}^xi$. $\qquad\square$

**Lemma 1.21** *Suppose that $char(k) = p$. Let $A$ be a $G$-algebra and $b$ a primitive idempotent of $A^G$. There exists a $p$-subgroup $P$ of $G$ and a primitive idempotent $i$ of $A^P$ such that*

$$i \leq b, \ \mathrm{Br}_P^A(i) \neq 0 \ and \ \ b \in \mathrm{Tr}_P^G(A^P i A^P).$$

*Proof.* Let $P$ be minimal amongst subgroups $G$ such that $b \in \mathrm{Tr}_P^G(A^P)$. By Lemma 1.19, there exists a primitive idempotent $i$ of $A^P$ with $i \leq b$ and $b = \mathrm{Tr}_P^G(a)$ for some $a \in A^P i A^P$. We will show that $\mathrm{Br}_P^A(i) \neq 0$. This will prove the lemma, since by Proposition 1.17, $\mathrm{Br}_P^A(A^P)$ being non-zero implies that $P$ is a $p$-group. Suppose if possible that $\mathrm{Br}_P^A(i) = 0$, that is $i \in A_{<P}^P$. The idempotent $i$ is primitive in $A^P$ and $A_{<P}^P$ is a sum of ideals of the form $A_Q^P$, $Q$ a proper subgroup of $G$. Hence by Lemma 1.5, $i \in \mathrm{Tr}_Q^P(A^Q)$ for some proper subgroup $Q$ of $P$. By Proposition 1.14 (g), $\mathrm{Tr}_Q^P(A^Q)$ is an ideal of $A^P$, hence $a \in A^P i A^P \subseteq \mathrm{Tr}_Q^P(A^Q)$ . But then by Proposition 1.14(d), $b \in \mathrm{Tr}_Q^P(A^Q)$, a contradiction to the minimality of $P$. $\qquad\square$

We will also need the following result.

**Lemma 1.22** *Assume that $char(k) = p$ and let $A$ be a $G$-algebra. Let $R$ be a subgroup of $G$ and let $C$ be a normal subgroup of $G$. Suppose that $1_A \in \mathrm{Tr}_R^G(A^R)$ and $1_A$ is primitive in $A^C$. Then, $RC/C$ contains a Sylow $p$-subgroup of $G/C$.*

*Proof.* Let $a \in A^R$ be such that

$$1_A = \mathrm{Tr}_R^G(a) = \mathrm{Tr}_{RC}^G(\mathrm{Tr}_R^{RC}(a)),$$

and set $u := \mathrm{Tr}_R^{RC}(a)$. Then, $u \in A^{RC} \subseteq A^C$. By hypothesis, the identity $1_A = 1_{A^C}$ of $A^C$ is the only idempotent of $A^C$. In other words, $A^C$ is a local algebra which means that $J(A^C)$ has co-dimension 1 in $A^C$ (see Proposition 1.7 (f)). Thus, $u = \lambda 1_A + v$ for some $\lambda \in k$ and $v \in J(A^C)$. This gives

$$1_A = \mathrm{Tr}_{RC}^G(\lambda 1_A + v) = |G{:}RC|\lambda 1_A + \mathrm{Tr}_{RC}^G(v).$$

Now, since $C$ is normal in $G$, $A^C$ is invariant under the action of $G$. In other words, the restriction of the action of $G$ on $A$ to the subalgebra $A^C$ gives $A^C$ the structure of an $G$-algebra. Since the Jacobson radical is an invariant of algebra automorphisms, it follows that $J(A^C)$ is $A$-invariant. In particular, $\mathrm{Tr}^G_{RC}(v) \in J(A^C)$. But $1_A \notin J(A^C)$. Hence, it follows from the above displayed equation that $|G{:}RC|\lambda 1_A \neq 0$, and consequently that $|G{:}RC|$ is not divisible by $p$. $\qquad\square$

## 2. $p$-PERMUTATION ALGEBRAS, BRAUER PAIRS AND FUSION SYSTEMS

In [AB], Alperin and Broué introduced the poset of Brauer pairs associated to a block of finite group algebra. A prototype of these pairs had been considered earlier by Brauer [Br4] and were also considered by Olsson[Ols2]. The results of [AB] were reinterpreted and extended to (certain) idempotents of a more general class of $G$-algebras in [BP2], $G$ a finite group. The aim in this section is to describe the main properties of Brauer pairs and to show how fusion systems arise from them (Proposition 2.22). For the moment, we will work in the general setting of $p$-permutation $G$-algebras; however the case of most interest for block theory is $A = kN$ for $N$ a normal subgroup of $G$, considered as $G$-algebra through the conjugation action of $G$; and even more particularly the sub-case $A = kG$. While reading the present section it might be helpful for the reader to occasionally refer ahead to subsection 3.4, where we discuss how the concepts introduced in this section may be understood in the above special case.

Throughout this section $k$ will denote an algebraically closed field of characteristic $p > 0$, $G$ a finite group and $A$ a $G$-algebra over $k$, finite dimensional as $k$-vector space.

### 2.1. $p$-permutation algebras and the Brauer homomorphisms.

**Definition 2.1**  • A subset $\mathcal{B}$ of $A$ is *P-stable* for a subgroup $P$ of $G$ if $^x a \in \mathcal{B}$ for all $a \in \mathcal{B}$ and all $x \in P$.

• The $G$-algebra $A$ is a *p-permutation algebra* if for any $p$-subgroup $P$ of $G$, $A$ has a $k$-basis which is $P$-stable.

If $\mathcal{B}$ is an $S$-stable $k$-basis of $A$, for $S$ a subgroup of $G$, then $\mathcal{B}$ is $P$-stable for any subgroup $P$ of $S$, and $^g\mathcal{B}$ is a $^gS$-stable $k$-basis of $A$ for any $g \in G$. So $A$ is a $p$-permutation algebra if and only if some Sylow $p$-subgroup of $G$ stabilises a $k$-basis of $A$. The group algebra $kG$ is a $p$-permutation algebra since the $k$-basis consisting of the elements of $G$ is $P$-stable for any subgroup $P$ of $G$. Similarly, the algebra $kN$, for $N$ a normal subgroup of $G$ (see Example 1.12 (b)) is a $p$-permutation $G$-algebra.

The existence of a $P$-stable basis, for $p$-subgroups $P$ of $G$ yields convenient alternate descriptions of relative trace maps and Brauer homomorphisms. Recall from Proposition 1.17 that if $H \leq G$ is not a $p$-group, then $A(H) = 0$.

**Lemma 2.2**  *Let $P$ be a $p$-subgroup of $G$ and suppose that $\mathcal{B}$ is a $P$-stable $k$-basis of $A$.*

(a)  *The set $\{C^+ : C \in Orb_P(\mathcal{B})\}$ is a $k$-basis of $A^P$.*

(b)  *For $Q \leq P$, $\{C^+ : C \in Orb_P(\mathcal{B})$ s.t. $C_P(x) \leq Q$ for $x \in C\}$ is a $k$-basis of $\mathrm{Tr}_Q^P(A^Q)$.*

(c)  *The set $\{C^+ : C \in Orb_P(\mathcal{B})$ s.t. $C \cap \mathcal{B}^P = \varnothing\} = \{C^+ : C \in Orb_P(\mathcal{B})$ s.t. $|C| > 1\}$ is a basis of $A^P_{<P}$.*

(d)  *For $Q \leq P$, the set $\{C^+ : C \in Orb_P(\mathcal{B})$ s.t. $C \cap \mathcal{B}^Q = \varnothing\}$, is a $k$-basis of $A^Q_{<Q} \cap A^P$. In particular, $A^P \cap A^Q_{<Q} \subseteq A^P_{<P}$.*

(e)  *The set $\{x + A^P_{<P} : x \in \mathcal{B}^P\}$ is a $k$-basis of $A(P)$.*

(f)  *For any element $a = \sum_{x \in \mathcal{B}} \alpha_x x$ of $A^P$, $\mathrm{Br}_P^A(a) = \sum_{x \in \mathcal{B}^P} \alpha_x x + A^P_{<P}$.*

*Proof.* (a) is clear. Before proceeding with the other parts of the proof, we record an elementary fact. If $V_1$ and $V_2$ are $k$-vector spaces with basis $A_1$ and $A_2$ respectively and $f : V_1 \to V_2$ is a linear map such that $f(A_1) \subseteq f(A_2)$, then $f(A_1)$ is a basis of the image $\mathrm{Im}(f)$ of $f$ and $\{f(a) + \mathrm{Im}(f) : a \in A_2 - f(A_1)\}$ is a basis of the cokernel $V_2/\mathrm{Im}(f)$ of $f$.

Let $Q \leq P$, and let $C \in Orb_Q(\mathcal{B})$, $x \in C$. Clearly, $C^+ = \mathrm{Tr}_{C_Q(x)}^Q(x)$, hence by Proposition 1.14(c),

$$\mathrm{Tr}_Q^P(C^+) = \mathrm{Tr}_Q^P \mathrm{Tr}_{C_Q(x)}^Q(x)$$

$$= \mathrm{Tr}_{C_P(x)}^P \mathrm{Tr}_{C_Q(x)}^{C_P(x)}(x) = |C_P(x):C_Q(x)| \mathrm{Tr}_{C_P(x)}^P(x).$$

Since $k$ has characteristic $p$, the above is 0 if $C_Q(x)$ is a proper subgroup of $C_P(x)$ and is the $P$-orbit sum of $x$ if $C_P(x) \leq Q$. By (a), the $Q$ (respectively $P$)-orbit sums of $\mathcal{B}$ form a basis of $A^Q$ (respectively $A^P$). Thus (b) follows by the observation above, applied with $V_1 = A^Q$, $V_2 = A^P$ and $f = \mathrm{Tr}_Q^P$. Assertion (c) is immediate from (b). Next, we prove (d). Let $Q \leq P$ and let $a \in A^Q_{<Q} \cap A^P$. By (a), we may write $a = \sum_{C \in Orb_P(\mathcal{B})} \alpha_C C^+$. Suppose that $C \in \mathcal{B}$ is such that $\alpha_C \neq 0$. Since $a \in A^Q_{<Q}$, by part (c), applied to $Q$ instead of $P$, it follows that for any $x \in C$, the $Q$-orbit of $x$ has size greater than 1, that is $x \notin \mathcal{B}^Q$. Thus, $A^Q_{<Q} \cap A^P$ is contained in the $k$-span of $\{C^+ : C \in Orb_P(\mathcal{B})$ s.t. $C \cap \mathcal{B}^Q = \varnothing\}$. Conversely, let $C$ be a $P$-orbit of $\mathcal{B}$ with $C \cap \mathcal{B}^Q = \varnothing$. Then $C$ is a union of $Q$-orbits each of length greater than 1, hence by (c) applied to $Q$, $C^+ \in A^Q_{<Q}$. This proves the first

assertion of (d). The second follows from the first and (c). Part (e) follows from (c), by applying the observation on linear transformations above to the inclusion of $A^P_{\leq P}$ in $A^P$. Finally, (f) is immediate from (e) and the definition of Brauer homomorphisms.          $\square$

Recall from the remarks after Proposition 1.17, that if $P \leq G$ is a $p$-group with $A(P) \neq 0$, then $A(P)$ is a $N_G(P)$-algebra. The following is an easy consequence of Lemma 2.2 (e).

**Proposition 2.3**  *If $A$ is a $p$-permutation $G$-algebra, then $A(P)$ is a $p$-permutation $N_G(P)$-algebra for any $p$-subgroup $P$ of $G$ such that $A(P) \neq 0$.*

*Proof.* Let $P \leq G$ be such that $A(P) \neq 0$ and let $S$ be a Sylow $p$-subgroup of $N_G(P)$. By the remark following Definition 2.1, it suffices to prove that $A(P)$ has an $S$-stable $k$-basis. So, let $\mathcal{B}$ be an $S$-stable (and hence $P$-stable) $k$-basis of $A$. By Lemma 2.2 (e), $\{x + A^P_{<P} : x \in \mathcal{B}^P\}$ is a $k$-basis of $A(P)$. This basis is $S$-stable since $P$ being normal in $S$ implies that $\mathcal{B}^P$ is $S$-stable.          $\square$

If $A$ is a $p$-permutation $G$-algebra, and $Q \leq P$ are $p$-subgroups of $G$, then by Lemma 2.2 (c), $\mathrm{Ker}(\mathrm{Br}_Q^A) \cap A^P$ is contained in $\mathrm{Ker}(\mathrm{Br}_P^A)$. Thus, $\mathrm{Br}_P^A$ factors through the restriction of $\mathrm{Br}_Q^A$ to $A^P$.

**Definition 2.4**  Let $A$ be a $p$-permutation $G$-algebra and let $Q \leq P$ of $G$ be $p$-subgroups of $G$. Define

$$\mathrm{Br}_{P,Q}^A : \mathrm{Br}_Q^A(A^P) \to A(P)$$

by $\mathrm{Br}_{P,Q}^A(\mathrm{Br}_Q^A(a)) = \mathrm{Br}_P^A(a)$ for $a \in A^P$.

Clearly, if $A(P) \neq 0$, then $\mathrm{Br}_{P,Q}^A$ is a surjective homomorphism of $k$-algebras. In particular, if $A(P) \neq 0$ then $A(Q) \neq 0$. Also, note that while the fact that $\mathrm{Br}_{P,Q}^A$ is well defined depends on the existence of a $P$-stable $k$-basis of $A$, the definition of $\mathrm{Br}_{P,Q}^A$ is independent of the choice of such a basis.

**Proposition 2.5**  *Let $A$ be a $p$-permutation $G$-algebra and let $Q \trianglelefteq P$ be $p$-subgroups of $G$ such that $A(P) \neq 0$. Then,*

$$A(Q)^P = \mathrm{Br}_Q^A(A^P) \quad and \quad \mathrm{Ker}(\mathrm{Br}_{P,Q}^A) = \mathrm{Ker}(\mathrm{Br}_P^{A(Q)}).$$

*Proof.* The inclusion $\mathrm{Br}_Q^A(A^P) \subseteq A(Q)^P$ is clear since $\mathrm{Br}_Q^A$ is a homomorphism of $N_G(Q)$ algebras. For the reverse inclusion, let $\mathcal{B}$ be a $P$-stable basis of $A$. By Lemma 2.2 (c) applied to $Q$, $\{\mathrm{Br}_Q^A(x) : x \in \mathcal{B}^Q\}$ is a $k$-basis of $A(Q)$. Since $Q$ is normal in $P$, this basis is $P$-stable, and it follows by Lemma 2.2 (a), that the $P$-orbit sums of $\{\mathrm{Br}_Q^A(x) : x \in \mathcal{B}^Q\}$ is a $k$-basis of $A(Q)^P$, that is the set $\{\mathrm{Br}_Q^A(C^+)\}$ as $C$ runs through the $P$-orbits of $\mathcal{B}^Q$

is a $k$-basis of $A(Q)^P$. Now since $Q$ is normal in $P$, any $P$-orbit $C$ of $\mathcal{B}^Q$ is also a $P$-orbit of $\mathcal{B}$ and consequently $C^+$ is an element of $A^P$. This shows that $A(Q)^P \subseteq \mathrm{Br}_Q^A(A^P)$. Again by Lemma 2.2 (c) applied to the $P$-stable basis $\{\mathrm{Br}_Q^A(x) : x \in \mathcal{B}^Q\}$ of $A(Q)$, $\mathrm{Ker}(\mathrm{Br}_P^{A(Q)})$ has basis $\{\mathrm{Br}_Q^A(C^+)\}$, where $C$ runs over the $P$-orbits of $\mathcal{B}^Q$ of size greater than 1. But by Lemma 2.2 (c) and (f) applied to $P$ and $Q$ respectively, this set is precisely the image of the basis $\{C^+ : C \in Orb_P\mathcal{B}, |C| > 1\}$ of $\mathrm{Ker}(\mathrm{Br}_P^A)$ under $\mathrm{Br}_Q^A$. Thus, $\mathrm{Ker}(\mathrm{Br}_P^{A(Q)}) = \mathrm{Br}_Q^A(\mathrm{Ker}(\mathrm{Br}_P^A)) = \mathrm{Ker}(\mathrm{Br}_{P,Q}^A)$. $\qquad\square$

It follows from the above that if $A$ is a $p$-permutation $G$-algebra, then for $p$-subgroups of $G$, $Q \trianglelefteq P$ of $G$, $\mathrm{Br}_{P,Q}^A$ induces a $k$-algebra isomorphism between $A(Q)(P)$ and $A(P)$.

**Definition 2.6**   With the notation and hypothesis of the above proposition, let $b_{P,Q}^A : A(Q)(P) \to A(P)$ denote the isomorphism induced by $\mathrm{Br}_{P,Q}^A$.

Note that for $P$ a $p$-subgroup of $N_G(Q)$ containing $Q$, $b_{P,Q}^A$ satisfies and is completely determined by the condition

$$b_{P,Q}^A(\mathrm{Br}_P^{A(Q)}(\mathrm{Br}_Q^A(x))) = \mathrm{Br}_{P,Q}^A(\mathrm{Br}_P^A(x)) = \mathrm{Br}_P^A(x) \quad \text{for all} \ \ x \in A^P.$$

The map $b_{P,Q}$ is $G$-equivariant in the sense that for all $g \in G, w \in A(Q)(P)$, and $z \in A^P$

$$b_{{}^gP,\,{}^gQ}^A({}^gw) = {}^g(b_{P,Q}^A(w)).$$

In particular, $b_{P,Q}^A$ is an isomorphism of $N_G(Q) \cap N_G(P)$-algebras.

## 2.2. $(A, G)$-Brauer pairs and inclusion.

For the rest of this section $A$ will denote a $p$-permutation $G$-algebra.

**Definition 2.7**   An $(A, G)$-Brauer pair is a pair $(P, e)$, where $P$ is a $p$-subgroup of $G$ such that $A(P) \neq 0$ and $e$ is a block of $A(P)$. If $(P, e)$ is an $(A, G)$-Brauer pair and $i$ is an idempotent of $A^P$, we say that $i$ is associated to $(P, e)$ or more simply that $i$ is associated to $e$ if $\mathrm{Br}_P^A(i) \neq 0$ and $\mathrm{Br}_P^A(i) \leq e$. In other words, $i$ is associated to $e$ if and only if

$$e\mathrm{Br}_P^A(i) = \mathrm{Br}_P^A(i)e = \mathrm{Br}_P^A(i) \neq 0.$$

**Lemma 2.8**   *Let $P$ be a $p$-subgroup of $G$. For a primitive idempotent $i$ of $A^P$ and a block $e$ of $A(P)$, $i$ is associated to $(P, e)$ if and only if $\mathrm{Br}_P^A(i)e \neq 0$. Moreover, for any block $e$ of $A(P)$, there exists a primitive idempotent of $A^P$ associated to $(P, e)$ and for any primitive idempotent $i$ of $A^P$ such that $\mathrm{Br}_P^A(i) \neq 0$, there is a unique block $e$ of $A(P)$ such that $i$ is associated to $(P, e)$.*

*Proof.* If $i$ is a primitive idempotent of $A(P)$ such that $\mathrm{Br}_P^A(i) \neq 0$, then by Proposition 1.4 (a), $\mathrm{Br}_P^A(i)$ is a primitive idempotent of $A(P)$. Hence, if $\mathrm{Br}_P^A(i) \neq 0$, then by Proposition 1.9(c), there is a unique block $e$ of of $A(P)$ such that $\mathrm{Br}_P^A(i)e \neq 0$ and this $e$ contains $\mathrm{Br}_P^A(i)$. This proves the first and third assertions. For the second, let $(P, e)$ be an $(A, G)$-Brauer pair and let $I$ be a primitive idempotent decomposition of $1_A$ in $A^P$. Since

$$0 \neq e = \mathrm{Br}_P^A(1_A)e = \sum_{i \in I} \mathrm{Br}_P^A(i)e,$$

$\mathrm{Br}_P^A(i)e \neq 0$ for some $i \in I$, and by the first assertion $i$ is asociated to $(P, e)$. $\qquad\square$

**Definition 2.9**   Let $(Q, f)$ and $(P, e)$ be $(A, G)$-Brauer pairs. We say that $(Q, f)$ is contained in $(P, e)$ and write $(Q, f) \leq (P, e)$ if $Q \leq P$ and if any primitive idempotent $i$ of $A^P$ which is associated to $e$ is also associated to $f$.

Note that if $Q \leq P$ then $A^P \leq A^Q$, hence if $i \in A^P$ is a primitive idempotent then $i$ is a (not necessarily primitive) idempotent of $A^Q$. Further since $\mathrm{Br}_P^A$ factors through the restriction of $\mathrm{Br}_Q^A$ to $A^P$, if $\mathrm{Br}_P^A(i) \neq 0$, then $\mathrm{Br}_Q^A(i) \neq 0$. Thus, $(Q, f) \leq (P, e)$ if and only if $\mathrm{Br}_Q^A(i)f = \mathrm{Br}_Q^A(i)$ for any primitive idempotent $i$ of $A^P$ associated to $e$.

The next result gives the fundamental properties of inclusion of Brauer pairs. Recall from Definition 2.4 and Proposition 2.5 the map $\mathrm{Br}_{P,Q}^A \colon A(Q)^P \to A(P)$ for $p$-subgroups $Q \trianglelefteq P$, of $G$.

**Theorem 2.10**   *Let $(P, e)$ be an $(A, G)$-Brauer pair and let $Q \leq P$.*

(a)   *There exists a unique block $f$ of $A(Q)$ such that $(Q, f) \leq (P, e)$.*

(b)   *If $Q$ is normal in $P$, then the block $f$ of (a) is the unique block of $A(Q)$ which is $P$-stable and such that $\mathrm{Br}_{P,Q}^A(f)e = e$.*

(c)   *Inclusion of $(A, G)$-pairs is a transitive relation.*

The above theorem is due to Alperin-Broué [AB] for the case of group algebras and to Broué-Puig [BP2] in the general case. The proof is essentially in the book [Th, Section 40]. We give details here since [Th] treats only the case $A = kG$ and uses the language of pointed groups which we have not introduced.

**Lemma 2.11**   *Let $(P, e)$ and $(Q, f)$ be $(A, G)$-Brauer pairs such that $(Q, f) \leq (P, e)$.*

(a)   *If $f'$ is a block of $A(Q)$ different from $f$, then for any primitive idempotent $i \in A^P$ associated to $e$, we have*

$$\mathrm{Br}_Q^A(i)f' = 0.$$

*In particular, $f$ is the unique block of $A(Q)$ with $(Q, f) \leq (P, e)$.*

(b)  *Let $(S, d)$ and $(S, d')$ be $(A, G)$-Brauer pairs such that $(S, d') \leq (Q, f)$ and $(S, d) \leq (P, e)$. Then $d = d'$.*

*Proof.* Let $f' \neq f$ be as in (a) and let $i \in A^P$ be a primitive idempotent associated to $e$. By hypothesis, $(Q, f) \leq (P, e)$, hence

$$0 \neq \mathrm{Br}_Q^A(i) = \mathrm{Br}_Q^A(i)f.$$

On the other hand, $f'f = 0$, so

$$\mathrm{Br}_Q^A(i)f' = \mathrm{Br}_Q^A(i)ff' = 0.$$

This proves (a).

Now let $S$, $d$ and $d'$ be as in (b) and let $i$ be a primitive idempotent of $A^P$ associated to $e$. We will show that $\mathrm{Br}_S^A(i)d' \neq 0$ and by (a), applied with $(Q, f)$ replaced by $(S, d)$ and $(Q, f')$ replaced by $(S, d')$, it will follow that $d' = d$, thus proving (b). Let $I$ be a primitive idempotent decomposition of $i$ in $A^Q$. Since $(Q, f) \leq (P, e)$, $i$ is also associated to $f$. Hence,

$$\sum_{j \in I} \mathrm{Br}_Q^A(j)f = \mathrm{Br}_Q^A(i)f = \mathrm{Br}_Q^A(i) \neq 0$$

and it follows that $\mathrm{Br}_Q^A(j)f \neq 0$ for some $j \in I$. By Lemma 2.8, $j$ is associated to $(Q, f)$. Since $(S, d') \leq (Q, f)$, $j$ is also associated to $d'$. So, $\mathrm{Br}_S^A(i)d' \neq 0$ as required.  $\square$

Recall from Proposition 1.9 that any idempotent of a commutative $k$-algebra has a unique primtitive decomposition in the algebra and that this is a subset of the unique primitive decomposition of the identity of the algebra.

**Lemma 2.12**  *Let $H$ be a finite group, $B$ a commutative $H$-algebra, finite dimensional as a vector space over $k$ and let $I$ be the unique primitive idempotent decomposition of $1_B$ in $B$. The set $\{J^+\}$ as $J$ runs over the set of $H$-orbits of $I$ is the unique primitive idempotent decomposition of $1_B$ in $B^H$.*

*Proof.* The statement makes sense since by Proposition 1.9, $I$ is the set of primitive idempotents of $B$ and is therefore $H$-stable. Let $e$ be an idempotent of $B^H$, and let $J \subseteq I$ be the primitive decomposition of $e$ in $B$. Since $^h e = e$, $^h J$ is also a primitive decomposition of $e$ in $B$. Since $J$ is the unique primitive decomposition of $e$ in $B$, it follows that $J$ is a union of $H$-orbits of $I$ and $e = J^+$. On the other hand, if $J$ is an $H$-orbit of $I$, then $J^+$ is an idempotent of $B^H$. Thus, the set of primitive idempotents of $B^H$ is precisely the set of orbit sums of $H$ on $I$.  $\square$

We now prove Theorem 2.10.

*Proof.* We will first prove (b) and (a) in the case that $Q$ is normal in $P$. Then, we will prove (a) in the general case. Part (c) will follow easily part (a) and Lemma 2.11 (b). Before proceeding, note that once the existence of a pair $(Q, f)$ such that $(Q, f) \leq (P, e)$ is proved, uniqueness follows from Lemma 2.11 (a). Suppose that $Q$ is normal in $P$. Then $A(Q)$ and hence $Z(A(Q))$ is a $P$-algebra and $P$ acts on the blocks of $A(Q)$. We will show that there exists a $P$-stable block $f$ of $A(Q)$ such that $\mathrm{Br}^A_{P,Q}(f)e = e$. The map $\mathrm{Br}^A_{P,Q} : A(Q)^P \to A(P)$ is surjective hence maps $Z(A(Q)^P)$ into $Z(A(P))$ and since $Z(A(Q))^P \subseteq Z(A(Q)^P)$, the image of $Z(A(Q))^P$ under $\mathrm{Br}^A_{P,Q}$ also lies in $Z(A(P))$. Let $I$ be the unique primitive decomposition of $1_{A(Q)}$ in $Z(A(Q))^P$. The set of non-zero elements of the image of $I$ under $\mathrm{Br}^A_{P,Q}$ is a decomposition of $1_{A(P)}$ in $Z(A(P))$ and $e$ is primitive in $Z(A(P))$. Thus there is a unique primitive idempotent $f$ of $Z(A(Q))^P$ such that $\mathrm{Br}^A_{P,Q}(f)e = e$. By Lemma 2.12 applied to the commutative $P$-algebra $Z(A(Q))$, $f$ is the sum of a $P$-orbit of blocks of $A(Q)$, that is $f = \mathrm{Tr}^P_R(f')$ for some block $f'$ of $A(Q)$ and some subgroup $R$ of $P$. Suppose, if possible, that $R$ is a proper subgroup of $P$. Then $f \in \mathrm{Ker}(\mathrm{Br}^{A(Q)}_P)$ and by Proposition 2.5, $\mathrm{Ker}(\mathrm{Br}^{A(Q)}_P) = \mathrm{Ker}(\mathrm{Br}^A_{P,Q})$. So $\mathrm{Br}^A_{P,Q}(f) = 0$, whence $0 = \mathrm{Br}^A_{P,Q}(f)e = e$, a contradiction. So $R = P$ and $f = f'$ is a $P$-stable block of $A(Q)$ as required.

Thus, in order to prove part (b) (and part (a) when $Q \trianglelefteq P$), it suffices to prove that if $f$ is a $P$-stable block of $A(Q)$ such that $\mathrm{Br}^A_{P,Q}(f)e = e$, then $(Q, f) \leq (P, e)$. So, let $f$ be a $P$-stable block of $A(Q)$ with $\mathrm{Br}^A_{P,Q}(f)e = e$, and let $i$ be a primitive idempotent of $A^P$ associated to $e$. Then,

$$\mathrm{Br}^A_{P,Q}(\mathrm{Br}^A_Q(i)f) = \mathrm{Br}^A_P(i)\mathrm{Br}^A_{P,Q}(f)$$
$$= \mathrm{Br}^A_P(i)e\mathrm{Br}^A_{P,Q}(f) = \mathrm{Br}^A_P(i)e = \mathrm{Br}^A_P(i) \neq 0,$$

and in particular, $\mathrm{Br}^A_Q(i)f \neq 0$. By Lemma 2.8, $i$ is associated to $f$, hence $(Q, f) \leq (P, e)$.

Next we prove (a) in the general case. We proceed by induction on $|P{:}Q|$. Let $Q$ be a proper subgroup of $P$ and let $N = N_P(Q)$. By the inductive hypothesis, for every subgroup $R$ of $P$ which properly contains $Q$, there is a unique block, say $e_R$ of $A(R)$ such that $(R, e_R) \leq (P, e)$. Since $Q$ is normal in $N$, by (b) there exists a block $f$ of $Z(A(Q))$ such that $(Q, f) \leq (N, e_N)$. Note that $e_N$ is defined as $Q$ is proper in $P$ and hence in $N_P(Q)$. We claim that for any pair $R$ and $T$ such that $Q < R \leq T \leq P$, $(R, e_R) \leq (T, e_T)$. Indeed, since $|T{:}R| < |P{:}Q|$, by the inductive hypothesis, there exists a block $e'_R$ of $A(R)$ such that $(R, e'_R) \leq (T, e_T)$. By Lemma 2.11 (b), applied to the groups $R, T, P$, $e'_R = e_R$ and $(R, e_R) \leq (T, e_T)$ as claimed. A similar argument to the one just given shows that for any $R$ such that $Q < R \leq N$, $(Q, f) \leq (R, e_R)$.

We will show that $(Q, f) \leq (P, e)$. So, let $i$ be a primitive idempotent of $A^P$ associated to $e$. We must show that $\mathrm{Br}_Q^A(i)f = \mathrm{Br}_Q^A(i)$ or equivalently that $\mathrm{Br}_Q^A(i)(1 - f) = 0$. As a first step, we show that

$$\mathrm{Br}_Q^A(i)(1 - f) \in \cap_{Q < R \leq N}\mathrm{Ker}(\mathrm{Br}_R^{A(Q)}) \cap A(Q)^N.$$

Let $R$ be a subgroup of $N$ properly containing $Q$. Then, $(Q, f) \leq (R, e_R)$ and $Q$ is normal in $R$. So, by (b), $f$ is $R$-stable and $\mathrm{Br}_{R,Q}^A(f)e_R = e_R$. Also, $(R, e_R) \leq (P, e)$, so $\mathrm{Br}_R^A(i)e_R = \mathrm{Br}_R^A(i)$ and

$$\mathrm{Br}_R^A(i)\mathrm{Br}_{R,Q}^A(f) = \mathrm{Br}_R^A(i)e_R\mathrm{Br}_{R,Q}^A(f) = \mathrm{Br}_R^A(i)e_R = \mathrm{Br}_R^A(i).$$

Since $\mathrm{Br}_R^A(i) = \mathrm{Br}_{R,Q}^A(\mathrm{Br}_Q^A(i))$, the above gives that $\mathrm{Br}_Q^A(i)(1 - f) \in \mathrm{Ker}(\mathrm{Br}_{R,Q}^A)$ and by Proposition 2.5, $\mathrm{Ker}(\mathrm{Br}_{R,Q}^A) = \mathrm{Ker}(\mathrm{Br}_R^{A(Q)})$. Since $i \in A^P$ and $f \in A(Q)^N$, it follows that

$$\mathrm{Br}_Q^A(i)(1 - f) \in \cap_{Q < R \leq N}\mathrm{Ker}(\mathrm{Br}_R^{A(Q)}) \cap A(Q)^N.$$

Next, we claim that $\cap_{Q < R \leq N} \mathrm{Ker}(\mathrm{Br}_R^{A(Q)}) \cap A(Q)^N = \mathrm{Br}_Q^A(\mathrm{Tr}_Q^P(A^Q))$. For this, let $\mathcal{B}$ be an $N$-stable basis of the $N_G(Q)$-algebra $A(Q)$ (see Proposition 2.3). Since $Q$ acts trivially on $A(Q)$, $Q \leq C_N(x)$ for all $x \in \mathcal{B}$. Thus, by Lemma 2.2 (d), $\cap_{Q < R \leq N}\mathrm{Ker}(\mathrm{Br}_R^{A(Q)}) \cap A(Q)^N$ is spanned by the set $\{C^+\}$, where $C$ runs over the set of those $N$-orbits of $\mathcal{B}$ for which $C_N(x) = Q$ for $x \in C$. Consequently, by Lemma 2.2 (b), $\cap_{Q < R \leq N}\mathrm{Ker}(\mathrm{Br}_R^{A(Q)}) \cap A(Q)^N = \mathrm{Tr}_Q^N(A(Q))$. The claim follows as by Lemma 1.18(b), applied with $G = P$, $\mathrm{Tr}_Q^N(A(Q)) = \mathrm{Br}_Q^A(\mathrm{Tr}_Q^P(A^Q))$.

By what we have shown so far, $\mathrm{Br}_Q^A(i)(1 - f) \in \mathrm{Br}_Q^A(\mathrm{Tr}_Q^P(A^Q))$. Since $\mathrm{Br}_Q^A(i)$ is an idempotent of $A(Q)$ and $(1 - f)$ is a central element of $A(Q)$, we have $\mathrm{Br}_Q^A(i)(1 - f) = \mathrm{Br}_Q^A(i)\mathrm{Br}_Q^A(i)(1 - f)\mathrm{Br}_Q^A(i)$, hence

$$\mathrm{Br}_Q^A(i)(1 - f) \in \mathrm{Br}_Q^A(i)(\mathrm{Tr}_Q^P(A^Q))\mathrm{Br}_Q^A(i) = \mathrm{Br}_Q^A(\mathrm{Tr}_Q^P(iA^Qi)).$$

Here, the last equality follows by Lemma 1.14(g). We claim that $\mathrm{Br}_Q^A(\mathrm{Tr}_Q^P(iA^Qi))$ contains only nilpotent elements. Since $(\mathrm{Br}_Q^A(i)(1-f))^2 = \mathrm{Br}_Q^A(i)(1-f)$, this will prove that $\mathrm{Br}_Q^A(i)(1-f) = 0$ and complete the proof of part (a). In order to prove the claim, note that since $iA^Qi = (iAi)^Q$, by 1.14 (g), applied to the $P$-algebra $iAi$, $\mathrm{Tr}_Q^P(iA^Qi)$ is an ideal of $iA^Pi$. Further, $\mathrm{Tr}_Q^P(iA^Qi)$ is a proper ideal of $iA^Pi$, since $\mathrm{Br}_P(i) \neq 0$ implies that $i \notin \mathrm{Tr}_Q^P(A^Q)$. Since $i$ is primitive in $A^P$, $iA^Pi$ is a local ring and any proper ideal of $iA^Pi$ is nilpotent (see Proposition 1.7). In particular, every element of $\mathrm{Tr}_Q^P(iA^Qi)$ is nilpotent and hence so is every element of $\mathrm{Br}_Q^A(\mathrm{Tr}_Q^P(iA^Qi))$. This proves the claim and completes the proof of (a).

It remains to prove (c). Let $(Q, f) \leq (P, e)$ and let $(S, d) \leq (Q, f)$. By (a), $(S, d') \leq (P, e)$ for some block $d'$ of $A(S)$. By part (b) of the lemma, $d = d'$, hence $(S, d) \leq (P, e)$ as required. $\qquad\square$

**Definition 2.13**  Let $(Q, f)$ and $(P, e)$ be $(A, G)$-Brauer pairs with $Q \leq P$. We write that $(Q, f) \trianglelefteq (P, e)$ if $Q$ is a normal subgroup of $P$, $f$ is $P$-stable and $\mathrm{Br}_{P,Q}^A(f)e = e$ .

Thus, by Theorem 2.10 (b), $(Q, f) \trianglelefteq (P, e)$ if $Q$ is normal in $P$ and $(Q, f) \leq (P, e)$.

**Proposition 2.14**  *Let $(Q, f)$ and $(P, e)$ be $(A, G)$-Brauer pairs. The following are equivalent.*

(a)  $(Q, f) \leq (P, e)$.

(b)  *There exist primitive idempotents $i \in A^P$ and $j \in A^Q$ such that $j \leq i$, $i$ is associated to $e$ and $j$ is associated to $f$.*

(c)  *There exist $(A, G)$-Brauer pairs $(S_i, d_i)$, $1 \leq i \leq n$ such that*
$$(Q, f) \trianglelefteq (S_1, d_1) \trianglelefteq (S_2, d_2) \trianglelefteq \cdots \trianglelefteq (S_n, d_n) \trianglelefteq (P, e).$$

*Proof.* **(a $\Rightarrow$ b)**  First note that for any primitive idempotent $j$ of $A^Q$, either $\mathrm{Br}_Q^A(j) = 0$ or $\mathrm{Br}_Q^A(j)$ is a primitive idempotent of $A(Q)$. In particuar, $j$ is associated to $f$ if and only if $\mathrm{Br}_Q^A(j)f \neq 0$. Suppose that $(Q, f) \leq (P, e)$. By Lemma 2.8, there exists a primitive idempotent, say $i$, of $A^P$ associated to $e$. Let $I$ be a primitive decomposition of $i$ in $A^Q$. Since $\mathrm{Br}_Q^A(i)f = \mathrm{Br}_Q^A(i)$, $\mathrm{Br}_Q^A(j)f \neq 0$ for some $j \in I$. Thus, as pointed out above, $j$ is associated to $f$ and since $j \in I$, $j \leq i$. So, (b) holds.

**(b $\Rightarrow$ a)**  Let $i$ and $j$ be as in (b). By part (a) of Theorem 2.10, there exists a block $d$ of $A(Q)$ such that $(Q, d) \leq (P, e)$. Since $ij = j = ji$ and $\mathrm{Br}_Q^A(j)f \neq 0$, $\mathrm{Br}_Q^A(i)f \neq 0$. So by part (a) of Lemma 2.11, $d = f$, hence (a) holds.

**(a $\Leftrightarrow$ c)**  This is clear from Theorem 2.10 and the fact that any inclusion of finite $p$-groups can be refined to a chain of normal inclusions. $\qquad\square$

If $(P, e)$ is an $(A, G)$-Brauer pair and $x \in G$, then $({}^x P, {}^x e)$ is again an $(A, G)$-Brauer pair.

**Definition 2.15**  Let $(P, e)$ be an $(A, G)$-Brauer pair and let $x \in G$. The $x$-conjugate of $(P, e)$ is the $(A, G)$-pair $({}^x P, {}^x e)$; we set ${}^x(P, e) = ({}^x P, {}^x e)$.

An easy "transport of structure" argument shows that if $(Q, f) \leq (P, e)$, then ${}^x(Q, f) \leq {}^x(P, e)$ for any $x \in G$. Thus, by Theorem 2.10 we have the following.

**Theorem 2.16**  *The set of $(A, G)$-Brauer pairs is a $G$-poset via the map sending an $(A, G)$-Brauer pair $(P, e)$ and an element $x \in G$ to the $(A, G)$-Brauer pair ${}^x(P, e)$.*

### 2.3. $(A, b, G)$-Brauer pairs and inclusion.

As before $A$ is a $p$-permutation $G$-algebra.

**Definition 2.17**  Let $b$ be a primitive idempotent of $A^G$. An $(A, b, G)$-Brauer pair is an $(A, G)$-Brauer pair $(P, e)$ such that $\mathrm{Br}_P^A(b) \neq 0$ and $\mathrm{Br}_P^A(b)e \neq 0$.

$(A, b, G)$-Brauer pairs are called $(b, G)$-Brauer pairs in [BP2]. We have adopted the more cumbersome notation as we will need to simultaneously consider Brauer pairs for different $k$-algebras. Note that if $A = kG$ then the primitive idempotents of $A^G$ are the blocks of $kG$. However, the above definition does not require $b$ to be central in $A$ in general.

**Lemma 2.18**  *Let $(P, e)$ be an $(A, G)$-Brauer pair and let $b$ be a primitive idempotent of $A^G$. If $(P, e)$ is an $(A, b, G)$-Brauer pair, then there exists a primitive idempotent $i$ of $A^P$ such that $i \leq b$ and $i$ is associated to $e$. Conversely, if $i \in A^P$ is an idempotent such that $i \leq b$ and $i$ is associated to $e$, then $(P, e)$ is an $(A, b, G)$-Brauer pair.*

*Proof.* The proof of the first part is as in Lemma 2.8 with $1_A$ replaced by $b$. Now let $i \in A^P$ be an idempotent such that $i \leq b$ and $i$ is associated to $e$. By the first condition, $\mathrm{Br}_P^A(i) = \mathrm{Br}_P^A(i)\mathrm{Br}_P^A(b)$ and by the second, $\mathrm{Br}_P^A(i) = \mathrm{Br}_P^A(i)e$. Hence,

$$0 \neq Br_P^A(i) = \mathrm{Br}_P^A(i)e\mathrm{Br}_P^A(b),$$

from which it follows that $\mathrm{Br}_P^A(b)e \neq 0$ and therefore that $(P, e)$ is an $(A, b, G)$-Brauer pair. $\square$

The set of $(A, b, G)$-Brauer pairs is a downwardly closed sub-$G$-poset of the set of $(A, G)$-Brauer pairs. More precisely,

**Lemma 2.19**  *Let $(Q, f) \leq (P, e)$ be $(A, G)$-Brauer pairs and let $x \in G$. If $(P, e)$ is an $(A, b, G)$-Brauer pair then so are $(Q, f)$ and ${}^x(P, e)$.*

*Proof.* By the previous lemma, there exists an idempotent $i \in A^P$ such that $i \leq b$ and $i$ is associated to $e$. Since $(Q, f) \leq (P, e)$, $i$ is also associated to $f$. But then by the previous lemma applied to $Q$, $(Q, f)$ is an $(A, b, G)$-Brauer pair as claimed. Since $b \in A^G$ (see also Defintion 1.16),

$$0 \neq {}^x(\mathrm{Br}_P^A(b)e) = \mathrm{Br}_{{}^xP}^A({}^xb)\,{}^xe = \mathrm{Br}_{{}^xP}^A(b)\,{}^xe,$$

hence ${}^x(P, e)$ is also an $(A, b, G)$-Brauer pair. $\square$

The crucial difference between the $G$-poset of $(A, G)$-Brauer pairs and the sub-poset of $(A, b, G)$-Brauer pairs is the nature of maximal objects. We have the following result, proved in [AB, Theorem 3.10] and [BP2, Theorem 1.14].

**Theorem 2.20**  *Let $A$ be a $p$-permutation $G$-algebra and let $b$ be a primitive idempotent of $A^G$. Then,*

(a)  *The group $G$ acts transitively on the set of maximal $(A, b, G)$-Brauer pairs.*

(b)  *Let $(P, e)$ be an $(A, b, G)$-Brauer pair. The following are equivalent.*

(i)  *$(P, e)$ is a maximal $(A, b, G)$-Brauer pair.*

(ii)  *$\mathrm{Br}^A_P(b) \neq 0$ and $P$ is maximal amongst subgroups $H$ of $G$ with the property that $\mathrm{Br}^A_H(b) \neq 0$.*

(iii)  *$b \in \mathrm{Tr}^G_P(A^P)$ and $P$ is minimal amongst subgroups $H$ of $G$ such that $b \in \mathrm{Tr}^G_H(A^H)$.*

*Proof.* (a) By Lemma 1.21, there exist a $p$-subgroup $P$ of $G$ and a primitive idempotent $i$ of $A^P$ such that $i \leq b$, $\mathrm{Br}^A_P(i) \neq 0$ and $b \in \mathrm{Tr}^G_P(A^P i A^P)$. By Lemma 2.8, there is a unique block, say $e$, of $A(Q)$ to which $i$ is associated, and by Lemma 2.18, $(P, e)$ is an $(A, b, G)$-Brauer pair. We will show that for any $(A, b, G)$-Brauer pair $(Q, f)$, $(Q, f) \leq {}^x(P, e)$ for some $x \in G$. This will prove part (a) of the theorem since by Theorem 2.19, the set of $(A, b, G)$-Brauer pairs is a $G$-poset. So, let $(Q, f)$ be an $(A, b, G)$-Brauer pair. By Lemma 2.18, there exists a primitive idempotent $j$ of $A^Q$ which is associated to $f$ and such that $j \leq b$. But $j$ being associated to $e$ means in particular that $\mathrm{Br}^A_Q(j) \neq 0$. Hence, by Lemma 1.20, $Q \leq {}^xP$ and ${}^uj \leq {}^xi$ for some $x \in G$ and some $u \in (A^Q)^*$. Now, since $i$ is associated to $e$, ${}^xi$ is associated to ${}^xe$. Further, since $u \in (A^Q)^*$ and $f$ is central in $A(Q)$, the fact that $j$ is associated to $f$ implies that

$$\mathrm{Br}^A_Q({}^uj)f = \mathrm{Br}^A_Q(u)\mathrm{Br}^A_Q(j)\mathrm{Br}^A_Q(u)^{-1}f = \mathrm{Br}^A_Q(u)\mathrm{Br}^A_Q(j)f\mathrm{Br}^A_Q(u)^{-1}$$
$$= \mathrm{Br}^A_Q(u)\mathrm{Br}^A_Q(j)\mathrm{Br}^A_Q(u)^{-1},$$

and hence also that $\mathrm{Br}^A_Q({}^uj)f \neq 0$. Thus, ${}^uj$ is associated to $f$ and ${}^xi$ is associated to ${}^xe$. By Proposition 2.14, $(Q, f) \leq {}^x(P, e)$.

We now prove the equivalences of part (b) of the theorem. First of all note by the proof of (a), that there is a maximal $(A, b, G)$-Brauer pair $(P', e')$ with the following property: there is a primitive idempotent of $i'$ of $A^{P'}$ such that $i' \leq b$, $i'$ is associated to $e'$ and $b \in \mathrm{Tr}^G_{P'}(A^{P'} i' A^{P'})$. Further, if $x \in G$, then ${}^xi'$ is a primitive idempotent of $A^{{}^xP'}$, ${}^xi' \leq b$, ${}^xi'$ is associated to ${}^xe'$ and $b \in \mathrm{Tr}^G_{{}^xP'}(A^{{}^xP'} {}^xi' A^{{}^xP'})$. Since, as has been just proved, all maximal $(A, b, G)$-Brauer pairs are $G$-conjugate, it follows that every maximal $(A, b, G)$-Brauer pair has this property. In particular, $\mathrm{Br}_{P'}^A(b) \neq 0$ and $b \in \mathrm{Tr}_{P'}^G(A^{P'})$ for any maximal $(A, b, G)$-Brauer pair $(P', e')$ and the implications, $((ii) \Rightarrow (i))$ and $((iii) \Rightarrow (i))$ are immediate.

Assume that (i) holds and let $i \in A^P$ be a primitive idempotent such that $i \leq b$, $i$ is associated to $e$ and $b \in \mathrm{Tr}_P^G(A^P i A^P)$ . Suppose $Q \geq P$ is such that $\mathrm{Br}_Q^A(b) \neq 0$. Then, by Lemma 1.19 (a) there is a primitive idempotent, say $j$ of $A^Q$ such that $j \leq b$ and $\mathrm{Br}_Q^A(j) \neq 0$. By Lemma 1.20, $Q \leq {}^x P$ for some $x \in G$. But $P \leq Q$. So, $P = Q$ and (ii) holds. Now, let $R \leq P$ be such that $b \in \mathrm{Tr}_R^G(A^R)$. By Lemma 1.19, there exists a primitive idempotent $t$ in $A^R$ such that $t \leq b$ and $b \in \mathrm{Tr}_R^G(A^R t A^R)$. Thus, $\mathrm{Br}_P^A(i) \neq 0$, $b \in \mathrm{Tr}_R^G(A^R t A^R)$ and $i \leq b$, $t \leq b$. So, by Lemma 1.20, ${}^x P \leq R$ for some $x \in G$. But $R \leq P$. Hence $P = R$ and (iii) holds. $\qquad\square$

### 2.4. $(A, b, G)$-Brauer pairs and fusion systems.

For each maximal $(A, b, G)$-Brauer pair, fusion of subpairs gives rise to a fusion category on the ambient $p$-group. More precisely, for $(A, b, G)$-Brauer pairs $(Q, f)$, $(R, u)$ and $X$ a subgroup of $G$, we define

$$\mathrm{Hom}_X((Q, f), (R, u)) = \{\varphi \in \mathrm{Hom}(Q, R) : \varphi = c_g \text{ for some } g \in X$$
$$\text{such that } {}^g(Q, f) \leq (R, u)\}.$$

**Definition 2.21**  Let $A$ be a $p$-permutation $G$-algebra, $b$ a primitive idempotent of $A^G$ and $(P, e_P)$ a maximal $(A, b, G)$-Brauer pair. For each $Q \leq S$, let $e_Q$ denote the unique block of $A(Q)$ such that $(Q, e_Q) \leq (P, e_P)$. The fusion category of $(A, b, G)$ over $(P, e_P)$ is the category $\mathcal{F}_{(P, e_P)}(A, b, G)$ whose objects are the subgroups of $P$ and which has morphism sets

$$\mathrm{Mor}_{\mathcal{F}_{(P, e_P)}(A, b, G)}(Q, R) = \mathrm{Hom}_G((Q, e_Q), (R, e_R))$$

for $Q, R \leq P$, and where composition of morphisms is the usual composition of functions.

We will write $\mathrm{Hom}_{(A, G, e_P)}(Q, R)$ to denote $\mathrm{Mor}_{\mathcal{F}_{(P, e_P)}(A, b, G)}(Q, R)$. Recall the definition of fusion systems from Definition I.2.1.

**Proposition 2.22**  *Let $A$ be a $p$-permutation $G$-algebra, $b$ a primitive idempotent of $A^G$ and $(P, e_P)$ a maximal $(A, b, G)$-Brauer pair. Then*

(a)  $\mathcal{F}_{(P, e_P)}(A, b, G)$ *is a fusion system over $P$.*

(b)  *If $(P', e_{P'})$ is another maximal $(A, b, G)$-Brauer pair, then*

$$\mathcal{F}_{(P', e_P)}(A, b, G) = {}^\varphi \mathcal{F}_{(P, e_P)}(A, b, G)$$

*for some group isomorphism $\varphi : P' \to P$ in $\mathrm{Hom}_G(P', P)$.*

*Proof.* **(a)** Let $\mathcal{F} = \mathcal{F}_{(P,e_P)}(A, b, G)$ and for each $Q \leq P$, let $e_Q$ be the unique block of $Z(A(Q))$ such that $(Q, e_Q) \leq (P, e_P)$. If $Q, R, S \leq P$ and $g, h \in G$ are such that $^g(Q, e_Q) \leq (R, e_R)$ and $^h(R, e_R) \leq (S, e_S)$, then by Proposition 2.16

$$^{hg}(Q, e_Q) \leq {}^h(R, e_R) \leq (S, e_S),$$

so the composition of $\mathcal{F}$-morphisms is an $\mathcal{F}$-morphism. Since the identity homomorphism is clearly an $\mathcal{F}$-morphism for any subgroup $Q$ of $P$, $\mathcal{F}$ is a category.

By Theorem 2.10, if $Q, R \leq P$ and $g \in G$ are such that $^gQ \leq R$, then $^g(Q, e_Q) \leq (R, e_R)$ if and only if $^g(Q, e_Q) \leq (P, e_P)$ and $^g(Q, e_Q) \leq (P, e_P)$ if and only if $^ge_Q = e_{^gQ}$. In particular, if $^g(Q, e_Q) \leq (R, e_R)$, then $c_g : Q \to {}^gQ$ is a morphism in $\mathcal{F}$. This shows that each $\mathcal{F}$-morphism is the composition of an $\mathcal{F}$-isomorphism followed by an inclusion. Now let $g \in P$ be such that $^gQ \leq R$. Since $g \in P$, $^g(P, e_P) = (P, e_P)$. So by Theorem 2.16

$$^g(Q, e_Q) \leq {}^g(P, e_P) = (P, e_P).$$

So, by the remark above $^g(Q, e_Q) \leq {}^g(R, e_R)$ and it follows that $c_g : Q \to R$ is a morphism in $\mathcal{F}$. This shows that $\mathrm{Hom}_P(Q, R) \leq \mathrm{Hom}_{\mathcal{F}}(Q, R)$. This proves that $\mathcal{F}$ is a fusion system on $P$.

**(b)** This is immediate from the fact that all maximal $(A, b, G)$-Brauer pairs are $G$-conjugate (see Theorem 2.20). $\qquad\square$

## 3. $p$-PERMUTATION ALGEBRAS AND SATURATED FUSION SYSTEMS

Throughout this section, $A$ will denote a $p$-permutation $G$-algebra, and $b$ a primitive idempotent of $A^G$. As seen above the category $\mathcal{F}_{(P,e_P)}(A, b, G)$ is a fusion system on $P$ for any maximal $(A, b, G)$-Brauer pair $(P, e_P)$. This section presents a criterion for $\mathcal{F}_{(P,e_P)}(A, b, G)$ to be saturated.

For an $(A, b, G)$-Brauer pair, $(Q, e)$, let $C_G(Q, e)$ denote the subgroup of $C_G(Q)$ stabilising the pair $(Q, e)$ and let $N_G(Q, e)$ denote the subgroup of $N_G(Q)$ stabilising the pair $(Q, e)$.

### 3.1. **Saturated triples.**

By definition, if $(Q, e)$ is an $(A, b, G)$-Brauer pair, then $e$ is a block of $A(Q)$. Now $A(Q)$ is an $N_G(Q)$-algebra, hence a $C_G(Q, e)$-algebra and by definition of $C_G(Q, e_Q)$, $e \in A(Q)^{C_G(Q,e)}$.

**Definition 3.1** The triple $(A, b, G)$ is *a saturated triple* if the following two conditions hold.

- $b$ is a central idempotent of $A$.

- For each $(A, b, G)$-Brauer pair $(Q, e)$ the idempotent $e$ is primitive in $A(Q)^{C_G(Q,e)}$.

The reason for the above terminology is the following result.

**Theorem 3.2**  *Suppose that $(A, b, G)$ is a saturated triple. Then $\mathcal{F}_{(P, e_P)}(A, b, G)$ is a saturated fusion system on $P$ for any maximal $(A, b, G)$-Brauer pair $(P, e_P)$.*

The proof requires some lemmas.

**Lemma 3.3**  *Let $H$ be a finite group, $B$ a p-permutation $H$-algebra and $e$ a primitive idempotent of $B^H$. If $e \in Z(B)$, then for a p-subgroup $Q$ of $H$ and a block $f$ of $B(Q)$, $(Q, f)$ is an $(B, e, H)$-Brauer pair if and only if $\mathrm{Br}_Q^B(e)f = f$.*

*Proof.*  Suppose that $e \in Z(B)$ and let $Q$ be a $p$-subgroup of $H$. Since

$$Z(B) \cap B^H \subseteq Z(B) \cap B^Q \subseteq Z(B^Q),$$

$e$ is a central idempotet of $B^Q$. Hence, either $\mathrm{Br}_Q^B(e) = 0$ or $\mathrm{Br}_Q^B(e)$ is a central idempotent of $B(Q)$ and for any block $f$ of $B(Q)$, either $\mathrm{Br}_Q^B(e)f = f$, or $\mathrm{Br}_Q^B(e)f = 0$. The result follows. $\qquad\square$

Recall that for an $(A, b, G)$-Brauer pair $(Q, e)$, $A(Q)$ is a $N_G(Q)$-algebra and $e$ is an idempotent of $A(Q)^{N_G(Q,e)}$. Thus, if $e$ is primitive in $A(Q)^{C_G(Q,e)}$, then $e$ is a primitive idempotent of $A(Q)^H$ for any $H$ such that $C_G(Q, e) \leq H \leq N_G(Q, e)$ and it makes sense to speak of $(A(Q), e, H)$-Brauer pairs. The next result compares $(A(Q), e, H)$-Brauer pairs for different $H$.

**Lemma 3.4**  *Suppose that $(Q, e)$ is an $(A, b, G)$-Brauer pair such that $e$ is primitive in $A(Q)^{C_G(Q,e)}$ and let $H$ be a subgroup of $G$ with $C_G(Q, e) \leq H \leq N_G(Q, e)$.*

(a)  *The $H$-poset of $(A(Q), e, H)$-Brauer pairs is the $H$-subposet of $(A(Q), e, N_G(Q, e))$-Brauer pairs consisting of those pairs whose first component is contained in $H$.*

(b)  *The map*

$$(R, \alpha) \to (QR, \alpha)$$

*is an $H$-poset map from the set of $(A(Q), e, H)$-Brauer pairs to the set of $(A(Q), e, QH)$-Brauer pairs and induces a bijection between the set of $(A(Q), e, H)$-Brauer pairs whose first component contains $Q \cap H$ and the set of $(A(Q), e, QH)$-Brauer pairs whose first component contains $Q$.*

(c)  *If $Q \leq H$, then $(Q, e)$ is the unique $(A(Q), e, H)$-Brauer pair with first component $Q$ and $(Q, e)$ is contained in every maximal $(A(Q), e, H)$-Brauer pair.*

*Proof.* **(a)** This is immediate from the definitions.

**(b)** Since $Q$ acts trivially on $A(Q)$, for any $p$-subgroup $R$ of $H$, $A(Q)^R = A(Q)^{QR}$ and $\mathrm{Br}_R^{A(Q)} = \mathrm{Br}_{QR}^{A(Q)}$. This proves the first assertion. The second follows from the first and the fact that $R \to QR$ is a bijection between subgroups of $H$ containing $Q \cap H$ and subgroups of $QH$ containing $Q$.

**(c)** By definition, $A(Q)^Q = A(Q)$, $A(Q)_{<Q}^Q = 0$, and $\mathrm{Br}_Q^{A(Q)}$ is the identity map on $A(Q)$. Thus, the set of $(A(Q), e, H)$-Brauer pairs with first component $Q$ consists precisely of the pairs $(Q, \alpha)$, where $\alpha$ is a block of $A(Q)$ such that $e\alpha \neq 0$. Since $e$ itself is a block of $A(Q)$ and any two distinct blocks of $A(Q)$ are orthogonal, it follows that $(Q, e)$ is an $(A(Q), e, H)$-Braur pair and that it is the unique one with first component $Q$. Since $^h(Q, e) = (Q, e)$ for all $h \in H$ and by Theorem 2.20(a) $H$ acts transitively on the set of maximal $(A(Q), e, H)$-Brauer pairs, $(Q, e)$ is contained in every maximal $(A(Q), e, H)$-Brauer pair.                    $\square$

The proof of Thoerem 3.2 follows the lines of the proof of Theorem I.2.3 with Brauer pairs playing the role of $p$-subgroups and Theorem 2.20 and Lemma 1.22 playing the role of Sylow's theorem and Lemma A.3. The isomorphisms $b_{R,Q}^A : A(Q)(R) \to A(R)$ for $p$-subgroups $Q \trianglelefteq R$ of $G$ introduced in Definition 2.6 allow us to pass back and forth between $(A, b, G)$-Brauer pairs and $(A(Q), e, H)$-Brauer pairs for subgroups $H$ of $G$ such that $QC_G(Q, e) \leq H \leq N_G(Q, e)$.

**Lemma 3.5**  *Suppose that $(Q, e)$ is an $(A, b, G)$-Brauer pair such that $e$ is primitive in $(A(Q))^{C_G(Q,e)}$ and let $H$ be a subgroup of $G$ with $QC_G(Q, e) \leq H \leq N_G(Q, e)$.*

*The map*

$$(R, \alpha) \to (R, b_{R,Q}^A(\alpha))$$

*is an $H$-poset equivalence between the subset of $(A(Q), e, H)$-Brauer pairs consisting of those pairs whose first component contains $Q$, and the subset of $(A, b, G)$-Brauer pairs containing $(Q, e)$ and whose first component is contained in $H$.*

*In particular, $H$ acts transitively on the subset of $(A, b, G)$-Brauer pairs which are maximal among those containing $(Q, e)$ and having first component contained in $H$.*

*Proof.* Let $\mathcal{P}_1$ be the subset of $(A(Q), e, H)$-Brauer pairs consisting of those pairs whose first component contains $Q$, and let $\mathcal{P}_2$ be the subset of $(A, b, G)$-Brauer pairs containing $(Q, e)$ and whose first component is contained in $H$. Since $H \leq N_G(Q, e) \leq N_G(Q)$, $\mathcal{P}_1$ and $\mathcal{P}_2$ are $H$-posets. Now let $Q \leq R \leq H$, and let $\alpha$ be a block of $A(Q)(R)$. By Lemma 3.3, $e = \mathrm{Br}_Q^A(b)e$, hence

$$\mathrm{Br}_{R,Q}^A(e) = b_{R,Q}^A(\mathrm{Br}_R^{A(Q)}(e)) = b_{R,Q}^A(\mathrm{Br}_R^{A(Q)}(\mathrm{Br}_Q^A(b)e)) = \mathrm{Br}_R^A(b)\mathrm{Br}_{R,Q}^A(e).$$

Suppose first that $(R, \alpha)$ is an $(A(Q), e, H)$-Brauer pair. By Lemma 3.3, $\alpha = \mathrm{Br}_R^{A(Q)}(e)\alpha$. Applying $b_{R,Q}^A$ to both sides of this equation, and using the displayed equation above, we get that

$$b_{R,Q}^A(\alpha) = \mathrm{Br}_{R,Q}^A(e)b_{R,Q}^A(\alpha) = \mathrm{Br}_R^A(b)\mathrm{Br}_{R,Q}^A(e)b_{R,Q}^A(\alpha).$$

In particular, $\mathrm{Br}_R^A(b)b_{R,Q}^A(\alpha) \neq 0$, whence $(R, b_{R,Q}^A(\alpha))$ is an $(A, b, G)$-Brauer pair. By Theorem 2.10(b) and the first equality above, $(Q, e) \leq (R, b_{R,Q}^A(\alpha))$ as $(A, b, G)$-Brauer pairs.

Conversely, if $(Q, e) \leq (R, b_{R,Q}^A(\alpha))$, then by Theorem 2.10(b), $b_{R,Q}^A(\alpha) = \mathrm{Br}_{R,Q}^A(e)b_{R,Q}^A(\alpha)$. Applying the inverse of $b_{R,Q}^A$ yields that $\alpha = \mathrm{Br}_R^{A(Q)}(e)\alpha$, hence that $(R, \alpha)$ is an $(A(Q), e, H)$-Brauer pair. This shows that $(R, \alpha) \to (R, b_{R,Q}^A(\alpha))$ is a bijection between $\mathcal{P}_1$ and $\mathcal{P}_2$. Since $Q$ is normal in $H$,

$$b_{{}^hR,Q}^A({}^h\alpha) = b_{{}^hR, {}^hQ}^A({}^h\alpha) = {}^hb_{R,Q}^A(\alpha)$$

for all $h \in H$, all $p$-subgroups $R$ of $G$ containing $Q$ as normal subgroup and all $\alpha \in A(Q)(R)$, hence the bijection is compatible with the $H$-action on $\mathcal{P}_1$ and $\mathcal{P}_2$.

We show that the bijection is inclusion preserving. Let $(R, \alpha)$ and $(S, \beta)$ be $(A(Q), e, H)$-Brauer pairs with $Q \leq R \leq S$. By Proposition 2.14, it suffices to consider the case that $R \trianglelefteq S$. By the equation displayed above, $\alpha$ is $S$-stable if and only if $b_{R,Q}^A(\alpha)$ is $S$-stable. Further, the restrictions of the maps $b_{S,Q}^A \circ \mathrm{Br}_{S,R}^{A(Q)} \circ \mathrm{Br}_R^{A(Q)} \circ \mathrm{Br}_Q^A$ and $\mathrm{Br}_{S,R}^A \circ b_{R,Q}^A \circ \mathrm{Br}_R^{A(Q)} \circ \mathrm{Br}_Q^A$ to $A^S$ both equal $\mathrm{Br}_S^A$. Since $\mathrm{Br}_R^{A(Q)} \circ \mathrm{Br}_Q^A(A^S) = [A(Q)(R)]^S$, it follows that $b_{S,Q}^A \circ \mathrm{Br}_{S,R}^{A(Q)}$ is equal to the restriction of $\mathrm{Br}_{S,R}^A \circ b_{R,Q}^A$ to $[A(Q)(R)]^S$. Since $b_{S,Q}^A$ is invertible, it follows that $\mathrm{Br}_{S,R}^A(b_{R,Q}^A(\alpha))b_{S,Q}^A(\beta) = b_{S,Q}^A(\beta)$ if and only if $\mathrm{Br}_{S,R}^{A(Q)}(\alpha)\beta = \beta$ Thus, by Theorem 2.10 $(R, b_{R,Q}^A(\alpha)) \leq (S, b_{S,Q}^A(\beta))$ if and only if $(R, \alpha) \leq (S, \beta)$. This proves that the bijection is an $H$-poset equivalence between $\mathcal{P}_1$ and $\mathcal{P}_2$.

The given map induces a bijection between the set of maximal elements of $\mathcal{P}_1$ and $\mathcal{P}_2$. But by Lemma 3.4 (c), the set of maximal elements in $\mathcal{P}_1$ is precisely the set of maximal $(A(Q), e, H)$-Brauer pairs. The final assertion follows from this and from the fact that $H$ acts transitively on the set of maximal $(A(Q), e, H)$-pairs (see 2.20 (a)).     $\square$

**Lemma 3.6**  *Suppose that $(A, b, G)$ is a saturated triple and let $(P, e_P)$ be a maximal $(A, b, G)$-Brauer pair. For each $Q \leq P$ let $e_Q$ be the unique* *block of $A(Q)$ such that $(Q, e_Q) \le (P, e_P)$ and let $\mathcal{F} = \mathcal{F}_{(P, e_P)}(A, b, G)$. If $Q \le P$ is such that $(N_P(Q), e_{N_P(Q)})$ is maximal amongst $(A, b, G)$-Brauer pairs $(R, f)$ with $(Q, e_Q) \le (R, f)$ and $R \le N_G(Q, e_Q)$, then $Q$ is fully $\mathcal{F}$-automised and $\mathcal{F}$-receptive.*

*Proof.* Suppose that $(N_P(Q), e_{N_P(Q)})$ is maximal amongst $(A, b, G)$-Brauer pairs $(R, f)$ such that $(Q, e_Q) \le (R, f)$ and $R \le N_G(Q, e_Q)$. Let $\alpha = b^A_{N_P(Q), Q}(e_{N_P(Q)})$. By Lemma 3.5, $(N_P(Q), \alpha)$ is a maximal $(A(Q), e_Q, N_G(Q, e_Q))$-Brauer pair. Thus, by Theorem 2.20 (b), $e_Q \in \mathrm{Tr}^{N_G(Q, e_Q)}_{N_P(Q)}(A(Q)^{N_P(Q)})$. Since $e_Q$ is central in $A(Q)$, idempotent and an element of $N_G(Q, e_Q)$, multiplying on both sides by $e_Q$ gives that

$$e_Q \in \mathrm{Tr}^{N_G(Q, e_Q)}_{N_P(Q)}((e_Q A(Q) e_Q)^{N_P(Q)}).$$

Now, $C_G(Q, e_Q)$ is a normal subgroup of $N_G(Q, e_Q)$ and since $(A, b, G)$ is a saturated triple $e_Q$ is a primitive idempotent of $(A(Q))^{C_G(Q, e_Q)}$ and hence of $(e_Q A(Q) e_Q)^{C_G(Q, e_Q)}$. Thus, by Lemma 1.22 applied to the $N_G(Q, e_Q)$-algebra $e_Q A(Q) e_Q$, and with $R = N_P(Q)$ and $C = C_G(Q, e_Q)$, we have that $N_P(Q)C_G(Q, e_Q)/C_G(Q, e_Q)$ is a Sylow $p$-subgroup of $N_G(Q, e_Q)/C_G(Q, e_Q)$. Since $N_P(Q)C_G(Q, e_Q)/C_G(Q, e_Q) \cong N_P(Q)/C_P(Q) \cong \mathrm{Aut}_P(Q)$ and $N_G(Q, e_Q)/C_G(Q, e_Q) \cong \mathrm{Aut}_{\mathcal{F}}(Q)$, it follows that $Q$ is fully $\mathcal{F}$-automised.

It remains to show that $Q$ is $\mathcal{F}$-receptive. For this, we first observe that the hypothesis on $Q$ implies that $(N_P(Q), e_{N_P(Q)})$ is also maximal amongst $(A, b, G)$-Brauer pairs $(R, f)$ such that $(Q, e_Q) \le (R, f)$ and $R \le N_P(Q)C_G(Q, e_Q)$. Hence, by Lemma 3.5, now applied with $H = N_P(Q)C_G(Q, e_Q)$, $(N_P(Q), e_{N_P(Q)})$ contains an $N_P(Q)C_G(Q, e_Q)$ conjugate of any $(A, b, G)$-Brauer pair which contains $(Q, e_Q)$ and whose first component is contained in $N_P(Q)C_G(Q, e_Q)$. Now let $\varphi : R \to Q$ be an isomorphism in $\mathcal{F}$, and let $g \in G$ induce $\varphi$, that is, ${}^g(R, e_R) = (Q, e_Q)$ and $\varphi(x) = gxg^{-1}$ for all $x \in R$. Then, it is an easy check that $N_\varphi = N_P(R) \cap {}^{g^{-1}}N_P(Q)C_G(Q, e_Q)$. Set $N' = {}^g N_\varphi = {}^g N_P(R) \cap N_P(Q)C_G(Q, e_Q)$, set $e'_{N'} = {}^g e_{N_\varphi}$ and consider the $(A, b, G)$-Brauer pair $(N', e'_{N'})$. Since $(R, e_R) \le (N_\varphi, e_{N_\varphi})$, $(Q, e_Q) \le {}^g(N_\varphi, e_{N_\varphi}) = (N', e'_{N'})$. Also, $N' \le N_P(Q)C_G(Q, e_Q)$. Thus, by what was pointed out above, ${}^h(N', e'_{N'}) \le (N_P(Q), e_{N_P(Q)})$ for some $h \in N_P(Q)C_G(Q, e_Q)$. Multiplying by some element of $N_P(Q)$ if necessary, we may assume that $h \in C_G(Q, e_Q)$. Since ${}^{hg}(N_\varphi, e_{N_\varphi}) \le (P, e_P)$, $\bar{\varphi} := c_{hg} : N_\varphi \to P$ is a morphism in $\mathcal{F}$ and since $h \in C_G(Q, e_Q)$, $\bar{\varphi}$ extends $\varphi$. Thus $Q$ is $\mathcal{F}$-receptive.  $\square$

We now give the proof of Theorem 3.2.

*Proof.* Keep the notation of the theorem, set $\mathcal{F} = \mathcal{F}_{(P, e_P)}(A, b, G)$ and for each $Q \le P$, let $e_Q$ be the unique block of $A(Q)$ such that $(Q, e_Q) \le (P, e_P)$. We have shown in Proposition 2.22 that $\mathcal{F}$ is a fusion system on $P$. Thus, by

Lemma 3.6 it suffices to show that each subgroup of $P$ is $\mathcal{F}$-conjugate to a subgroup $Q$ of $P$ such that $(N_P(Q), e_{N_P(Q)})$ is maximal amongst $(A, b, G)$-Brauer pairs $(R, f)$ with $(Q, e_Q) \leq (R, f)$ and $R \leq N_G(Q, e_Q)$. So, let $Q' \leq P$, and let $(T, \alpha)$ be a maximal $(A(Q'), e_{Q'}, N_G(Q', e_{Q'}))$-Brauer pair. By Lemma 3.4 (c), $Q' \leq T$. Set $f = b_{R,Q}^{A^{-1}}(\alpha)$. By Lemma 3.5, $(T, f)$ is an $(A, b, G)$-Brauer pair with $(Q', e_{Q'}) \leq (T, f)$. Since $(P, e_P)$ is a maximal $(A, b, G)$-Brauer pair, we have

$$^g(Q', e_{Q'}) \leq {}^g(T, f) \leq (P, e_P)$$

for some $g \in G$. Set $Q = {}^g Q'$. By the above, $c_g : Q' \to Q$ is a morphism in $\mathcal{F}$, so $Q$ is $\mathcal{F}$-conjugate to $Q'$. We will show that $(N_P(Q), e_{N_P(Q)})$ has the required maximality property. Note that by Lemma 3.5, $(T, f)$ is maximal amongst $(A, b, G)$-Brauer pairs which contain $(Q', e_{Q'})$ and whose first component is contained in $N_G(Q', e_{Q'})$. Thus, by transport of structure $^g(T, f)$ is maximal amongst $(A, b, G)$-Brauer pairs which contain $(Q, e_Q)$ and whose first component is contained in $N_G(Q, e_Q)$. Since $^g(T, f) \leq (P, e_P)$, $^g T \leq N_P(Q)$ and $^g f = e_{{}^g T}$. Consequently, $^g(T, f) \leq (N_P(Q), e_{N_P(Q)})$. Since $(N_P(Q), e_{N_P(Q)})$ contains $(Q, e_Q)$ and $N_P(Q)$ is contained in $N_G(Q, e_Q)$, the maximality of $^g(T, f)$ forces $^g(T, f) = (N_P(Q), e_{N_P(Q)})$. This completes the proof of the theorem. $\qquad\square$

### 3.2. Normaliser systems and saturated triples.

The notion of saturated triples is compatible with passage to local subgroups and normal subgroups. Just as in the previous subsection, the arguments to prove the results below are essentially the same as for the analogous results for fusion systems of finite groups, with $p$-subgroups replaced by Brauer pairs. Lemmas 3.4 and 3.5 again form the technical backbone.

**Proposition 3.7** *If $(A, b, G)$ is a saturated triple, then so is $(A(Q), e, H)$ for any $(A, b, G)$-Brauer pair $(Q, e)$ and any subgroup $H$ of $G$ with $C_G(Q, e) \leq H \leq N_G(Q, e)$.*

*Proof.* Suppose that $(A, b, G)$ is a saturated triple, let $(Q, e)$ be an $(A, b, G)$-Brauer pair and let $H \leq G$ be such that $C_G(Q, e) \leq H \leq N_G(Q, e)$. Note that by the remarks before Lemma 3.4, it makes sense to speak of $(A(Q), e, H)$-Brauer pairs. Also, the first condition of Definition 3.1 holds as by definition of Brauer pairs, $e$ is central in $A(Q)$. It remains to show that for any $(A(Q), e, H)$-Brauer pair $(R, \alpha)$, $\alpha$ is primitive in $A(Q)(R)^{C_H(R,\alpha)}$. By Lemma 3.4(b), $(QR, \alpha)$ is an $(A(Q), e, QH)$-Brauer pair. Further,

$$C_H(R, \alpha) \geq C_H(QR, \alpha) = C_{QH}(QR, \alpha),$$

the last equation holding since $H \geq C_{QH}(Q)$. Hence

$$[A(Q)(R)]^{C_H(R,\alpha)} \subseteq [A(Q)(QR)]^{C_{QH}(QR,\alpha)}.$$

This shows that we may assume that $Q \leq H$ and that $Q \leq R$. Set $f = b^A_{R,Q}(\alpha)$. By Lemma 3.5, $(R, f)$ is an $(A, b, G)$-Brauer pair and by hypothesis $(A, b, G)$ is a saturated triple. So $f$ is primitive in $(A(R))^{C_G(R,f)}$, and since $C_G(R, f) = C_H(R, f)$, $f$ is primitive in $(A(R))^{C_H(R,f)}$. Since $b_{R,Q}$ is an $N_G(Q) \cap N_G(R)$-algebra homomorphism and $C_H(R) \leq N_G(Q) \cap N_G(R)$, it follows that $\alpha$ is primitive in $[A(Q)(R)]^{C_H(R,\alpha)}$ as required. $\qquad\square$

If $(A, b, G)$ is a saturated triple, then by Theorem 3.2 and the above proposition, for any $(A, b, G)$-Brauer pair $(Q, e)$, any subgroup $H$ of $G$ with $C_G(Q, e) \leq H \leq N_G(Q, e)$, and any maximal $(A(Q), e, H)$-Brauer pair $(N, \alpha)$, $\mathcal{F}_{(N,\alpha)}(A(Q), e, H)$ is a saturated fusion system on $N$. The following is an analogue of Proposition I.5.4.

**Proposition 3.8**  *Assume that $(A, b, G)$ is a saturated triple. Let $(P, e_P)$ be a maximal $(A, b, G)$-Brauer pair and let $\mathcal{F} = \mathcal{F}_{(P,e_P)}(A, b, G)$. For any $R \leq P$, let $e_R$ be the unique block of $A(R)$ with $(R, e_R) \leq (P, e_P)$. Let $Q$ be a subgroup of $P$, $H$ a subgroup of $G$ with $C_G(Q, e_Q) \leq H \leq N_G(Q, e_Q)$ and $K = \mathrm{Aut}_H(Q)$. Let $\alpha$ be the block of $A(Q)(QN^K_P(Q))$ which satisfies $b^A_{QN^K_P(Q),Q}(\alpha) = e_{QN^K_P(Q)}$. Then $Q$ is fully $K$-normalised if and only if $(N^K_P(Q), \alpha)$ is a maximal $(A(Q), e_Q, H)$-Brauer pair. If this is the case, then*

$$N^K_{\mathcal{F}}(Q) = \mathcal{F}_{(N^K_P(Q),\alpha)}(A(Q), e_Q, H).$$

*Proof.* By Lemma 3.4 (b), $(N^K_P(Q), \alpha)$ is a maximal $(A(Q), e_Q, H)$-Brauer pair if and only if $(QN^K_P(Q), \alpha)$ is a maximal $(A(Q), e_Q, QH)$-Brauer pair. On the other hand, by Proposition I.5.2 (equivalence of (a) and (b)), $Q$ is fully $K$-normalised if and only if $Q$ is fully $\mathrm{Aut}_Q(Q)K$-normalised. Since clearly, $\mathrm{Aut}_Q(Q)K = \mathrm{Aut}_{QH}(Q)$, it follows that in order to prove the proposition, we may (and will) assume that $Q \leq H$ and therefore also that $Q \leq N^K_P(Q)$.

By Lemma 3.5, it suffices to prove that $Q$ is fully $K$-normalised if and only if $(N^K_P(Q), e_{N^K_P(Q)})$ is maximal amongst $(A, b, G)$-Brauer pairs which contain $(Q, e_Q)$ and whose first component is contained in $H$.

Amongst the $(A, b, G)$-Brauer pairs which are maximal with respect to containing $(Q, e_Q)$ and having first component contained in $H$, choose one, say $(R, f)$ containing $(N^K_P(Q), e_{N^K_P(Q)})$. Then, it suffices to prove that $Q$ is fully $K$-normalised if and only if $(N^K_P(Q), e_{N^K_P(Q)}) = (R, f)$. Since

$$(Q, e_Q) \leq (N^K_P(Q), e_{N^K_P(Q)}) \leq (R, f),$$

by Theorem 2.20 there exists $g \in G$ such that

$$^g(Q, e_Q) \leq {}^g(N_P^K(Q), e_{N_P^K(Q)}) \leq {}^g(R, f) \leq (P, e_P).$$

If we set $\tau = c_g : Q \to P$, then $\tau$ is a morphism in $\mathcal{F}$ and $^g R \leq N_P^{\tau K}(\tau(Q))$. So, if $Q$ is fully $K$-normalised then $|N_P^K(Q)| \geq |N_P^{\tau K}(\tau(Q))|$ and it follows from the above that $(N_P^K(Q), e_{N_P^K(Q)}) = (R, f)$. Conversely, let $t \in G$ be such that $^t(Q, e_Q) \leq (P, e_P)$ and set $\varphi = c_t : Q \to P$, a morphism in $\mathcal{F}$. Then,

$$(\varphi(Q), e_{\varphi(Q)}) = ({}^tQ, e_{{}^tQ}) = {}^t(Q, e_Q) \leq {}^t(N_P^K(Q), e_{N_P^K(Q)}) \leq {}^t(R, f)$$

and $^t(R, f)$ is maximal amongst $(A, b, G)$-Brauer pairs which contain $(^tQ, e_{{}^tQ})$ and whose first component is contained in $^tH$. Now, $\varphi(Q) \leq N_P^{\varphi K}(\varphi(Q)) \leq P$, hence there is an $(A, b, G)$-Brauer pair which contains $(\varphi(Q), e_{\varphi(Q)})$ and whose first component is $N_P^{\varphi K}(\varphi(Q))$. Since

$$N_P^{\varphi K}(\varphi(Q)) = N_P(\varphi(Q)) \cap {}^tH$$

and $^t(R, f)$ is maximal amongst $(A, b, G)$-Brauer pairs which contain $(^tQ, e_{{}^tQ})$ and whose first component is contained in $^tH$, by Lemma 3.5 $|N_P^{\varphi K}(\varphi(Q))| \leq |{}^tR| = |R|$. Thus, if $(N_P^K(Q), e_{N_P^K(Q)}) = (R, f)$, then $Q$ is fully $K$-normalised.

Now assume that $(N_P^K(Q), \alpha)$ is a maximal $(A(Q), e_Q, H)$-Brauer pair. For each $R \leq N_P^K(Q)$, let $\alpha_R$ be the unique block of $A(R)$ such that $(R, \alpha_R) \leq (N_P^K(Q), \alpha)$, and let $\mathcal{N} = \mathcal{F}_{(N_P^K(Q), \alpha)}(A(Q), e_Q, H)$ be the corresponding saturated fusion system. Note that $\alpha = \alpha_{N_P^K(Q)}$. For subgroups, $R, T \leq N_P^K(Q)$, $\mathrm{Hom}_{\mathcal{N}}(R, T) = \mathrm{Hom}_H((R, \alpha_R), (T, \alpha_T))$, and it is an easy check that also $\mathrm{Hom}_{N_{\mathcal{F}}^K(Q)}(R, T) = \mathrm{Hom}_H((R, e_R), (T, e_T))$. Thus, by Lemma 3.4 (a) in order to prove the second assertion of the lemma it suffices to show that if $R, T \leq N_P^K(Q)$ and $h \in N_G(Q, e_Q)$ are such that $^h R \leq T$, then $^h(R, e_R) \leq (T, e_T)$ if and only $^h(R, \alpha_R) \leq (T, \alpha_T)$.

So, let $R, T$ and $h$ be as above. Then, $^h(R, e_R) \leq (T, e_T)$ if and only if $^h(QR, e_{QR}) \leq (RT, e_{RT})$ and similarly $^h(R, \alpha_R) \leq (T, \alpha_T)$ if and only if $^h(QR, \alpha_{QR}) \leq (RT, \alpha_{RT})$. So we may assume that $Q \leq R, T$. By definition $e_{N_P^K(P)} = \mathrm{Br}_{N_P^K(P), Q}^A(\alpha)$. Hence by Lemma 3.5, $e_X = b_{X, Q}^A(\alpha_X)$ for $X = R, T$ and $^h(R, e_R) \leq (T, e_T)$ if and only $^h(R, \alpha_R) \leq (T, \alpha_T)$ as required. $\qquad \square$

### 3.3. Saturated triples and normal subgroups.

Suppose that $N$ is a subgroup of $G$ such that $b$ is also primitive in $A^N$. Then, it makes sense to speak of $(A, b, N)$-Brauer pairs. Further, the $N$-poset of $(A, b, N)$-Brauer pairs is just the $N$ sub-poset of $(A, b, G)$-Brauer pairs consisting of those pairs whose first component is contained in $N$, and this leads to an obvious inclusion of the corresponding fusion systems. The next result shows that If $(A, b, G)$ and $(A, b, N)$ are both saturated triples, and if $N$ is normal in $G$, then this inclusion is normal.

**Proposition 3.9** *Suppose that $N$ is a normal subgroup of $G$, $b$ is primitive in $A^N$ and $(A, b, G)$ and $(A, b, N)$ are both saturated triples. Let $(P, e_P)$ be a maximal $(A, b, G)$-Brauer pair and for each $Q \leq P$, let $e_Q$ be the unique block of $A(Q)$ such that $(Q, e_Q) \leq (P, e_P)$. Then $(P \cap N, e_{P \cap N})$ is a maximal $(A, b, N)$-Brauer pair. Further, if $\mathcal{F} = \mathcal{F}_{(P, e_P)}(A, b, G)$ and $\mathcal{E} = \mathcal{F}_{(P \cap N, e_{P \cap N})}(A, b, N)$, then $\mathcal{E}$ is a normal subsystem of $\mathcal{F}$.*

*Proof.* Set $T = P \cap N$. Clearly, $(T, e_T)$ is a $(A, b, N)$-Brauer pair. Suppose if possible that $(T, e_T) \leq (R, f)$ for some $(A, b, N)$-Brauer pair such that $R$ contains $T$ properly. By Theorem 2.20 (a), there exists $g \in G$ such that ${}^g(R, f) \leq (P, e_P)$. Since $N$ is normal in $G$ and $R \leq N$, ${}^g R \leq P \cap N = T$, a contradiction since $|R|$ is strictly larger than $|T|$. Thus, $(T, e_T)$ is a maximal $(A, b, G)$-brauer pair and this proves the first assertion.

Clearly, $T$ is strongly $\mathcal{F}$-closed and ${}^\alpha \mathcal{E} = \mathcal{E}$ for any $\alpha \in \mathrm{Aut}_{\mathcal{F}}(T)$. Also, $G$ acts on the set of maximal $(A, b, N)$-Brauer pairs and by Theorem 2.20 (a), $N$ acts transitively on the set of maximal $(A, b, N)$-Brauer pairs. Thus, by the Frattini argument, $G = N_G(T, e_T) \cdot N$. Let $g \in G$ be such that ${}^g(Q, e_Q) \leq (R, e_R)$ for some $Q, R \leq T$. Then $g = xn$, for some $n \in N$ and some $x \in N_G(T, e_T)$. Since ${}^x(Q, e_Q) \leq {}^x(T, e_T) = (T, e_T)$, it follows that ${}^n(Q, e_Q) \leq (T, e_T)$. Thus, $c_n : Q \to T$ is a morphism in $\mathcal{E}$ and $c_g : Q \to R$ factors as a morphism in $\mathcal{E}$ followed by one in $\mathrm{Aut}_{\mathcal{F}}(T)$. This proves the Frattini condition in Definition I.6.1.

Since $T$ is strongly $\mathcal{F}$-closed in $P$, $T$ is fully $\mathcal{F}$-centralised. Hence by Proposition 3.8, $(C_P(T), \alpha)$ is a maximal $(A(T), e_T, C_G(T, e_T))$-Brauer pair, where $\alpha$ is defined by $e_{TC_P(T)} = b^A_{TC_P(T), T}(\alpha)$. Consequently, $(C_T(P), \alpha)$ is also a maximal $(A(T), e_T, C_N(T, e_T)C_P(T))$-Brauer pair. Now by the argument used above for the Frattini condition applied to $C_N(T, e_T)C_P(T) \trianglelefteq N_N(T, e_T)C_P(T)$ and the set of maximal $(A(T), e_T, C_N(T, e_T)C_P(T))$-Brauer pairs,

$$N_N(T, e_T)C_P(T) = C_N(T, e_T)C_P(T) \cdot N_{N_N(T, e_T)}(C_P(T), \alpha).$$

Thus for each $g \in N_N(T, e_T)$, $g = an$, where $a \in C_{NP}(T, e_T)$ and $n \in N_{N_N(T, e_T)}(C_P(T), \alpha)$. But then $n$ is also clearly in $N_{N_N(T, e_T)}(TC_P(T), \alpha)$ and it follows by Lemma 3.5 that $n \in N_{N_N(T, e_T)}(TC_P(T), e_{TC_P(T)})$. Thus, $c_g \in \mathrm{Aut}_{\mathcal{E}}(T)$ extends to $c_n \in \mathrm{Aut}_{\mathcal{F}}(TC_P(T))$ where $[c_n, C_P(T)] \in C_P(T) \cap N = Z(T)$ as $T = P \cap N$. This proves the extension condition of Definition I.6.1, and finishes the proof. $\qquad\square$

### 3.4. **Block fusion systems.**

Let $N$ be a normal subgroup of $G$. Consider $kN$ as a $G$-algebra as in Example 1.12. The action of $G$ on $kN$ induces an action of $G$ on $Z(kN)$ and since $N \leq G$, $(kN)^G = Z(kN)^G$. The following is an immediate consequence of Lemma 2.12, applied with $H = G$ and $B = Z(kN)$.

**Proposition 3.10**   *The set of primitive idempotents of $(kN)^G$ is equal to the set $\{J^+\}$ as $J$ runs over the orbits of $G$ on the set of blocks of $kN$. In particular, the primitive idempotents of $(kG)^G$ are precisely the blocks of $kG$.*

Since the $k$-basis of $kN$ consisting of the elements of $N$ is $G$-stable, $kN$ is a $p$-permutation $G$-algebra. Applying Lemma 2.2 with $\mathcal{B} = N$ allows us to identify the homomorphism $\mathrm{Br}_P^{kN}$, for $P$ a $p$-subgroup of $G$, with the Brauer homomorphism as orginally defined by Brauer.

**Definition 3.11**   Let $P$ be a $p$-subgroup of $G$. Let $\mathrm{br}_P\colon (kG)^P \to kC_G(P)$ be the map which sends an element $\sum_{x \in G} \alpha_x x$ of $kG$ to the element $\sum_{x \in C_G(P)} \alpha_x x$ of $kC_G(P)$.

The map $\mathrm{br}_P$ is clearly $k$-linear and surjective and should be thought of as the truncation map which cuts off the support of an element of $kG$ outside of $C_G(P)$. The next result identifies $\mathrm{Br}_P^{kN}$ with $\mathrm{br}_P$. Note that for any subgroup $P$ of $G$, $C_N(P) \subseteq (kN)^P$, and since $C_N(P)$ is normalized by $N_G(P)$, $kC_N(P)$ is a $kN_G(P)$-subalgebra of $(kN)^P$.

**Proposition 3.12**   *Let $P$ be a $p$-subgroup of $G$.*

(a)   *The composite of the inclusion $kC_N(P) \to (kN)^P$ with $\mathrm{Br}_P^{kN}\colon (kN)^P \to kN(P)$ induces an isomorphism of $kN_G(P)$-algebras between $kC_N(P)$ and $kN(P)$.*

(b)   *If $kN(P)$ is identified with $kC_N(P)$ via the isomorphism of (a), then the Brauer homomorphism $\mathrm{Br}_P^{kN}$ is identified with the restriction of $\mathrm{br}_P$ to $(kN)^P$.*

*Proof.* Consider the $G$-invariant basis $N$ of $kN$. The orbits of $P$ on $N$ are the $P$-conjugacy classes of $N$. A $P$-conjugacy class of $N$ is a singleton $\{x\}$ if and only if $x \in C_N(P)$. In other words, the set of fixed points of $P$ on $N$ is $C_N(P)$ and all $P$-class sums of $N$ outside of $C_N(P)$ are in the kernel of $\mathrm{Br}_P^A$. The result follows from Lemma 2.2.   $\square$

From now on, for any $p$-subgroup $P$ of $G$, we shall identify $kN(P)$ with $kC_N(P)$ and $\mathrm{Br}_P^{kN}$ with $\mathrm{br}_P$. Further, if $Q$ is a normal subgroup $P$, then applying the above discussion to the normal subgroup $C_N(Q)$ of $N_G(Q)$ we may, and will, identify $kC_N(Q)(P)$ with $kC_N(P)$, and consequently

$\mathrm{Br}_P^{kC_N(Q)}$ with the restriction of $\mathrm{br}_P$ to $(kC_N(Q))^P$ and $\mathrm{Br}_{P,Q}^{kN}$ with the identity map. These identifications allow for an alternate, easier description of the poset of Brauer pairs, which is given in the next proposition. Note that since by Proposition 2.14, inclusion of Brauer pairs is just the transitive extension of the "subnormality" relation $\trianglelefteq$, part (b) below describes inclusion of Brauer pairs completely.

**Proposition 3.13**    *Let $b$ be a primitive idempotent of $(kN)^G$.*

(a)   *A $(kN, G)$-Brauer pair is a pair $(P, e)$, where $P$ is a $p$-subgroup of $G$ and $e$ is a block of $kC_N(P)$.*

(b)   *If $(Q, f)$ and $(P, e)$ are $(kN, G)$-Brauer pairs with $Q \trianglelefteq P$, then $(Q, f) \leq (P, e)$ if and only if $f \in (kC_N(Q))^P$ and $\mathrm{br}_P(f)e = e$.*

(c)   *A $(kN, G)$-Brauer pair $(P, e)$ is a $(kN, b, G)$-Brauer pair if and only if $\mathrm{br}_P(b)e \neq 0$, and this in turn is equivalent to $\mathrm{br}_P(b)e = e$.*

(d)   *Any $(kN, G)$-Brauer pair is a $(kN, b, G)$-Brauer pair for a unique primitive idempotent of $(kN)^G$.*

(e)   *The pair $(\{1\}, b)$ is a $(kG, G)$-Brauer pair and a $(kG, b)$-Brauer pair $(P, e)$ is a $(kG, b, G)$-Brauer pair if and only if $(\{1\}, b) \leq (P, e)$ as $(kG, G)$-Brauer pairs.*

*Proof.*   **(a)** and **(b)** and the first part of **(c)** are immediate given the identifications specified above. Since $(kN)^G \subseteq Z(kN)$, the second assertion of **(c)** follows from Lemma 3.3. By Proposition 3.10 any two distinct primitive idempotents of $(kN)^G$ are orthogonal whence **(d)**. Clearly, $b$ is a block of $kC_G(\{1\}) = kG$, so $(\{1\}, b)$ is a $(kG, G)$-Brauer pair. Further by part **(b)**, applied with $N = G$, $(\{1\}, b) \leq (P, e)$ if and only if $\mathrm{br}_P(b)e = e$ and by part **(c)**,$(P, e)$ is a $(kG, G)$-Brauer pair if and only if $\mathrm{br}_P(b)e = e$. This proves **(e)**. $\qquad\square$

**Proposition 3.14**    *Let $b$ be a primitive idempotent of $(kN)^G$. The triple $(kN, b, G)$ is of saturated type.*

*Proof.* By Proposition 3.10, $b$ is a central idempotent of $kN$, hence the first condition of Definition 3.1 holds. Let $(Q, e)$ be a $(kN, G)$-Brauer pair. By definition, $e$ is a primitive idempotent of $Z(kC_N(Q))$. Therefore $e$ is primitive in any subalgebra of $Z(kC_N(Q))$ which contains $e$. Since clearly $C_N(Q) \leq C_G(Q, e)$,

$$Z(kC_N(Q)) = kC_N(Q)^{C_N(Q)} \supset (kC_N(Q))^{C_G(Q,e)}$$

and it follows that $e$ is primitive in $(kC_N(Q))^{C_G(Q,e)}$. Thus the second condition of Definition 3.1 holds. $\qquad\square$

By the above proposition and Theorem 3.2, for a primitive idempotent $b$ of $(kN)^G$ and a maximal $(kN, b, G)$-Brauer pair $(P, e_P)$, the category $\mathcal{F}_{(P,e_P)}(kN, b, G)$ is a saturated fusion system on $P$. In the special case that $N = G$, the primitive idempotents of $(kG)^G$ are just the blocks of $kG$.

**Definition 3.15** Let $b$ be a block of $kG$.

- A $p$-subgroup $P$ of $G$ is a *defect group* of $kGb$ (or more simply of $b$) if there exists a maximal $(kG, b, G)$-Brauer pair with first component $P$.

- Fix a maximal $(kG, b, G)$-Brauer pair $(P, e_P)$. The *fusion system* of $kGb$ (or of $b$) over $(P, e_P)$ is the category $\mathcal{F}_{(P,e_P)}(kG, b, G)$.

By Theorem 2.20, any two maximal $(kG, b, G)$-Brauer pairs are $G$-conjugate, hence $P$ is a defect group of $kGb$ if and only if $\mathrm{br}_P(b) \neq 0$ and every $(kG, b, G)$-Brauer pair with first component $P$ is maximal. By Theorem 2.22 (b), the fusion systems corresponding to any two maximal $(kG, b, G)$-Brauer pairs are isomorphic. For this reason, we often refer to $\mathcal{F}_{(P,e_P)}(G, b)$ in the definition above as *the* fusion system of the block $b$.

### 3.5. Fusion systems of blocks of local subgroups.

For the rest of the section, we concentrate on block fusion systems, that is the case $A = kG$. The aim of this subsection is to show that if $\mathcal{F}$ is the fusion system of a block $b$ of $kG$ with respect to some maximal $(kG, b, G)$-Brauer pair $(P, e_P)$, then for a fully-$\mathcal{F}$ normalised subgroup $Q$ of $P$, $N_{\mathcal{F}}(Q)$ is the fusion system of a block of $kN_G(Q)$ and if $Q$ is fully-$\mathcal{F}$ centralised, then $C_{\mathcal{F}}(Q)$ is the fusion system of a block of $kC_G(Q)$. The results of Subsection 3.2 come very close to what we will prove here but the point of view is different.

**Lemma 3.16** *Suppose that $Q$ is a normal $p$-subgroup of $G$. Then $x - 1 \in J(kG)$ for all $x \in Q$ and $\mathrm{Tr}_R^Q((kG)^R) \subseteq J(kG)$ for any proper subgroup $R$ of $Q$.*

*Proof.* The first statement is equivalent to showing that for any simple $kG$-module $V$, $(x - 1)V = 0$ for all $x \in Q$. Proceed by induction on $|Q|$. Suppose first that $|Q| = p$ and let $W \subseteq V$ be defined by $W = \{v \in V : xv = v, \forall x \in Q\}$. Since $Q$ is normal in $G$, $W$ is a $kG$-submodule of $V$. Hence either $W = 0$ or $W = V$ and it suffices to show that $W \neq 0$. Let $z$ be a generator of $Q$. Since $z$ has order $p$, $z^p v = v$ for all $v \in V$ which means that

$$(z - 1)^p V = (z^p - 1)V = 0.$$

In particular, for some $n$, $0 \leq n < p$ and for some $v \in V$, we have that $(z - 1)^n v \neq 0$ and $(z - 1)^{n+1} v = 0$. Set $w = (z - 1)^n v$. By definiton $zw = w$, hence $z^i w = w$ for any positive integer $i$, that is $w \in W$ and

$W \neq 0$ as required. Now suppose that $|Q| > p$, and let $U$ be a proper non-trivial normal subgroup of $Q$. By induction $U$ acts as the identity on $V$. Thus, $V$ is a (simple) $kG/U$-module and $Q/U$ is a normal subgroup of $G/U$. By induction again, $Q/U$ acts as the identity on $V$. Thus, $Q$ acts as the identity on $V$, proving the first statement.

Now, let $R < Q$ and let $a \in (kG)^R$. Then by the first part, for any $x \in Q$, $xax^{-1} - a \in J(kG)$, hence

$$\mathrm{Tr}_R^Q(a) = \sum_{x \in Q/R} xax^{-1} - |Q{:}R|a \in J(kG),$$

thus proving the second assertion. $\qquad\square$

**Lemma 3.17** *Suppose that $Q$ is a normal p-subgroup of $G$. The set of blocks of $kG$ is the set of sums of $G$-orbits of blocks of $kC_G(Q)$ under the conjugation action of $G$.*

*Proof.* Since the orbit sum of any $G$-orbit of blocks of $kC_G(Q)$ is a central idempotent of $kG$, it suffices to show that any central idempotent of $kG$ is an element of $kC_G(Q)$.

So, let $d$ be an idempotent of $Z(kG)$. The class sums of $G$ form a basis of $Z(kG)$. Further, since $Q$ is normal in $G$, any conjugacy class of $G$ is either contained in $C_G(Q)$ or intersects $C_G(Q)$ trivially. Thus, we may write $d = d_1 + d_2$, where $d_1 \in Z(kG) \cap kC_G(Q)$ and $d_2 \in (kG)_{<Q}^Q$. By Lemma 3.16, $d_2 \in J(kG)$, hence $d_2^{p^n} = 0$ for some natural number $n$. Since $d_1$ and $d_2$ commute, $d^{p^n} = d_1^{p^n} + d_2^{p^n} = d_1^{p^n}$. Since $d$ is idempotent, this means that $d = d_1^{p^n} \in kC_G(Q)$, as required. $\qquad\square$

**Lemma 3.18** *Let $Q$ be a p-subgroup of $G$ and $H$ a subgroup of $G$ with $QC_G(Q) \leq H$. Let $e$ be a block of $kC_G(Q)$ and $b$ and $c$ be blocks of $kG$ and $kH$ respectively such that $\mathrm{br}_Q(b)e \neq 0 \neq \mathrm{br}_Q(c)e$.*

(a) *The $H$-poset of $(kH, c, H)$-Brauer pairs containing $(Q, e)$ is equal to the $H$-poset of $(kG, b, G)$-Brauer pairs containing $(Q, e)$ and whose first component is contained in $H$.*

(b) *Suppose that $Q$ is normal in $G$ and $H = N_G(Q, e)$. Then $e = c$ and the set of maximal $(kH, e, H)$-Brauer pairs is the set of maximal $(kG, b, G)$-Brauer pairs containing $(Q, e)$. Further, for any maximal $(kH, e, H)$-Brauer pair $(R, f)$*

$$\mathcal{F}_{(R,f)}(kH, e, H) = \mathcal{F}_{(R,f)}(kG, b, G).$$

*Proof.* (a) Since $C_G(Q) \leq H$, $C_G(R) = C_H(R)$ for any $R$ such that $Q \leq R \leq H$. Thus, by Proposition 3.13, the $H$-poset of $(kH, H)$-Brauer pairs whose first component contains $Q$ equals the $H$-poset of $(kG, G)$-Brauer pairs whose first component contains $Q$ and is contained in $H$. Note that by Proposition 3.13(b), and the remark just before Proposition 3.13, for such pairs the inclusion as $(kH, H)$-Brauer pairs is the same as inclusion as $(kG, G)$-Brauer pairs. Also by Proposition 3.13, any $(kG, G)$-Brauer pair which contains $(Q, e)$ is a $(kG, b, G)$-Brauer pair and similarly any $(kH, H)$-Brauer pair which contains $(Q, e)$ is a $(kH, c, H)$-Brauer pair.

**(b)** The assertion that $e = c$ is immediate from Lemma 3.17. Let $(R, f)$ be a $(kG, b, G)$-Brauer pair containing $(Q, e)$ and let $x \in R$. Then ${}^{x}(Q, e) \leq (R, f)$ and ${}^{x}Q = Q$. Hence, by the uniqueness of inclusion of Brauer pairs, ${}^{x}(Q, e) = (Q, e)$, that is $x \in H$. So, the first component of any $(kG, b, G)$-Brauer pair which contains $(Q, e)$ is contained in $H$. By part (a), it follows that the $H$-poset of $(kH, e, H)$-Brauer pairs containing $(Q, e)$ is equal to the $H$-poset of $(kG, b, G)$-Brauer pairs containing $(Q, e)$. Since $H = N_H(Q, e)$, $(Q, e)$ is contained in any maximal $(kH, e, H)$-Brauer pair, hence the set of maximal $(kH, e, H)$-Brauer pairs equals the set of maximal $(kG, b, G)$-Brauer pairs containing $(Q, e)$.

Now let $(R, f)$ be a maximal $(kH, e, H)$-Brauer pair and set $\mathcal{E} = \mathcal{F}_{(R,f)}(kH, e, H)$, $\mathcal{F} = \mathcal{F}_{(R,f)}(kG, b, G)$. Since $H = N_G(Q, e)$, $\mathcal{E} = N_{\mathcal{F}}(Q)$. Thus, in order to show that $\mathcal{E} = \mathcal{F}$, it suffices to show that $Q$ is normal in $\mathcal{F}$. Let $U$ be a $\mathcal{F}$-centric-radical subgroup of $R$. Since $Q$ is normal in $G$, $\mathrm{Aut}_{QU}(U)$ is a normal $p$-subgroup of $\mathrm{Aut}_{\mathcal{F}}(U)$, hence

$$\mathrm{Aut}_{QU}(U) \leq O_p(\mathrm{Aut}_{\mathcal{F}}(U)) = \mathrm{Aut}_U(U).$$

So,

$$N_{QU}(U) \leq UC_G(U) \cap R = UC_R(U) = U.$$

Thus, $Q \leq U$. Since $Q$ is clearly stongly $\mathcal{F}$-closed, the result follows by Proposition I.4.5. $\qquad\square$

**Theorem 3.19** *Let $b$ a block of $kG$ and $(P, e_P)$ a maximal $(kG, b, G)$-Brauer pair. For each $Q \leq P$, let $e_Q$ be the unique block of $kC_G(Q)$ such that $(Q, e_Q) \leq (P, e_P)$ and set $\mathcal{F} = \mathcal{F}_{(P,e_P)}(kG, b, G)$.*

(a) *Let $Q \leq P$, let $N$ be a subgroup of $G$ such that $N_G(Q, e_Q) \leq N \leq N_G(Q)$ and let $c$ be the unique block of $kN$ with $\mathrm{br}_Q(c)e_Q = e_Q$. Then $Q$ is fully $\mathcal{F}$-normalized if and only if $(N_P(Q), e_{N_P(Q)})$ is a maximal $(kN, c, N)$-Brauer pair. If this is the case, then*

$$N_{\mathcal{F}}(Q) = \mathcal{F}_{(N_P(Q), e_{N_P(Q)})}(kN, c, N).$$

(b) *A subgroup $Q \leq P$ is fully centralised if and only if $(C_P(Q), e_{C_P(Q)})$ is a maximal $(kC_G(Q), e_Q, C_G(Q))$-Brauer pair. If this is the case, then*

$$C_{\mathcal{F}}(Q) = \mathcal{F}_{(C_P(Q), e_{C_P(Q)})}(kC_G(Q), e_Q, C_G(Q)).$$

*Proof.* **(a)** By Lemma 3.17 applied with the group $G$ replaced by $N$ and $H$ replaced by $N_G(Q, e_Q)$, we may assume that $N = N_G(Q, e_Q)$ and

$c = e_Q$. By Lemma 3.5 applied to the Brauer pair $(Q, e_Q)$ of the saturated triple $(kN, e_Q, N)$ and noting that $C_N(Q) = C_G(Q)$, the $N$-poset of $(kN, e_Q, N)$-Brauer pairs which contain $(Q, e_Q)$ is equal to the $N$-poset of $(kC_G(Q), e_Q, N)$-Brauer pairs which contain $(Q, e_Q)$. In particular, if $(R, f)$ is a maximal $(kC_G(Q), e_Q, N)$-Brauer pair containing $(Q, e_Q)$, then $(R, f)$ is a maximal $(kN, e_Q, N)$-Brauer pair containing $(Q, e_Q)$ and $\mathcal{F}_{(R,f)}(kN, e_Q, N) = \mathcal{F}_{(R,f)}(kC_G(Q), e_Q, N)$. Now the result is immediate from Proposition 3.8.

**(b)** Since $C_G(Q, e_Q) = C_G(Q)$, this is immediate from Proposition 3.8. $\qquad\square$

As pointed out in the remarks after Definition 3.15, the condition that $(N_P(Q), e_{N_P(Q)})$ is a maximal $(kN, c, N)$-Brauer pair in the above theorem is equivalent to $N_P(Q)$ being a defect group of the block $c$ of $kN$, and similarly the condition that $(C_P(Q), e_{C_P(Q)})$ is a maximal $(kC_G(Q), e_Q, C_G(Q))$-Brauer pair is equivalent to $C_P(Q)$ being a defect group of the block $e_Q$ of $kC_G(Q)$. The next result is a characterization of centric subgroups.

**Theorem 3.20**   *Keep the notation of Theorem 3.19. A subgroup $Q$ of $P$ is $\mathcal{F}$-centric if and only if $Z(Q)$ is a defect group of the block $e_Q$ of $kC_G(Q)$.*

*Proof.* If $Q$ is $\mathcal{F}$-centric, then $Q$ is fully $\mathcal{F}$-centralized and by Theorem 3.19, $Z(Q) = C_P(Q)$ is a defect group of the block $e_Q$. Conversely, suppose that $Z(Q)$ is a defect group of the block $e_Q$ of $kC_G(Q)$. Let $R \leq P$ be $\mathcal{F}$-conjugate to $Q$ and let $g \in G$ be such that ${}^g(Q, e_Q) = (R, e_R)$. By Theorem 2.20, $Z(Q)$ is maximal amongst $p$-subgroups of $C_G(Q)$ such that $\mathrm{br}_{Z(Q)}(e_Q) \neq 0$. So, $Z(R) = {}^g Z(Q)$ is maximal amongst $p$-subgroups of $C_G(R) = {}^g C_G(Q)$ such that $\mathrm{br}_{Z(R)}(e_R) = \mathrm{br}_{Z(R)}({}^g e_Q) \neq 0$. By theorem 2.20, $Z(R)$ is a defect group of the block $e_R$ of $kC_G(R)$. On the other hand, there is a $(kC_G(R), e_R, C_G(R))$-Brauer pair with first component $C_P(R)$, namely $(C_P(R), e_{C_P(R)})$ and $Z(R) \leq C_P(R)$. Thus, $Z(R) = C_P(R)$ for all $R \leq P$ which are $\mathcal{F}$-conjugate to $Q$. $\qquad\square$

## 4. Background on finite group representations

We have seen in the previous section that every block of a finite group algebra gives rise to an isomorphism class of saturated fusion systems on a defect group of the block. The central problem of modular representation theory is the connection between a block fusion system and the representation theory of the corresponding block algebra. In order to be able to discuss this connection, we need to invoke some of the standard language and notions of modular group representation theory, and this is what we will do in this section. As in Section 1, we will be very brief — many books on group representation theory give an extensive treatement of the material presented here. Some of these are [CuR1], [CuR2], [NT] and [Fe].

## 4.1. **Ordinary and modular representations.**

Let $F$ be a field and let $G$ be a finite group. In order to avoid complications of rationality we will assume that $F$ is a splitting field for $G$, that is $\mathrm{End}_{FG}(V) \cong F$ for any simple $FG$-module $V$. Recall that by the convention of Section 1, all modules are finitely generated left modules.

**Definition 4.1**   Let $H$ be a subgroup of $G$. An $FG$-module $V$ is *relatively $H$-projective* if the following holds:

For any surjective homomorphism

$$\varphi : W \to V$$

of $FG$-modules, if $\varphi$ splits as a map of $FH$-modules, then $\varphi$ splits as a map of $FG$-modules.

It is an easy exercise that a $FG$-module $V$ is relatively $\{1\}$-projective if and only if $V$ is a projective $FG$-module. Recall from Example 1.12 that if $V$ is a $FG$-module, then $\mathrm{End}_F(V)$ is naturally a $G$-algebra. For a subgroup $H$ of $G$, the $H$-fixed point subalgebra $(\mathrm{End}_F(V))^H$ is equal to the subalgebra of $FH$-module endomorphisms $\mathrm{End}_{FH}(V)$.

**Theorem 4.2**   *(Higman's criterion). Let $G$ be a finite group, $H \leq G$ and let $V$ be a $FG$-module. Then $V$ is relatively $H$-projective if and only if the identity $\mathrm{Id}_V : V \to V$ of $\mathrm{End}_k(V)$ is a relative trace $\mathrm{Tr}_H^G(\pi)$ for some $\pi \in \mathrm{End}_k(V)^H$.*

**Theorem 4.3**   *Let $H \leq G$. The following are equivalent.*

(a)   $1_{FG} = \mathrm{Tr}_H^G(a)$ *for some $a \in (FG)^H$.*

(b)   *Any finitely generated $FG$-module is relatively $H$-projective.*

The above is a standard fact, but the statement does not appear often in this form in the literature-we provide a proof for the convenience of the reader.

*Proof.* **(a $\Rightarrow$ b)**   Suppose that $1_{FG} = \mathrm{Tr}_H^G(a)$, $a \in (FG)^H$ and let $V$ be an $FG$-module. Let $\pi : V \to V$ be the map given by $\pi(v) = av$ for $v \in V$. Since $a \in (FG)^H$, we have

$$^h\pi(v) = h\pi(h^{-1}v) = hah^{-1}v = av = \pi(v), \quad \text{for all} \ \ h \in H, v \in V,$$

so $\pi \in \mathrm{End}_{FH}(V)$. Also, we have

$$\mathrm{Tr}_H^G(\pi(v)) = \sum_{g \in [G/H]} g\pi(g^{-1}v) = \sum_{g \in [G/H]} gag^{-1}v = v \quad \text{for all} \ \ v \in V.$$

So, by Higman's criterion, $V$ is relatively $H$-projective and (b) holds.

**(b $\Rightarrow$ a)** Suppose that every $FG$-module is relatively $H$-projective. Consider $FG$ as an $FG$-module via the conjugation action of $G$ on itself. We will prove that (a) holds by applying (b) to $FG$. Note that since $FG$ is being considered as a $FG$-module via conjugation, for $\pi \in \mathrm{End}_F(FG)$, $g \in G$, the map ${}^g\pi : FG \to FG$ satisfies ${}^g\pi(a) = g\pi(g^{-1}ag)g^{-1}$ for $a \in FGb$. In particular, $\pi \in \mathrm{End}_{FH}(FG)$ if and only if $h\pi(a)h^{-1} = \pi(hah^{-1})$ for all $a \in FG$ and all $h \in H$.

By hypothesis, $FG$ is relatively $H$-projective. Hence by Higman's criterion, the identity map $\mathrm{Id}_{FG} : FG \to FG$ is a relative trace $\mathrm{Tr}_H^G(\pi)$ for some $\pi \in \mathrm{End}_{FH}(FG)$. Let $a = \pi(1_{FG})$. Since $\pi \in \mathrm{End}_{FH}(FG)$,

$$hah^{-1} = h\pi(1_{FG})h^{-1} = \pi(1_{FG}) = a \text{ for all } h \in H,$$

so $a \in (FG)^H$. We have,

$$\mathrm{Tr}_H^G(a) = \sum_{g\in[G/H]} gag^{-1} = \sum_{g\in[G/H]} g\pi(1_{FG})g^{-1} = \sum_{g\in[G/H]} {}^g\pi(1_{FG})$$
$$= \mathrm{Id}_{FG}(1_{FG}) = 1_{FG}.$$

Hence (a) holds. $\qquad\square$

**Proposition 4.4** *The identity element $1_{FG}$ belongs to $\mathrm{Tr}_1^G(FG)$ if and only if $char(F)$ does not divide $|G|$.*

*Proof.* Let $g \in G$ and let $C$ be the $G$-conjugacy class of $g$. Then $\mathrm{Tr}_1^G(g) = |C_G(g)|C^+$, where $C^+$ denotes the sum of elements of $C$ in $FG$. So, the elements $C^+$, where $C$ runs over those conjugacy classes of $G$ for which $|C_G(x)|$ is non-zero in $F$ form a basis of $\mathrm{Tr}_1^G(FG)$. The result is immediate from this. $\qquad\square$

The results above show the essential difference between representation theory of finite groups over fields of characteristic 0 (ordinary representation theory) and fields of positive characteristic (modular representation theory). For, if $char(F)$ does not divide $|G|$, then by the above every finitely generated every $FG$-module is projective (in case $char(F) = 0$, this fact is known as Maschke's theorem). Thus the Artin-Wedderburn theory gives that if $char(F)$ does not divide $|G|$, then $FG$ is a product of matrix algebras. In particular, every block of $FG$ is a matrix algebra. On the other hand, if $char(F)$ does divide $|G|$, then there exist simple $kG$-modules which are not projective, hence $FG$ has blocks which are not matrix algebras:

**Theorem 4.5** *Suppose that $char(F)$ does not divide $|G|$. Then, $J(FG) = 0$ and*

$$FG = \prod_{i=1}^{r} \Lambda_i,$$

*where each $\Lambda_i \cong \mathrm{Mat}_{n_i}(F)$ for some natural number $n_i$. Further, the following holds.*

(a) *For each $i$, $1 \le i \le r$, let $e_i$ be a primitive idempotent of $\Lambda_i$, and regard $\Lambda_i e_i$ as a $FG$-module through the projection of $FG$ onto $\Lambda_i$. Then $\{e_i : 1 \le i \le r\}$ is a set of representatives of conjugacy classes of primitive idempotents of $FG$ and $\{\Lambda_i e_i : 1 \le i \le r\}$ is a set of representatives of isomorphism classes of simple $FG$-modules.*

(b) *The set $\{1_{\Lambda_i} : 1 \le i \le r\}$ is the block decomposition of $1_{FG}$, and each block of $FG$ is a matrix algebra over $F$.*

*If $char(F)$ divides $|G|$, then there exist blocks $b$ of of $FG$ such that $J(FGb) \ne 0$ (and hence $FGb$ is not a matrix algebra).*

## 4.2. $p$-modular systems.

**Definition 4.6**  A *$p$-modular system* is a triple $(K, \mathcal{O}, k)$ where $\mathcal{O}$ is a local principal ideal domain, $K$ is the field of quotients of $\mathcal{O}$ and $k = \mathcal{O}/\mathfrak{p}$, where $\mathfrak{p}$ is the unique maximal ideal of $\mathcal{O}$ such that the following hold:

- $\mathcal{O}$ is complete with respect to the natural topology induced by $\mathfrak{p}$,

- $K$ has characteristic 0; and

- $k$ has characteristic $p$.

**Example 4.7**  Let $L$ be a number field, $R$ the ring of algebraic integers in $L$, $\mathfrak{p}$ a prime ideal of $R$ lying over $p$. Let $R_\mathfrak{p}$ be the localization of $A$ with respect to $\mathfrak{p}$ and let $\mathcal{O} = \varprojlim R_\mathfrak{p}/\mathfrak{p}^n$, the completion of $A$ with respect to the $\mathfrak{p}$-adic topology. Then $\mathcal{O}$ is a local principal ideal domain, and the corresponding triple $(K, \mathcal{O}, k)$ is a $p$-modular system.

For the rest of the article, we will fix a $p$-modular system $(K, \mathcal{O}, k)$ which satisfies the following additional properties:

- $k$ is algebraically closed and perfect; and

- $K$ is a splitting field for any finite group appearing in the sequel.

The existence of $p$-modular systems satisfying the above properties relies on two results. The first is that given any natural number $n$ there exists a $p$-modular system $(K, \mathcal{O}, k)$ such that $k$ is algebraically closed and perfect and such that $K$ contains a primitive $n$-th root of unity (see [Se1, Chapter 2, Section 3, Theorem 1 and Chapter 2, Section 5, Theorem 3]). The second is Brauer's theorem that any field of characteristic 0 containing a primitive $n$-th root of unity is a splitting field for any finite group of order $n$, which in turn is a consequence of Brauer's characterisation of characters (see [CuR1, Theorem 41.1]).

We record the following fact. Note that the natural surjective ring homomorphism $\mathcal{O} \to k$ induces a group homomorphism $\mathcal{O}^* \to k^*$.

**Proposition 4.8**  *Let $\mathfrak{p}$ denote the maximal ideal of $\mathcal{O}$. The group homomorphism $\mathcal{O}^* \to k^*$ is surjective with kernel $1 + \mathfrak{p}$. Further, the short exact sequence*

$$1 \to 1 + \mathfrak{p} \to \mathcal{O}^* \to k^* \to 1$$

*splits uniquely. In particular, the surjection $\mathcal{O}^* \to k^*$ restricts to an isomorphism between the torsion subgroup of $\mathcal{O}^*$ and $k^*$.*

### 4.3. Cartan and decomposition maps.

The extension of the natural surjection of $\mathcal{O}$ onto $k$, and of the inclusion of $\mathcal{O}$ in $K$ to the corresponding group rings, yields the following diagram of ring homomorphisms

$$KG \hookleftarrow \mathcal{O}G \twoheadrightarrow kG,$$

where the map on the left is inclusion and the map on the right sends an element $a := \sum_{g \in G} \alpha_g g$ of $\mathcal{O}G$ to the element $\bar{a} := \sum_{g \in G} (\alpha_g + \mathfrak{p})g$ of $kG$.

Since conjugacy class sums of elements of $G$ form a basis for the center of the group ring $RG$ of $G$ over any commutaitive ring $R$ with 1, restricting the above maps to $Z(\mathcal{O}G)$ gives corresponding homomorphisms

$$Z(KG) \hookleftarrow Z(\mathcal{O}G) \twoheadrightarrow Z(kG).$$

The above maps lead to a two-way traffic between the characteristic 0 representations and characteristic $p$ representations of $G$. Note that if $M$ is an $\mathcal{O}G$-module, then by extension of scalars $K \otimes_{\mathcal{O}} M$ is a $KG$-module and similarly $k \otimes_{\mathcal{O}} M \cong M/\mathfrak{p}M$ is a $kG$-module.

The completeness of $\mathcal{O}$ with respect to the $\mathfrak{p}$-adic topology has as consequence that the idempotent lifting results of Subsection 1.1 carry over to $\mathcal{O}$-algebras. More precisely, Propositions 1.2, 1.3, 1.4, 1.8, 1.9 and Lemma 1.5 hold for $\mathcal{O}$-algebras which are finitely generated as $\mathcal{O}$-modules. In particular, we have the following.

**Proposition 4.9**  *The map $a \to \bar{a}$ induces a bijection between $\mathcal{O}G$-conjugacy classes of primitive idempotents of $\mathcal{O}G$ and $kG$-conjugacy classes of primitive idempotents of $kG$ and also a bijection between the blocks of $\mathcal{O}G$ and the blocks of $kG$.*

There is an analogous result for modules.

**Proposition 4.10**  *Let $I$ be a set of representatives of $\mathcal{O}G$-conjugacy classes of primitive idempotents of $\mathcal{O}G$.*

(a)  *The set $\{\mathcal{O}Gi : i \in I\}$ is a set of representatives of the isomorphism classes of projective indecomposable $\mathcal{O}G$-modules.*

(b)  *The set $\{kG\bar{i} \cong k \otimes_{\mathcal{O}} \mathcal{O}G : i \in I\}$ is a set of representatives of the isomorphism classes of projective indecomposable $kG$-modules.*

(c)  *The set $\{kG\bar{i}/J(kG)\bar{i} : i \in I\}$ is a set of representatives of the isomorphism classes of simple $kG$-modules.*

The following definition introduces notation for various Grothendieck groups and subgroups associated to the categories of $KG$, $\mathcal{O}G$ and $kG$-modules.

**Definition 4.11**  Let $F$ be a field.

- Denote by $R_F(G)$ the Grothendieck group (with respect to short exact sequences) of finitely generated $FG$-modules. For a finitely generated $FG$-module $V$, denote by $[V]$ the image of $V$ in $R_F(G)$.

- Denote by $\mathrm{Irr}_F(G)$ the set of images of the simple $FG$-modules in $R_F(G)$.

- Denote by $\mathrm{IPr}_F(G)$ the set of the projective indecomposable $FG$-modules in $R_F(G)$ and by $Pr_F(G)$ the subgroup of $R_F(G)$ generated by $\mathrm{IPr}_F(G)$.

- Denote by $\mathrm{IPr}_{\mathcal{O}}(G)$ the set of images of modules of the form $K \otimes_{\mathcal{O}} U$, where $U$ is a finitely generated projective indecomposable $\mathcal{O}G$-module and by $Pr_{\mathcal{O}}(G)$ the subgroup of $R_K(G)$ generated by $\mathrm{IPr}_{\mathcal{O}}(G)$.

Note that for $F = K$ or $k$, $R_F(G)$ is a finitely generated free abelian group with basis $\mathrm{Irr}_F(G)$. Contrary to what might be expected, the above list does not include the subgroup of $R_K(G)$ generated by the scalar extensions of simple $\mathcal{O}G$-modules. The reason for this is the following. If $M$ is a non-zero $\mathcal{O}G$-module, finitely generated as $\mathcal{O}$-module, then $\mathfrak{p}M$ is a proper $\mathcal{O}G$-submodule of $M$ (this is a special case of what is known as Nakyama's lemma). So, if $M$ is a simple $\mathcal{O}G$-module, then $\mathfrak{p}M = 0$ and consequently $K \otimes_{\mathcal{O}} M = 0$. Also, note that by Theorem 4.3 and Proposition 4.4, every finitely generated $KG$-module is projective and hence $\mathrm{IPr}_K(G) = \mathrm{Irr}_K(G)$ and $Pr_K(G) = R_K(G)$. On the other hand, if $p$ divides the order of $G$, then $Pr_k(G)$ is a proper subgroup of $R_k(G)$.

**Definition 4.12**  The *Cartan map* of $FG$ is the inclusion map $c : P_F(G) \to R_F(G)$. For $\Phi \in \mathrm{IPr}_F(G)$ and $\varphi \in \mathrm{Irr}_F(G)$ the *Cartan number* $c_{\Phi,\varphi}$ of $\Phi$ with respect to $\varphi$ is the number defined by the equation

$$c(\Phi) = \sum_{\varphi \in \mathrm{Irr}_F(G)} c_{\Phi,\varphi} \varphi.$$

The matrix $(c_{\Phi,\varphi})_{\Phi \in \mathrm{IPr}_F(G), \varphi \in \mathrm{Irr}_F(G)}$ is the *Cartan matrix* of $FG$.

The Cartan numbers of $kG$ have the following alternative descriptions. For a projective indecomposable $kG$-module $P$ and a simple $kG$-module $S$, the Cartan number $c_{[P],[V]}$ is the multiplicity of $V$ as a composition factor of $P$ and this in turn is equal to $\dim_k(jkGi)$, for any primitive idempotents $i$ and $j$ of $kG$ such that $P \cong kGi$ and $V \cong kGj/J(kG)j$.

Since $\mathcal{O}$ is a principal ideal domain, if $V$ is a finitely generated $KG$-module. $V$ has an $\mathcal{O}$-form, i.e., there exists an $\mathcal{O}G$-module $V_0$ which is $\mathcal{O}$-free and such that $V = K \otimes_{\mathcal{O}} V_0$. Then, $V_0/\mathfrak{p}V_0$ is a $kG$-module of the same $k$-dimension as the $K$-dimension of $V$. Neither the isomorphism class of $V_0$ nor that of $V_0/\mathfrak{p}V_0$ is determined by the isomorphism class of $V$ in general. However, the image $[V_0/\mathfrak{p}V_0]$ of $V_0/\mathfrak{p}V_0$ in $R_k(G)$ is uniquely determined by the image $[V]$ of $V$ in $R_K(G)$. More precisely,

**Proposition 4.13** *The map* $d: R_K(G) \to R_k(G)$ *which sends the image* $[V]$ *of a finitely generated* $\mathcal{O}G$-*module to the element* $[V_0/\mathfrak{p}V_0] \in R_k(G)$, *where* $V_0$ *is an* $\mathcal{O}$-*free* $\mathcal{O}G$-*module such that* $V = K \otimes_{\mathcal{O}} V_0$ *is a well defined group homomorphism.*

**Definition 4.14** With the notation above,

- The map $d$ is called the *decomposition map* of $\mathcal{O}G$.

- For each $\chi \in \mathrm{Irr}_K(G)$, write

$$d(\chi) = \sum_{\varphi \in \mathrm{Irr}_k(G)} d_{\chi,\varphi}\varphi.$$

The number $d_{\chi,\varphi}$ is the *decomposition number* of $\chi$ with respect to $\varphi$ and the matrix $(d_{\chi,\varphi})_{\chi \in \mathrm{Irr}_K(G), \varphi \in \mathrm{Irr}_K(G)}$ is the *decomposition matrix* of $G$.

It is easy to see that for $V$ a simple $KG$-module and $S$ a simple $kG$-module, the decomposition number $d_{[V],[S]}$ is the multiplicity of $S$ as a composition factor of $V_0/\mathfrak{p}V_0$, where $V_0$ is an $\mathcal{O}$-free $\mathcal{O}G$-module such that $V = K \otimes_{\mathcal{O}} V_0$.

For the next theorem, note that by Proposition 4.10, if $I$ is a set of representatives of conjugacy classes of primitive idempotents of $\mathcal{O}G$, then $\mathrm{IPr}_k(G) = \{[kG\bar{i}] : i \in I\}$ and $\mathrm{Irr}_k(G) = \{[kG\bar{i}/J(kG)\bar{i}] : i \in I\}$.

**Theorem 4.15** *Let* $d: R_K(G) \to R_k(G)$ *be the decomposition map of* $\mathcal{O}G$ *and let* $c: Pr_K(G) \to R_k(G)$ *be the Cartan map of* $kG$. *Let* $I$ *be a set of representatives of conjugacy classes of primitive idempotents of* $\mathcal{O}G$ *and for each* $i \in I$, *let* $\Phi_i = [kG\bar{i}]$ *and* $\varphi_i = [kG\bar{i}/J(kG)\bar{i}]$.

(a) *The map* $d$ *induces an isomorphism between the subgroups* $Pr_{\mathcal{O}}(G)$ *and* $Pr_k(G)$.

(b) *Let* $e: Pr_k(G) \to R_K(G)$ *be the group homomorphism which for each* $i$ *in* $I$ *sends the element* $[kG\bar{i}]$ *of* $Pr_k(G)$ *to the element* $[K \otimes_{\mathcal{O}} \mathcal{O}Gi]$ *of* $R_K(G)$. *Then* $de = c$, *and*

$$e(\Phi_i) = \sum_{\chi \in \mathrm{Irr}_K(G)} d_{\chi,\varphi_i}\chi,$$

*for all $i \in I$. In particular, if $D$ is the decomposition matrix of $\mathcal{O}G$ and $C$ is the Cartan matrix of $kG$, then*

$$C = D^t D.$$

(c)   *The map $d$ is surjective.*

The above notions are compatible with block decompositions. Let $R$ denote one of the rings $K, \mathcal{O}$, or $k$ and let $b$ be a central idempotent of the group ring $RG$. We say that an $RG$-module $M$ *belongs to $b$*, if $bM = M$. If $M$ is an $RG$-module which belongs to $b$, and if $m \in M$, then writing $m = bn$ for $n \in M$, we have that

$$bm = bbn = bn = m.$$

So, if $M$ belongs to $b$, then $b$ acts as the identity on $M$ and $M$ is therefore an $RGb$-module. Conversely, if $N$ is an $RGb$-module, then $N$ is an $RG$-module via $a.n = abn$, for $a \in RG$ and $m \in M$. Thus $RGb$-modules are precisely $RG$-modules which belong to $b$. Further, if $\{b_i : 1 \leq i \leq r\}$ is an orthogonal decomposition of $1_{RG}$ in $Z(RG)$, and $M$ is an $RG$-module, then since $b_i$ is central in $RG$, $b_iM$ is an $RG$-module belonging to $b_i$ for $1 \leq i \leq r$ and we have a direct sum decomposition

$$M = b_1 M \oplus \cdots \oplus b_r M,$$

into $RG$-modules. In particular, if $M$ is indecomposable, then $M$ belongs to $b_i$ for a unique $i$, $1 \leq i \leq r$.

For the next definition, note that by Proposition 4.10, for any central idempotent $b$ of $kG$, there is a unique central idempotent $\hat{b}$ of $\mathcal{O}G$ such that $\bar{\hat{b}} = b$ and if $b$ is a block of $kGb$, then $\hat{b}$ is a block of $\mathcal{O}Gb$. The central idempotent $\hat{b}$ of $\mathcal{O}G$ is called the *lift* of $b$. Also, since $Z(\mathcal{O}G)$ is a subring of $Z(KG)$, any central idempotent of $\mathcal{O}G$ is a central idempotent of $KG$.

**Definition 4.16**   Let $b$ be a central idempotent of $kG$ and let $\hat{b} \in \mathcal{O}G$ be the lift of $b$.

(a)   Denote by $R_k(G, b)$ the subgroup of $R_k(G)$ generated by images of finitely generated $kG$-modules which belong to $b$; by $\mathrm{Irr}_k(G, b)$ the images of simple $kG$-modules which belong to $b$; by $\mathrm{Pr}_k(G, b)$ the group $\mathrm{Pr}_k(G) \cap R_k(G, b)$; and by $\mathrm{IPr}_k(G, b)$ the set $\mathrm{IPr}_k(G) \cap R_k(G, b)$.

(b)   Denote by $R_K(G, b)$ the subgroup of $R_K(G)$ generated by images of finitely generated $KG$-modules which belong to $\hat{b}$; by $\mathrm{Irr}_K(G, b)$ the images of simple $KG$-modules which belong to $\hat{b}$; by $\mathrm{Pr}_{\mathcal{O}}(G, b)$ the group $\mathrm{Pr}_{\mathcal{O}}(G) \cap R_K(G, b)$; and by $\mathrm{IPr}_{\mathcal{O}}(G, b)$ the set $\mathrm{IPr}_{\mathcal{O}}(G, b) \cap R_K(G, b)$.

For a central idempotent $b$ of $kG$, $|\mathrm{Irr}_k(G,b)|$ is the number of isomorphism classes of simple $kGb$-modules and similarly $|\mathrm{Irr}_K(G,b)|$ is the number of ismorphism classes of $KGb$-modules. Also, if $b$ and $c$ are orthogonal idempotents of $Z(kGb)$, then for any $\varphi \in \mathrm{Irr}_k(G,b)$, $\chi \in \mathrm{Irr}_K(G,c)$ and $\Phi \in \mathrm{IPr}_k(G,c)$, the corresponding Cartan and decomposition numbers satisfy

$$d_{\chi,\varphi} = 0 = c_{\Phi,\varphi}.$$

**Definition 4.17**   Let $b$ be a central idempotent of $kG$ and let $\hat{b}$ be the central idempotent of $\mathcal{O}G$ with $\bar{\hat{b}} = b$. The *Cartan matrix* of $kGb$ is the submatrix of the Cartan matrix of $kG$ whose entries correspond to modules which belong to $b$ and the *decomposition matrix* of $\mathcal{O}Gb$ is the submatrix of the decomposition matrix of $\mathcal{O}G$ whose entries correspond to modules which belong to $b$.


## 4.4.  Ordinary and Brauer characters.

Let $R$ be an integral domain and let $M$ be a finitely generated $RG$-module such that $M$ is free as $R$-module. Choose an $R$-basis of $M$ and identify $\mathrm{End}_R(M)$ with $\mathrm{Mat}_n(R)$ through the chosen basis, where $n$ is the cardinality of the basis.

**Definition 4.18**   With the above notation, the character $\chi_M : RG \to R$ of $M$ is the function defined by $\chi_M(a) = \mathrm{trace}(\phi(a))$, $a \in RG$, where $\phi(a) \in \mathrm{End}_R(M)$ is defined by

$$\phi(a)(m) = am, \quad m \in M.$$


The character $\chi_M$ is independent of the choice of $R$-basis of $M$ and is completely determined by its values on the elements of the group $G$. So it is customary to identify $\chi_M$ with its restriction to $G$. As such, $\chi_M : G \to R$ is a *class function* on $G$, i.e., $\chi_M$ has constant value on $G$-conjugacy classes.


Over fields, characters separate simple modules:

**Proposition 4.19**   *Let $F = K$ or $k$ and let $M_1, M_2$ be simple $FG$-modules. Then $\chi_{M_1} = \chi_{M_2}$ if and only if $M_1 \cong M_2$ as $FG$-modules.*

The character of a $KG$-module is called an *ordinary character*, and the character of a simple $KG$-module is an *ordinary irreducible character*. The character of a $kG$-module is a *modular character* and the character of a simple $kG$-module is a *modular irreducible character*. By the above result, we may identify $\mathrm{Irr}_K(G)$ with the set of ordinary irreducible characters of $G$ and it is customary to do so. This can of course also be done for $\mathrm{Irr}_k(G)$

but there is another subtle and very useful interpretation of $\mathrm{Irr}_k(G)$ due to Brauer which we now describe.

Let $M$ be a $kG$-module, and let $X$ be an abelian subgroup of $G$. Since $k$ is algebraically closed, by the standard "simultaneous eigenvector" argument in linear algebra, there exists a $k$-basis of $M$ such that identifying $\mathrm{End}_k(M)$ with $\mathrm{Mat}_n(k)$ through this basis, the matrix representing the action of $x$ on $M$ is upper triangular, for all $x \in X$. For each $x \in X$, the diagonal entries, say $\lambda_1, \cdots, \lambda_n$ of the corresponding matrix are the eigenvalues of $x$ on $M$ and in particular are $t$-th roots of unity, where $t$ is the order of $x$ in $G$. Since $char(k) = p$, the only $p$-power root of unity in $k$ is 1. Thus, if $x$ and $y$ are a pair of commuting elements of $G$ with $x$ a $p$-element, then $\chi_M(xy) = \chi_M(y)$. On the other hand, every element $x$ of $G$ is a product $x = x_p x_{p'}$, for uniquely determined elements $x_p, x_{p'}$ of $G$ such that $x_p$ and $x_{p'}$ commute, $x_p$ is a $p$-element and $x_{p'}$ is a $p'$-element. So, $\chi_M(x) = \chi_M(x_{p'})$ for all $x \in G$ and $\chi_M$ is completely determined by its values on the $p'$-elements of $G$. Now, let $x \in G$ be a $p'$-element and let $\lambda_1, \cdots, \lambda_n$ be the eigenvalues of $x$ on $M$. For each $i$, $1 \leq i \leq n$, $\lambda_i$ is a $p'$-root of unity, so by Proposition 4.8, there exists a unique root of unity, say $\hat{\lambda}_i$ in $\mathcal{O}^*$ such that $\overline{\hat{\lambda}}_i = \lambda_i$. Let $G_{p'}$ denote the set of $p'$-elements of $G$ and define $\hat{\chi}_M : G_{p'} \to K$ by $\hat{\chi}_M(x) = \hat{\lambda}_1 + \cdots + \hat{\lambda}_n$ for $x \in G_{p'}$. So, $\overline{\hat{\chi}_M(x)} = \chi_M(x)$, for all $x \in G_{p'}$.

**Definition 4.20**   With the notation above the map $\hat{\chi}_M : G_{p'} \to K$ is called the *Brauer character* of $M$. The Brauer character of a simple $kG$-module is called an *irreducible Brauer character*.

By the discussion preceding the above definition, for any $kG$-module $M$ the modular character $\chi_M$ of $M$ is completeley determined by the Brauer character $\hat{\chi}_M$ of $M$. So, by Proposition 4.19, two simple $kG$-modules $M_1$ and $M_2$ are isomorphic if and only they have the same Brauer characters. Thus $\mathrm{Irr}_k(G)$ may be identified with the set of irreducible Brauer characters of $G$, and it is customary to do so.

The functional, i.e., character theoretic approach is a big part of the representation theory of finite groups. We describe below a few properties of characters-the choice of statements is dictated by the material of the next sections. The first assertion of the next theorem forms part of the "orthogonality relations" of characters.

**Theorem 4.21**   *With the above identifications, $\mathrm{Irr}_K(G)$ is a basis of the $K$-vector space of $K$-valued class functions on $G$ and $\mathrm{Irr}_k(G)$ is a basis of the $K$-vector space of $K$-valued class functions on $G_{p'}$.*

By the above, the Grothendieck group $R_K(G)$ becomes identified with the subgroup of the $K$-vector space of $K$-valued class functions on $G$ generated by $\mathrm{Irr}_K(G)$ and $R_k(G) \cong \mathbb{Z}\mathrm{Irr}_k(G)$ with the subgroup of the $K$-vector space of $K$-valued class functions on $G_{p'}$ generated by $\mathrm{Irr}_k(G)$.

If $\chi : G \to K$ is a $K$-valued class function on $G$, then the restriction $\mathrm{Res}^G_{G_{p'}} \chi$ of $\chi$ to $G_{p'}$ is a $K$-valued class function on $G_{p'}$. The map $\chi \to \mathrm{Res}^G_{G_{p'}} \chi$ is a surjective $K$-linear map from the space of $K$-valued class functions on $G$ to the space of $K$-valued class functions on $G_{p'}$; restriction of this map to the group $R_K(G)$ is just the character theoretic analogue of the decomposition map. This map has two obvious sections: either one can extend a class function $\Phi$ on $G_{p'}$ to the class function $\tilde{\Phi}$ on $G$ by setting $\tilde{\Phi}(x) = \Phi(x_{p'})$ for $x \in G$, or to the class function $\Phi^*$ on $G$ by setting $\Phi^*(x) = 0$ if $x \in G - G_{p'}$. The next result, which is related to the surjectivity of the decomposition map, shows that these sections are meaningful on restriction to Grothendieck groups. These proofs (see [NT, Chapter 3, Lemma 6.13, Lemma 6.31]) depend on Brauer's characterisation of characters. Recall that $\mathrm{Pr}_{\mathcal{O}}(G)$ is a subgroup of $R_K(G)$.

**Theorem 4.22**    *Keep the notation above and let $p^a$ be the p-part of the order of $G$.*

(a)   *$\mathrm{Pr}_{\mathcal{O}}(G)$ consists of the elements of $R_K(G)$ which vanish on $G - G_{p'}$.*

(b)   *For any $\Phi \in R_k(G)$, $\tilde{\Phi} \in R_K(G)$.*

(c)   *For any $\Phi \in R_k(G)$, $p^a \Phi^* \in \mathrm{Pr}_{\mathcal{O}}(G)$. Consequently, $p^a (\mathrm{Res}^G_{G_{p'}} \chi)^* \in \mathrm{Pr}_{\mathcal{O}}(G)$ for all $\chi \in R_K(G)$.*

By Theorem 4.5, if $F$ is a splitting field for $G$ such that $char(F)$ does not divide $|G|$, then the map which sends a simple $FG$-module to the block of $FG$ containing it is a bijection between $\mathrm{Irr}_F(G)$ and the set of blocks of $FG$. The next theorem is a formula which expresses a block of $FG$ as an element of $FG$ in terms of the values of the character of the unique simple $FG$-module contained in the block. This formula also encodes to some extent the orthogonality relations of characters.

**Theorem 4.23** (Fourier inversion formula)    *Let $F$ be a splitting field for $G$ whose characteristic does not divide $|G|$ and identify $\mathrm{Irr}_F(G)$ with the set of characters of the simple $FG$-modules. For each $\chi \in \mathrm{Irr}_F(G)$, let $e_\chi$ be the block of $FG$ containing $\chi$. Then,*

$$e_\chi = \frac{\chi(1)}{|G|} \sum_{g \in G} \chi(g) g^{-1}.$$

If $b$ is a block of $kG$, and $\hat{b}$ the lift of $b$ in $\mathcal{O}G$, then $\hat{b}$ is a central idempotent of $KG$ since $Z(\mathcal{O}G) \subseteq Z(KG)$, but $\hat{b}$ is not in general a block of $KG$.

**Theorem 4.24** *With the notation above,*

(a) *For any block $b$ of $kG$,*

$$\hat{b} = \sum_{\chi \in \mathrm{Irr}_K(G,b)} e_\chi.$$

(b) *For any $\chi \in \mathrm{Irr}_K(G)$,*

$$\frac{|G|\chi(g)}{|C_G(g)|\chi(1)} \in \mathcal{O} \quad \text{for all} \ \ g \in G.$$

*Further, if $\chi, \chi' \in \mathrm{Irr}_K(G)$, then $\chi$ and $\chi'$ are in $\mathrm{Irr}_K(G,b)$ for the same block $b$ of $kG$ if and only if for any $p'$-element $g \in G$,*

$$\frac{|G|\chi(g)}{|C_G(g)|\chi(1)} - \frac{|G|\chi'(g)}{|C_G(g)|\chi'(1)} \in \mathfrak{p}.$$

The above two results give a recipe for calculating block idempotents for $kG$ from the values of the ordinary irreducible characters of $G$. We illustrate this with a small example. Examples 5.43 and 6.5 illustrate how this recipe can be extended to also calculate Brauer pairs and the corresponding fusion systems.

**Example 4.25** Let $G = S_3$, the symmetric group on three letters. By Theorem 4.24, $|\mathrm{Irr}_K(G)| = \dim_K Z(KG)$. Also, $\dim_K Z(KG)$ equals the number of conjugacy classes of $G$. Thus, up to isomorphism $KG$ has three simple modules. There are two simple modules of dimension 1, corresponding to the trivial and sign representation respectively, and one of dimension 2: the quotient of the natural permutation module for $G$ by the submodule of $G$-fixpoints. So,

$$KG = K \times K \times Mat_2(K).$$

Denoting by $\chi_1$ the character of the trivial $kG$-module, by $\chi_2$ the character of the sign representation and by $\chi_3$ the character of the two-dimensional simple $KG$-module, by the Fourier Inversion formula, the primitive central idempotents of $KG$ are

$$e_{\chi_1} = \frac{1}{6}\sum_{g \in G} g, \quad e_{\chi_2} = \frac{1}{6}\sum_{g \in G} sgn(g)g, \quad e_{\chi_3} = \frac{2}{3}1_G - \frac{1}{3}\sum_{g \in A_3 - 1_G} g.$$

Theorem 4.24 gives the $kG$ and hence the $\mathcal{O}G$-block distribution of $\mathrm{Irr}_K(G)$ for different values of $p$.

- If $p = 2$, then $\chi_1$ and $\chi_2$ are in the same block of $kG$, but not in the same block as $\chi_3$. Thus $\mathcal{O}G$ has two blocks:

$$b_1 := e_{\chi_1} + e_{\chi_2} = \frac{1}{3} \sum_{g \in A_3} g$$

and

$$b_2 = e_{\chi_3} = \frac{2}{3} 1_G - \frac{1}{3} \sum_{g \in A_3 - 1_G} g,$$

where $A_3$ is the alternating subgroup of $S_3$.

- If $p = 3$, then all three irreducible characters of $KG$ are in the same block of $kG$. Thus, $\mathcal{O}G$ has a unique block, i.e. $1_{\mathcal{O}G}$ is primitive in $Z(\mathcal{O}G)$, and $\mathcal{O}G$ is indecomposable.

- If $p \geq 5$, then $e_{\chi_i}$ is a block idempotent of $\mathcal{O}G$, for each $i$, $1 \leq i \leq 3$.

Another useful fact about blocks of $\mathcal{O}G$ and of $kG$ is that their support is contained in the $p'$-part of $G$ (see [NT, Chapter 3, Theorem 6.22]):

**Proposition 4.26**   *Let $R = \mathcal{O}$ or $k$. Any block of $RG$ is an $R$-linear combination of $p'$-elements of $G$.*


## 5. FUSION AND STRUCTURE

Let $(K, \mathcal{O}, k)$ be a $p$-modular system satisfying the conditions of Subsection 4.2. For a $p$-subgroup $P$ of $G$, let $\mathrm{br}_P : (kG)^P \to kC_G(P)$ be the Brauer homomorphism with respect to $P$, as in Subsection 3.4.

### 5.1. **The three main theorems of Brauer.**

The three main theorems are the bedrock of block theory. All involve the Brauer homomorphism, Brauer pairs and hence fusion systems. Of the three the second makes an explicit link between representation theory and Brauer pairs. The ideas of the proof of the first main theorem are also the main ideas behind the existence of saturated fusion systems associated to blocks. The third main theorem describes the fusion system of a particular block (the principal block) of $kG$.

**Theorem 5.1** (Brauer's First Main Theorem)   *Let $P$ be a $p$-subgroup of $G$. The map $\mathrm{br}_P$ induces a bijection between the set of blocks of $G$ with defect group $P$ and the set of blocks of $N_G(P)$ with defect group $P$.*

*Proof.* The main ingredients of the proof are Theorem 2.20, Lemma 3.17 and Lemma 3.18. We indicate how the pieces are put together to yield the theorem. Let $H = N_G(P)$, let $e$ be a block for $kC_G(P)$ and let $b$ and

$c$ be blocks of $kG$ and $kH$ respectively such that $\mathrm{br}_P(b)e \neq 0 \neq \mathrm{br}_P(c)e$. Then $(P,e)$ is a $(kG,b,G)$-Brauer pair as well as a $(kH,c,H)$-Brauer pair. By Lemma 3.18, $(P,e)$ is a maximal $(kG,b,G)$-Brauer pair if and only if $(P,e)$ is a maximal $(kH,c,H)$-Brauer pair since if $(P,e)$ is not maximal as $(kG,b,G)$-Brauer pair, then there exists a $(kG,b,G)$-Brauer pair containing $(P,e)$ and whose first component normalises $P$. Since for any block $e$ of $kC_G(P)$, the blocks $b$ and $c$ of $kG$ and $kH$ are uniquely determined by the condition $\mathrm{br}_P(b)e \neq 0$ and $\mathrm{br}_P(c)e \neq 0$, it suffices to prove that if $b$ is a block of $kG$ with defect group $P$, then $\mathrm{br}_P(b)$ is a block of $kH$. So, let $b$ be a block of $kG$ with defect group $P$, and let $(P,e)$ be a maximal $(kG,b,G)$-Brauer pair. Let $c$ be the unique block of $kH$ such that $\mathrm{br}_P(c)e \neq 0$. By Lemma 3.17, $c$ is the sum of $H$-conjugates of $e$. But by Theorem 2.20, $H$ acts transitively on the set of $(kG,b,G)$-Brauer pairs with first component $P$. In other words, $H$ acts transitively on the set of blocks of $kC_G(P)$ which appear in the block decomposition of $\mathrm{br}_P(b)$ in $kC_G(P)$. Thus $c = \mathrm{br}_P(b)$. $\qquad\square$

**Definition 5.2**  Let $P$ be a $p$-subgroup of $G$ and let $b$ be a block of $kG$ with $P$ as a defect group. The *Brauer correspondent* of $b$ in $N_G(P)$ is the block $\mathrm{br}_P(b)$ of $kN_G(P)$.

An immediate consequence of Theorem 3.19 and Brauer's First Main Theorem is that the fusion system of the Brauer correspondant of a block is the normaliser subsystem (on the underlying $p$-group) of the fusion system of the block. More precisely:

**Corollary 5.3**  *Suppose that $b$ is a block of $kG$, $P \leq G$ is a defect group of $kGb$ and $c$ is the Brauer correspondent of $b$ in $kN_G(P)$. Then for any maximal $(kG,b,G)$-Brauer pair of the form $(P,e)$, $(P,e)$ is a maximal $(kN_G(P),c,N_G(P))$-Brauer pair and setting $\mathcal{F} = \mathcal{F}_{(P,e)}(kG,b,G)$, we have that $\mathcal{F}_{(P,e)}(kN_G(P),c,N_G(P)) = N_{\mathcal{F}}(P)$.*

**Theorem 5.4** (Brauer's Second Main Theorem)   *Let $b$ be a block of $kG$, $x$ a $p$-element of $G$ and $e$ a block of $kC_G(x)$. Let $\hat{b}$ (respectively $\hat{e}$) be the block of $\mathcal{O}G$ (respectively $\mathcal{O}C_G(x)$ ) lifting $b$ (respectively $e$). If $(\langle x\rangle, e)$ is not an $(kG,b,G)$-Brauer pair, then $\chi_V(\hat{e}xy) = 0$ for any $\mathcal{O}Gb$-module $V$ which is free and of finite rank as $\mathcal{O}$-module and any $p'$-element $y$ of $C_G(x)$.*

*Proof.* We sketch the ideas behind the proof given in [NT, Chapter 5, Theorem 4.1]. Let $V$ be as in the theorem, an $\mathcal{O}$-free finitely generated $\mathcal{O}Gb$-module and set $H = C_G(x)$. Let

$$V = W_1 \oplus \cdots \oplus W_t,$$

be a decomposition of $\mathrm{Res}_{\mathcal{O}H}^{\mathcal{O}G} V$ into indecomposable $\mathcal{O}H$-modules and for each $i$, $1 \leq i \leq t$, let $e_i$ be the block of $kH$ such that $W_i$ belongs to the lift , $\hat{e}_i$, of $e_i$ to a block of $\mathcal{O}H$. Then for $1 \leq i \leq t$ one proves:

- If $\mathrm{br}_P(b)e_i = 0$, then $W_i$ is relatively $Q_i$-projective for some proper subgroup $Q_i$ of $\langle x \rangle$, where the definition of relative projectivity for $\mathcal{O}H$-modules is as in Definition 4.1 with $F$ replaced by $\mathcal{O}$. The proof uses Green's theory of vertices of indecomposable modules (see [NT, Chapter 4]).

- If $W_i$ is relatively $Q_i$-projective for some proper subgroup $Q_i$ of $\langle x \rangle$, then $\chi_{W_i}(xy) = 0$ for any $p'$-element $y$ of $H$. This is a consequence of a deep result known as Green's indecomposability theorem (see [NT, Chapter 4, Theorem 7.4]).

Now if $e$ is a block of $kH$, then multiplying the above decomposition of $V$ with $e$ gives

$$eV = \oplus_{i \in I} W_i,$$

where $I \subset \{1, \cdots, t\}$ consists of those $i$ for which $e_i = e$. This is because for $1 \leq i \leq t$, $W_i = e_i W_i$ and $ee_i = 0$ if $e \neq e_i$. Thus, if $\mathrm{br}_P(b)e = 0$,

$$\chi_V(exy) = \chi_{eV}(xy) = \sum_{i \in I} \chi_{W_i}(xy) = 0. \qquad \square$$

The second main theorem is often stated in terms of generalized decomposition numbers. We indicate briefly the connection with the version we have given above. keeping in mind the identification of modules and characters laid out in Section 4.4, let $x$ be a $p$-element of $G$ and let $\chi \in \mathrm{Irr}_K(G)$. The restriction $\mathrm{Res}_{C_G(x)}^{G} \chi$ of $\chi$ to $KC_G(x)$ is a sum of ordinary irreducible $C_G(x)$ characters, say

$$\mathrm{Res}_{C_G(x)}^{G} \chi = \sum_{\tau \in \mathrm{Irr}_K(C_G(x))} n_{\chi,\zeta} \tau.$$

Since $x$ is central in $C_G(x)$, and $K$ is a splitting field of $C_G(x)$, by Schur's Lemma (see [NT, Chapter 3, Lemma 3.5]), $x$ acts as a scalar linear transformation on any simple $KC_G(x)$-module. Since $x$ is a $p$-element, this means that for any $\zeta \in \mathrm{Irr}_K(C_G(x))$, there exists an element $\lambda_{x,\zeta}$ of $K$ of $p$-power order such that $\zeta(xy) = \lambda_{x,\zeta}\zeta(y)$ for all $y \in C_G(x)$. Let $d^x : R_K(C_G(x)) \to R_k(C_G(x))$ be the decomposition map of $\mathcal{O}C_G(x)$, and for each $\zeta \in \mathrm{Irr}_K(C_G(x))$ and each $\tau \in \mathrm{Irr}_k(C_G(x))$, let $d_{\zeta,\tau}^{(x)}$ be the corresponding decomposition number. For each $\chi \in \mathrm{Irr}_K(G)$ and each $\tau \in \mathrm{Irr}_k(C_G(x))$, the *generalised decomposition number* is defined by

$$d_{\chi,\tau}^x = \sum_{\zeta \in \mathrm{Irr}_K(C_G(x))} n_{\chi,\zeta} \lambda_{x,\zeta} d_{\zeta,\tau}^{(x)}.$$

Then, it is easy to see that for all $\chi \in \mathrm{Irr}_K(G)$, $\tau \in \mathrm{Irr}_k(C_G(x))$ and any $p'$-element $y$ of $C_G(x)$,

$$\chi(xy) = \sum_{\tau \in \mathrm{Irr}_k(C_G(x))} d^x_{\chi,\tau} \tau(y),$$

where each $\tau \in \mathrm{Irr}_k(C_G(x))$ is identified to its Brauer character. From this and using the fact that the set of irreducible Brauer characters on $C_G(x)$ is a linearly independent subset of the set of $K$-valued functions on the set of $p'$-elements of $C_G(x)$(see Theorem 4.21), it can be shown that Theorem 5.4 is equivalent to the following statement.

**Theorem 5.5** *Let $x$ be a $p$-element of $G$, $b$ a block of $kG$, $e$ a block of $kC_G(x)$. If $(\langle x\rangle, e)$ is not a $(kG, b, G)$-Brauer pair, then for any $\chi \in \mathrm{Irr}_K(G, b)$ and any $\tau \in \mathrm{Irr}_k(C_G(x), e)$, $d^x_{\chi,\tau} = 0$.*

A consequence of Theorem 5.4 is that the number of ordinary irreducible characters in a block is "locally" determined, in a sense that is made precise below.

**Definition 5.6** Let $P$ be a finite $p$-group and let $\mathcal{F}$ be a saturated fusion system on $P$. Two elements $x, y \in P$ are $\mathcal{F}$-*isomorphic* if there is a morphism $\varphi: \langle x\rangle \to \langle y\rangle$ in $\mathcal{F}$ such that $\varphi(x) = y$. Denote by $P/\mathcal{F}$ a set of representatives of the $\mathcal{F}$-isomorphism classes of elements of $P$.

**Theorem 5.7** *Let $b$ be a block of $kG$ and let $\mathcal{F} = \mathcal{F}_{(P,e_P)}(kG, b, G)$ for some maximal $(kG, b, G)$-Brauer pair $(P, e_P)$. For each $Q \leq P$, let $e_Q$ be the unique $(kG, b, G)$-Brauer pair such that $(Q, e_Q) \leq (P, e_P)$. Then,*

$$|\mathrm{Irr}_K(G, b)| = \sum_{x \in P/\mathcal{F}} |\mathrm{Irr}_k(C_G(x), e_{\langle x\rangle})|.$$

*Proof.* See [NT, Chapter 5, Theorem 4.13, Theorem 9.4]. $\square$

Examples 4.25 and 5.43 show how to calcuate the poset of Brauer pairs using character theory. Brauer's Second Main Theorem can also used for this. In fact, the second main theorem is one of the tools for understanding the fusion systems of blocks of families of groups such as the finite groups of Lie type and their Weyl groups. To explain this properly would require an exposition of the representation theory of these classes of groups, but a rough idea of how the theorem is used may be gleaned from the following situation. Suppose that $\chi \in \mathrm{Irr}_K(G, b)$, and $x$ is a $p$-element of $G$ such that $\chi(xy) \neq 0$ for some $p'$-element of $C_G(x)$ and such that $\mathrm{Res}^{C_G(x)}_G \chi \in \mathrm{Irr}_K(C_G(x))$. Then letting $e$ be the block of $kC_G(x)$ to which $\mathrm{Res}^{C_G(x)}_G \chi$ belongs (with the obvious abuse of notation), by Theorem 5.4, $(\langle x\rangle, e)$ is a $(kG, b, G)$-Brauer pair.

The above applies to the the character of the trivial $KG$-module, and can be used to give a proof Brauer's Third Main Theorem which is stated below.

**Definition 5.8**   Let $G$ be a finite group. The *principal block* of $kG$ is the block $b$ of $kG$ containing the trivial $kG$-module.

**Theorem 5.9** (Brauer's Third Main Theorem)   *Suppose that $b$ is the principal block of $kG$. Then, for any $p$-subgroup $Q$ of $G$, $\mathrm{br}_Q(b)$ is the principal block of $kC_G(Q)$. Consequently, for any maximal $(kG, b, G)$-Brauer pair $(P, e)$, $P$ is a Sylow $p$-subgroup of $G$ and $\mathcal{F}_{(P,e)}(kG, b, G) = \mathcal{F}_P(G)$.*

*Proof.* We sketch a proof of the above given in [AB, Theorem 3.13]. For any finite group $H$, let $p_H$ denote the principal block of $kH$. Applying the discussion before Definition 5.8 repeatedly (or by using the fact that the principal block is the unique block of $kG$ not contained in the augmentation ideal of $kG$) one proves that

- For any $p$-element $x$ of $G$, $(\{1\}, p_G) \leq (\langle x \rangle, p_{C_G(x)})$; and

- For a $p$-subgroup $Q$ of $G$ and $z$ a $p$-element of $N_G(Q)$, $(Q, p_{C_G(Q)}) \leq (\langle z, Q \rangle, p_{C_G(\langle z, Q \rangle)})$. Then by induction,

- For $p$-subgroups $Q \leq Q'$ of $G$ $(Q, p_{C_G(Q)}) \leq (Q', p_{C_G(Q')})$.

The theorem follows from the uniqueness of inclusion of Brauer pairs and the fact that the principal block of $kH$ for any finite group $H$ is invariant under any automorphism of $H$. $\qquad\square$


## 5.2. **Relative projectivity and representation type.**

The next result gives various characterisations of defect groups, each of which has been used as a definition of defect groups in the literature. We only give versions over $k$, but there are analogous characterisations over $\mathcal{O}$. Note that if $b$ is a central idempotent of $kG$, then $kGb$ is invariant under left and right multiplication by $G$ and hence $kGb$ is a $k[G \times G]$-module via $(x, y).a = xay^{-1}$ for $x, y \in G$ and $a \in kGb = bkGb$. For any $H \leq G$, let $\Delta H = \{(x, x) : x \in G\} \leq G \times G$ be the diagonally embedded copy of $H$ in $G \times G$.

**Proposition 5.10**   *Let $b$ be a block of $kG$ and let $P$ be a $p$-subgroup of $G$. The following are equivalent.*

(a)   *$P$ is a defect group of $b$.*

(b)   *$\mathrm{br}_P(b) \neq 0$ and $P$ is maximal amongst $p$-subgroups of $G$ with this property.*

(c)  $b \in \mathrm{Tr}_P^G((kG)^P)$ *and $P$ is minimal amongst subgroups with this property.*

(d)  *The $k[G \times G]$-module $kGb$ is relatively $\Delta P$-projective and $\Delta P$ is minimal amongst subgroups of $G \times G$ with this property.*

(e)  *Any $kGb$-module is relatively $P$-projective and $P$ is minimal amongst subgroups of $G$ with this property.*

*Proof.* The equivalence of (a), (b) and (c) is immediate from Theorem 2.20 (recall the identification of the $\mathrm{Br}^{kG}P$ with $\mathrm{br}_P$ as given in Proposition 3.12). The equivalence of (c) and (d) is given in [Be1, Proposition 6.1.2], and the equivalence of (d) and (e) is in [Be1, Proposition 6.3.3]. The subtle part of the equivalence of (d) and (e) is to deduce, from (d), the existence of a $kGb$-module which is not relatively $Q$-projective for any proper subgroup $Q$ of $G$.  $\qquad\square$

Combining the equivalence of (a) and (e) above for $P = \{1\}$ with the structure theory of finite dimensional algebras gives the following.

**Theorem 5.11** *Let $b$ be a block of $kG$ and let $P$ be a defect group of $kGb$. If $P = 1$, $kGb$ is isomorphic to a full matrix algebra over $k$. If $P$ is non-trivial, then $J(kGb) \neq 0$.*

By the equivalence of (a) and (e) in Proposition 5.10 we see that the defect groups of a block $b$ of $kG$ measure how far away $kGb$-modules are from being projective, and this in turn is a measure of the complexity of the module category of the block algebra. The *representation type* of a finite dimensional algebra is another measure of the complexity of the module category of the algebra. For the definition of representation type, we refer the reader to [Be1, Section 4.4]. Here we just note that there are three possibilities for representation type: *finite*, *tame*, and *wild*. Roughly speaking, algebras of wild type have the most complex module categories, while algebras of finite type have the least complicated structure.

**Theorem 5.12**  *([Bre], [BonD], [H]) Let $b$ be a block of $kG$ and let $P$ be a defect group of $kGb$. The block algebra $kGb$ and the group algebra $kP$ have the same representation type. Further,*

(a)  *$kP$ is of finite representation type if $P$ is cyclic.*

(b)  *$kP$ is of tame representation type if $p = 2$ and $P$ is the Klein 4-group, or a generalized quaternion, dihedral or semi-dihedral group.*

(c)  *In all other cases $kP$ is of wild representation type.*

Blocks of finite representation type, i.e., blocks with cyclic defect groups, are well understood by the work of many authors, notably Brauer, Thompson, Dade, Janusz, Green and Linckelmann. Blocks of tame representation type are also well understood by the work of Erdmann. By contrast, blocks of wild representation type are little understood. The differences in the state of knowledge for the various types will become apparent in the discussion regarding the status of various block theoretic conjectures. Many of the standard reference books on representation theory contain an exposition of the theory of cyclic blocks; Erdmann's book [Er] gives an account of tame blocks.

### 5.3. Finiteness conjectures.

**Definition 5.13** Let $P$ be a finite group. A *P-block* is a block $b$ of $kH$, for $H$ some finite group, such that the defect groups of $kHb$ are isomorphic to $P$.

One of the themes of block theory is this : Given a finite $p$-group $P$, to what extent are the representation theoretic invariants of $P$-blocks bounded by a function of $|P|$?

**Theorem 5.14** (Brauer) *Let $b$ be a block of $kG$ and let $P$ be a defect group of $kGb$. The group $R_k(G,b)/Pr_k(G,b)$ has exponent $|P|$.*

*Proof.* We sketch the idea behind the proof given in [NT, Chapter 3, Theorem 6.35]. Let $|P| = p^d$. One proves the following block-wise version of Theorem 4.22 (b):

- With the notation of Theorem 4.22 (b), for any $\Phi \in R_k(G,b)$, $p^d\Phi^* \in \mathrm{Pr}_{\mathcal{O}}(G,b)$.

The result follows from this and the fact that the decomposition map induces an isomorphism between $\mathrm{Pr}_{\mathcal{O}}(G,b)$ and $\mathrm{Pr}_k(G,b)$ (Theorem 4.15 (a)). $\qquad\square$

**Theorem 5.15** (Brauer-Feit) *Let $b$ be a block of $kG$ and let $P$ be a defect group of $kGb$. Then,*
$$|\mathrm{Irr}_K(G,b)| \leq |P|^2.$$

*Proof.* We sketch the idea behind the proof given in [NT, Chapter 3, Theorem 6.39]. The starting point is the result highlighted in the proof of the previous theorem. So, in particular, $p^d(\mathrm{Res}^G_{G_{p'}}\chi)^* \in R_K(G,b)$ for all $\chi \in \mathrm{Irr}_K(G,b)$. Then one proves:

- For any $\chi \in \mathrm{Irr}_k(G,b)$, writing
$$p^d(\mathrm{Res}^G_{G_{p'}}\chi)^* = \sum_{\tau \in \mathrm{Irr}_K(G,b)} n_\tau \tau,$$

we have

$$\sum_{\tau \in \mathrm{Irr}_K(G,b)} n_\tau^2 \le p^{2d}.$$

This is an easy consequence of the orthogonality relations for characters.

- There exists a $\chi \in \mathrm{Irr}_k(G, b)$ such that writing $p^d(\mathrm{Res}^G_{G_{p'}} \chi)^*$ as above, we have that $n_\tau \ne 0$ for all $\tau \in \mathrm{Irr}_K(G, b)$.

Since each $n_\tau$ above is an integer, the proof follows. $\square$

Actually, the proof above yields the better upper bound of $\frac{1}{4}|P|^2 + 1$. This is still quadractic in $|P|$. The following conjecture, due to Brauer, predicts that $|P|$ itself is an upper bound for $|\mathrm{Irr}_K(G, b)|$.

**Conjecture 5.16** (Brauer's $\mathbf{k}(b)$ conjecture) *Let $b$ be a block of $kG$ and let $P$ be a defect group of $kGb$. Then,*

$$|\mathrm{Irr}_K(G, b)| \le |P|.$$

By Theorem 4.21, $|\mathrm{Irr}_k(G, b)| \le |\mathrm{Irr}_K(G, b)|$, so it follows from Theorem 5.15 that the Cartan matrix of a $P$-block has size at most $|P|^2$. By Theorem 5.14, the elementary divisors of the Cartan matrix of a $P$-block are of the form $p^m \le |P|$. Thus there are only finitely many possibilities for the size and the elementary divisors of the Cartan matrix of a $P$-block. The following conjecture says that the same is true of the entries of the Cartan matrix.

**Conjecture 5.17** (Weak Donovan Conjecture) [Al2] *Let $P$ be a finite $p$-group. There are only finitely many possibilities for Cartan numbers of $P$-blocks.*

For a commutative ring $R$ with identity, and $A$ an $R$-algebra, denote by $\mathrm{mod}(A)$, the $R$-linear category of finitely generated $A$-modules. Two $R$-algebras $A$ and $B$ are said to be *Morita equivalent* if $\mathrm{mod}(A)$ and $\mathrm{mod}(B)$ are equivalent as $R$-linear categories. The above conjecture is implied by the following conjecture, also due to Donovan. For the statement, we identify a block with the corresponding block algebra.

**Conjecture 5.18** (Strong Donovan conjecture) [Al2] *Let $P$ be a finite $p$-group. Up to Morita equivalence, there are only finitely many $P$-blocks.*

Conjecture 5.18 is know to be true if $P$ is cyclic by results of Brauer, Dade, Green and Janusz. Erdmann settled the case $P \cong C_2 \times C_2$ and $P \cong Q_8$. For the other tame cases, Erdmann's work nearly settles the conjecture, but there are some outstanding rationality questions; Conjecture 5.17 is known to hold for all tame blocks. By Puig's nilpotent block theorem (see

Theorem 5.40), Conjecture 5.18 also holds for those $p$-groups $P$ with the property that the only saturated fusion system on $P$ is the system $\mathcal{F}_P(P)$. Both conjectures are open for all other $P$.

The isomorphism class of the center $Z(A)$ of a finite dimensional $k$-algebra is an invariant of the Morita equivalence class of $A$. In this context, we have the following result.

**Theorem 5.19** *Let $P$ be a finite $p$-group. Up to isomorphism, there are only finitely many commutative $k$-algebras that occur as centers of $P$-blocks.*

*Proof.* Let $b$ be a block of $kG$ with defect group $P$. Since $\dim_k(Z(kGb)) = \dim_K(Z(KGb)) = |\mathrm{Irr}_K(G, b)|$, it follows from Theorem 5.15, that the $k$-dimension of $Z(kGb)$ is bounded by $a: = |P|^2$. By a result of Cliff, Plesken and Weiss [CPW], $Z(kGb)$ has an $\mathbb{F}_p$-form, that is $Z(kGb)$ has a $k$-basis such that the multiplicative constants of $Z(kGb)$ with respect to this basis are all in $\mathbb{F}_p$. Thus there are at most $p^{a^3}$-possibilities for the isomorphism type of $Z(kGb)$. $\qquad\square$

### 5.4. **Source algebras and Puig's conjecture.**

The concept of source algebras is due to Puig, who also proved their main properties. For a detailed account see [Th, Section 38].

**Definition 5.20**   Let $b$ be a block of $kG$ and let $P$ be a defect group of $kGb$.

- A *source idempotent* of $kGb$ (or of $b$) with respect to $P$ is a primitive idempotent $i$ of $(kG)^P$ such that $i \le b$ and $\mathrm{br}_P(i) \ne 0$.

- A *source algebra* of $kGb$ (or of $b$) is an interior $P$-algebra $(ikGi, \iota)$ where $i$ is a source idempotent of $kGb$ and $\iota$ is defined by $\iota(x) = ixi$ for $x \in P$.

By Lemma 1.19 (a), it is clear that source algebras exist. They are unique up to $G$-conjugation, as we now explain. Let $P$ be a defect group of $b$ and let $i$ be a source idempotent of $b$ with respect to $P$. Let $e(i)$ be the unique block of $kC_G(P)$ such that $i$ is associated to $e(i)$ (see Lemma 2.8). For any $u \in (A^P)^*$, $^u e$ is a source idempotent of $kGb$ and $e(\,^u i) = e(i)$.

**Proposition 5.21**   *With the notation above, the mapping $(P, i) \to (P, e(i))$, induces a bijection between pairs $(P, \gamma)$ such that $P$ is a defect group of $b$ and $\gamma$ is an $(A^P)^*$-conjugacy class of source idempotents of $b$ with respect to $P$ and the set of maximal $(kG, b, G)$-Brauer pairs.*

*Consequently, if $ikGi$ and $jkGj$ are source algebras of $b$, where $i$ and $j$ are source idempotents with respect to defect groups $P$ and $Q$ respectively,* *then there exists $g \in G$ and an isomorphism of $k$-algebras $f : jkGj \to ikGi$ such that $P = {}^gQ$ and $f(jxj) = i{}^gxi$ for all $x \in Q$.*

*Proof.* See [Th, Proposition 18.3]. $\square$

Thus, for a given defect group $P$, a source algebra of $b$ is unique up to twisting the interior $P$-algebra structure by an outer automorphism of $P$.

**Proposition 5.22** *Let $b$ be a block of $kG$ and let $i$ be a source idempotent of $b$ with respect to some defect group of $kGb$. The algebras $ikGi$ and $kGb$ are Morita equivalent.*

*Proof.* See [Th, Proposition 38.2]. $\square$

The $k$-dimension of a source algebra $ikGi$ of $kGb$ is usually much smaller than that of $kGb$, so the above result reduces the study of the representation theory of $kGb$ to a smaller algebra. A source algebra of $kGb$ is not in any sense "the smallest" amongst subalgebras of $kGb$ which are Morita equivalent to $kGb$. For instance, if $I$ is a primitive decomposition of a source idempotent $i$ in $ikGi$, and $J \subseteq I$ is such that $J$ contains at least one $kG$-conjugate of each element of $I$, then setting $j = J^+$, the algebra $jkGj$ is also Morita equivalent to $kGb$. The advantage of source algebras is that they contain information about blocks which is finer than that captured by Morita equivalence. For instance, the next result shows that the fusion system of $kGb$ can be read off from a source algebra of $kGb$.

If $b$ is a block of $kG$ and $i$ is a source idempotent of $b$ with respect to $P$, then since $xi = ix$ for all $x \in P$, $ikGi$ is invariant under left and right multilplication by elements of $P$. Thus, $ikGi$ is a $k[P \times P]$-module via $(x, y).a = xay^{-1}$ for $x, y \in P$ and $a \in ikGi$. This $k[P \times P]$-module structure determines the fusion system of $b$. We make this more precise. For a finite group $H$, and an automorphism $\varphi : H \to H$, we denote by ${}_\varphi kH$ the $k[H \times H]$-module defined by ${}_\varphi kH = kH$ as $k$-space and where the $k[H \times H]$-module structure is given by $(g, h).m = \varphi(g)mh$ for $g, h \in H$ and $m \in kH$.

**Proposition 5.23** *Let $b$ be a block of $kG$, let $P$ be a defect group of $kGb$ and let $i$ be a source idempotent of $b$ with respect to $P$. Let $e_P$ be the block of $kC_G(P)$ such that $i$ is associated to $e_P$ and let $\mathcal{F} = \mathcal{F}_{(P,e_P)}(kG, b, G)$. Then $\mathcal{F}$ is generated by the set of inclusions between subgroups of $P$ and those automorphisms $\varphi$ of subgroups $Q$ of $P$ such that ${}_\varphi kQ$ is isomorphic to a direct summand of $ikGi$ as $k[Q \times Q]$-module.*

*Proof.* This is essentially proved in [P2]. In the the formulation above, the result appears in [Li5, Corollary 5.3]. $\square$

Source idempotents and source algebras are also defined over $\mathcal{O}$. Let $\hat{b}$ be the block of $\mathcal{O}G$ lifting $b$ and let $P$ be a defect group of $kGb$. A *source idempotent of* $\mathcal{O}G\hat{b}$ (or of $\hat{b}$) with respect to a defect group $P$ of $b$ is a primitive idempotent $i$ of $(\mathcal{O}G)^P$ such that $i \leq \hat{b}$ and $\mathrm{br}_P(i + \mathfrak{p}) \neq 0$ and the corresponding source algebra is $i\mathcal{O}G\hat{b}i$ considered as interior $P$-algebra via the map $x \to ixi$, for $x \in P$. A primitive idempotent $i$ of $(\mathcal{O}G)^P$ is a source idempotent if and only if $\bar{i} := i + \mathfrak{p}$ is a source idempotent of $kGb$ and in that case $\bar{i}kG\bar{i} = i\mathcal{O}Gi/\mathfrak{p}i\mathcal{O}Gi$. Conversely, by idempotent lifting, for any source idempotent $i$ of $kGb$, there exists a source idemotent $\hat{i}$ of $\mathcal{O}Gb$ with $\hat{i} + \mathfrak{p} = i$. From the point of view of representation theory it is better to study source algebras over $\mathcal{O}$ since this allows us to keep track of the ordinary character theory of blocks as well. On the other hand, the structure of a source algebra over $\mathcal{O}$ is determined largely by the structure over $k$.

**Proposition 5.24**  *Let $b$ be a block of $kG$ and $\hat{b}$ the lift of $b$ in $\mathcal{O}G$. Let $P$ be a defect group of $kGb$, let $i$ be a source idempotent of $\mathcal{O}Gb$ and let $\bar{i} = i + \mathfrak{p}$. Then, $i\mathcal{O}Gi$ has an $\mathcal{O}$-basis stable under left and right multiplication by $P$ and $\bar{i}kG\bar{i} = i\mathcal{O}Gi/\mathfrak{p}i\mathcal{O}Gi$. Further, if $B$ is a interior $P$-algebra over $\mathcal{O}$ having an $\mathcal{O}$-basis which is stable under left and right multiplication by $P$ and such that $\bar{i}kG\bar{i}$ is isomorphic to $B/\mathfrak{p}B$ as interior $P$-algebras, then $B$ is isomorphic to $i\mathcal{O}Gi$ as interior $P$-algebras.*

*Proof.* See [Th, Proposition 38.8].                                     □

**Definition 5.25**  Let $G$ and $H$ be finite groups, $b$ a block of $kG$ and $c$ a block of $kH$. We say that $b$ and $c$ are *source algebra equivalent* if $b$ and $c$ have a common defect group $P$ and there are source idempotents $i$ of $b$ and $j$ of $c$ with respect to $P$ such that there is an isomorphism of interior $P$-algebras algebras $ikGi \cong jkHj$.

From Propositions 5.22 and 5.23 it is immediate that two source algebra equivalent blocks have equivalent module categories and isomorphic fusion systems. We describe one well-known situation of a source algebra equivalence between blocks of different finite groups.

**Proposition 5.26**  *Let $H$ be a subgroup of $G$. Suppose that $c$ is a central idempotent of $kH$ such that $\mathrm{Tr}_H^G(c)$ is a central idempotent of $kG$ and $c^g c = 0$ for all $g \in G - H$. Set $d = \mathrm{Tr}_H^G(c)$. Then the map $e \to \mathrm{Tr}_H^G(e)$ is a bijection between the set of blocks $e$ of $kH$ such that $e \leq c$ and the set of blocks $f$ of $kG$ such that $f \leq d$. Further, if $e$ is a block of $kH$ such that $e \leq c$, then the block algebras $kHe$ and $kG\mathrm{Tr}_H^G(e)$ are source algebra equivalent.*

*Proof.* See [Th, Lemma 16.1, Proposition 16.6, Proposition 16.9].        □

**Conjecture 5.27** (Puig's conjecture) *Let $P$ be a finite $p$-group. Up to source algebra equivalence, there are only finitely many $P$-blocks.*

The above conjecture implies the Donovan conjectures. The conjecture is known to be true when $P$ is cyclic ([Li1]). Recently, using the classification of finite simple groups it has been settled for the case $P \cong C_2 \times C_2$ ([CrEKL]). It is open for all other $p$-groups.

## 5.5. Külshammer-Puig classes.

Let $n$ be a natural number and $A = Mat_n(k)$. We recall the following standard facts about $A$.

- (Noether-Skolem theorem) $\mathrm{Aut}(A) = \mathrm{Inn}(A)$.

- $Z(A)$ consists of scalar matrices. Hence, under the natural identification of the group of invertible scalar matrices with the group of non-zero elements $k^*$ of $k$, we have a group isomorphism $\mathrm{Inn}(A) \cong A^*/k^*$.

Now suppose $N$ is a finite group and that $A = Mat_n(k)$ is an $N$-algebra. Suppose further that $C$ is a normal subgroup of $N$ and that the restriction of the $N$-algebra structure of $A$ to $C$ is interior, that is suppose that there is a group homomorphism $\iota : C \to A^*$ such that for all $x \in C$, the action of $x$ on $A$ is given by conjugation by $\iota(x)$. We use this data to describe a certain distinguished element $\alpha_{A,N,C}$ of $H^2(N/C, k^\times)$. For each $y \in N$, the map $a \to {}^y a$ for $a \in A$ defines a $k$-algebra automorphism of $A$ and by the Noether-Skolem theorem every algebra automorphism of $A$ is an inner automorphism. Define a function $s : N \to A^*$ as follows. Choose a set of coset representatives, say $I$ of $C$ in $N$ containing 1. For each $x \in N$, set $s(x) = \iota(x)$ and for each $z \in I - N$ choose $s(z) \in A^*$ such that ${}^y a = s(y) a s(y)^{-1}$ for all $a \in A$. Then, for $y \in N - C$, set $s(y)$ to be equal to $\iota(x)s(z)$, where $y = xz$, $x \in C, z \in I$.

For all $y \in N$, ${}^y a = s(y) a s(y)^{-1}$ for all $a \in A$. So, for any $y, z \in N$, and any $a \in A$,

$$s(yz)as(yz)^{-1} = {}^{yz}a = s(y)({}^z a)s(y)^{-1} = s(y)s(z)as(z)^{-1}s(y)^{-1}.$$

Hence, $s(z)^{-1}s(y)^{-1}s(yz) \in Z(A) \cap A^*$. Since $Z(A)$ consists of scalar matrices, it follows that $s(yz) = \alpha(y, z)s(y)s(z)$ for some element $\alpha(y, z) \in k^*$. It is a straightforward check that the map $\alpha : N \times N \to k^*$ sending an element $(y, z)$ to $\alpha(y, z)$ is a 2-cocycle. Further, for any $y, z \in N$ and any $x, x' \in C$, $\alpha(yx, zx') = \alpha(y, z)$. Hence $\alpha$ induces a 2-cocycle, say $\bar{\alpha}$ of $N/C$. If $\alpha, \alpha'$ are 2-cocycles of $N$ resulting from different choices of $s$ above, then the corresponding 2-cocycles $\bar{\alpha}$ and $\bar{\alpha}'$ of $N/C$ are cohomologous. Thus, we can make the following definition.

**Definition 5.28**   With the notation above, $\alpha_{A,N,C}$ is the class in $H^2(N/C, k^*)$ of the 2-cocycle $N/C \times N/C \to k^*$ which sends an element $(yC, zC)$ of $N/C \times N/C$ to the element $\alpha(y, z)$ of $k^*$.

In the special case that $A = k$, the $N$-action on $A$ is trivial and the interior $C$-algebra structure on $A$ is defined by a homomorphism $\iota : C \to k^*$.

**Proposition 5.29**  *Suppose that $A = k$, and let $N$, $C$ and $\iota$ be as above. Denote also by $\iota$ the induced map from $C/\ker(\iota) \to k^*$ and let $\iota_* : H^2(N/C, C/\ker(\iota)) \to H^2(N/C, k^*)$ denote the map induced by $\iota$. Suppose that $\ker(\iota)$ is normal in $N$ and that $C/\ker(\iota)$ is central in $N/\ker(\iota)$. Then, $\alpha_{A,N,C} = \iota_*(\beta)$ where $\beta \in H^2(N/C, C/\ker(\iota))$ corresponds to the central extension*

$$1 \to C/\ker(\iota) \to N/\ker(\iota) \to N/C \to 1.$$

*Proof.* In the recipe given for constructing $\alpha_{A,N,C}$ chooose $s(z) = 1_A$ for every element of the chosen system of coset representatives of $C$ in $N$.   □

Before proceeding we record the following fact.

**Proposition 5.30**  *Let $Q$ be a central $p$-subgroup of a finite group $H$. Then the canonical $k$-algebra surjection*

$$kH \to k(H/Q)$$

*induces a bijection between blocks of $kH$ and blocks of $kH/Q$. If $R$ is a defect group of a block of $kH$, then $Q \leq R$ and $R/Q$ is a defect group of the corresponding block of $kH/Q$.*

*Proof.* This is in [NT, Chapter 5, Theorem 8.11].   □

Let $b$ be a block of $kG$, $(P, e_P)$ a maximal $(kG, b, G)$-Brauer pair and $\mathcal{F} = \mathcal{F}_{(P,e_P)}(kG, b, G)$. Suppose that $Q$ is an $\mathcal{F}$-centric subgroup of $P$. Let $\overline{C_G(Q)} = C_G(Q)/Z(Q)$ and let $\overline{e_Q}$ be the image of $e_Q$ in $k\overline{C_G(Q)}$ under the canonical surjection of $kC_G(Q) \to k\overline{C_G(Q)}$. By Theorem 3.20, $Z(Q)$ is a defect group of the block $e_Q$. Since $Z(Q)$ is central in $C_G(Q)$, by Proposition 5.30, $\overline{e_Q}$ is a block of $k\overline{C_G(Q)}$ with trivial defect groups. In particular, $k\overline{C_G(Q)}\,\overline{e_Q}$ is a full matrix algebra over $k$. Set $A = k\overline{C_G(Q)}\,\overline{e_Q}$. Regard $\overline{C_G(Q)}$ as a subgroup of $N_G(Q, e_Q)/Q$ through the natural isomorphism $\overline{C_G(Q)} \cong QC_G(Q)/Q$. Regard $A$ as an $N_G(Q, e_Q)/Q$ algebra through the conjugation action of $N_G(Q, e_Q)$ on $kC_G(Q)$ and regard $A$ as interior $\overline{C_G(Q)}$-algebra through the map $\iota{:}\overline{C_G(Q)} \to A^*$ which sends an element $x$ of $\overline{C_G(Q)}$ to $x\overline{e_Q}$. Let $\alpha_Q$ be the element $\alpha_{A,N,C,\iota}$ of Definition 5.28 with $N = N_G(Q, e_Q)/Q$ and $C = \overline{C_G(Q)}$. We regard $\alpha_Q$ as an element of $H^2(\mathrm{Out}_{\mathcal{F}}(Q), k^*)$ through the identification $\mathrm{Out}_{\mathcal{F}}(Q) =$

$N_G(Q, e_Q)/QC_G(Q) = N/C$. We will denote also by $\alpha_Q$ the element of $H^2(\mathrm{Aut}_{\mathcal{F}}(Q), k^*)$ obtained from $\alpha_Q$ through the canonical homomorphism $\mathrm{Aut}_{\mathcal{F}}(Q) \to \mathrm{Out}_{\mathcal{F}}(Q)$.

**Definition 5.31**  Let $b$ be a block of $kG$, $(P, e_P)$ a maximal $(kG, b, G)$-Brauer pair, and $\mathcal{F} = \mathcal{F}_{(P, e_P)}(kG, b, G)$. Let $Q$ be an $\mathcal{F}$-centric subgroup of $P$. With the notation above, the *Külshammer-Puig class* at $(Q, e_Q)$ is the element $\alpha_Q$ of $H^2(\mathrm{Out}_{\mathcal{F}}(Q), k^*)$.

We describe Külshammer-Puig classes in some cases.

**Proposition 5.32**  *With the notation of Definition 5.31, if $b$ is the principal block of $kG$, then $\alpha_Q$ is trivial for all $Q \in \mathcal{F}^c$.*

*Proof.* By Brauer's Third Main Theorem (see Theorem 5.9) $P$ is a Sylow $p$-subgroup of $G$, $\mathcal{F} = \mathcal{F}_P(G)$ and $e_Q$ is the principal block of $kC_G(Q)$ for all $Q \leq P$. Thus $\bar{e}_Q$ is the principal block of $k\overline{C_G(Q)}$ for all $Q \leq P$.

Suppose that $Q$ is $\mathcal{F}$-centric. Since $\mathcal{F} = \mathcal{F}_P(G)$, $Z(Q)$ is a Sylow $p$-subgroup of $C_G(Q)$, $C_G(Q) = Z(Q) \times O_{p'}(C_G(Q))$ and $\overline{C_G(Q)} \simeq O_{p'}(C_G(Q))$. Since $k$ is of characteristic $p$, by the Fourier inversion formula, the principal block idempotent, $\overline{e_Q}$ of $k\overline{C_G(Q)}$ is given by the formula

$$\bar{e}_Q = \frac{1}{|k\overline{C_G(Q)}|} \sum_{x \in k\overline{C_G(Q)}} x.$$

Hence, $x\bar{e}_Q = \bar{e}_Q$ for all $x \in \overline{C_G(Q)}$. It follows that $k\overline{C_G(Q)}\bar{e}_Q$ is a one-dimensional algebra, and $\iota : \overline{C_G(Q)} \to k^*$ is the trivial map. So $\alpha_Q$ is trivial by Proposition 5.29. $\qquad\square$

Certain defect groups also support only trivial Külshammer-Puig classes.

**Proposition 5.33**  *With the notation of Definition 5.31, suppose that either $P$ is cyclic or $p = 2$ and $P$ is a Klein 4-group, or a dihedral, semi-dihedral or quaternion group. Then $\alpha_Q$ is trivial for all $Q \in \mathcal{F}^c$.*

*Proof.* Let $Q$ be a subgroup of $P$. We will show that $H^2(\mathrm{Out}_F(Q), k^\times) = 0$ for any subgroup $Q$ of $P$. Since $H^2(H, k^\times) = 0$ if $H$ is cyclic or if $H$ is a $p$-group, we may assume that $\mathrm{Out}_{\mathcal{F}}(Q)$ is neither cyclic nor a $p$-group and hence that $\mathrm{Aut}(Q)$ is neither cyclic nor a $p$-group. Since the automorphism group of a cyclic group of odd prime power is cyclic, we may assume that $p = 2$ and that $\mathrm{Aut}(Q)$ is not a 2-group. By Example I.3.8, $Q$ is isomorphic to one of $Q_8$ or $C_2^2$. Now $\mathrm{Aut}(C_2^2)$ is cyclic, so we may assume that $Q \cong Q_8$. In this case, $\mathrm{Out}_{\mathcal{F}}(Q)$ is isomorphic to one of $C_3$, $C_2$ or $S_3$ and $H^2(C_3, k^*) = H^2(C_2, k^*) = H^2(S_3, k^*) = 0$. $\qquad\square$

In contrast to the above two results, we describe one source for non-trivial Külshammer-Puig classes. Recall the definition of constrained fusion systems from before Theorem I.4.9.

**Definition 5.34**  For a saturated fusion system $\mathcal{F}$ on a finite $p$-group and a block $b$ of $kG$, we say that $b$ is an *$\mathcal{F}$-block* if $\mathcal{F} \cong \mathcal{F}_{(P,e_P)}(kG, b, G)$ for some maximal $(kG, b, G)$-Brauer pair $(P, e_P)$.

**Proposition 5.35**  *Let $\mathcal{F}$ be a constrained fusion system on a finite $p$-group $P$ and let $Q \trianglelefteq \mathcal{F}$ be an $\mathcal{F}$-centric subgroup of $P$. Then for any element $\alpha \in H^2(\mathrm{Out}_{\mathcal{F}}(Q), k^*)$, there exists a finite group $G$ and an $\mathcal{F}$-block $b$ of $kG$ such that the corresponding Külshammer-Puig cocycle at $Q$ is $\alpha$.*

*Proof.* Let $L$ be a model for $\mathcal{F}$ (see Theorem I.4.9) and identify $L/Q$ with $\mathrm{Out}_{\mathcal{F}}(Q)$. Let $C$ be the cyclic subgroup of order $|L|_{p'}$ of $k^*$ and let $\iota : C \to k^*$ be the inclusion of $C$ in $k^*$. The group $C$ exists since $k$ is algebraically closed. Since $k$ is of characteristic $p$, the induced map $\iota_* : H^2(L/Q, C) \to H^2(L/Q, k^*)$ is surjective. Let $\beta \in H^2(L/Q, C)$ be such that $\alpha = \iota_*(\beta)$. Denote also by $\beta$ the element of $H^2(L, C)$ obtained from $\beta$ through the canonical homomorphism $L \to L/Q$ and let

$$1 \to C \to G \to L \to 1$$

and

$$1 \to C \to G/Q \to L/Q \to 1$$

be the corresponding central extensions. For each $p$-subgroup $R$ of $G$, identify $R$ with the Sylow $p$-subgroup of the full inverse image of $R$ in $G$. So $P$ is a Sylow $p$-subgroup of $G$ and $\mathcal{F}_P(G) = \mathcal{F}_P(L) \cong \mathcal{F}$.

Set

$$b = \frac{1}{|C|} \sum_{z \in C} \iota(z)^{-1} z.$$

So by the Fourier inversion formula, $b$ is the block of $kC$ corresponding to the linear representation $\iota$ of $C$. We claim that $b$ is a block of $kG$. Indeed, since $C$ is central in $G$, $b$ is a central idempotent of $kG$. On the other hand, since $Q$ is a normal $p$-subgroup of $G$, by Lemma 3.17 any central idempotent of $kG$ is an element of $kC_G(Q)$ and by Proposition 4.26, any central idempotent of $kG$ is a $k$-linear combination of $p'$-elements of $G$. Thus any central idempotent of $kG$ is a $k$-linear combination of $p'$-elements of $C_G(Q)$. Since $L$ is strongly $p$-constrained, $C_G(Q) = C \times Z(Q)$. Thus any central idempotent of $kG$ is in $kC$. Since by definition, $b$ is primitive in $Z(kC)$, it follows that $b$ is primitive in $Z(kG)$, proving the claim.

Let $R$ be a $p$-subgroup of $G$. Since $C \leq Z(G) \leq C_G(R)$, and $b \in kC$, we have $\mathrm{br}_R(b) = b$ and $(R, b)$ is a $(kG, b, G)$-Brauer-pair. In particular, $(P, b)$ is a maximal $(kG, b, G)$-Brauer-pair, $(R, b) \leq (P, b)$ for any subgroup $R$ of $G$

and $N_G(R, b) = N_G(R)$ for any $R \leq P$. In particular, $\mathcal{F}_{(P,b)}(kG, b, G) = \mathcal{F}$, that is $b$ is an $\mathcal{F}$-block.

Now we show that the Külshammer-Puig class $\alpha_Q$ equals $\alpha$. Note that $N_G(Q, b) = G$ and $C_G(Q) = C \times Z(Q)$. Now

$$k\overline{C_G(Q)}\bar{b} = k[C \times Z(Q)/Z(Q)]\bar{b} \cong kCb \cong k.$$

Identify $\overline{C_G(Q)}$ with $C$ through the canonical isomorphism $\overline{C_G(Q)} \cong C$. Then the interior $\overline{C_G(Q)}$-algebra structure of $k\overline{C_G(Q)}\bar{b}$ corresponds to $\iota\!:\!C \to k^*$, $\iota$ is faithful and $C$ is central in $G/Q$. Thus Proposition 5.29 applies. By definition the central extension

$$1 \to C \to G/Q \to L/Q$$

corresponds to the element $\beta \in H^2(C, k^*)$. Thus, by Proposition 5.29, $\alpha_Q = \iota_*(\beta)$. This proves the result as by definition $\iota_*(\beta) = \alpha$. $\qquad\square$

**Definition 5.36**  Let $R$ be a a commutative ring with 1 and let $\alpha \in H^2(G, R^*)$. Let $\tilde{\alpha}\!:\!G \times G \to R^*$ be a 2-cocycle in the class of $\alpha$. The *twisted group algebra* of $G$ with respect to $\alpha$ is the $R$-algebra $R_\alpha G$ which is equal to $RG$ as $R$-module and with multiplication $*$ satisfying $g * g' = \tilde{\alpha}(g, g')gg'$, for $g, g' \in G$.

It is easy to check that up to isomorphism, the $R$-algebra $R_\alpha G$ is independent of the choice of 2-cocycle $\tilde{\alpha}$. Also, if $\alpha$ is the trivial class then $R_\alpha G$ is isomorphic to the usual group algebra $RG$.

A twisted group algebra of $G$ over $k$ corresponds to a factor of the untwisted group algebra of a central extension of $G$ by a $p'$-group:

**Proposition 5.37**  *Let $\alpha \in H^2(G, k^*)$. Then there exists a central extension of finite groups*

$$1 \to Z \to H \to G \to 1,$$

*with $Z$ cyclic of order prime to $p$ such that $k_\alpha G$ is isomorphic to $kHd$ for some central idempotent $d$ of $kH$.*

*Proof.* This is very similar to the first part of the proof of Proposition 5.35. Let $Z$ be a cyclic group of order $|G|_{p'}$ and let $\iota\!:\!Z \to k^*$ be an injective group homomorphism. The induced map $\iota_*\!:\!H^2(G, Z) \to H^2(G, k^*)$ is surjective. Choose $\beta \in H^2(G, Z)$ with $\iota_*(\beta) = \alpha$ and let

$$1 \to Z \to G \to H \to 1$$

be the central extension corresponding to $\beta$. Set

$$d = \frac{1}{|Z|} \sum_{z \in Z} \iota(z)^{-1} z.$$

Then $d$ is a block of $kZ$ and hence a central idempotent of $kH$. Choose a section $s : G \to H$ and let $\varphi : k_\alpha G \to kHd$ be the $k$-linear map which sends $g$ to $s(g)d$ for each $g \in G$. Then, $\varphi$ is an isomorphism of $k$-algebras. $\qquad\square$

## 5.6. Nilpotent blocks and extensions.

Let $G$ be a finite group, $b$ a block of $kG$ and let $(P, e_P)$ be a maximal $(kG, b, G)$-Brauer pair. Set $\mathcal{F} := \mathcal{F}_{(P, e_P)}(G, b)$ and let $\hat{b}$ denote the block of of $\mathcal{O}G$ lifting $b$, that is $b$ is the image of $\hat{b}$ under the surjection $\mathcal{O}G \to kG$.

Recall Frobenius's theorem: For a finite group $G$ and $S$ a Sylow $p$-subgroup of $G$, $\mathcal{F}_S(G) = \mathcal{F}_S(S)$ if and only if $G$ has a normal $p$-complement, that is if and only if $G = O_{p'}(G) \rtimes S$.

**Definition 5.38**   The block $b$ is *nilpotent* if $\mathcal{F} = \mathcal{F}_P(P)$.

The above definition, due to Broué and Puig [BP1] is a natural generalization of Frobenius's condition as by Brauer's Third Main Theorem, if $b$ is the principal block of $kG$, then $b$ is nilpotent if and only if $\mathcal{F}_S(G) = \mathcal{F}_S(S)$ for any Sylow $p$-subgroup $S$ of $G$. It was shown in [BP1] that the ordinary character theory of a nilpotent block is analogous to that of a $p$-nilpotent group. In [P3], Puig gave a structure theorem for nilpotent blocks, which we state below. For a detailed exposition, we refer the reader to [Th, Chapter 7]. Recall that if $V$ is a $kP$-module then $\mathrm{End}_k(V)$ is naturally an interior $P$-algebra (see Example 1.12). We will refer to this interior $P$-algebra structure below. Recall also that any interior $P$-algebra inherits a natural $P$-algebra structure. Further, note that if $(A, \iota)$ and $(B, \kappa)$ are interior $P$-algebras over $k$, then $A \otimes_k B$ is naturally an interior $P$-algebra via the map which sends an element $x$ of $P$ to the element $\iota(x) \otimes \kappa(x)$ of $A \otimes_k B$.

**Definition 5.39**   A *Dade P-algebra* is an interior $P$-algebra $\mathcal{S}$ such that

- As $k$-algebra $\mathcal{S} \cong Mat_n(k)$; and

- $\mathcal{S}$ is a $p$-permutation $P$-algebra with $\mathrm{Br}_P^{\mathcal{S}}(\mathcal{S}^P) \neq 0$.

A *primitive Dade P-algebra* is a Dade $P$-algebra $\mathcal{S}$ such that $1_{\mathcal{S}}$ is a primitive idempotent of $\mathcal{S}^P$.

**Theorem 5.40**   *Suppose that $b$ is a nilpotent block of $kG$, and let $i$ be a source idempotent of $b$ with respect to $P$. There exists a primitive Dade $P$-algebra $\mathcal{S}$ such that*

$$ikGi \cong \mathcal{S} \otimes_k kP$$

*as interior $P$-algebras.*

*In particular, the $k$-algebra $kGb$ is Morita equivalent to $kP$ and up to isomorphism, $kGb$ has a unique simple module.*

*Proof.* This is the main result of [P3]. In fact, [P3] gives a structure theorem over $\mathcal{O}$. The lifting of the above result from $k$ to $\mathcal{O}$ is a deep result, even given the rigidity of source algebras over $\mathcal{O}$ described in Proposition 5.24. The main difficulty is in proving that the $kP$-module $V$ above has a lift to an $\mathcal{O}$-free $\mathcal{O}P$-module. The paper [KulOW] gives a simplification of Puig's proof over $k$. $\qquad\square$

In [KulP] Külshammer and Puig proved an extension theorem for nilpotent blocks. We describe a special case of their result which gives the structure of the normaliser block algebras of centric Brauer correspondents. Suppose that $Q \in \mathcal{F}^c$ is fully $\mathcal{F}$-normalized. By Proposition 3.17, $e_Q$ is a block of $kN_G(Q, e_Q)$, and by Theorem 3.19, $(N_P(Q), e_{N_P(Q)})$ is a maximal $(kN_G(Q, e_Q), e_Q, N_G(Q, e_Q))$-Brauer pair, and the corresponding fusion system is $N_{\mathcal{F}}(Q)$. Since $Q$ is $\mathcal{F}$-centric, $QC_{\mathcal{F}}(Q) = \mathcal{F}_Q(Q)$ and hence $N_{\mathcal{F}}(Q)$ is constrained. Consequently, $N_{\mathcal{F}}(Q)$ has a model (see Theorem I.4.9). In [KulP], Külshammer and Puig described the source algebra of $kN_G(Q, e_Q)e_Q$ in terms of a model for $N_{\mathcal{F}}(Q)$ and the Kulshammer-Puig cohomology class $\alpha_Q$ (the existence and uniqueness of the model for $N_{\mathcal{F}}(Q)$ in this context is also a part of their proof).

**Theorem 5.41** *With the notation above, let $Q \in \mathcal{F}^c$ be fully $\mathcal{F}$-normalized and let $\alpha_Q$ be as in Definition 5.31. There exists a strongly p-constrained group $L_Q$, unique up to isomorphism, with $N_P(Q)$ as Sylow p-subgroup, $Q = O_p(L_Q)$, $\mathcal{F}_{N_P(Q)}(L_Q) \cong N_{\mathcal{F}}(Q)$ and a primitive Dade P-algebra $\mathcal{S}$, such that denoting also by $\alpha_Q$ the element of $H^2(L_Q, k^*)$ obtained by restriction through the surjection of $L_Q$ onto $L_Q/Q \cong \mathrm{Out}_{\mathcal{F}}(Q)$, there is an isomorphism of interior P-algebras*

$$ikN_G(Q, e_Q)i \cong \mathcal{S} \otimes_k k_{\alpha_Q} L_Q,$$

*where $i$ is a source idempotent of $kN_G(Q, e_Q)e_Q$. In particular, the k-algebra $kN_G(Q, e_Q)e_Q)$ is Morita equivalent to $k_{\alpha_Q} L_Q)$*

*Proof.* See [KulP, Theorem 1.12, 1.20.3]. $\qquad\square$

Theorems 5.40 and 5.41 show that the module categories of certain blocks are determined to a large extent by the corresponding fusion systems. Another example of such control is the following consequence of Glauberman's $Z^*$-theorem (see I.1):

**Theorem 5.42** *Suppose that $p = 2$, that $b$ is the principal block of $kG$ and that $Z(\mathcal{F})$ contains a subgroup $Q$ of order 2. Then $kGb$ and $kC_G(Q)e_Q$ are isomorphic k-algebras.*

*Proof.* By Brauer's Third Main Theorem, $e_Q$ is the principal block of $kC_G(Q)$, $P$ is a Sylow $p$-subgroup of $G$ and $\mathcal{F} = \mathcal{F}_P(G)$. So, by Glauberman's $Z^*$-theorem, $G = C_G(Q)O_{p'}(G)$. The isomorphism of block algebras follows by combining Glauberman's theorem with the following fact about principal blocks (see [NT, Chapter 5, Theorem 8.1]):

- If $b$ is the principal block of $kG$, then $xa = a$ for all $x \in O_{p'}(G)$ and all $a \in kGb$.

$\square$

## 5.7. Counting Conjectures.

Let $G$ be a finite group, $b$ a block of $kG$ and let $(P, e_P)$ be a maximal $(kG, b, G)$-Brauer pair. Set $\mathcal{F}: = \mathcal{F}_{(P, e_P)}(G, b)$ and let $\hat{b}$ denote the block of $\mathcal{O}G$ lifting $b$. For a finite dimensional algebra $A$ over $k$, let $l(A)$ denote the number of isomorphism classes of simple $A$-modules. So, $l(kGb) = |\mathrm{Irr}_k(G, b)|$. Let $z(A)$ denote the number of isomorphism classes of $A$-modules which are both simple and projective. In this subsection, we consider the question: to what extent does $\mathcal{F}$ determine $|\mathrm{Irr}_k(G, b)|$ and $|\mathrm{Irr}_K(G, b)|$? For instance, if $\mathcal{F} = \mathcal{F}_P(P)$, then by Theorem 5.40, $|\mathrm{Irr}_k(G, b)| = 1$. If $kP$ has tame representation type, then also the number of isomorphism classes of simple $kGb$-modules is completely determined by $\mathcal{F}$:

- If $P$ is cyclic, a Klein 4-group or a quaternion group of order 8, then $|\mathrm{Irr}_k(G, b)| = |\mathrm{Out}_{\mathcal{F}}(P)|$.

- If $P$ is a dihedral, semidihedral or quaternion group of order $2^n \geq 16$ as in Examples I.2.7 and I.3.8, then
    - $|\mathrm{Irr}_k(G, b)| = 1$ if $\mathcal{F} = \mathcal{F}_{00}$,
    - $|\mathrm{Irr}_k(G, b)| = 2$ if $\mathcal{F} = \mathcal{F}_{10}$ or $\mathcal{F} = \mathcal{F}_{01}$; and
    - $|\mathrm{Irr}_k(G, b)| = 3$ if $\mathcal{F} = \mathcal{F}_{11}$.

The above results are due to Brauer ([Br1]) and Dade ([Da1]) when $P$ is cyclic, to Brauer ([Br2], [Br3] ) when $P$ is a Klein 4 group or dihedral and to Olsson ([Ols]) when $P$ is semidihedral or quaternion.

**Example 5.43**   Let $p = 3$ and let $P$ be an elementary abelian group of order 9 with generators $x$, $y$. Let $E = \langle \sigma, \tau : \sigma^4 = \tau^4 = 1, \sigma\tau\sigma^{-1} = \tau^{-1} \rangle$ be a quaternion group of order 8. The group $E$ acts on $P$ via

$$^{\sigma}x = x^2 \quad {}^{\tau}x = x, \, {}^{\sigma}y = y \quad {}^{\tau}y = y^2.$$

Set

$$G: = P \rtimes E, \quad b_0 = \frac{1}{2}(1 + \sigma^2), b_1 = \frac{1}{2}(1 - \sigma^2).$$

By the Fourier inversion formula, $b_0$ is the principal block of $k\langle\sigma^2\rangle$ and $b_1$ is the block of $k\langle\sigma^2\rangle$ containing the non-trivial one dimensional $k\langle\sigma^2\rangle$-module. Now, $\langle\sigma^2\rangle = C_G(P)$ and is a central subgroup of $G$; thus by Lemma 3.17, $b_0$ and $b_1$ are blocks of $kG$.

Since $\mathrm{br}_P(b_0) = b_0$, $\mathrm{br}_P(b_1) = b_1$ and $b_0$ and $b_1$ are also blocks of $kC_G(P)$, it follows that $(P, b_0)$ is a maximal $b_0$-Brauer pair, $(P, b_1)$ is a maximal $b_1$-Brauer pair. So, $N_G(P, b_0) = N_G(P, b_1) = N_G(P)$. Since $P$ is abelian it follows that $\mathcal{F}_{(P,b_0)}(G, b_0) = \mathcal{F}_{(P,b_1)}(G, b_1) = \mathcal{F}_P(G)$.

We count the number of simple modules in $b_0$ and $b_1$. For this, note that as $b_0 + b_1 = 1_{kG}$, $b_0$ and $b_1$ are the only blocks of $kG$ and hence any simple $kGb$-module belongs to exactly one of $b_0$ or $b_1$. Also, note that if $N$ is a normal subgroup of a finite group $H$, then a $k[H/N]$-module $V$ is a $kH$-module through restriction along the natural map $H \to H/N$ and $V$ is simple as $k[H/N]$-module if and only if $V$ is simple as $kH$-module. Further, a $kH$-module $W$ is the inflation of a $k[H/N]$-module as above if and only if $xw = w$ for all $x \in N$ and all $w \in W$.

Since $P$ is normal in $G$, by the first part of the proof of Lemma 3.16, $P$ acts trivially on any simple $kG$-module. It follows that simple $kG$-modules are just inflations of simple $k[G/P]$-modules. Now $G/P$ is isomorphic to $E$, and $|E|$ is relatively prime to $3 = char(k)$. Thus, by Theorem 4.5, $|\mathrm{Irr}_k(G)| = |\mathrm{Irr}_k(G/P)|$ equals the $k$-dimension of $E$, which in turn equals the number of conjugacy classes of $E$, that is $|\mathrm{Irr}_k(G/P)| = 5$. Now, if $V$ is a simple $kE$-module, it is easy to see that $V$ regarded as $kG$-module belongs to the block $b_0$ if and only if $\sigma^2$ acts trivially on $V$, that is if and only if $V$ is the inflation of a simple $kE/\langle\sigma^2\rangle$-module. Since $E/\langle\sigma^2\rangle = C_2^2$ and $\mathrm{Irr}_k(C_2^2) = 4$, it follows that $|\mathrm{Irr}_k(G, b_0)| = 4$ and $|\mathrm{Irr}_k(G, b_1)| = 1$.

The above example shows that $|\mathrm{Irr}_k(G, b)|$ is not in general completely determined by $\mathcal{F}$. However, in [Al3], Alperin gave a conjectural formula for $|\mathrm{Irr}_k(G, b)|$ which implies that $|\mathrm{Irr}_k(G, b)|$ is described completely by $\mathcal{F}$ and the Külshammer-Puig classes associated to $\mathcal{F}$-centric-radical subgroups of $P$.

For a $p$-subgroup $Q$ of $G$, and an element $a \in kN_G(Q)$, denote by $\bar{a}$ the image of an element $a$ of $kN_G(Q)$ under the canonical surjection

$$kN_G(Q) \to kN_G(Q)/Q.$$

Recall that either $\mathrm{br}_Q(b) = 0$ or $\mathrm{br}_Q(b)$ is a central idempotent of $N_G(Q)$. Consequently, either $\overline{\mathrm{br}_Q(b)} = 0$ or $\overline{\mathrm{br}_Q(b)}$ is a central idempotent of $kN_G(Q)/Q$.

**Definition 5.44**  A weight of $kGb$ is a pair of the form $(Q, w)$, where $Q$ is a $p$-subgroup of $G$, and $w$ is a block of $kN_G(Q)/Q$ such that $w$ has trivial defect group and such that $w\overline{\mathrm{Br}_Q(b)} = w$.

Note that $G$ acts on the set of $kGb$-weights by conjugation.

**Conjecture 5.45**  *(Alperin's Weight Conjecture)* [Al3] *The number of isomorphism classes of simple $kGb$-modules equals the number of $G$-orbits of $kGb$-weights.*

The number of $G$-orbits of $kGb$-weights can be expressed in terms of the fusion system $\mathcal{F}$ and the Külshammer-Puig classes. Recall from Definition I.3.1 that $\mathcal{F}^{cr}$ denotes the full subcategory of $\mathcal{F}$ whose objects are the $\mathcal{F}$-centric-radical subgroups of $P$.

**Proposition 5.46**  *For $Q \in \mathcal{F}^{cr}$, let $\alpha_Q$ denote the element of $H^2(\mathrm{Out}_{\mathcal{F}}(Q), k^*)$ as in Definition 5.31 and let $\mathcal{F}/\mathcal{F}^{cr}$ denote a set of representatives of the $\mathcal{F}$- isomorphism classes of objects of $\mathcal{F}^{cr}$.*

*The number of $G$-orbits of $kGb$-weights is equal to*

$$\sum_{Q \in \mathcal{F}^{cr}/\mathcal{F}} z(k_{\alpha_Q} \mathrm{Out}_{\mathcal{F}}(Q)).$$

*Thus, conjecture 5.45 is equivalent to the statement*

$$|\mathrm{Irr}_k(G, b)| = \sum_{Q \in \mathcal{F}^{cr}/\mathcal{F}} z(k_{\alpha_Q} \mathrm{Out}_{\mathcal{F}}(Q)).$$

*Proof.* See [K1, Proposition 5.4]. $\square$

If $\mathcal{F} = \mathcal{F}_P(P)$, then $P$ is the only centric-radical subgroup of $P$, and $\mathrm{Out}_{\mathcal{F}}(P) = 1$. Hence in this case, by Proposition 5.46, Conjecture 5.45 is equivalent to the equality $|\mathrm{Irr}_k(G, b)| = 1$, which is known to hold by Theorem 5.40.

By Proposition 5.33, if $P$ is cyclic or $p = 2$ and $P$ is a Klein 4-group, or a dihedral, semi-dihedral or quaternion group, then $\alpha_Q$ is trivial for all $Q$ in $\mathcal{F}^c$. Then, using Proposition 5.46, it is an easy exercise to check that the results of Brauer, Dade and Olsson described at the beginning of this subsection are consistent with Alperin's weight conjecture. Hence, the conjecture is true if $kP$ is of finite or tame representation type. It has also been proved in the case that $P$ is an abelian 2-group of rank 2 ([PUs], [Sa1]). Using the classification of finite simple groups, Conjecture 5.45 has been proved when $P$ is an elementary abelian 2-group of order 8 [KKoL].

We consider the statement of Conjecture 5.45 in some special cases. First suppose that $b$ is the principal block of $kG$. By Proposition 5.32, $\alpha_Q$ is trivial for all $Q \in \mathcal{F}^c$, hence by Proposition 5.46, Alperin's conjecture in this case is the equality

$$|\mathrm{Irr}_k(G, b)| = \sum_{Q \in \mathcal{F}^{cr}/\mathcal{F}} z(k \mathrm{Out}_{\mathcal{F}}(Q)).$$

In particular, two principal blocks with isomorphic fusion systems have conjecturally the same number of simple modules.

Now suppose $P$ is abelian. Then $P$ is the only $\mathcal{F}$-centric subgroup of $P$ and the sum in the right hand side of the equality in Proposition 5.46 reduces to a single term. By Proposition 5.37, the twisted group algebra $k_{\alpha_P}\mathrm{Out}_F(P)$ is isomorphic to $kHd$, where $H$ is a central extension of $\mathrm{Out}_{\mathcal{F}}(P)$ by a cyclic $p'$-group, and $d$ is a central idempotent of $kH$. Since $\mathrm{Out}_{\mathcal{F}}(P)$ is a $p'$-group, $H$ is a $p'$-group, and it follows from Theorem 4.5, that every simple $k_{\alpha_P}\mathrm{Out}_F(P)$-module is projective. Thus, if $P$ is abelian, Conjecture 5.45 is the equality

$$|\mathrm{Irr}_k(G,b)| = l(k_{\alpha_P}\mathrm{Out}_{\mathcal{F}}(P)).$$

Let $L_P$ be a model for $N_{\mathcal{F}}(P)$ and denote by $\alpha_P$ the element of $H^2(L_P, k^*)$ obtained from the restriction of $\alpha_P$ through the canonical surjection of $L_P$ onto $\mathrm{Out}_{\mathcal{F}}(P)$. Since $L_P$ is an extension of $\mathrm{Out}_{\mathcal{F}}(P)$ by a $p$-group, using Proposition 5.37, and the fact that a normal $p$-subgroup of a finite group $H$ acts trivially on each simple $kH$-module (see the first part of the proof of Lemma 3.16), it follows that $l(k_{\alpha_P}L_P) = l(k_\alpha\mathrm{Out}_{\mathcal{F}}(P))$. Hence, if $P$ is abelian, then Conjecture 5.45 is the equality

$$|\mathrm{Irr}_k(G,b)| = l(k_{\alpha_P}L_P).$$

The number of isomorphism classes of simple modules of a finite dimensional $k$-algebra is an invariant of the bounded derived category $D^b(\mathrm{mod}(A))$ of the module category of $A$ (see [KoZ] for an account of derived categories in the context of group representation theory). In the case that $P$ is abelian, Conjecture 5.45 is therefore a consequence of the following.

**Conjecture 5.47** (Broué's Abelian Defect Group Conjecture) [B2] *Suppose that $P$ is abelian. Let $\alpha_P \in H^2(\mathrm{Out}_{\mathcal{F}}(P), k^*)$ be as in Definition 5.31 and let $L_P$ be a model for $N_{\mathcal{F}}(P)$. Denote also by $\alpha_P$ the element of $H^2(L_P, k^*)$ obtained by restriction of the surjection $L_P$ onto $L_P/P \simeq \mathrm{Out}_{\mathcal{F}}(P)$. Then, there is an equivalence of triangulated categories $D^b(\mathrm{mod}(kGb)) \cong D^b(\mathrm{mod}(k_{\alpha_P}L_P))$.*

By Theorem 5.41, the block algebra $kN_G(P, e_P)e_P$ is Morita equivalent to $k_{\alpha_P}L_P$. On the other hand, by Proposition 5.26 and by Brauer's First Main Theorem it follows that $kN_G(P, e_P)e_P$ is source algebra equivalent to $kN_G(P)\mathrm{br}_P(b)$. Since source alegbra equivalence implies Morita equivalence and Morita equivalence implies derived equivalence, the above conjecture has the following reformulation, which is the version in which it was originally stated.

**Conjecture 5.48** *Suppose that $P$ is abelian. Then, there is an equivalence of triangulated categories $D^b(\mathrm{mod}(kGb)) \cong D^b(\mathrm{mod}(kN_G(P)\mathrm{br}_P(b)))$.*

Conjecture 5.47 is known to be true when $P$ is cyclic ([Ri1, Theorem 4.2]) or $P \cong C_2 \times C_2$([Ri2, Section 3]).

In [KR], Knörr and Robinson showed that Conjecture 5.45 can be reformulated in terms of alternating sums indexed by certain chains of $p$-subgroups of $G$. Using that $\mathrm{Irr}_K(G, b) - \mathrm{Irr}_k(G, b)$ is locally determined (see Theorem 5.7), they also gave a version of the conjecture which involves ordinary irreducible characters. Various refinements of the weight conjecture have appeared since (see for instance [Da2], [Da3], [Da4] , [Ro1], [Ro2] [Bo], [Li4], [Li2], [Un]), [Wb2]). We describe one such version, the ordinary weight conjecture, due to Robinson.

It is a standard fact of ordinary character theory that if $V$ is a simple $KG$-module, then $\dim_K(V)$ is a divisor of $|G|$ (see [NT, Chapter 3, Theorem 2.4]); the $p$-*defect* of $V$ is defined to be the non-negative integer $d$ such that $p^d$ is the highest power of $p$ dividing $\frac{|G|}{\dim_K(V)}$. For a non-negative integer $d$, we denote by $\mathrm{Irr}_K^d(G)$ the subset of $\mathrm{Irr}_K(G)$ consisting of the images $[V]$ in $R_K(G)$ of those simple $KG$-modules $V$ whose $p$-defect equals $d$ and we denote by $\mathrm{Irr}_K^d(G, b)$ the set $\mathrm{Irr}_K^d(G) \cap \mathrm{Irr}_K(G, b)$.

Let $Q \leq P$ be $\mathcal{F}$-centric and $\mathcal{F}$-radical. Let $\mathcal{N}_Q$ denote the set of strictly increasing chains

$$\sigma : 1 < X_1 < \cdots < X_l$$

of $p$-subgroups of $\mathrm{Out}_{\mathcal{F}}(Q)$ such that for each $i$, $1 \leq i \leq l$, $X_i$ is normal in $X_l$. For each $\sigma$ as above, let $|\sigma| = l + 1$ (so if $\sigma$ is the chain consisting solely of the trivial subgroup, then $|\sigma| = 1$).

Clearly, $\mathrm{Out}_{\mathcal{F}}(Q)$ acts on $\mathcal{N}_Q$ by conjugation. For $\sigma \in \mathcal{N}_Q$, denote by $I(\sigma)$ the stabilizer in $\mathrm{Out}_{\mathcal{F}}(Q)$ of $\sigma$. There is a natural action of $\mathrm{Aut}(Q)$ on $\mathrm{Irr}_K(Q)$, where for $\varphi \in \mathrm{Aut}(Q)$ and a simple $KQ$-module $V$, $^{\varphi}V$ is the $kQ$-module which equals $V$ as $K$-vector space and where $x.v := \varphi(x)v$ for $x \in Q$, and $v \in V$. If $\varphi \in \mathrm{Inn}(Q)$, then for any simple $KQ$-module $V$, $^{\varphi}V \cong V$. Further, $\dim_K(^{\varphi}V) = \dim_K(V)$ for any simple $KQ$-module $V$ and any $\varphi \in \mathrm{Aut}(Q)$. Thus the action of $\mathrm{Aut}(Q)$ on $\mathrm{Irr}_K(Q)$ induces an action of $\mathrm{Out}_{\mathcal{F}}(Q)$ on $\mathrm{Irr}_K(Q)$ and on $\mathrm{Irr}_K^d(Q)$ for any non-negative integer $d$. For $\sigma \in \mathcal{N}_Q$, and $\mu \in \mathrm{Irr}_K(Q)$, let $I(\sigma, \mu)$ denote the intersection of $I(\sigma)$ with the stabilizer in $\mathrm{Out}_{\mathcal{F}}(Q)$ of $\mu$; denote by $\alpha_Q$ also the restriction of the Külshammer-Puig class $\alpha_Q$ to $I(\sigma, \mu)$ (see definition 5.31) and define the integer $\omega(Q, \sigma, \mu)$ by

$$\omega(Q, \sigma, \mu) := z(k_{\alpha_Q} I(\sigma, \mu)).$$

For a non-negative integer $d$, set

$$\mathbf{w}(Q, d) := \sum_{\sigma \in \mathcal{N}_Q / \mathrm{Out}_{\mathcal{F}}(Q)} (-1)^{|\sigma|+1} \sum_{\mu \in Irr_K^d(Q)/I(\sigma)} \omega(Q, \sigma, \mu).$$

**Conjecture 5.49** (Ordinary Weight Conjecture)  [Ro2] *Let $d$ be a non-negative integer. Then,*

$$|\mathrm{Irr}_K^d(G,b)| = \sum_{Q \in \mathcal{F}^{cr}/\mathcal{F}} \mathbf{w}(Q,d).$$

Summing over all non-negative integers in the above gives a formula for the number, $|\mathrm{Irr}_K(G,b)|$, of simple $KGb$-modules.

The Ordinary Weight Conjecture is known to hold for all blocks of tame representation type (see Example after [Ro2]), for all $P$ which admit only $\mathcal{F}_P(P)$ as saturated fusion system and for a few other 2-groups ([KKoL, Sa1, Sa2, Sa3]).

We state two other famous counting conjectures-these are related to the conjectures stated above and have interesting refinements (see [IsNa], [Tu]). We say that an element $\chi \in \mathrm{Irr}_K(G,b)$ is a *height zero* character if the $p$-defect of $\chi$ is maximal amongst all elements of $\mathrm{Irr}_K(G,b)$.

**Conjecture 5.50** (Brauer's height zero conjecture)  *All elements of $\mathrm{Irr}_K(G,b)$ are of height zero if and only if $P$ is abelian.*

**Conjecture 5.51** (Alperin-McKay Conjecture)  *The number of height zero elements in $|\mathrm{Irr}_K(G,b)|$ and in $|\mathrm{Irr}_K(N_G(P),\mathrm{br}_P(b))|$ is the same.*

**Remark.** The conjectures of this section as well as the finiteness conjectures of the previous sections have been established for only a few isomorphism types of defect groups or fusion systems. However, work on four families of finite groups: $p$-solvable groups, Weyl groups and their covers, finite groups of Lie type, and sporadic simple groups has yielded a huge amount of evidence for these conjectures. Understanding blocks in the different families requires the use of very different and often deep techniques and we cannot give an exposition here which would do the justice to the state of the art. There has also been a lot of progress, especially recently, on reducing the counting conjectures to statements about quasi-simple groups. Again it is not possible to give a proper account of these results here.

## 6. BLOCK FUSION SYSTEMS AND NORMAL SUBGROUPS.

Let $N$ be a normal subgroup of $G$ and consider $kN$ as a $G$-algebra via the conjugation action of $G$. As noted in Section 3.4, the $G$-action on $kN$ induces an action of $G$ on the set of blocks of $kN$. For a block $c$ of $kN$, we let $I(c)$ be the stabilizer in $G$ of $c$. By Proposition 3.10, if $c$ is a block of $kN$, then $\mathrm{Tr}_{I(c)}^G(c)$ is an idempotent contained in $(kN)^G$ and the set $\{\mathrm{Tr}_{I(c)}^G(c)\}$ as $c$ runs over a set of representatives of the $G$-orbits of the blocks of $kN$ is the primitive idempotent decomposition of $1$ in $(kN)^G$.

Since $(kN)^G = kN \cap Z(kG) \subseteq Z(kG)$, the set $\{\mathrm{Tr}^G_{I(c)}(c)\}$ is an idempotent decomposition of 1 in $Z(kG)$. By the uniqueness of block decompositions, it follows that for each block $b$ of $kG$, there exists a unique $G$-orbit, say $X$, of blocks of $kN$ such that $b\mathrm{Tr}^G_{I(c)}(c) \neq 0$ for some (any) $c \in X$ and for this $c$, $b\mathrm{Tr}^G_{I(c)}(c) = b = \mathrm{Tr}^G_{I(c)}(c)b$.

**Definition 6.1**    Let $b$ be a block of $kG$ and let $c$ be a block of $kN$. We say $b$ covers $c$ if $b\mathrm{Tr}^G_{I(c)}(c) \neq 0$.

It is immediate from the definition that for any block $b$ of $kG$, the set of blocks of $kN$ covered by $b$ is a $G$-orbit of blocks of $kN$. The following is an easy characterization of covering blocks.

**Proposition 6.2**    *Let $b$ be a block of $kG$ and let $c$ be a block of $kN$. Then $b$ covers $c$ if and only if $bc \neq 0$.*

*Proof.* Let $X$ be the $G$-orbit of $c$. Then, $\mathrm{Tr}^G_{I(c)}(c) = \sum_{d\in X} d$ and the elements of $X$ are pairwise orthogonal idempotents of $kN$ and hence of $kG$. Since $b$ is a central idempotent of $kG$, it follows that $b\mathrm{Tr}^G_{I(c)}(c) = 0$ if and only if $bd = 0$ for all $d \in X$. Let $d \in X$. By definition, $d = {}^x c$ for some $x \in G$. Since $b$ is central in $kG$, we have

$$bd = b\,{}^x c = {}^x(bc).$$

Thus, $bc = 0$ if and only if $bd = 0$ for all $d \in X$. The result follows.    $\square$

If $b$ is a block of $kG$ and $V$ is a $kG$-module lying in the block $b$, then the restriction of $V$ to $kN$ is a direct sum of $kN$-modules, $W_d$, where $d$ runs over the set of blocks of $kN$ covered by $b$. The relationship between the representation theory of a block and of the blocks it covers, often referred to as Clifford theory, is an important part of block theory. So, it makes sense to investigate what happens to fusion systems on passage to covered blocks. The following proposition shows that for this (as for Clifford theory), one may always reduce to the case where there is a unique covered block.

**Proposition 6.3**    *Let $N$ be a normal subgroup of $G$, and $c$ a block of $kN$. Set $I = I(c)$, the stabilizer of $c$ in $G$. The map $e \to \mathrm{Tr}^G_H(e)$ is a bijection between the set of blocks of $kI$ covering $c$ and the set of blocks of $kG$ covering $c$. Further, if $e$ is a block of $kI$ covering $c$, then the block algebras $kIe$ and $kG\mathrm{Tr}^G_H(e)$ are source algebra equivalent. In particular, if $b$ is a block of $kG$ covering $c$, then the fusion system of $kGb$ is isomorphic to the fusion system of some block of $kI$.*

*Proof.* By definition of $I(c)$, $c \in (kN)^I \subseteq Z(kI)$. Thus, $c$ is a central idempotent of $kI$. If $g \in G - I$, then again by definition of $I$, ${}^g c$ is a block of $kN$ different from $c$, hence $c\,{}^g c = 0$. Also note that since $N$ is normal in $I$, the condition $e \leq c$ for a block of $e$ of $kI$ is equivalent to the condition that $e$ covers $c$, and similarly for blocks of $kG$. The first part of the statement is now immediate from Proposition 5.26. The second part follows from the fact that source algebra equivalent blocks have isomorphic fusion systems (see Proposition 5.23). $\qquad\square$

Now let $b$ be a block of $kG$ covering $c$ and assume that $I(c) = G$, so that $bc = b$ and $c \in (kN)^G$. Then, by Proposition 3.14, $(kN, c, G)$, $(kN, c, N)$ and $(kG, b, G)$ are triples of saturated type and thus by Theorem 3.2, the corresponding fusion systems are saturated, the latter two triples yielding the fusion systems of $kNc$ and $kGb$ respectively. Proposition 3.9 describes the relationship between the fusion systems associated to the triples $(kN, c, G)$ and $(kN, c, N)$. From the point of view of block theory however, one wants to understand the relationship between the fusion systems associated to the triples $(kN, c, G)$ and $(kG, b, G)$.

**Theorem 6.4**  *Let $N$ be a normal subgroup of $G$ and $c$ a block of $kN$ which is $G$-stable. Let $b$ be a block of $kG$ covering $c$ and let $(P, e_P)$ be a maximal $(kG, b, G)$-Brauer pair.*

(a)  *There exists a maximal $(kN, c, G)$-Brauer pair $(S, e'_S)$ such that $P \leq S$ and*

$$\mathcal{F}_{(P,e_P)}(kG, b, G) \subseteq \mathcal{F}_{(S,e'_S)}(kN, c, G).$$

(b)  *For each $Q \leq S$, let $(Q, e'_Q)$ be the unique $(kN, c, G)$-Brauer pair contained in $(S, e'_S)$. Then, $P \cap N = S \cap N$ and $(S \cap N, e'_{S \cap N})$ is a maximal $(kN, c, N)$-Brauer pair. Further, $\mathcal{F}_{(S \cap N, e'_{S \cap N})}(kN, c, N)$ is a normal subsystem of $\mathcal{F}_{(S, e'_S)}(kN, c, G)$.*

*Proof.* Part (a) is proved in [KS, Theorem 3.5]. Part (b) is just Proposition 3.9, applied with $A = kG$, $S$ replacing $P$ and $c$ replacing $b$. A version of Part (b) with "weakly normal" replacing normal is proved in [KS, Theorem 3.5]. $\qquad\square$

The above theorem gives an indirect relationship between a fusion system of a block and a covered block through a third saturated fusion system. By analogy with finite group fusion systems, one might expect a more direct relationship, namely that a fusion system of a covered block is a normal subsystem of the fusion system of the covering block. The next example shows that such a straightforward relationship does not hold and it can even be the case that a fusion system of the covering block is a subsystem of the covered block.

**Example 6.5**  The details of the arguments for this example are similar to those used in Example 5.43 and have been omitted. Let $p = 3$ and let $P$ be the cyclic group of order 3 on generator $r$. Let $T = \langle x, y \rangle$ be a quaternion group of order 8. Let $T$ act on $P$ via

$$^x r = r^2 \quad ^y r = r.$$

Let $G = P \rtimes T$ and let $N$ be the normal subgroup $P\langle x \rangle$ of $G$. Consider

$$b = \frac{1}{2}(1 - x^2) = \frac{1}{2}(1 - y^2).$$

Then $b$ is a block of $kG$ as well as of $kN$. Now $C_G(P) = P \times \langle y \rangle$. Thus, $\mathrm{br}_P(b)$ is a sum of two blocks, $\frac{1}{4}(1 + iy - y^2 - iy^3)$ and $\frac{1}{4}(1 - iy - y^2 + iy^3)$ of $kC_G(P)$, where $i$ is a primitive fourth root of unity. Set $e_P := \frac{1}{4}(1 + iy - y^2 - iy^3)$. Then $(P, e_P)$ is a maximal $(kG, b, G)$-Brauer pair.

On the other hand, $C_N(P) = P \times \langle x^2 \rangle$ whence $\mathrm{br}_P(b)$ is a block of $kC_N(P)$ and $(P, \mathrm{br}_P(b))$ is a maximal $(kN, b, N)$-Brauer pair. Let $\mathcal{F} = \mathcal{F}_{(P,e_P)}(kG, b, G)$ and $\mathcal{F}' = \mathcal{F}_{(P,\mathrm{br}_P(b))}(kG, b, G)$. We have $\mathrm{Aut}_{\mathcal{F}}(P) = 1$ and $\mathrm{Aut}_{\mathcal{F}'}(P)$ is cyclic of order 2.

If $N$ is a normal $p'$-subgroup of $G$, and $S$ is a Sylow $p$-subgroup of $G$, then $S$ is a Sylow $p$-subgroup of $G/N$ under the natural identification of an element $x$ of $S$ with its coset $xN$ in $G/N$ and $\mathcal{F}_S(G) = \mathcal{F}_S(G/N)$. In block theory, the analogue of a normal $p'$-subgroup is a covered block having trivial defect group. The equality of fusion systems in the group theoretic situation has an analogue in the block theory setting. Note that if $N$ is a normal subgroup of $G$ and $c$ is a block of $kN$ which is $G$-stable and such that $kNc$ has trivial defect groups, then $kNc$ is a matrix algebra, which is both an interior $kN$-algebra and a $kG$-algebra.

**Theorem 6.6** *Let $N$ be a normal subgroup of $G$ and $c$ a $G$-stable block of $kN$. Suppose that $\{1\}$ is a defect group of $c$. Let $\alpha = \alpha_{kNc,N,G}$ be as in Definition 5.28 and let $s : G \to (kNc)^*$ induce $\alpha$ as in Definition 5.28.*

(a) *Let*

$$\varphi : kNc \otimes_k k_\alpha G/N \to kGc$$

*be the $k$-linear function satisfying $x \otimes gN \to xs(g)^{-1}g$ for $x \in kNc$ and $g \in G$. Then $\varphi$ is a $k$-algebra isomorphism, with $\varphi^{-1}(gc) = s(g) \otimes gN$ for $g \in G$.*

(b) *Let $b$ be a block of $G$ covering $c$ and let $(P, e_P)$ be a maximal $(kG, b, G)$-Brauer pair. Then $N \cap P = 1$ and there exists a central extension*

$$1 \to Z \to H \to G/N \to 1,$$

*where $Z$ is a cyclic $p'$-group and a block $d$ of $kH$ such that identifying $P$ with the Sylow $p$-subgroup of the inverse image in $H$ of $PN/N$, there is a maximal $(kH, d, H)$-Brauer pair of the form $(P, f_P)$ with $\mathcal{F}_{(P,e_P)}(kG, b, G) = \mathcal{F}_{(P,f_P)}(kH, d, H)$.*

*Proof.* The first assertion is a straight forward verification. The second is deduced from the first through Proposition 5.37 and relies on some properties of endo-permutation modules. The result is implicit in the [P3]; the details may be found in [Ke1, Section 3]. $\qquad\square$

By Proposition I.6.2, if $\mathcal{F}$ is a simple saturated fusion system and $\mathcal{F} = \mathcal{F}_P(G)$, for a finite group $G$ and a Sylow $p$-subgroup $P$ of $G$, then $\mathcal{F} = \mathcal{F}_S(H)$ for some simple group $H$. In other words, any realizable simple saturated fusion system (see remarks after Theorem I.2.3) is realizable by a finite simple group, i.e., is the fusion system of a finite simple group.

**Definition 6.7**   A saturated fusion system $\mathcal{F}$ over a $p$-group $P$ is *block-realizable* if there exists a finite group $G$ and a block $b$ of $kG$ such that there is a maximal $(kG, b, G)$-Brauer pair of the form $(P, e_P)$ with $\mathcal{F} \cong \mathcal{F}_{(P,e_P)}(kG, b, G)$; otherwise, $\mathcal{F}$ is said to be *block-exotic*.

It is not known whether any block-realizable simple saturated fusion system is realizable by a block of a finite simple or quasi-simple group. Recall that a finite group $G$ is quasi-simple if $G/Z(G)$ is simple and $G = [G, G]$. Combining Proposition 6.3, Theorem 6.4 and Theorem 6.6 yields the following partial result.

**Theorem 6.8**   *Let $\mathcal{F}$ be a simple saturated fusion system on a finite $p$-group $P$. Suppose further that $P$ contains no proper, non-trivial strongly $\mathcal{F}$-closed subgroup and that there does not exist a saturated fusion system $\mathcal{G}$ on $P$ containing $\mathcal{F}$ as a proper subsystem. If $\mathcal{F}$ is block-realizable, then there exists a quasi-simple group $G$ with $Z(G)$ a $p'$-group and a block $b$ of $kG$ having a maximal $(kG, b, G)$-Brauer pair of the form $(P, e_P)$ such that $\mathcal{F} \cong \mathcal{F}_{(P,e_P)}(kG, b, G)$.*

*Proof.* This is a special case of [KS, Theorem 4.2]. $\qquad\square$

By Brauer's Third Main Theorem, a realizable saturated fusion system is block realizable. The converse is open. The above theorem (or some variant of it) reduces the question of the block-exoticity of certain saturated fusion systems to the question of whether these occur as fusion systems of blocks of finite quasi-simple groups. Even though the fusion systems of blocks of quasi-simple groups are not completely understood, it is possible to show that some of the known exotic saturated fusion systems are block-exotic.

**Theorem 6.9**   *Let $\mathcal{F}$ be the exotic saturated fusion system $\mathcal{F}_{\mathrm{Sol}}(q)$ at $p = 2$ (see Theorem III.6.7) or one of the three Ruiz-Viruel fusion systems at $p = 7$ (see Section III.6.2). Then $\mathcal{F}$ is block-exotic.*

*Proof.* The case $\mathcal{F} = \mathcal{F}_{Sol}(3)$ is proved in [Ke1], the case $\mathcal{F}_{Sol}(q)$ for general $q$ is in [Cr2] and the case that $\mathcal{F}$ is a Ruiz-Viruel system is in [KS]. To check that these systems do not occur as block fusion systems in quasi-simple groups, one relies heavily on the modular versions of Deligne-Lusztig theory of representations of finite groups of Lie type developed by many authors, notably Bonnafe, Broué, Cabanes, Dignes, Enguehard, Fong, Geck, Hiss, Malle, Michel, Rouquier and Srinivasa. $\square$

## 7. OPEN PROBLEMS

The block theory conjectures mentioned in the previous sections are the big open problems in modular representation theory. We list additionally a few questions related to block fusion systems.

**1. Are all exotic fusion systems also block-exotic?** At the moment, other than in the situation of principal blocks, there seems no way to use directly the information that a given saturated fusion system is a block fusion system in order to conclude that it must be the fusion system of a finite group. However, Theorem 6.9 seems to suggest that it may be possible to use the classification of finite simple groups to answer the above question, at least partially. The next two questions in our list form part of this approach.

**2. Improve the reduction result of Theorem 6.8.** Is any block-realizable simple saturated fusion system realizable by a block of a quasi-simple group?

**3. Describe the fusion systems of blocks quasi-simple groups.** In particular, does there exist a block of a quasi-simple group whose fusion system is exotic?

**4. Does there exist a solution to the gluing problem for blocks?** Recall from Definition 5.31 that if $\mathcal{F}$ is the saturated fusion system of a block $b$ of $kG$, then to each $\mathcal{F}$-centric subgroup $Q$ of $P$ is associated an element $\alpha_Q$ of $H^2(\mathrm{Aut}_{\mathcal{F}}(Q), k^*)$. The gluing problem for blocks, first stated in [Li2, Conjecture 4.2]) is:

Does there exist a second cohomology class $\alpha \in H^2(\mathcal{F}^c, k^*)$ whose restriction to $\mathrm{Aut}_{\mathcal{F}}(Q)$ equals $\alpha_Q$ for each centric subgroup $Q$ of $P$, and if so is there a canonical choice for $\alpha$? For instance if $b$ is the principal block of $kG$, then by Proposition 5.32, $\alpha$ may be taken to be the zero class. As another example, if $\mathcal{F}$ is constrained with $Q = O_p(\mathcal{F})$, then restriction induces an isomorphism $H^2(\mathcal{F}^c, k^*) \cong H^2(\mathrm{Aut}_{\mathcal{F}}(Q), k^*)$, and hence there is a unique choice for $\alpha$ (see [Li6, Proposition 6.4 and 6.5]). In [Pa1]

S.Park has shown that $\alpha$ is not unique in general. If the gluing problem
has a solution, then one obtains interesting structural reformulations of
Alperin's weight conjecture ([Li2, 4.5, 4.7], [Li4, Theorem 4.3]) involving
centric linking systems.


**5. Find interesting classes of $p$-permutation $G$-algebras to which
Theorem 3.2 applies.** Block algebras are a particular class of $p$-permu-
tation algebras, and it is possible that there are other interesting classes
which might provide a source of saturated fusion systems. In [KKuM], saturated triples arising from $p$-permutation modules are considered. Finding
other classes of triples of saturated type might be interesting for exoticity
questions. For instance the following question is open: Given a saturated
fusion system on a finite $p$-group $P$, does there exist a finite group $G$, a
$p$-permutation algebra $A$, and a primitive idempotent $b$ of $A^G$ such that
$(A, b, G)$ is a saturated triple and such that the corresponding fusion system
is isomorphic to $\mathcal{F}$?

## Appendix A. Background facts about groups

We collect here, for reference, some of the standard results in group theory which have been used throughout the book. We begin with some of the standard elementary properties of $p$-groups and Sylow subgroups.

**Lemma A.1**  *For any pair of $p$-groups $P \le Q$, if $N_Q(P) = P$, then $Q = P$.*

*Proof.* This is a general property of nilpotent groups (finite or not). If $G$ is nilpotent and $H < G$, then by definition, there is $K > H$ such that $[G, K] \le H$. Hence $H < K \le N_G(H)$. That all $p$-groups are nilpotent is shown, e.g., in [A4, 9.8] or [G1, Theorem 2.3.3].  $\square$

**Lemma A.2**  *Fix a $p$-group $P$, and an automorphism $\alpha \in \mathrm{Aut}(P)$ of order prime to $p$. Assume $1 = P_0 \trianglelefteq P_1 \trianglelefteq \cdots \trianglelefteq P_m = P$ is a sequence of subgroups all normal in $P$, such that for each $1 \le i \le m$, $\alpha|_{P_i} \equiv \mathrm{Id}_{P_i} \pmod{P_{i-1}}$. Then $\alpha = \mathrm{Id}_P$.*

*Proof.* See, for example, [G1, Theorem 5.3.2]. It suffices by induction to prove this when $m = 2$, and when the order of $\alpha$ is a prime $q \ne p$. In this case, for each $g \in P$, $\alpha$ acts on the coset $gP_1$ with fixed subset of order $\equiv |gP_1| \pmod{q}$. Since $|gP_1| = |P_1|$ is a power of $p$, this shows that $\alpha$ fixes at least one element in $gP_1$. Thus $\alpha$ is the identity on $P_1$ and on at least one element in each coset of $P_1$, and so $\alpha = \mathrm{Id}_P$.  $\square$

**Lemma A.3**  *Assume that $H \trianglelefteq G$ are finite groups and $S \le G$ is a $p$-subgroup. Then $S \in \mathrm{Syl}_p(G)$ if and only if $S \cap H \in \mathrm{Syl}_p(H)$ and $SH/H \in \mathrm{Syl}_p(G/H)$.*

*Proof.* Since $|S| = |S \cap H| \cdot |SH/H|$, $[G{:}S]$ is prime to $p$ if and only if $[H{:}S \cap H]$ and $[G/H{:}SH/H]$ are both prime to $p$.  $\square$

**Proposition A.4** (Frattini argument)  *Fix a group $G$ and a normal subgroup $H \trianglelefteq G$.*

(a)  *If $H$ is finite, then for any $T \in \mathrm{Syl}_p(H)$, $G = H \cdot N_G(T)$.*

(b)  *More generally, if $G$ acts on a set $X$ in such a way that $H$ acts transitively on $X$, then for any $x \in X$ with stabilizer subgroup $\mathrm{Stab}_G(x)$, $G = H \cdot \mathrm{Stab}_G(x)$.*

*Proof.* Fix $g \in G$. In the situation of (b), there is $h \in H$ such that $gx = hx$ (since $H$ acts transitively), and so $g = h(h^{-1}g)$ where $h^{-1}g \in \mathrm{Stab}_G(x)$. Point (a) now follows by setting $X = \mathrm{Syl}_p(H)$.  $\square$

We next look at minimal normal subgroups.

**Lemma A.5** *If $M$ is a minimal nontrivial normal subgroup of $G$, then $M$ is a product of simple groups which are pairwise isomorphic to each other.*

*Proof.* This is shown in [A4, 8.2–8.3], but we also sketch the proof here. Let $1 \neq L \trianglelefteq M$ be a minimal normal subgroup, and let $L = L_1, \ldots, L_k$ be the distinct subgroups of $M$ which are $G$-conjugate to $L$. For $i \neq j$, $L_i \cap L_j$ is also normal in $M$, and hence $[L_i, L_j] \leq L_i \cap L_j = 1$ by the minimality of $L$. Also, $\langle L_1, \ldots, L_k \rangle \leq M$ is normal in $G$, and hence equal to $M$ since $M$ is a minimal normal subgroup of $G$. In particular, any subgroup normal in $L$ is also normal in $M$ (since the other $L_i$ commute with $L$), and so the $L_i$ are simple by the minimality of $L$ again.

If $|L| = p$ for some prime $p$, then $M$ is an elementary abelian $p$-group. If not, then the $L_i$ are nonabelian and commute pairwise. Hence the product of the inclusions is a surjective homomorphism $L_1 \times \cdots \times L_k \longrightarrow M$, whose kernel intersects trivially with each factor. We leave it as an exercise to check that this kernel must be trivial, and hence that $M = L_1 \times \cdots \times L_k$. $\square$

We next look at strongly embedded subgroups of a finite group.

**Definition A.6** Fix a finite group $G$ and a prime $p$. A proper subgroup $H$ of $G$ is *strongly $p$-embedded* if $p \mid |H|$, and for each $x \in G \smallsetminus H$, $H \cap {}^x H$ has order prime to $p$.

The properties of strongly embedded subgroups which were used to prove Theorem I.3.5 (the version of Alperin's fusion theorem stated here) are listed in the following proposition.

**Proposition A.7** *Fix a finite group $G$. For each $S \in \mathrm{Syl}_p(G)$, set*

$$H_S = \langle x \in G \mid {}^x S \cap S \neq 1 \rangle .$$

*Then the following hold.*

(a) *Each strongly $p$-embedded subgroup $H < G$ contains some $S \in \mathrm{Syl}_p(G)$.*

(b) *For each $S \in \mathrm{Syl}_p(G)$, either $H_S = G$ and $G$ contains no strongly $p$-embedded subgroups; or else $H_S < G$, $H_S$ is strongly $p$-embedded, and each strongly $p$-embedded subgroup $H < G$ which contains $S$ also contains $H_S$.*

(c) *If $G$ contains a strongly $p$-embedded subgroup, then $O_p(G) = 1$.*

*Proof.* **(a)** Assume $H < G$ is strongly $p$-embedded. Since $p \mid |H|$ by assumption, there is $g \in H$ of order $p$. Choose any $S \in \mathrm{Syl}_p(G)$ such that $g \in S$. For $x \in Z(S)$ of order $p$, $g \in H \cap {}^x H$ implies $x \in H$. Hence for all $y \in S$, $x \in H \cap {}^y H$ implies $y \in H$. Thus $S \leq H$.

**(b)** Assume $S \le H < G$, where $S \in \mathrm{Syl}_p(G)$ and $H$ is strongly $p$-embedded. For each $x \in G$ such that $S \cap {}^x S \ne 1$, $p \mid |H \cap {}^x H|$, and hence $x \in H$. Thus $H_S \le H$, and $H_S < G$. Clearly, if $H_S < G$, then $H_T < G$ for each $T \in \mathrm{Syl}_p(G)$. In other words, by (a), if $H_S = G$ for some $S \in \mathrm{Syl}_p(G)$, then $G$ contains no strongly $p$-embedded subgroup.

To prove (b), it remains to show that $H_S$ is strongly $p$-embedded if it is a proper subgroup of $G$. So assume this holds, and fix $x \in G$ such that $p \mid |H_S \cap {}^x H_S|$. Choose $g \in H_S \cap {}^x H_S$ of order $p$. Since $g, {}^{x^{-1}} g \in H_S$ and $S \in \mathrm{Syl}_p(H_S)$, there are elements $y, z \in H_S$ such that ${}^y g \in S$ and ${}^{zx^{-1}} g \in S$. Then ${}^y g \in S \cap {}^{yxz^{-1}} S$, so $yxz^{-1} \in H_S$, and $x \in H_S$ since $y, z \in H_S$ by assumption. Thus $H_S$ is strongly $p$-embedded.

**(c)** Set $Q = O_p(G)$. Assume $H < G$ is strongly $p$-embedded, and fix $S \in \mathrm{Syl}_p(G)$ such that $S \le H$. Thus $Q \le S$ and $Q \trianglelefteq G$. Fix $x \in G \smallsetminus H$. Then $H \cap {}^x H$ contains $Q$ and has order prime to $p$, so $Q = 1$. $\square$

**Theorem A.8** (Burnside's Fusion Theorem)  *Let $G$ be a finite group, $p$ a prime, and assume a Sylow $p$-subgroup $S$ of $G$ is abelian. Then $N_G(S)$ controls fusion in $S$.*

*Proof.* This is a result of Burnside, appearing in [Bu]. Or see for example [A4, 37.6]. Here, "control of fusion" is interpreted in the strong sense of the first paragraph of the introduction; equivalently, $\mathcal{F}_S(G) = \mathcal{F}_S(N_G(S))$ by Exercise II.2.2. $\square$

**Definition A.9**  Fix a group $G$.

- For any $x, y \in G$, the *commutator* of $x$ and $y$ is the element
$$[x, y] = x^{-1} y^{-1} xy \in G$$
  (cf. Section 8 in [A4]).

- For subgroups $X$ and $Y$ of $G$, the *commutator* of $X$ and $Y$ is the subgroup
$$[X, Y] = \langle [x, y] \mid x \in X,\ y \in Y \rangle \le G \,.$$

- The *commutator subgroup of $G$* is the subgroup $[G, G]$. This subgroup is also called the *derived subgroup of $G$*.

- The *derived series of $G$* is defined recursively by $G^{(0)} = G$, and for $n > 0$, $G^{(n)} = [G^{(n-1)}, G^{(n-1)}]$.

- $G$ is solvable if and only if $G^{(n)} = 1$ for some positive integer $n$ (cf. [A4, 9.1]).

- $G$ is *perfect* if $G = [G, G]$.

Observe that $[X, Y] = 1$ if and only if each element of $X$ commutes with each element of $Y$.

Observe that the Puig series of a saturated fusion system, defined in Definition II.12.2, is analogous to the derived series of a finite group, so that Puig's definition of a "solvable fusion system" (which we call *Puig solvable*) is quite natural.

Later we will need:

**Lemma A.10** (3-Subgroups Lemma)  *Let $A, B, C$ be subgroups of a group $G$, and set $[A, B, C] = [[A, B], C]$. If two of the subgroups $[A, B, C]$, $[B, C, A]$, and $[C, A, B]$ are trivial, then so is the third.*

*Proof.* This follows from the relation

$$[[a, b^{-1}], c]^b \cdot [[b, c^{-1}], a]^c \cdot [[c, a^{-1}], b]^a = 1$$

for any triple of elements $a, b, c$ in a group $G$. See [A4, 8.7] for more details.  □

**Definition A.11**  Let $G$ be a finite group.

- A subgroup $H$ of $G$ is *subnormal* in $G$ if there exists a series
  $$H = H_0 \trianglelefteq H_1 \trianglelefteq \cdots \trianglelefteq H_n = G$$
  (cf. [A4, Section 7]). Thus subnormality is the transitive extension of the normality relation.

- $G$ is *quasisimple* if $G$ is perfect (cf. Definition A.9) and $G/Z(G)$ is simple.

- The *components* of $G$ are its nontrivial subnormal quasisimple subgroups. Also, $E(G)$ is the subgroup of $G$ generated by the components of $G$ (cf. [A4, Section 31]).

- The *Fitting subgroup* of $G$, denoted $F(G)$, is the largest nilpotent normal subgroup of $G$.

- $F^*(G) = E(G)F(G)$ is the *generalized Fitting subgroup* of $G$.

Recall (cf. [A4, 9.11]) that a finite group is nilpotent if and only if it is the direct product of its Sylow subgroups. Hence $F(G)$ is the product of the groups $O_p(G)$, taken over all primes $p \big| |G|$.

**Lemma A.12**  *If $L$ is a component of $G$ and $H$ is a subnormal subgroup of $G$, then either $[L, H] = 1$, or $L$ is a component of $H$.*

*Proof.* Let $G$ be a minimal counterexample. Thus $L \not\leq H$ and $[L, H] \neq 1$. Let $X = \langle L^G \rangle$ and $Y = \langle H^G \rangle$ be the normal closures of $L$ and $H$, respectively, in $G$. If $G = L$, then since $L \neq H$ and each proper normal subgroup of $L$ is contained in $Z(L)$, $H \leq Z(L)$, contradicting the assumption $[L, H] \neq 1$. Thus $L < G$, and so $X < G$. Also, $H \neq G$ since $L \not\leq H$, so $Y < G$.

Set $K = X \cap Y$. If $L \leq K$, then $L, H \leq Y < G$, contrary to the minimality of $G$. Thus $L$ is a component of $X < G$, $K \trianglelefteq X$, and $L \not\leq K$, which by the minimality of $G$ implies $[L, K] = 1$. Now, $[L, Y] \leq [X, Y] \leq K$, so $[L, Y, L] = [Y, L, L] \leq [K, L] = 1$. Then by the three subgroups lemma (Lemma A.10), $[L, L, Y] = 1$, and so $[L, Y] = 1$ as $L$ is perfect. Thus $[L, H] \leq [L, Y] = 1$, contradicting the assumption that $G$ is a counterexample.  $\square$

**Theorem A.13**  *Let $G$ be a finite group. Then the following hold.*

(a)  *Distinct components of $G$ commute elementwise. Thus $E(G)$ is the central product of the components of $G$, and these are permuted by $G$ via conjugation.*

(b)  $[F(G), E(G)] = 1.$

(c)  $C_G(F^*(G)) = Z(F^*(G)).$

*Proof.* This is shown, for example, in 31.5, 31.12, and 31.13 in [A4], but we will reproduce the proof here.

As $F(G)$ is solvable, it contains no component of $G$, so point (b) follows from Lemma A.12. Similarly if $L$ and $K$ are components of $G$ then by Lemma A.12, either $[K, L] = 1$, or $K \leq L \leq K$. This proves point (a).

To prove (c), let $H = C_G(F^*(G))$. Then $H \cap F^*(G) = Z(F^*(G)) \leq Z(H)$ is abelian. If $L$ is a component of $H$, then $L$ is also a component of $G$ (since $H \trianglelefteq G$), so $L \leq H \cap F^*(G)$, which is impossible since $L$ cannot be both abelian and quasisimple. Thus $F^*(H) = F(H)$ is nilpotent and normal in $G$, so $F^*(H) \leq F(G)$, and hence $F^*(H) = Z(H) \leq F(G)$ (since $F(H) \geq Z(H)$ for any $H$). If $H$ is abelian, then $H = C_G(F^*(G)) \leq F(G)$, and we are done.

So assume $H$ is nonabelian, and choose a minimal (nontrivial) normal subgroup $M/Z(H)$ of $H/Z(H)$. Then by Lemma A.5, $M/Z(H) = S_1 \times \cdots \times S_n$ is the direct product of pairwise isomorphic simple groups. If $S_1$ is of prime order, then $[M, M] \leq Z(H)$, so $M$ is nilpotent and hence $M \leq F^*(H) = Z(H)$, a contradiction. Thus $S_1$ is nonabelian. Let $X$ be the preimage of $S_1$ in $M$, and set $Y = [X, X]$. Then $Y$ is perfect with $Y/Z(Y) \cong S_1$, so $Y$ is a component of $G$, $Y \leq H \cap F^*(G) \leq Z(H)$, which is a contradiction.  $\square$

Note in particular the following consequence of Theorem A.13. We say that a subgroup $H \leq G$ is *centric* in $G$ if $C_G(H) \leq H$.

**Corollary A.14** *For any finite group $G$, $F(G)$ is centric in $G$ if and only if $F(G) = F^*(G)$ (equivalently, $E(G) = 1$). For any prime $p\,|\,|G|$, $O_p(G)$ is centric in $G$ if and only if $O_p(G) = F(G) = F^*(G)$.*

**Definition A.15** Let $S$ be a $p$-group for some prime $p$.

- The *Thompson subgroup* of $S$ is the subgroup generated by the elementary abelian subgroups (ie. abelian subgroups of exponent $p$) of $S$ of maximal order (cf. [A4, Section 32]). We write $J(S)$ for the Thompson subgroup of $S$.

- The *Baumann subgroup* of $S$ is $C_S(\Omega_1(Z(J(S))))$. We denote the Baumann subgroup of $S$ by $\mathrm{Baum}(S)$.

Recall that for any $p$-group $X$, $\Omega_1(X) = \langle x \in X \,|\, x^p = 1 \rangle$ (cf. [A4, Section 1]).

For more information about the Baumann subgroup, see [ASm, Section B.2].

## References


[Ad]     J. F. Adams, *Stable Homotopy and Generalized Homology*, Univ. Chicago Press (1974).

[Al1]    J. Alperin, Sylow intersections and fusion, *J. Algebra*, **6** (1967), 222–241.

[Al2]    J. Alperin, Local Representation Theory, The Santa Cruz conference on finite groups, *Proc. Symp. Pure Math.*, **37**, Amer. Math. Soc., Providence (1980), 369–375.

[Al3]    J. L. Alperin, Weights for finite groups, The Arcata Conference on Finite Groups, *Proc. Sympos. Pure Math.*, **47**, Amer. Math. Soc., Providence (1987), 369–379.

[ABG]    J. Alperin, R. Brauer, and D. Gorenstein, Finite groups with quasi-dihedral and wreathed Sylow 2-subgroups, *Trans. Amer. Math. Soc.*, **151** (1970), 1–261.

[AB]     J. L. Alperin and M. Broué, Local methods in block theory, *Ann. of Math.*, **110** (1979), 143–157.

[ABC]    T. Altinel, A. Borovik, & G. Cherlin, Simple groups of finite Morley rank, *Math. Surveys & Monogr.*, **145**, American Math. Soc. (2008).

[AOV]    K. Andersen, B. Oliver, & J. Ventura, Reduced, tame, and exotic fusion systems, preprint.

[A1]     M. Aschbacher, A characterization of Chevalley groups over fields of odd order, *Annals of Math.*, **106** (1975), 353-468.

[A2]     M. Aschbacher, On finite groups of component type, *Illinois J. Math.*, **19** (1975), 87-113.

[A3]     M. Aschbacher, On finite groups of Lie type and odd characteristic, *J. Algebra*, **66** (1980), 400-424.

[A4]     M. Aschbacher, *Finite Group Theory*, Cambridge Univ. Press (1986).

[A5]     M. Aschbacher, Normal subsystems of fusion systems, *Proc. London Math. Soc.*, **97** (2008), 239–271.

[A6]     M. Aschbacher, The generalized Fitting subsystem of a fusion system, *Memoirs Amer. Math. Soc.*, **209** (2011), nr. 986.

[A7]     M. Aschbacher, Generation of fusion systems of characteristic 2-type, *Invent. Math.*, **180** (2010), 225–299.

[A8]     M. Aschbacher, $S_3$-free 2-fusion systems, *Proc. Edinburgh Math. Soc.*, (proceedings of the 2009 Skye conference on algebraic topology, group theory and representation theory, to appear).

[A9]     M. Aschbacher, N-groups and fusion systems, preprint.

[AC]     M. Aschbacher & A. Chermak, A group-theoretic approach to a family of 2-local finite groups constructed by Levi and Oliver, *Annals of Math.*, **171** (2010), 881–978.

[ASm]    M. Aschbacher and S. Smith, The Classification of the Quasithin Groups, American Mat. Soc., (2004).

[Ben]    H. Bender, Finite groups with dihedral Sylow 2-subgroups, *J. Algebra*, **70** (1981), 216–228.

[BG]     H. Bender and G. Glauberman, Characters of finite groups with dihedral Sylow 2-subgroups, *J. Algebra*, **70** (1981), 200–215.

[Be1]    D. Benson, *Representations and Cohomology I: Cohomology of Groups and Modules*, Cambridge Univ. Press (1991).

[Be2]    D. Benson, *Representations and Cohomology II: Cohomology of Groups and Modules*, Cambridge Univ. Press (1991).

[Be3]    D. Benson, Cohomology of sporadic groups, finite loop spaces, and the Dickson invariants, Geometry and cohomology in group theory, *London Math. Soc. Lecture notes ser.* **252**, Cambridge Univ. Press (1998), 10–23.

[Bo] R. Boltje, Alperin's weight conjecture in terms of linear source modules and trivial source modules, *Modular representation theory of finite groups (Charlottesville, VA, 1998)*, de Gruyter, Berlin (2001), 147–155.

[BonD] V.M. Bondarenko, J.A. Drozd, The representation type of finite groups *Zap. anchn. Sem. Leningrad. Otdel. Mat. Inst. Steklov*, **57** (1977) 24-41. English translation: J. Soviet Math., 20(1982), 2515–2528.

[Bf] P. Bousfield, On the $p$-adic completions of nonnilpotent spaces, *Trans. Amer. Math. Soc.*, **331** (1992), 335–359.

[BK] P. Bousfield & D. Kan, Homotopy limits, completions, and localizations, *Lecture notes in math.*, **304**, Springer-Verlag (1972).

[Br1] R. Brauer, Investigations on group characters, *Ann. of Math.*, (2) **42** (1941), 936–958.

[Br2] R. Brauer, Some applications of the theory of blocks of characters of finite groups IV, *J. Algebra*, **17** (1971), 489–521.

[Br3] R. Brauer, On 2-blocks with dihedral defect groups, *Symposia Mathematica*, Vol. XIII (Convegno di Gruppi e loro Rappresentazioni, INDAM, Rome, 1972), Academic Press, London (1974), 367–393.

[Br4] R. Brauer, On the structure of blocks of characters of finite groups, Proc. Second Intern. Conf. on Theory of Groups, *Lecture Notes in Mathematics* **372**, Springer-Verlag, (1974).

[Bre] S. Brenner, Modular representations of p-groups *J. Algebra*, **15** (1970), 89-102.

[5a1] C. Broto, N. Castellana, J. Grodal, R. Levi, & B. Oliver, Subgroup families controlling $p$-local finite groups, *Proc. London Math. Soc.*, **91** (2005), 325–354.

[5a2] C. Broto, N. Castellana, J. Grodal, R. Levi, & B. Oliver, Extensions of $p$-local finite groups, *Trans. Amer. Math. Soc.*, **359** (2007), 3791-3858.

[BL] C. Broto & R. Levi, On spaces of self homotopy equivalences of $p$-completed classifying spaces of finite groups and homotopy group extensions, *Topology*, **41** (2002), 229–255.

[BLO1] C. Broto, R. Levi, & B. Oliver, Homotopy equivalences of $p$ completed classifying spaces of finite groups, *Invent. Math.*, **151** (2003), 611–664.

[BLO2] C. Broto, R. Levi, & B. Oliver, The homotopy theory of fusion systems, *J. Amer. Math. Soc.*, **16** (2003), 779–856.

[BLO3] C. Broto, R. Levi, & B. Oliver, Discrete models for the $p$-local homotopy theory of compact Lie groups and p-compact groups, *Geometry and Topology*, **11** (2007), 315-427.

[BLO4] C. Broto, R. Levi, & B. Oliver, A geometric construction of saturated fusion systems, An alpine anthology of homotopy theory (proceedings Arolla 2004), Contemp. math. **399** (2006), 11-39.

[BLO] C. Broto, R. Levi, & B. Oliver, The theory of $p$-local groups: A survey, Homotopy theory (Northwestern Univ. 2002), *Contemp. math.*, **346**, Amer. Math. Soc. (2004), 51–84.

[BM] C. Broto & J. Møller, Chevalley $p$-local finite groups, *Algebr. & Geom. Topology*, **7** (2007), 1809–1919.

[BMO] C. Broto, J. Møller, & B. Oliver, Equivalences between fusion systems of finite groups of Lie type, preprint.

[B2] M. Broué, Isométries parfaites, types de blocs, catégories dérivées, *Astérisque*, **181-182** (1990) 61–92.

[BP1] M. Broué and L. Puig, A Frobenius theorem for blocks, *Invent. Math.*, **56** (1980), no. 2, 117–128.

[BP2] M. Broué and L. Puig, Characters and Local structures in $G$-Algebras, *Journal of Algebra*, **63**, (1980), 51–59.

[Br] K. Brown, *Cohomology of Groups*, Springer-Verlag (1982).

[Bu]     W. Burnside, *The Theory of Groups of Finite Order*, Cambridge Univ. Press (1897).

[Cm]     N. Campbell, *Pushing Up in Finite Groups*, Thesis, Cal. Tech., (1979).

[Ca]     G. Carlsson, Equivariant stable homotopy and Sullivan's conjecture, *Invent. Math.*, **103** (1991), 497–525.

[CE]     H. Cartan & S. Eilenberg, *Homological Algebra*, Princeton Univ. Press (1956).

[CL]     N. Castellana & A. Libman, Wreath products and representations of $p$-local finite groups, *Advances in Math.*, **221** (2009), 1302–1344.

[COS]    A. Chermak, B. Oliver, & S. Shpectorov, The linking systems of the Solomon 2-local finite groups are simply connected, *Proc. London Math. Soc.*, **97** (2008), 209–238.

[CP]     M. Clelland & C. Parker, Two families of exotic fusion systems, *J. Algebra*, **323** (2010), 287–304.

[CPW]    G. Cliff, W. Plesken, A. Weiss, Order-theoretic properties of the center of a block, The Arcata Conference on Representations of Finite Groups (Arcata, Calif., 1986), *Proc. Sympos. Pure Math.*, **47**, Part 1, *Amer. Math. Soc.*, Providence, RI (1987), 413–420.

[Cr1]    D. Craven, Control of fusion and solubility in fusion systems, *J. Algebra*, **323** (2010), 2429–2448.

[Cr2]    D. Craven, *The Theory of Fusion Systems: an Algebraic Approach*, Cambridge Univ. Press (2011).

[Cr3]    D. Craven, Normal subsystems of fusion systems, *Journal London Math. Soc.* (to appear).

[CG]     D. Craven & A. Glesser, Fusion systems on small $p$-groups, *Trans. Amer. Math. Soc.* (to appear).

[CrEKL]  D. Craven, C.Eaton, R. Kessar, M.Linckelmann, The structure of blocks with a Klein 4 defect group, *Math. Z.* (to appear).

[Cu]     E. Curtis, Simplicial homotopy theory, *Adv. in Math.*, **6** (1971), 107–209.

[CuR1]   C. W. Curtis and I. Reiner. *Representation theory of finite groups and associative algebras*, Wiley-Interscience (1962).

[CuR2]   C. W. Curtis and I. Reiner. *Methods in representation theory*, Vol. I, J. Wiley and Sons (1981).

[CuR3]   C. W. Curtis and I. Reiner. *Methods in representation theory*, Vol. II, J. Wiley and Sons (1987).

[Da1]    E.C. Dade, Blocks with cyclic defect groups, *Ann. of Math.*, **84** (1966), 20–48.

[Da2]    E.C. Dade, Counting characters in blocks I, *Invent. Math.*, **109** (1992), no. 1, 187–210.

[Da3]    E.C. Dade, Counting characters in blocks II, *J. Reine Angew. Math.*, **448** (1994), 97–190.

[Da4]    E.C. Dade, Counting characters in blocks, II.9, in *Representation Theory of Finite Groups*, Ohio State University Math Research Institute Publications, Vol. **6**, de Gruyter, Berlin (1997), 45–59.

[DGMP1]  A. Díaz, A. Glesser, N. Mazza, & S. Park, Control of transfer and weak closure in fusion systems, *J. Algebra*, **323** (2010), 382–392.

[DGMP2]  A. Díaz, A. Glesser, N. Mazza, & S. Park, Glauberman's and Thompson's theorems for fusion systems, *Proc. Amer. Math. Soc.*, **137** (2009), 495–503.

[DGPS]   A. Díaz, A. Glesser, S. Park, & R. Stancu, Tate's and Yoshida's theorem for fusion systems, *Journal London Math. Soc.* (to appear).

[DN1]    A. Díaz & A. Libman, Segal's conjecture and the Burnside ring of fusion systems, *J. London Math. Soc.*, **80** (2009), 665–679.

[DN2]    A. Díaz & A. Libman, The Burnside ring of fusion systems, *Adv. Math.*, **222** (2009), 1943–1963.

[DRV]    A. Díaz, A. Ruiz & A. Viruel, All $p$-local finite groups of rank two for odd prime $p$, *Trans. Amer. Math. Soc.*, **359** (2007), 1725–1764.

[Dr]     A. Dress, Induction and structure theorems for orthogonal representations of finite groups, *Annals of Math.*, **102** (1975), 291–325.

[Dw]     W. Dwyer, Homology decompositions for classifying spaces of finite groups, *Topology*, **36** (1997), 783–804.

[DK1]    W. Dwyer & D. Kan, Realizing diagrams in the homotopy category by means of diagrams of simplicial sets, *Proc. Amer. Math. Soc.*, **91** (1984), 456–460.

[DK2]    W. Dwyer & D. Kan, Centric maps and realizations of diagrams in the homotopy category, *Proc. Amer. Math. Soc.*, **114** (1992), 575–584.

[Er]     K. Erdmann, Blocks of tame representation type and related algebras, *Lecture Notes in Mathematics*, **1428**, Springer-Verlag, Berlin (1990).

[Fe]     W. Feit, *The Representation Theory of Finite Groups*, North Holland (1982).

[FT]     W. Feit and J. Thompson, Solvability of groups of odd order, *Pacific J. Math.*, **13** (1963), 775–1029; 218–270; 354–393.

[FF]     R. Flores and R. Foote, Strongly closed subgroups of finite groups, *Adv. in Math.*, **222** (2009), 453-484.

[F]      R. Foote, A characterization of finite groups containing a strongly closed 2-subgroup, *Comm. Alg.*, **25** (1997), 593–606.

[Fr]     E. Friedlander, *Étale Homotopy of Simplicial Schemes*, Annals of Mathematics Studies, vol. **104**, Princeton University Press (1982).

[Gl1]    G. Glauberman, Central elements in core-free groups, *J. Algebra*, **4** (1966), 403–420.

[Gl2]    G. Glauberman, Factorizations in local subgroups of finite groups, *Regional Conference Series in Mathematics*, **33**, Amer. Math. Soc. (1977).

[GN]     G. Glauberman & R. Niles, A pair of characteristic subgroups for pushing-up in finite groups, *Proc. London Math. Soc.*, **46** (1983), 411–453.

[GZ]     P. Gabriel & M. Zisman, *Calculus of Fractions and Homotopy Theory*, Springer-Verlag (1967).

[GJ]     P. Goerss & R. Jardine, *Simplicial Homotopy Theory*, Birkhäuser Verlag (1999).

[Gd1]    D. Goldschmidt, A conjugation family for finite groups, *J. Algebra*, **16** (1970), 138–142.

[Gd2]    D. Goldschmidt, Strongly closed 2-subgroups of finite groups, *Annals of Math.*, **102** (1975), 475–489.

[Gd3]    D. Goldschmidt, 2-fusion in finite groups, *Annals of Math.*, **99** (1974), 70-117.

[G1]     D. Gorenstein, *Finite groups*, Harper & Row (1968).

[G2]     D. Gorenstein, *The Classification of the Finite Simple Groups*, I, Plenum (1983).

[GH]     D. Gorenstein and M. Harris, Finite groups with product fusion, *Annals of Math.*, **101** (1975), 45-87.

[GL]     D. Gorenstein and R. Lyons, Nonsolvable finite groups with solvable 2-local subgroups, *J. Algebra*, **38** (1976), 453–522.

[GLS3]   D. Gorenstein, R. Lyons, and R. Solomon, *The Classification of the Finite Simple Groups*, Number 3, Mathematical Surveys and Monographs, vol. **40**, Amer. Math. Soc. (1998).

[GLS6]   D. Gorenstein, R. Lyons, and R. Solomon, *The Classification of the Finite Simple Groups*, Number 6, Mathematical Surveys and Monographs, vol. **40**, Amer. Math. Soc. (2005).

[GW1]    D. Gorenstein and J. Walter, The characterization of finite simple groups with dihedral Sylow 2-subgroups, *J. Algebra*, **2** (1964), 85–151; 218–270; 354–393.

[GW2]    D. Gorenstein and J. Walter, Balance and generation in finite groups, *J. Algebra*, **33** (1975), 224-287.

[GHL]    D. Green, L. Héthelyi, & M. Lilienthal, On Oliver's $p$-group conjecture, *Algebra Number Theory*, **2** (2008), 969–977.

[GHM]   D. Green, L. Héthelyi, & N. Mazza, On Oliver's $p$-group conjecture: II, *Math. Annalen*, **347** (2010), 111–122.

[Gr]     J. Grodal, Higher limits via subgroup complexes, *Annals of Math.*, **155** (2002), 405–457.

[Ht]     A. Hatcher, *Algebraic Topology*, Cambridge Univ. Press (2002).

[H]      D. Higman, Indecomposable representations at characteristic $p$, *Duke J. Math.*, **21** (1954), 377–381.

[HV]     J. Hollender & R. Vogt, Modules of topological spaces, applications to homotopy limits and $E_\infty$ structures, *Arch. Math.*, **59** (1992), 115–129.

[IsNa]   I.M. Isaacs and G. Navarro, New refinements of the McKay conjecture for arbitrary finite groups, *Annals of Math.*, **156** (2002), 333–344.

[JM]     S. Jackowski & J. McClure, Homotopy decomposition of classifying spaces via elementary abelian subgroups, *Topology*, **31** (1992), 113–132.

[JMO]    S. Jackowski, J. McClure, & B. Oliver, Homotopy classification of self-maps of $BG$ via $G$-actions, *Annals of Math.*, **135** (1992), 183–270.

[JS]     S. Jackowski & J. Słomińska, $G$-functors, $G$-posets and homotopy decompositions of $G$-spaces, *Fundamenta Math.*, **169** (2001), 249–287.

[J]      Z. Janko, Nonsolvable finite groups all of whose 2-local subgroups are solvable, I, *J. Algebra*, **21** (1972), 458–517.

[K1]     R. Kessar, Introduction to block theory, *Group Representation Theory*, EPFL Press, Lausanne (2007) 47–77.

[Ke1]    R. Kessar, The Solomon system $\mathcal{F}_{\mathrm{Sol}}(3)$ does not occur as fusion system of a 2-block *J. Algebra*, **296**, no. 2 (2006), 409–425.

[KL]     R. Kessar & M. Linckelmann, $ZJ$-theorems for fusion systems, *Trans. Amer. Math. Soc.*, **360** (2008), 3093–3106.

[KS]     R. Kessar, R. Stancu, A reduction theorem for fusion systems of blocks, *J. Algebra*, **319** (2008), 806–823.

[KKoL]   R. Kessar, S.Koshitani, M. Linckelmann, Conjectures of Alperin and Broué for 2-blocks with elementary abelian defect groups of order 8, *J. Reine Angew. Math.* (to appear).

[KKuM]   R. Kessar, N. Kunugi, N. Mitsuhashi, On saturated fusion systems and Brauer indecomposability of Scott modules, *J. Algebra* (to appear).

[KR]     R. Knörr and G. R. Robinson, Some remarks on a conjecture of Alperin, *J. London Math. Soc.* (2), **39** (1989), no. 1, 48–60.

[KoZ]    S. Koenig, A. Zimmermann, Derived equivalences for group rings, *Lecture Notes in Mathematics*, **1685**, Springer-Verlag, Berlin (1998).

[Ku]     B. Külshammer, *Lectures on block theory*, London Mathematical Society Lecture Note Series **161**, Cambridge Univ. Press, Cambridge (1991).

[KulP]   B. Külshammer, L. Puig, Extensions of nilpotent blocks, *Invent. Math.*, **102**, no. 1 (1990), 17–71.

[KulOW]  B. Külshammer, A. Watanabe, and T. Okuyama, A lifting theorem with applications to blocks and source algebras, *J. Algebra*, **232**, no. 1 (2000), 299–309.

[La]     J. Lannes, Sur les espaces fonctionnels dont la source est le classifiant d'un $p$-groupe abélien élémentaire, *Publ. Math. I.H.E.S.*, **75** (1992), 135–244.

[LS]     I. Leary & R. Stancu, Realising fusion systems, *Algebra & Number Theory*, **1** (2007), 17–34.

[LO]     R. Levi & B. Oliver, Construction of 2-local finite groups of a type studied by Solomon and Benson, *Geometry & Topology*, **6** (2002), 917–990.

[LO2]    R. Levi & B. Oliver, Correction to: Construction of 2-local finite groups of a type studied by Solomon and Benson, *Geometry & Topology*, **9** (2005), 2395–2415.

[LR]     R. Levi & K. Ragnarsson, $p$-local finite group cohomology, *Homotopy, Homology, Appl.* (to appear).

[Lb]     A. Libman, The normaliser decomposition for $p$-local finite groups, *Alg. Geom. Topology*, **6** (2006), 1267–1288.

[LbS]    A. Libman & N. Seeliger, Homology decompositions and groups inducing fusion systems, preprint.

[LV]     A. Libman & A. Viruel, On the homotopy type of the non-completed classifying space of a $p$-local finite group, *Forum Math.*, **21** (2009), 723–757.

[Li1]    M. Linckelmann, The isomorphism problem for cyclic blocks and their source algebras. *Invent. Math.*, **125** (1996), 265-283.

[Li2]    M. Linckelmann, Fusion category algebras, *J. Algebra*, **277**, no. 1 (2004), 222–235.

[Li3]    M. Linckelmann, Simple fusion systems and the Solomon 2-local groups, *J. Algebra* , **296**, no. 2 (2006), 385–401.

[Li4]    M. Linckelmann, Alperin's weight conjecture in terms of equivariant Bredon cohomology, *Math. Z.*, **250**, no. 3 (2005), 495–513.

[Li5]    M. Linckelmann, Trivial source bimodule rings for blocks and $p$-permutation equivalences, *Trans. Amer. Math. Soc.*, **361** (2009), 1279–1316.

[Li6]    M. Linckelmann, On $H^*(\mathcal{C}; k^\times)$ for fusion systems, *Homology, Homotopy Appl.*, **11**, no. 1 (2009), 203–218.

[LP]     J. Lynd & S. Park, Analogues of Goldschmidt's thesis for fusion systems, *J. Algebra*, **324** (2010), 3487–3493.

[McL]    S. MacLane, *Homology*, Springer-Verlag (1975).

[MP1]    J. Martino & S. Priddy, Stable homotopy classification of $BG_p^\wedge$, *Topology*, **34** (1995), 633–649.

[MP2]    J. Martino & S. Priddy, Unstable homotopy classification of $BG_p^\wedge$, *Math. Proc. Cambridge Phil. Soc.*, **119** (1996), 119–137.

[May]    J. P. May, *Simplicial Objects in Algebraic Topology*, Univ. Chicago Press (1967).

[MSS]    U. Meierfrankenfeld, B. Stellmacher, and G. Stroth, Finite groups of local characteristic $p$: an overview, *Groups, Combinatorics, and Geometry (Durham 2001)*, World Sci. Publ. (2003), 155–192.

[Mi]     H. Miller, The Sullivan conjecture on maps from classifying spaces, *Annals of Math.*, **120** (1984), 39–87.

[Ms]     G. Mislin, On group homomorphisms inducing mod-$p$ cohomology isomorphisms, *Comment. Math. Helv.*, **65** (1990), 454–461.

[NT]     H. Nagao and Y. Tsushima, *Representations of Finite Groups*, Academic Press, Boston (1988).

[OW]     T. Okuyama and M. Wajima, Irreducible characters of $p$-solvable groups *Proc. Japan Acad. Ser. A Math. Sci.*, **55** (1979), no. 8, 309–312.

[O1]     B. Oliver, Higher limits via Steinberg representations, *Comm. in Algebra*, **22** (1994), 1381–1402.

[O2]     B. Oliver, Equivalences of classifying spaces completed at odd primes, *Math. Proc. Camb. Phil. Soc.*, **137** (2004), 321–347.

[O3]     B. Oliver, Equivalences of classifying spaces completed at the prime two, *Memoirs Amer. Math. Soc.*, **848** (2006).

[O4]     B. Oliver, Extensions of linking systems and fusion systems, *Trans. Amer. Math. Soc.*, **362** (2010), 5483–5500.

[O5]     B. Oliver, Splitting fusion systems over 2-groups, *Proc. Edinburgh Math. Soc.*, proceedings of the 2009 Skye conference on algebraic topology, group theory and representation theory (to appear).

[OV1]    B. Oliver & J. Ventura, Extensions of linking systems with $p$-group kernel, *Math. Annalen*, **338** (2007), 983-1043.

[OV2]   B. Oliver & J. Ventura, Saturated fusion systems over 2-groups, *Trans. Amer. Math. Soc.*, **361** (2009), 6661–6728.

[Ols]   J. B. Olsson, On 2-blocks with quaternion and quasidihedral defect groups *J. Algebra*, **36** (1975), 212-241.

[Ols2]  J. B. Olsson, On subpairs and modular representation theory, *J. Algebra*, **76** (1982), 261–279.

[OS]    S. Onofrei and R. Stancu, A characteristic subgroup for fusion systems, *J. Algebra*, **322** (2009), 1705–1718.

[Pa1]   S. Park, The gluing problem for some block fusion systems, *J. Algebra*, **323** (2010), 1690–1697.

[Pa2]   S. Park, Realizing a fusion system by a single finite group, *Arch. Math.*, **94** (2010), 405-410.

[P1]    L. Puig, Structure locale dans les groupes finis, *Bull. Soc. Math. France Suppl. Mém.*, **47** (1976).

[P2]    L. Puig, Local fusions in block source algebras, *J. Algebra* , **104**, no. 2 (1986), 358–369.

[P3]    L. Puig, Nilpotent blocks and their source algebras, *Invent. Math.*, **93**, no. 1 (1988), 77–116.

[P4]    L. Puig, The Nakayama conjecture and the Brauer pairs *Seminaire sur les groupes finis* **III**, Publications Matehmatiques De L'Université Paris VII, 171–189.

[P5]    L. Puig, The hyperfocal subalgebra of a block, *Invent. math.*, **141** (2000), 365–397.

[P6]    L. Puig, Frobenius categories, *J. Algebra*, **303** (2006), 309–357.

[P7]    L. Puig, *Frobenius Categories versus Brauer Blocks*, Birkhäuser (2009).

[PUs]   L. Puig and Y. Usami, Perfect isometries for blocks with abelian defect groups and cyclic inertial quotients of order 4, *J. Algebra*, **172** (1995), 205–213.

[Rg]    K. Ragnarsson, Classifying spectra of saturated fusion systems, *Algebr. Geom. Topol.*, **6** (2006), 195–252.

[RSt]   K. Ragnarsson & R. Stancu, Saturated fusion systems as idempotents in the double Burnside ring, preprint.

[Ri1]   J.Rickard, Derived categories and stable equivalence, *J. Pure Appl. Algebra*, **61**, no. 3 (1989), 303–317.

[Ri2]   J.Rickard, Splendid equivalences: derived categories and permutation modules, *Proc. London Math. Soc. (3)*, **72**, no. 2 (1996), 331–358.

[RS]    K. Roberts & S. Shpectorov, On the definition of saturated fusion systems, *J. Group Theory*, **12** (2009), 679–687.

[Ro1]   G. R. Robinson, Local structure, vertices and Alperin's conjecture, *Proc. London Math. Soc.*, **72** (1996), 312–330.

[Ro2]   G. R. Robinson, Weight conjectures for ordinary characters, *J. Algebra*, **276** (2004), 761–775.

[Ro3]   G. Robinson, Amalgams, blocks, weights, fusion systems, and finite simple groups, *J. Algebra*, **314** (2007), 912–923.

[Rz]    A. Ruiz, Exotic normal fusion subsystems of general linear groups, *J. London Math. Soc.*, **76** (2007), 181–196.

[RV]    A. Ruiz & A. Viruel, The classification of $p$-local finite groups over the extraspecial group of order $p^3$ and exponent $p$, *Math. Z.*, **248** (2004), 45–65.

[Sa1]   B. Sambale, 2-blocks with mimimal non-abelian defect groups, *J. Algebra* (to appear).

[Sa2]   B. Sambale, Blocks with defect group $D_{2^n} \times C_{2^n}$, *J. Pure Appl. Algebra* (to appear).

[Sa3]   B. Sambale, Fusion systems on metacyclic 2-groups, preprint.

[Sg]    G. Segal, Classifying spaces and spectral sequences, *Publ. Math. I.H.E.S.*, **34** (1968), 105–112.

[Se1]   J. P. Serre, *Corps Locaux*, Hermann (1968).

[Se2]   J.-P. Serre, *Trees*, Springer-Verlag (1980).

[Sm]    F. Smith, Finite simple groups all of whose 2-local subgroups are solvable, *J. Algebra*, **34** (1975), 481–520.

[So]    R. Solomon, Finite groups with Sylow 2-subgroups of type .3, *J. Algebra*, **28** (1974), 182–198.

[Sta1]  R. Stancu, Control of fusion in fusion systems, *J. Algebra Appl.*, **5** (2006), 817–837.

[Sta2]  R. Stancu, Equivalent definitions of fusion systems, preprint.

[Stn]   R. Steinberg, *Lectures on Chevalley Groups*, Yale Lecture Notes (1967).

[St1]   B. Stellmacher, A characteristic subgroup of $S_4$-free groups, *Israel J. Math.*, **94** (1996), 367–379.

[St2]   B. Stellmacher, An application of the amalgam method: the 2-local structure of N-groups of characteristic 2-type, *J. Algebra*, **190** (1997), 11–67.

[Sw]    R. Switzer, *Algebraic Topology*, Springer-Verlag (1975).

[Th]    J. Thévenaz, *G-Algebras and Modular Representation Theory*, Oxford Science Publications (1995).

[Th1]   J. Thompson, Nonsolvable finite groups all of whose local subgroups are solvable, I, *Bull. Amer. Math. Soc.*, **74** (1968), 383–437; II, *Pacific J. Math.*, **33** (1970), 451–536; III, *Pacific J. Math.*, **39** (1971), 483–534; IV, *Pacific J. Math.*, **48** (1973), 511–592; V, *Pacific J. Math.*, **50** (1974), 215–297; VI, *Pacific J. Math.*, **51** (1974), 573–630.

[Th2]   J. Thompson, Simple 3'-groups, *Symposia Math.*, **13** (1974), 517–530.

[Tu]    A. Turull, Strengthening the McKay Conjecture to include local fields and local Schur indices, *J. Algebra*, **319** (2008), 4853–4868.

[Un]    K. Uno, Conjectures on character degrees for the simple Thompson group, *Osaka J. Math.*, **41** (2004), 11–36.

[W]     J. Walter, The B-Conjecture; characterization of Chevalley groups, *Memoirs Amer. Math. Soc.*, **61** no. 345 (1986), 1–196.

[Wb1]   P. Webb, A split exact sequence of Mackey functors, *Comment. Math. Helv.*, **66** (1991), 34–69.

[Wb2]   P.J. Webb, Standard stratifications of EI categories and Alperin's weight conjecture, *J. Algebra*, **320** (2008), 4073–4091.

[Wei]   C. Weibel, *An Introduction to Homological Algebra*, Cambridge Univ. Press (1994).

[Wh]    G. Whitehead, *Elements of Homotopy Theory*, Springer-Verlag (1978).

[Wo]    Z. Wojtkowiak, On maps from holim F to Z (Algebraic topology, Barcelona, 1986), *Lecture notes in math.*, **1298**, Springer-Verlag (1987), 227–236.

[Zi]    K. Ziemiański, Homotopy representations of $SO(7)$ and Spin(7) at the prime 2, *Jour. Pure Appl. Algebra*, **212** (2008), 1525–1541.

## List of Notation